Y0-GAJ-392

# MATHEMATICAL ASPECTS OF STRING THEORY

# ADVANCED SERIES IN MATHEMATICAL PHYSICS

**Editors-in-charge**
D H PHONG (*Columbia University*)
S-T YAU (*University of California, San Diego*)

**Associate Editors**
L ALVAREZ-GAUMÉ (*CERN*)
J P BOURGUIGNON (*Ecole Polytechnique, Palaiseau*)
T EGUCHI (*University of Tokyo/CNRS*)
B JULIA (*CNRS*)
F WILCZEK (*University of California, Santa Barbara*)

***Forthcoming Title***

Volume 2: Bombay Lectures on Highest Weight Representations of Infinite-Dimensional Lie Algebras
by V G Kac (*MIT*) & A K Raina (*Tata Institute*)

Advanced Series in Mathematical Physics — Vol. 1

# MATHEMATICAL ASPECTS OF STRING THEORY

edited by **S T Yau**

Proceedings of the Conference held at
University of California, San Diego
July 21 – August 1, 1986

World Scientific
*Singapore • New Jersey • Hong Kong*

# TABLE OF CONTENTS

# Introduction to Quantum Field Theory, Path Integrals, and Strings

B. F. Hatfield

*C-012, University of California, La Jolla, CA 92093*

"Teichmüller" has become a new word in the physicist's vocabulary. The study of quantum strings has sent physicists running to ask the mathematicians to tell them all they know about these spaces, probably to the mathematician's surprise (what next?). Now we have conferences such as this one to discuss these matters but often we physicists forget to explain how Teichmüller spaces arise in the study of strings. I will try to atleast partially remedy this. Since most mathematicians don't spend their days doing quantum mechanics, I will begin by reviewing the basic ideas of quantum theory. Following this, I will show how the path integral arises in quantum mechanics. Finally, I briefly examine the case of the free bosonic string using "path" integrals for genus 0 surfaces. [1] The reduction of the string path integral for arbitrary genus to an integral over moduli space will be given by Polchinski.

## Quantum Mechanics

The object of interest to calculate in a quantum-mechanical process is the quantum amplitude of the process, a complex number.

$$Amplitude \in \mathbb{C}$$

The square of the amplitude is the probability that the process will occur.

$$|\, Amplitude \,|^2 = Prob$$

For example, the process might just be a particle starting at $x$ at time $t_0$ and travelling to $x'$ at time $t'$.

With this process we associate an amplitude, $A$. The probability that the particle starting at $(x,t_0)$ will be found at $(x',t')$ is $|\, A \,|^2$.

Quantum mechanics postulates that there are a set of state vectors, symbolized as $| \ >$, that describe all possible configurations of the system. For example, when the particle is at $(x, t_0)$ the state of the system is $| x, t_0>$. The state vectors are elements of a Hilbert space.

The probability amplitude that is associated with the above process is then the overlap between the initial state and the final state.

$$A = <x', t' | x, t_0>$$

$< | >$ means scalar product. The states are normalized so that $<x', t | x, t> = \delta(x' - x)$. So again, the probability of the process occuring is given by

$$Prob = |<x', t' | x, t_0>|^2$$

So far our states are time independent (even though our notation may make you think that they are not). The state $| x, t_0>$ just represents the particle being found at $(x, t_0)$, it doesn't evolve or change with time.

Let's switch to a state that does have time dependence and call it $| \psi(t)>$. Another postulate of quantum mechanics is that all physically observable quantities are associated with operators and the result of a measurement of an observable is one of the eigenvalues of the operator.

The position of the particle is an observable quantity so there is an operator associated with it, call it $\mathbf{X}$. When $\mathbf{X}$ operates on a state that is an eigenvector of $\mathbf{X}$ we get an eigenvalue, the result of the measurement. Often we label the eigenstate with its eigenvalue. For example,

$$\mathbf{X} | x> = x | x>$$

Now suppose we operate on $| \psi(t)>$ at $t = t_0$ with $\mathbf{X}$ and find that we get the state back again multiplied by $x$,

$$\mathbf{X} | \psi(t_0)> = x | \psi(t_0)>.$$

Then $| \psi(t = t_0)> = | x, t_0>$. If we got $x'$ instead or found that $| \psi(t_0)>$ wasn't an eigenvector then $| \psi(t_0)> \neq | x, t_0>$.

How does quantum mechanics provide for the time evolution of $| \psi(t)>$? We have a time-evolution operator, denoted $\mathbf{H}$, which time translates $| \psi(t)>$ from $t$ to $t + \epsilon$, $\epsilon$ small. In other words, $| \psi(t)>$ satisfies

$$i \frac{\partial | \psi(t)>}{\partial t} = \mathbf{H} | \psi(t)>,$$

the Schrödinger equation. $\mathbf{H}$ is the Hamiltonian operator. If we have a classical system that we want to quantize, then $x$ becomes an operator $\mathbf{X}$, $p$, the momentum, becomes an operator, $\mathbf{P}$, so the classical Hamiltonian, $H(p,x)$ becomes an operator through its dependence on $x$ and $p$. The quantization is completed by replacing i times the classical Poisson bracket by a commutator (Planck's constant here has been set to one) and by choosing an operator ordering. If we do not have a classical system (and sometimes we do not) then we must guess what form $\mathbf{H}$ should be.

The formal solution to the Schrödinger equation is

$$|\,\psi(t)> = e^{-i\mathbf{H}(t-t_0)}\,|\,\psi(t_0)>.$$

Suppose that we choose $|\,\psi(t_0)>$ to be an eigenfunction of $\mathbf{X}$ with eigenvalue $x$ so that $|\,\psi(t_0)> = |\,x>$. Then at $t\,'$ the particle is in the state $|\,\psi(t\,')>$. $|\,\psi(t\,')>$ may no longer be an eigenvector of $\mathbf{X}$ (most certainly it won't). The amplitude that the particle will be at $x\,'$ is the overlap of $|\,\psi(t\,')>$ with $|\,x\,'>$, that is, $|\,\psi(t\,')>$ projected onto the eigenvector $|\,x\,'>$. So $<x\,'\,|\,\psi(t\,')>$ is the amplitude that the particle will be found at $x\,'$ at $t\,'$ given that it started in state $|\,\psi(t_0)>$ which we chose to be $|\,x>$. But this is just $<x\,'t\,'\,|\,xt_0>$ hence

$$\begin{aligned} <x\,'t\,'\,|\,xt_0> &= <x\,'\,|\,\psi(t\,')> \\ &= <x\,'\,|\,e^{-i\mathbf{H}(t\,'-t_0)}\,|\,x> \end{aligned}$$

which is just the matrix element of the time-evolution operator in the "x" basis.

### Path Integrals

Let us find another expression for $<x\,'t\,'\,|\,xt_0>$, one in terms of a path integral.

$<x\,'t\,'\,|\,xt_0>$ is the amplitude to go from $(x,t_0)$ to $(x\,',t\,')$. We *do not know* how the particle got from $(x,t_0)$ to $(x\,',t\,')$. We do not know what path it took. In fact, if we don't try to measure which path it takes, then quantum mechanics tells us that it took all paths, or no paths, however you want to look at it. We can, however, calculate the amplitude that the particle took a certain path. The total amplitude, $<x\,'t\,'\,|\,xt_0>$, will then be the sum of all the amplitudes, one amplitude for each path. The total amplitude will be a sum over paths, weighting each path with its amplitude. This sum over paths is what we call a path integral.

What is the weight, the amplitude, for each path? We can construct this by noting that we could compute $<x't'|xt_0>$ by considering the amplitude to go from $(x,t_0)$ to $(x_1,t_1)$ and then from $(x_1,t_1)$ to $(x',t')$.

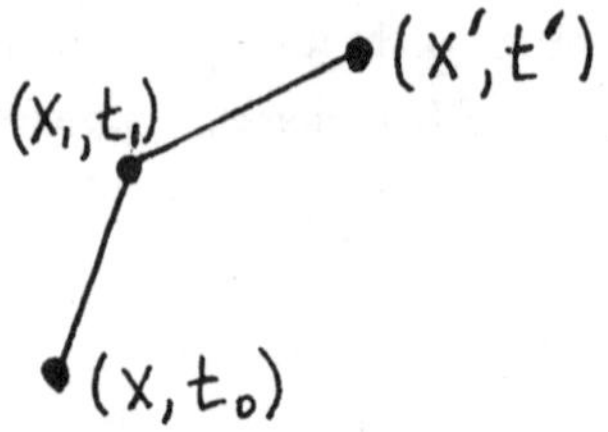


Then

$$<x't'|xt_0> = \int <x't'|x_1t_1><x_1t_1|xt_0>\,dx_1$$

Note that we must sum over all the intermediate values of $x$ at $t_1$ to get the total amplitude. Thus $<x't'|x_1t_1><x_1t_1|xt_0>$ is the amplitude for the particle to go from $(x,t_0)$ through the point $(x_1,t_1)$ to $(x',t')$.

We can also obtain the above expression for $<x't'|xt_0>$ in terms of the matrix elements of the time-evolution operator by noting that

$$1 = \int dx_1\,|x_1><x_1|$$

and inserting 1 into the matrix element of the time-evolution operator.

$$\begin{aligned} <x'|e^{-i\mathbf{H}(t'-t_0)}|x> &= <x'|e^{-i\mathbf{H}(t'-t_1)}e^{-i\mathbf{H}(t_1-t_0)}|x> \\ &= \int <x'|e^{-i\mathbf{H}(t'-t_1)}|x_1><x_1|e^{-i\mathbf{H}(t_1-t_0)}|x>\,dx_1 \\ &= \int <x't'|x_1t_1><x_1t_1|xt_0>\,dx_1 \end{aligned}$$

Now break up the interval between $t_0$ and $t'$ into $N+1$ equal pieces. Make $N$ large so that $t_{i+1}-t_i = \epsilon$ is small.

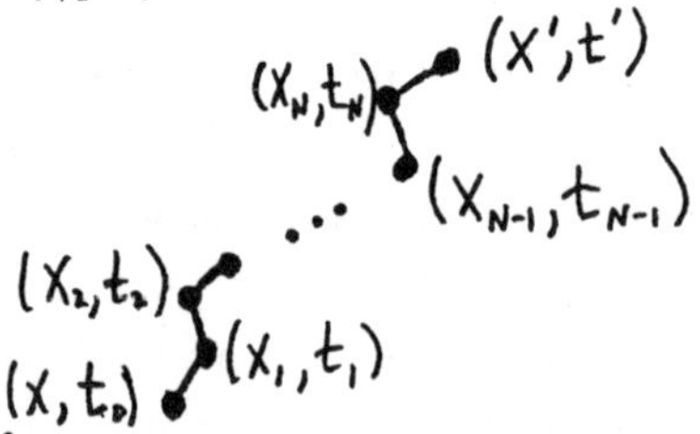


Inserting $1 = \int dx_i\,|x_i><x_i|$ into the matrix element $N$ times we obtain,

$$<x't'|xt_0> = \int <x't'|x_Nt_N> \cdots <x_2t_2|x_1t_1><x_1t_1|xt_0>\,dx_1 \cdots dx_N$$

Now let's look at $<x_{i+1}t_{i+1}|\, x_i t_i>$. Considering that $t_{i+1}-t_i = \epsilon$ is suppose to be small, we expand the expotential.

$$
\begin{aligned}
<x_{i+1}t_{i+1}|\, x_i t_i> &= <x_{i+1}|\, e^{-i\mathbf{H}\epsilon}|\, x_i> \\
&= <x_{i+1}|\, 1 - i\mathbf{H}\epsilon + O(\epsilon^2) + \cdots |\, x_i> \\
&= <x_{i+1}|\, x_i> - i\epsilon<x_{i+1}|\, \mathbf{H}|\, x_i> + \cdots
\end{aligned}
$$

Fourier transform the right-hand side above.

$$<x_{i+1}|\, e^{-i\mathbf{H}\epsilon}|\, x_i> = \int_{-\infty}^{\infty} \frac{dp_i}{2\pi} e^{ip_i(x_{i+1}-x_i)}(1 - i\epsilon H(p_i,(x_{i+1}+x_i)/2) + \cdots$$

Since $\epsilon$ is small, we make little error by dropping the terms of order $\epsilon^2$ and greater, then expotentiating $1 - i\epsilon H$ to obtain

$$<x_{i+1}t_{i+1}|\, x_i t_i> \approx \int_{-\infty}^{\infty} \frac{dp_i}{2\pi} e^{i(p_i(x_{i+1}-x_i) - \epsilon H(p_i,(x_{i-1}+x_i)/2)}$$

Plug this back into our last expression for $<x't'\,|\, xt_0>$ to get

$$
\begin{aligned}
<x't'\,|\, xt_0> &= \lim_{N\to\infty} \int \prod_{i=1}^{N} dx_i\, dp_i \exp(i\sum_{j=1}^{N} p_j(x_{j+1}-x_j) - \frac{t'-t_0}{N+1} H(p_j,(x_{j+1}+x_j)/2)) \\
&\equiv \int Dx\, Dp \exp(i\int_{t_0}^{t'} p\dot{x} - H(p,x)\, dt\,)
\end{aligned}
$$

which is what we call a path integral. Capital $D$ in "$Dx$" above means that the integral is a functional integral. You may have noticed that $x_{i+1}+x_i$ appeared in $H$ above instead of $x_{i+1}-x_i$. This corresponds to a choice of operator ordering (symmetric) while quantizing.

If $H$ is quadratic in $p$ with no funny extra $p$ dependence, then we can carry out the $N$ $p$ integrals (they are Gaussian) and take the limit. The result will be

$$<x't'\,|\, xt_0> = \int Dx\; e^{iS}$$

where $S$ is the classical action. Thus, the amplitude to go from $(x,t_0)$ to $(x',t')$ is found by summing all the paths between $x$ and $x'$ weighting each path by $e^{iS}$. The amplitude for the particle to take a certain path is thus $e^{iS}$.

So far I have been assuming that the particles are free, that they do not interact. Suppose they do self interact, say 1 particle can decay into 2 others.

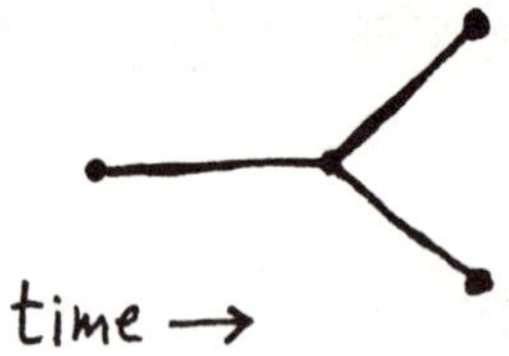


Now the amplitude to go from $(x,t_0)$ to $(x',t')$ is a sum over all paths including all possible interactions which we could characterize as a sum of all possibilities.

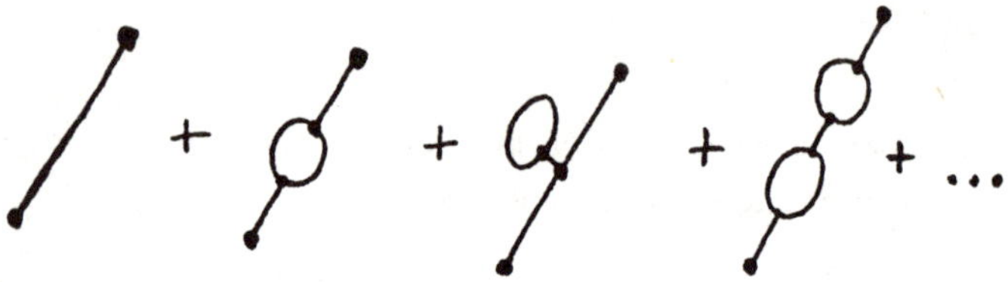

(This picture representation of the path integral is perturbative.) To calculate this sum over all intermediate states all we do is just place the interacting Hamiltonian or action in the path integral. Of course, the path integral will be a lot harder to compute.

### Strings

The idea behind strings is that while an electron looks like a point at $10^{-13}$ cm

if we maginfied it to scales of $10^{-33}$ cm then the electron will look like a tiny string. The idea of building a theory of strings such that all forces and particles are represented by the same string means that we take a single string, excite it one way to get an electron, excite it another way to get a graviton, etc. Now if we want to compute the probability amplitude for a *string* to go from $(x,t_0)$ to $(x',t')$ and again allow the string to interact we must sum over all possibilities.

In other words, we must sum over *surfaces*, the world sheet of the string with quantum fluctuations. We found that the quantum mechanical amplitude for the propagation of point particles was a sum over paths. For strings it will be a sum over surfaces. (The picture above representing the string propagator is perturbative.)

How do we write this sum? For point particles we had $\int Dx \exp(-iS)$ where $S$ was the classical action of the system of point particles so it was a functional of the path of the particle. So, to write this sum for strings, we begin by writing down a classical action for strings. Here is one choice for the bosonic string.

$$S = \frac{-T}{2} \int d^2\xi \sqrt{g}\, g^{ab}\, \partial_a x^\mu \partial_b x^\mu$$

where $T$ is the string tension, a number, $(\xi_1, \xi_2)$ are coordinates on the string world sheet surface, $g_{ab}$ is the metric on the surface, and $x^\mu$ is the map that imbeds the surface in $d$ dimensional Euclidean space.

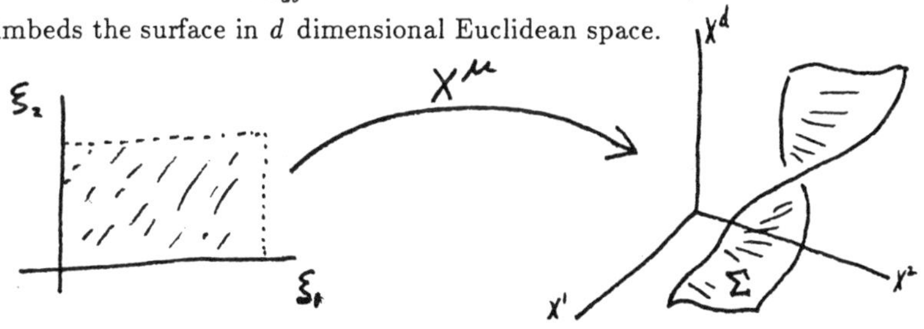


Now we sum over all surfaces which means integrating over all imbeddings, $x^\mu$, and all metrics, $g_{ab}$. Since the target space is Euclidean, the factor of $i$ multiplying $S$ in the path integral is absorbed in $S$ and disappears leaving a minus sign.

$$Z = \int Dg_{ab}\, Dx^\mu \exp -(S + \int d^2\xi \sqrt{g}\, (\lambda R + m_0^2))$$

where $S$ is the action given above, $R$ is the scalar curvature, and $\lambda$ and $m_0^2$ are constants. $Z$ is sometimes called a partition function. You will notice that I added a couple of more terms to the action. This was done for the purposes of renormalization, another topic in itself. But briefly, when we calculate physical quantities with quantum field theory we sometimes find divergences. These divergences are not physical so we want to remove them to leave only the finite physical part. To do this, we change the theory in some drastic way to make everything finite. This is called regularization. Sometimes it may mean cutting off the upper limit of an integral. This cutoff is parametrized, say, by $\Lambda$. Now all physical results will be finite, but will depend on this artificial cutoff. We must remove this cutoff dependence. To do this, we let the constants of the theory depend on this cutoff in such a way that the physical quantities will be independent of the cutoff. ( $\frac{d}{d\Lambda}$ (*physical quantity*) = 0.) To accomplish this we must change the original action by adding terms (counterterms). In anticipation of this, I have added some counterterms.

By the Gauss-Bonnet theorem the first counterterm can be rewritten as

$$\int d^2\xi\sqrt{g}\, R = 4\pi\chi(\Sigma)$$
$$= 4\pi(2 - 2h - b)$$

where $h$ is the genus of the surface $\Sigma$ and $b$ is the number of boundaries. To simplify things a little, let's only consider surfaces without boundaries so $b = 0$. This counterterm does not depend on the metric or the imbedding so we may pull it out of the integral. Thus

$$Z = \sum_{h=0}^{\infty} e^{-4\pi\lambda(2-2h)} \int_{genus=h} Dg_{ab}(\xi)\, Dx^{\mu}(\xi) \exp-(S + \int d^2\xi\sqrt{g}\, m_0^2)$$

Let's follow Polyakov [1], and consider only genus = 0. This is the first term in the expansion above, call it $Z_0$. The relative importance of the higher genus integrals depends on the size of the coupling $\lambda$. If $\lambda$ is "small", perturbation theory will be valid and we will only need to compute $Z$ for small genus to see what is going on. If $\lambda$ is large, we are in trouble. The size of $\lambda$ is a question of dynamics.

Our classical action is reparameterization invariant and conformally invariant. This presents a problem we have not discussed yet. Let's digress a moment to a simpler system to see what the problem is and what to do.

Consider free photons. They are described by the electromagnetic 4-potential, $A^{\mu}$. Since we have no charged particles present set the temporal component to 0, $A^0 = 0$. The classical action is

$$S = \int \vec{E}^2 - \vec{B}^2 \, d^3x \, dt$$

where $\vec{E} = -\dot{\vec{A}}$ is the electric field, and $\vec{B} = \nabla \times \vec{A}$ is the magnetic field. The partition function is

$$Z = \int DA_i(x) e^{iS}$$

The action is gauge invariant. For $A_i \to A_i + \nabla_i U(\vec{x})$, the electric field and magnetic field remain unchanged, $E_i \to E_i$ and $B_i \to B_i$, so the classical action doesn't change. Decompose the vector potential into longitudinal, $\vec{A}_L$, and transverse, $\vec{A}_T$, components. ($\nabla \cdot \vec{A}_T = 0$ and $\nabla \times \vec{A}_L = 0$). By applying a gauge transformation we can see that $\vec{A}_T$ is gauge invariant while $\vec{A}_L$ is not. The partition function can be rewritten as

$$Z = \int DA_L \int DA_T \, e^{iS[A_T]}$$

$\int DA_L$ represents a huge infinity. This is the problem that arises when symmetries are present. We want to remove this infinity (put it into the normalization). The physically relevant part is the integral over $A_T$. So the real partition function that we want to calculate is

$$Z = \int DA_T \, e^{iS[A_T]}$$

For the case of free photons, it is easy to seperate out the infinity due to the presence of the symmetry. For most theories, however, we do not have such a nic{ decomposition of the field into gauge invariant and gauge dependent parts. So we would like to develop some procedure to carry this out. This procedure was developed by Faddeev and Popov.

We begin by inserting a funny version of 1 into the functional integral. We write 1 as $\int D(\nabla \cdot A)\, \delta[\nabla \cdot A]$, a functional integral over $\nabla \cdot A$ which is equal to 1 by the definition of the delta functional.

$$Z = \int DA_L \, DA_T \, e^{iS[A_T]} = \int DA_L \, DA_T \, D(\nabla \cdot A)\, \delta[\nabla \cdot A] e^{iS[A_T]}$$

$$= \int D(\nabla \cdot A) \int DA_i \, \delta[\nabla \cdot A] \, e^{iS[A]}$$

Now we change variables from $\nabla \cdot \vec{A}(\vec{x})$ to $U(\vec{x})$, the arbitrary function in the gauge transform.

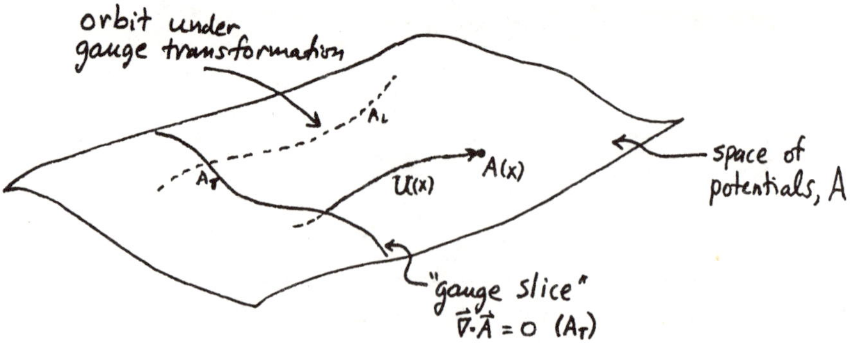


The change of variables requires us to include a Jacobian. Since this is a functional integral, the Jacobian is the determinant of an operator. Changing variables we have

$$Z = \int DA_i\, \delta[\nabla\cdot A]\, \det\left|\frac{\delta(\nabla\cdot A)}{\delta U}\right| e^{iS[A]}$$

Nothing in the integral depends on $U(\vec{x})$ (all of it is gauge invariant) so we can put $\int DU(\vec{x})$ into the normalization. $\int DU(\vec{x})$ is called the gauge volume. The partition function we want to calculate is

$$Z = \int DA_i\, \delta[\nabla\cdot A]\, \det\left|\frac{\delta(\nabla\cdot A)}{\delta U}\right| e^{iS[A]}$$

What we have done is called "choosing a gauge." The gauge choice is $\nabla\cdot A = 0$ (that is, $A_L = 0$). We must fix the gauge to quantize. The physical partition function is the functional integral over the gauge slice only, not over the entire space of 4-potentials. To accomplish this, we choose a gauge, and insert a delta functional to integrate over the gauge slice only. Next, change coordinates so that any 4-potential off the slice is written as the gauge transform, $U(\vec{x})$, it takes to go from the slice to the "off-slice" 4-potential on the gauge orbit and throw away the integral over the gauge transforms. Sometimes this procedure is symbolically written as

$$Z = \frac{\int DA\, e^{iS[A]}}{V}$$

where $V$ is the volume of the gauge group. For free photons the Faddeev-Popov determinant doesn't depend on $\vec{A}$ so it may also be absorbed into the normalization. ($\nabla\cdot A(U) = \nabla\cdot(A_i + \nabla_i U) = \nabla\cdot A + \nabla^2 U$ so $\det\left|\frac{\delta(\nabla\cdot A)}{\delta U}\right| = \det(\nabla^2)$.)

Now we must do the same procedure for strings. The "gauge" symmetry is the reparameterization invariance of the action. So first we must choose a "gauge." A popular choice is

$$g_{ab} = e^{\phi(\xi)}\delta_{ab},$$

the "conformally flat" gauge. So our gauge condition is $G_{ab} = g_{ab} - e^{\phi(\xi)}\delta_{ab} = 0$. Insert $\delta[G_{ab}]$ into the partition function. Change variables to $U(\xi)$, symbolic for diffeomorphisms that take the metric off the conformally flat slice without sliding along it.

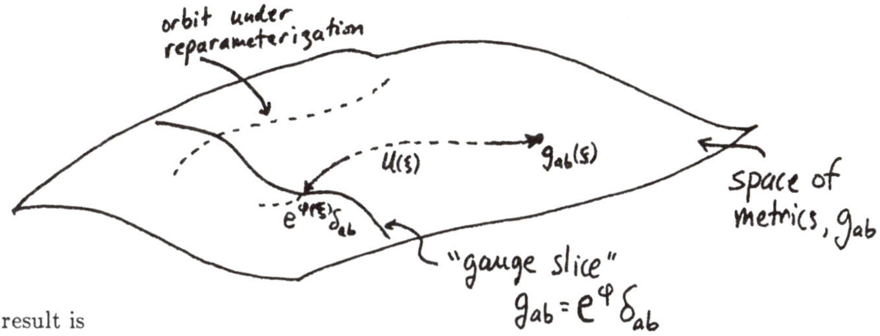


The result is

$$Z_0 = \int DU(\xi) \int Dg_{ab}\, Dx^\mu\, \delta[G_{ab}] \det\left|\frac{\delta G_{ab}}{\delta U(\xi)}\right| \exp-(S + \int d^2\xi\sqrt{g}\, m_0^2)$$

$Z_0$ is a gaussian integral so we can evaluate it explicitly. We will just state the result. The computation of the conformal anomaly is another topic in itself. The result is

$$Z_0 = \int D\phi(\xi) \exp(-(\frac{26-d}{48\pi})\int d^2\xi \frac{1}{2}(\partial_\mu\phi)^2 + m^2 e^\phi)$$

$d$ is the dimension of the Euclidean space-time. It enters the result above through the integral over the imbeddings. If $d \neq 26$, the partition function looks like the partition function for a 2 dimensional quantum field theory for the field $\phi(\xi)$, with the classical action given by the argument of the expotential. This action is called the Liouville model. It describes how these strings behave in non-critical ($d \neq 26$) dimensions.

If $d = 26$, $Z_0 = \int D\phi(\xi)$ which just represents the volume associated with the conformal group. It is absorbed into the normalization, too. So for $d = 26$, the effective action is conformally invariant (no anomaly) so we write the partition function as

$$Z = \frac{\int Dg_{ab}\, Dx^{\mu}\, e^{-S + c.t.}}{V_{diff_0} V_{conf}}$$

and we are integrating over the space of metrics divided by the group of diffeomorphisms connected to the identity, and the group of conformal transformations on the surface, i.e. Teichmüller space. If not all of the diffeomorphisms can be connected to the identity, which is the case for genus $\neq 0$, then we also divide by the mapping class group and integrate over moduli space. This will be discussed by Polchinski.

### Reference

[1] A.M. Polyakov, Phys. Lett. **103B** (1981) 207.

# FROM POLYAKOV TO MODULI

Joseph Polchinski*

*Theory Group*
*Department of Physics*
*University of Texas*
*Austin, TX 78712*

ABSTRACT

We review the reduction of the Polyakov functional integral to an integral over moduli space.

*Supported in part by the A. P. Sloan Foundation, the Robert A. Welch Foundation and NSF Grant Number 8605978.

I would like to explain the reduction of the Polyakov functional integral[1] to an integral over the moduli space of Riemann surfaces. The relevance of moduli in the Polyakov path integral was first discussed by Friedan[2]. Alvarez worked out most of the necessary formalism in ref. 3. I then worked out a specific example in some detail[4] and Moore and Nelson[5] and D'Hoker and Phong[6] completed the general formalism. Since these papers, a great deal of progress has been made in a number of directions. My talk will be introductory: I will outline the approach of refs. 2-6. Some of the more recent developments will be discussed here by Alvarez, Friedan, Giddings, Nelson and others.

The starting point is the Polyakov functional integral, described in the talk by Hatfield:

$$Z = \sum_g \int \frac{Dg_{ab}\ Dx^\mu}{V_{\text{diff x Weyl}}} e^{-S} \qquad (1)$$

where the action is

$$S = \int d^2\sigma \sqrt{g}\ (\frac{T}{2} g^{ab}\partial_a x^\mu \partial_b x^\mu + \lambda R). \qquad (2)$$

In (1), one is integrating over all embeddings $x^\mu$ of a two dimensional surface in a d-dimensional Euclidean spacetime, and over all metrics $g_{ab}$ on the surface. The constant T is the string tension. The constant $\lambda$ determines the relative weight of different topologies, as $\int \sqrt{g}\ R$ is proportional to the Euler number. For simplicity we restrict attention to closed, oriented surfaces, corresponding to the world-sheet of a closed oriented string. The $\sum_g$ indicates that we are to sum over all topologies. The action (2) has two local symmetries, (1) diffeomorphisms of the two dimensional coordinate $\sigma$, and (2) Weyl transformations, $g_{ab}(\sigma) \rightarrow e^{\phi(\sigma)} g_{ab}(\sigma)$. Metrics which are related under this diff x Weyl group are to be identified. Thus, we are actually integrating only over conformal structures on the surface. Heuristically, one eliminates the overcounting in $Dg_{ab}$ by dividing out the volume of the local symmetry group, as indicated in (1). This is the Fadeev-Popov procedure, described also in Hatfield's talk. Finally, since conformal structures on a 2-d surface are equivalent to complex structures, we are actually integrating in the end over the moduli space of complex structures on the surface.

We will define the functional measures and volumes in the integral (1) in a formal but physically well motivated way, and then show that we obtain a well defined measure on moduli space. Let moduli space be labelled by n real coordinates $t^i$ ($n = 6g-6$ for $g > 1$). In order to carry out the Fadeev-Popov procedure it is necessary to pick a gauge slice, a representative metric $g_{ab}(t)$ from each gauge equivalence class. Later we will discuss the dependence on the choice of slice. If we consider a general metric variation about a given such metric, it can be decomposed as

$$\delta g_{ab} = \text{gauge variation} + \text{modular variation}$$

$$= (\nabla_a \delta\xi_b + \nabla_b \delta\xi_a + \delta\phi\, g_{ab}) + g_{ab,i}\delta t^i. \qquad (3)$$

where $\delta\xi_a$ and $\delta\phi$ are infinitesimal coordinate and Weyl transformations. Thus we can change variables

$$Dg_{ab} = D\xi_a\, D\phi\, d^n t\, J \qquad (4)$$

where J is a Jacobian to be determined. The Fadeev-Popov procedure amounts to dropping the $\xi$ and $\phi$ integrals as gauge redundancies, being left with $d^n t\, J$ as the measure on moduli space. (Actually, for reasons to be discussed later, it is J times the result of Dx integration which should be regarded as a measure.).

We will define the functional measures in (1) and (4) by assigning values to certain Gaussian functional integrals. The Jacobian in (4) and the result of the x-integration can then be obtained by formal comparison of Gaussian integrals. Define first metrics for field variations

$$\begin{aligned} \langle\delta g,\delta g\rangle &= \int d^2\sigma \sqrt{g}\, g^{ac}g^{bd}\delta g_{ab}\delta g_{cd} \\ \langle\delta\xi,\delta\xi\rangle &= \int d^2\sigma \sqrt{g}\, g^{ab}\delta\xi_a\delta\xi_b \\ \langle\delta\phi,\delta\phi\rangle &= \int d^2\sigma \sqrt{g}\, \delta\phi^2 \\ \langle\delta x,\delta x\rangle &= \int d^2\sigma \sqrt{g}\, \delta x^\mu\delta x^\mu \end{aligned} \tag{5}$$

These variations live in the tangent space to field space (denoted Tg, Tξ,•••). Now use these metrics to define measures on the tangent spaces by setting

$$\int D\delta g_{ab}\; e^{-\langle\delta g,\delta g\rangle} \equiv 1 \tag{6a}$$

$$\int D\delta\xi_a\; e^{-\langle\delta\xi,\delta\xi\rangle} \equiv 1 \tag{6b}$$

$$\int D\delta\phi\; e^{-\langle\delta\phi,\delta\phi\rangle} \equiv 1 \tag{6c}$$

$$\int D\delta x\; e^{-\langle\delta x,\delta x\rangle} \equiv 1 \tag{6d}$$

Eqs. (6) define measures on the tangent space to the various field spaces, while eq. (4) refers to the measure on field space itself. However, there is a natural equivalence between these, so that we can rewrite (4) as

$$D\delta g_{ab} = D\delta\xi_a \, D\delta\phi \, d^n\delta t \, J. \tag{7}$$

By inserting (4) and (7) into equation (6a) and comparing to eq. (6b,c), we will obtain J in terms of determinants. First, using (4), we have

$$\langle\delta g,\delta g\rangle = 2\langle\delta\phi+\cdots, \delta\phi+\cdots\rangle + \langle P_1\delta\xi + \chi_i\delta t^i, P_1\delta\xi + \chi_i\delta t^i\rangle \tag{8}$$

The two terms on the right of (8) are respectively from the trace and the traceless part of (4). The terms denoted by ellipses are unimportant. The notation $P_1\delta\xi$ denotes the traceless part of $\nabla_a\xi_b + \nabla_b\xi_a$. Thus $P_1$ is the covariant derivative taking rank 1 tensors into traceless rank 2 tensors. $P_n$ would be defined analogously. Also, $\chi_i$ is the traceless part of $g_{ab,i}$. In order to simplify the second term in (8), we wish to eliminate the cross term between $\delta\xi$ and $\delta t^i$. The $\chi_i$ represent $(Tg)_{traceless}/\mathrm{range}(P_1)$. This is dual to the space $\ker(P_1^\dagger)$, for which we define a (real) basis $\psi_j$. We have the identity

$$\begin{aligned}\chi_i &= P_1 \frac{1}{P_1^\dagger P_1} P_1^\dagger \chi_i + \psi_j(\langle\psi,\psi\rangle^{-1})_{jk}\langle\psi_k,\chi_i\rangle \\ &\equiv P_1 V_i + \psi_j(\langle\psi,\psi\rangle^{-1})_{jk}\langle\psi_k,\chi_i\rangle \end{aligned} \tag{9}$$

where for convenience we assume $\ker(P_1) = 0$.

Thus

$$\langle\delta g,\delta g\rangle = 2\langle\delta\phi + \cdots,\ \delta\phi + \cdots\rangle$$

$$+ \langle P_1(\delta\xi + v_i\delta t^i), P_1(\delta\xi + v_i\delta t^i)\rangle \qquad (10)$$

$$+ \langle\chi_i\delta t^i,\psi_j\rangle(\langle\psi,\psi\rangle^{-1})_{jk}\langle\psi_k,\chi_\ell\delta t^\ell\rangle$$

and

$$1 = J\int d\delta\xi\ d\delta\phi\ d\delta t\ e^{-\langle\delta g,\delta g\rangle} \qquad (11)$$

with $\langle\delta g,\delta g\rangle$ in (10). The ellipses and the $v_i$ terms in (10) can be absorbed in shifts of $\delta\phi$ and $\delta\xi$ respectively. We can then evaluate (11) by comparison with (6b,c) to give*

$$J = \det{}^{1/2}(P_1{}^\dagger P_1)\ \det\langle\chi_i,\psi_j\rangle\ \det{}^{-1/2}\langle\psi_j,\psi_k\rangle \qquad (12)$$

In the x-integral, it is necessary to divide $x^\mu(\sigma)$ into a constant mode plus an orthogonal part, $x^\mu(\sigma) = x_0^\mu + x^{\mu\prime}(\sigma)$. The integral over $x^\mu$ is again Gaussian and gives

$$\det{}'^{-d/2}\ \nabla^2. \qquad (13)$$

---

*We are being a bit careless with factors like $2\pi$, but these can in principle be determined[4,7].

where the prime excludes the zero eigenvalue.

The integral over $x_0$ is unbounded; to define it we must take spacetime to have a finite volume V. Then we obtain

$$V(\int d^2\sigma \sqrt{g}\,)^{d/2} \tag{14}$$

In all, we have[5,6]

$$Z = V\int d^n t\ \det\langle\chi_i,\psi_j\rangle\left\{\left(\frac{\det P_1^{\dagger}P_1}{\det\langle\psi,\psi\rangle}\right)^{1/2}\left(\frac{\det'\nabla^2}{\int d^2\sigma\sqrt{g}}\right)^{-d/2}\right\} \tag{15}$$

We must discuss the range of integration in (15). Gauge equivalent configurations are to be counted only once. The gauge group has a piece $\mathrm{diff}_0 \times \mathrm{Weyl}$ which is connected to the identity. There remains the mapping class, or modular, group, $\mathrm{diff}/\mathrm{diff}_0$. If we integrate over $\{\mathrm{metrics}\}/\mathrm{diff}_0 \times \mathrm{Weyl}$, which is Teichmuller space, we overcount. The integration in (15) thus runs over moduli space, $\{\mathrm{metrics}\}/\mathrm{diff} \times \mathrm{Weyl}$. (An additional numerical constant must be divided out if some elements of the modular group act trivially on the metric).

The determinants of differential operators in (15) are renormalized as usual in quantum field theory. For example, one can use ζ-function renormalization

$$\ln\det\Delta \equiv \lim_{s\to 0}\frac{d}{ds}\left(\frac{1}{\Gamma(s)}\int_0^\infty dt\ t^{s-1}\sum_i e^{-t\lambda_i}\right) \tag{16}$$

where $\lambda_i$ are the eigenvalues of the operator Δ. For Re(s) sufficiently large and positive, the sum and integral on the right of (16) converge. The value as $s \to 0$ is then defined by analytic continuation.

Let us give the explicit form of (15) for genus 1. We take $\sigma^1$ and $\sigma^2$ to be periodic from 0 to 1. A good slice is given by metrics of the form

$$ds^2 = |d\sigma^1 + \tau\, d\sigma^2|^2 \tag{17}$$

with $\tau$ a complex parameter in the upper half plane. There is one additional subtlety at genus 1. We have assumed in deriving (15) that $\ker(P_1) = 0$. When $\ker(P_1)$ is nonzero, as occurs for genus < 2, some additional steps parallel to the treatment of $\ker(P_1^{\dagger})$ are needed, and (15) is slightly modified. All the factors in the resulting expression are readily evaluated, giving

$$Z = \int \frac{d^2\tau}{2\,\mathrm{Im}(\tau)^2} \left(2\pi\, \mathrm{Im}(\tau)\right)^{1-d/2} |\eta(\tau)|^{-2d+4} \tag{18}$$

where $\eta(\tau)$ is the Dedekind eta-function.

$$\eta(\tau) = e^{i\pi\tau/12} \prod_{n=1}^{\infty} (1 - e^{2\pi i n\tau}). \tag{19}$$

The $\tau$ integral runs over a fundamental region of the modular group, for example

$$|\mathrm{Re}(\tau)| < \frac{1}{2} \qquad |\tau| > 1. \tag{20}$$

We will now discuss a number of issues, starting with the procedure that led to the integrals (15) and (18) over moduli space, then some of the properties of these expressions, and some generalizations. (In the references I will not attempt to be exhaustive, which would be impossible, but will try to list some useful papers).

1. The guiding principle used to define the functional integrations in (1) is locality on the string world-sheet. Thus, in the definitions (6a–d) of the measures, the left-hand side is formally a product of independent integrals at each point of the world-sheet. The right-hand side must have this same structure, so it must be exponential of an integral of a local function. Requiring renormalizability, the most general such integral would have the form of the action itself, and can be absorbed into redefinitions of the constants T and $\lambda$; thus we can take the right-hand sides in equation (6) all to be 1. Actually, there is an additional operator, $\int d^2\sigma \sqrt{g}$, which is allowed by renormalizability, but forbidden by Weyl invariance; see point 2 below. Other choices made in the procedure, such as the normalizations of the metrics (5), the omission of a second tensor structure in $\langle \delta g, \delta g \rangle$, and the particular renormalization procedure (16) all can be absorbed in the definitions of T and $\lambda$.

2. In order to carry out the Fadeev-Popov procedure, we had to single out a gauge slice: is the final result independent of slice? The metrics (5) and renormalization (16) are explicitly invariant under the full diffeomorphism group. It is not possible, however, to simultaneously preserve Weyl invariance at these stages. One must check the Weyl dependence of the final expression. It was shown by Polyakov[1] that quantization destroys Weyl invariance except in the critical dimension, d=26 for the bosonic string theory considered here. (Some renormalization procedures will introduce a spurious Weyl violation even in d=26, which can be absorbed by adding a term of the form $\int d^2\sigma \sqrt{g}$ into the action.) Only the combination of the Jacobian J and the x-integration is independent of slice.

3. In (15), the slice independence is further obscured by the need to separate out the zero modes of $P_0$ and $P_1^{\dagger}$. As shown in refs. 3, 5, it is only the combination in braces, and not any of the separate factors, which is slice independent. Note that the object in braces is not precisely a function on moduli space, as it depends also on the basis $\psi_i$ chosen. This basis dependence cancels against that of $\det\langle\chi,\psi\rangle$. The matrix $\langle\chi_i,\psi_j\rangle$ is separately slice independent: In complex coordinates $ds^2 = 2g_{z\bar{z}}dzd\bar{z}$, the $\psi_j$ are quadratic differentials, $\partial_{\bar{z}}\psi_{jzz} = 0$. The $\chi_i$ are related to slice-independent Beltrami differentials $\mu_{i\bar{z}}{}^{z} = g^{\bar{z}z}\chi_{i\bar{z}\bar{z}}$. Then

$$\langle\chi_i,\psi_j\rangle = \int d^2z\left(\mu_{i\bar{z}}{}^{z}\psi_{jzz} + \mu_{iz}{}^{\bar{z}}\psi_{j\bar{z}\bar{z}}\right) \qquad (21)$$

does not depend on $g_{z\bar{z}}$. Finally $d^nt \det\langle\chi,\psi\rangle$ is a volume element on moduli space. Equivalent Beltrami differentials, $\mu_{i\bar{z}}{}^{z} \sim \mu_{i\bar{z}}{}^{z} + \partial_{\bar{z}}\xi_i{}^{z}$, are seen to give identical results after integration by parts. In applications, it is sometimes convenient to take the slice of constant curvature metrics. In this case, the combination $d^nt \det\langle\chi,\psi\rangle \det^{-1/2}\langle\psi,\psi\rangle$ becomes the Weil-Peterson measure[6]. However, the significance of this is sometimes overemphasized, as these factors by themselves have little meaning. Rather, it is the whole integrand in (15) which should be regarded as the measure, since, as we have discussed above, the requirements of slice-independence and $\psi$-basis independence unify this integrand into an indivisible whole[5].

4. The quantity Z has the physical interpretation of a vacuum-to-vacuum amplitude. Amplitudes between initial and final states containing particles are obtained by inserting into the integral (1) local functions of $x^{\mu}$ and $g_{ab}$ known as vertex operators. The term operator is used because when the functional integral is given a Hilbert space interpretation, such insertions become linear

operators on the string Hilbert space. These operators are required to be invariant under the Weyl group and, as with the measure, quantization introduces anomalous Weyl dependence; see, for example, refs. 1, 8, 9 for further discussion. Under diffeomorphisms they transform as densities, so that they form invariants after integration over the world sheet. One also considers conserved currents, which transform as one-forms and so can naturally be integrated along contours, and the stress tensor, which transforms as a symmetric rank two tensor and so can be naturally convoluted with a Beltrami differential. In a Hilbert space interpretation, moments of conserved currents and of the stress tensor generate affine Lie algebras, the Kac-Moody and Virasoro algebras respectively; see, for example, refs. 10, 11.

5. The genus 1 modular integrand (18) is seen to be a fairly simple function of $\tau$ times the square of a holomorphic function. A striking recent development is the realization that this structure holds for arbitrary genus[12]. This holomorphic factorization leads to many connections with algebraic geometry; refs. 13-17 give some recent applications.

In complex coordinates, the classical equation of motion for $x^\mu$ resulting from the action (2) is

$$\partial_z \partial_{\bar{z}} x^\mu = 0 \tag{22}$$

to which the general local solution is an arbitrary analytic function of z plus an arbitrary antianalytic function. In the Minkowski space string theory from which (1) is analytically continued, this separation corresponds to the independence of right-moving and left-moving modes propagating on the string. This separation is central to string theory and has a number of important consequences in the quantum theory. For example, the operator algebras which arise tend to separate into analytic and antianalytic parts. The holomorphic factorization of the modular integrand discussed above is a second consequence.

6. The string theory described in (1) can be generalized in a number of ways. Additional fields, of various types, can be added to the string world sheet. In particular, the theory can be supersymmetrized, so that the world-sheet becomes a supermanifold and the variables of integration superfields. The resulting functional integrand can be reduced to an integration over a supermoduli space[18] (the critical dimension is now $d=10$).

The most interesting generalizations are chiral string theories, in which the right- and left-moving strings differ, either in their field content or at least in their spin structure or periodicity conditions (unlike (22), where $x^\mu$ propagates both right and left). In this case, one can no longer define diffeomorphism invariant metrics as in (5), and so the procedure leading to the analog of (15) is not so straightforward. However, it is easy to guess what the answer must be. If we consider two different non-chiral strings, the modular integrand for each will be essentially the square of a holomorphic function. The chiral string obtained by combining the right-moving sector of one string with the left-moving sector of the other must correspond to the holomorphic part of one integrand times the antiholomorphic part of the other.

In non-chiral strings, diffeomorphism invariance of the metrics as in (5) guarantees independence of gauge slices both under $\mathrm{diff}_0$ and in fact under the full diff group. In the chiral strings, the process of taking holomorphic square roots leads to potential anomalous phases; absence of anomalies puts constraints on the theories. The connected gauge group $\mathrm{diff}_0 \times \mathrm{Weyl}$, and the remainder, the modular group, give different sorts of constraints. The group $\mathrm{diff}_0 \times \mathrm{Weyl}$ puts restrictions on the local field content. The modular group puts further restrictions on this content but also puts global restrictions.

For example, when there are spinors on the world sheet, modular invariance strongly constrains the manner in which the various spin structures are summed[14,19].

7. One important generalization that we should mention, although we will give no details, is the replacement of the flat d-dimensional Euclidean metric in (1) with a general Riemannian metric. In addition, other background fields may be present. Weyl invariance puts restrictions on the possible backgrounds: in essence, the background must satisfy the string equations of motion[20]. The x-integration is now no longer Gaussian, although for some special manifolds it can still be carried out exactly.

8. Our starting point, a functional integral over local fields (1), is physically well motivated but mathematically ill defined. Our result (15) is mathematically well defined but a key piece of physics, the locality, is obscured, as the Teichmuller parameters depend globally on the world sheet metric. Locality is important to assure a sensible quantum mechanical interpretation. The metrics (5) and Gaussian integrals (6) that we introduced were crutches to assure that we maintained locality. However, this procedure is not totally satisfactory. For example, as noted in point 6, it is difficult or impossible to carry out in chiral theories. It would be desirable either to develop some method for seeing locality in the modular integral (15), or to have a better defined starting point. Conformal field theory may be that starting point[10,17]. For some discussion of locality in conformal field theory, see ref. 9.

9. One consequence of locality is a factorization which occurs in the limit of a degenerating Riemann surface. When a handle pinches off, it leads to a Riemann surface, possibly disconnected, with two nodes. The asymptotics of the modular integrand for the pinching handle are given by an infinite series in

the pinch parameter $q$ where the coefficients are modular integrands for the Riemann surface with nodes, with vertex operators located at the nodes. For some discussion of this factorization, see refs. 14-17.

References


1. A.M. Polyakov, Phys. Lett. B103 (1981) 207, 211.

2. D. Friedan, at the Copenhagen Workshop on String Models (1981) unpublished.

3. O. Alvarez, Nucl. Phys. B216 (1983) 125.

4. J. Polchinski, Comm. Math. Phys. 104 (1986) 37.

5. G. Moore and P. Nelson, Nucl. Phys. B266 (1986) 58.

6. E. D'Hoker and D. Phong, Nucl. Phys. B269 (1986) 205.

7. W. Weisberger, "Normalization of the Path Integral Measure and the Coupling Constants for Bosonic Strings," Stony Brook preprint (1986).

8. S. Weinberg, Phys. Lett. B156 (1985) 309.

9. J. Polchinski, "Vertex Operators in the Polyakov Path Integral," Texas preprint UTTG-27-86 (1986).

10. A.A. Belavin, A.M. Polyakov, and A.B. Zamolodchikov, Nucl.Phys. B241 (1984) 333.
D. Friedan, E. Martinec, and S. Shenker, Nucl. Phys. B271 (1986) 93.

11. P. Goddard and D. Olive, in Vertex Operators in Mathematics and Physics, ed. J. Lepowsky (Springer-Verlag, New York, 1984). I. Bars, ibid. S. Mandelstam, ibid.

12. A. Belavin and V. Knizhnik, Phys. Lett. B168 (1986) 201; "Complex Geometry and Theory of Quantum Strings," Landau Inst. preprint (1986).
J. Bost and J. Jolicoeur, Phys. Lett. B174 (1986) 273.
R. Catenacci, M. Cornalba, M. Martinelli, and C. Reina, Phys. Lett. B172 (1986) 328.

13. Yu. I. Manin, JETP Lett. 43 (1986) 204.

14. L. Alvarez-Gaume, G. Moore and C. Vafa, Comm. Math. Phys. 106 (1986) 1.

15. P. Nelson, "Lectures on Strings and Moduli Space," to appear in Phys. Reports.

16. D. Friedan and S. Shenker, "The Integrable Analytic Geometry of String," Chicago preprint (1986).

17. E. Martinec, "Conformal Field Theory on a (Super) Riemann Surface," Princeton preprint (1986).

18. G. Moore, P. Nelson, and J. Polchinski, Phys. Lett. B169 (1986) 47.
E. D'Hoker and D. Phong, Nucl. Phys. B278 (1986) 225.
S. Chandhuri, H. Kawai, and S.H. Tye, Cornell preprint (1986).

19. N. Seiberg and E. Witten, Nucl. Phys B276 (1986) 272.
L. Alvarez-Gaume, P. Ginsparg, G. Moore and C. Vafa, Phys. Lett. B171 (1986) 155.
L. Dixon and J. Harvey, Nucl. Phys. B274 (1986) 93.

20. C. Callan, D. Friedan, E. Martinec and M. Perry, Nucl. Phys. B262 (1985) 593.

# GEOMETRY OF QUANTUM STRINGS*+

by

Eric D'Hoker
Physics Department
Princeton University
Princeton, N.J. 08544

D. H. Phong
Mathematics Department
Columbia University
New York, N.Y. 10027

---

* Contribution to Proceedings of "Mathematical Aspects of String Theory" Conference at San Diego, July 1986

+ Work supported in part by the National Science Foundation under Grants PHY-80-19754 and DMS-84-02710.

## I. Introduction

String theories are prime candidates for consistent unified theories of all interactions including gravity [1]. It is likely that string scattering amplitudes may not even require renormalization, they are just finite. These are indications obtained from analysis at the one-loop order, and it is an important issue to understand consistency and finiteness to all orders of string perturbation theory.

In this article we shall give a brief survey of some recent progresses in the study of higher string loop amplitudes, which have revealed an amazingly rich structure. The approach to string quantization developed here is that of Polyakov [2], which is the one where the role of (super) geometry and (super) moduli space is perhaps most transparent. Other approaches are the light cone gauge approach [3] and conformal field theory [4]. Some early work from dual models can be found in [5].

# I. QUANTIZATION OF STRINGS

## 1. Bosonic Strings

The action for the bosonic string in the Polyakov formulation is that of a sigma model with space-time as target manifold

$$I(X,g) = \frac{T}{2}\int d^2\xi\sqrt{g}\, g^{mn}\partial_m X^\mu \partial_n X^\nu G_{\mu\nu}(X) \qquad , \mu = 1,\dots, d = 26 \qquad (1.1)$$

Here g is a metric on the world sheet which is non-dynamical classically. At $\delta I/\delta g = 0$ g is conformal to the metric induced from space-time $\partial_m X^\mu \partial_n X^\nu G_{\mu\nu}(X)$ and I reduces to the area of the world sheet, which is the natural generalization of the action for point particles. To quantize the theory we must consider all configurations of X and g. A key ingredient of the quantum theory is the partition function

$$Z = \int DgDX^\mu \exp(-I(X,g))/\mathcal{N} \qquad (1.2)$$

which can be interpreted as the space-time integral of the cosmological constant. Here $\mathcal{N}$ is a normalization factor which corresponds to the volumes of the symmetry groups of the theory. For the bosonic string these are

(i) Weyl scalings: $g = \exp(2\sigma)\hat{g}$

(ii) Reparametrizations of the world sheet

We shall denote the second group (ii) by Diff(M). It will be convenient to separate it into $\text{Diff}_0(M)$ (the group of diffeomorphisms connected to the

m

identity, generated by vector fields $\delta V$ on M), and the mapping class group $\Gamma = \text{Diff}(M)/\text{Diff}_0(M)$.

We shall analyze the theory in the case of flat space-time $G_{\mu\nu} = \delta_{\mu\nu}$. In this case there will be an additional Poincaré symmetry on the $X^\mu$'s. It should also be noted that renormalization of (1.2) will introduce counterterms of the form of a two-dimensional cosmological constant, and of a correction to the string coupling constant. To simplify the notation we shall not write them down explicitly.

For flat space-time the integral over the matter fields $DX^\mu$ is Gaussian and (1.2) becomes

$$Z = \int Dg \left(\frac{2\pi \det{}'\Delta_g}{\int d^2\xi\sqrt{g}}\right)^{-13} / \mathcal{N}$$

In the absence of anomalies the integral over metrics should now reduce to an integral over cosets {(Metrics)/Weyl×Diff(M)}. This space is exactly the Moduli space $\mathfrak{M}_h$ of the surface M, i.e., the space of inequivalent complex structures on M. Actually the Weyl group is anomalous, but the conformal anomalies between matter fields and gauge fixing cancel when d = 26, so the above principle applies.

The physics of strings can be gathered from amplitudes for absorption and emission of general particles. Particles correspond to vertex operators

$$V_i(k) = \int d^2\xi\sqrt{g}\, U_i(\varepsilon, DX, R)\, \exp(ik_\mu X^\mu)$$

where $U_i$ are polynomials in the covariant derivatives of $X^\mu$ and of the scalar curvature R, and $\varepsilon$ is a polarization tensor. The $V_i$'s should be consistent with the symmetries of the theory upon quantization. Thus the above expression is dictated by space-time Poincaré invariance and world sheet reparametrization invariance. Weyl symmetry on the other hand is anomalous, and the dependence on conformal factors arising from various terms in the U's is required to cancel after inclusion of possible anomalies. It is this requirement which dictates in fact the mass spectrum and the physical states. Some examples for the bosonic string are (in Minkowski signature) [6][7]

$$k^2 = -8\pi \qquad V_{-1} = \int d^2\xi\sqrt{g}\ \exp(ik_\mu X^\mu)$$

$$k^2 = 0 \qquad V_0 = \int d^2\xi\sqrt{g}\ \exp(ik_\mu X^\mu)(\varepsilon_{\mu\nu}\partial_z X^\mu \partial_{\bar{z}} X^\nu + \beta R + \gamma)$$

$$k^\mu k^\nu \varepsilon_{\mu\nu} = 0,\ \beta = \frac{1}{2\pi}\ \varepsilon_\mu{}^\mu - \frac{1}{16\pi}(\varepsilon_1 + \varepsilon_2)\quad \gamma = -\frac{1}{8\pi\varepsilon}\ \varepsilon_\mu{}^\mu$$

$$k^2 = 8\pi \qquad V_1 = \int d^2\xi\sqrt{g}\ \exp(ik_\mu X^\mu)\varepsilon_{\mu\nu,\kappa\lambda}(D_z X^\mu D_z X^\nu D_{\bar{z}} X^\kappa D_{\bar{z}} X^\lambda + \frac{1}{2\pi}\delta^{\nu\lambda} R \partial_z X^\mu \partial_{\bar{z}} X^\lambda + \frac{1}{8\pi^2} R^2 \delta^{\mu\kappa}\delta^{\nu\lambda}) \qquad (1.3)$$

Observe that vertex operators are unchanged by addition of a total derivative operator. This is an interpretation of the origin of gauge invariance.

Scattering amplitudes are then given by

$$\langle V_1(k_1)\ldots V_p(k_p)\rangle = \sum_{h=0}^{\infty} \int DgDX\ \exp(-I(X,g))V_1(k_1)\ldots V_p(k_p)/\mathcal{N}$$

At the h-loop level, the worldsheet is a compact Riemann surface without boundary of genus h.

Again the integral with respect to DX is straightforward, and the symmetries of the theory show then that scattering amplitudes reduce as well to finite-dimensional integrals over moduli space.

We now exhibit the gauge fixed formula for Z [8][9].

$$Z = \int_{\text{Slice } S} d\mu_S \left(\frac{\det P_1^+ P_1}{\det \langle \kappa_j | \kappa_k \rangle}\right)^{1/2} \left(\frac{2\pi \det' \Delta_g}{\int d^2\xi \sqrt{g}\, \det \langle \phi_j | \phi_k \rangle}\right)^{-13}$$

$$d\mu_S = \det \langle \phi_j | \phi_k \rangle^{-13} \det \langle \kappa_j | f_\kappa \rangle \quad d\beta_1 \ldots d\beta_{6h-6} \tag{1.4}$$

where S is a 6h-6 dimensional slice imbedded in the space of metrics, $\beta_i$ are local coordinates for S, $f_i$ are the corresponding tangent vectors, and $\kappa_j$, $\phi_k$ are arbitrary bases of quadratic and abelian differentials. The operator $(P_1 \delta V)_{mn} = \nabla_m \delta V_n + \nabla_n \delta V_m - g_{mn} \nabla^P \delta V_P$ gives the traceless part of the infinitesimal deformation of the metric resulting from an infinitesimal reparametrization $\delta V^m$. Its determinant is a Faddeev-Popov determinant resulting from the Jacobian of the change of variables from the space of metrics to the slice S and the orbits of the gauge groups.

Similar gauge-fixed formulas hold for the scattering amplitudes (1.3) with the integral in DX producing additional factors of Green's functions.

## 2. Fermionic Strings

In the Neveu-Schwarz-Ramond model, the basic fields are now a zweibein $e_m{}^a$, the scalar fields $X^\mu$ $\mu=1,\ldots d=10$, and their supersymmetric partners, a gravitino field $\chi_m{}^\alpha$ and 10 Majorana fields $\psi^\mu$ . The action is [10]

$$I_{NSR} = \int d^2\xi\sqrt{g}\ (\frac{1}{2}\ g^{mn}\partial_m X^\mu \partial_n X_\mu - \frac{i}{2}\ \psi^\mu \gamma^m \partial_m \psi_\mu - \frac{1}{2}\ \psi^\mu \gamma^a \gamma^m \chi_a \partial_m X_\mu$$

$$+ \frac{1}{8}\ \psi^\mu\ \gamma^a \gamma^b \chi_a (\chi_b \psi_\mu))$$

with the following symmetries in addition to the symmetries of the bosonic string:

(iii) super Weyl scalings $\delta\chi_m = \gamma_m \delta\lambda$ , etc.

(iv) local N = 1 supersymmetry $\delta\chi_m = 2D_m \delta\xi$, etc.

(v) Lorentz symmetries.

The partition function

$$Z_{NSR} = \sum_{spin} \int DgD\chi D\psi DX \quad \exp(I_{NSR})/\mathcal{N}$$

reduces to (after cancellation of anomalies in the critical dimension d = 10 [11][12][13])

$$Z_{NSR} = \int_{SLICE\ S} d\mu_S \sum_{spin\ \nu,\bar{\nu}} \left( \frac{\det'\Delta_g}{\int d^2\xi\sqrt{g}\det\langle\phi_j|\phi_k\rangle} \right)^{-5} \left( \frac{\det P_1^+ P_1}{\det\langle\kappa_j|\kappa_k\rangle} \right)^{\frac{1}{2}} \quad (\det' D)^5$$

$$(\det\ P_{1/2}^+ P_{1/2})^{-1/2}\ \langle \mathrm{Pfaff} K(g)\rangle \qquad (1.5)$$

The origin of the various factors on (1.5) is roughly the following. The measure on the slice S and $\det P_1^+ P_1$ arise from the same gauging fixing procedure as in the bosonic string. Similarly modes of the gravitino field $\chi_m$ which can be gauged away by combined super Weyl and supersymmetry are those in $\{\gamma_m \delta\lambda\}$ + {Range $P_{1/2}$}, with $(P_{1/2}\delta\xi)_m = 2D_m\delta\xi - \gamma_m \gamma^n D_n \delta\xi$. The determinant of $P_{1/2}$ is the Faddeev-Popov determinant of this gauge theory. This leaves us with a 4h-4 finite-dimensional space of "supermoduli modes." Integration of the matter fields $X^\mu$ and $\psi^\mu$ then leads to the remaining determinant and an additional factor <Pfaff K(g)> which represents couplings with the supermoduli modes. An explicit formula for <Pfaff K(g)> can be found in [11]. Finally contributions from left and right movers (i.e. holomorphic and antiholomorphic factors on moduli space) are to be separated out in (1.5) and assigned independent spin structures $\nu$ and $\bar{\nu}$. Invariance under the mapping class group requires a summation over spin structures.

The heterotic string can be treated similarly in the fermionic representation. The resulting partition function is [12][14]

$$Z_{HET.} = \int_{SLICES} d\mu_s \sum_{\nu_1,\nu_2,\bar{\nu}} \left(\frac{2\pi \ \det'\Delta_g}{\int d^2\xi\sqrt{g}\ \det\langle\phi_j|\phi_k\rangle}\right)^{-5} \left(\frac{\det P_1^+ P_1}{\det\langle\kappa_j|\kappa_k\rangle}\right)^{1/2} \times$$
$$(\det_{\bar{\nu}}{}^{\bar{z}} D)^5 (\det^z_{\nu_1} D)^8 (\det^z_{\nu_2} D)^8 \ \langle \mathrm{Pfaff}_{\bar{\nu}}\ K^{\bar{z}}(g)\rangle \qquad (1.6)$$

Here we have kept the contributions of the right movers from the NSR string, and replaced the left movers by two families of 16 internal fermions each, with spin structures $\nu_1$ and $\nu_2$. Taking $\nu_1 = \nu_2$ leads to the Spin $32/Z_2$ string, while taking $\nu_1$ independent of $\nu_2$ leads to the $E_8 \times E_8$ heterotic string. Other (non-supersymmetric) theories can be obtained by altering the fermion projection [15].

The above expressions for fermionic quantum measures rely in an essential way on the existence of determinants of chiral operators. This is a non-trivial issue which will be discussed at greater length in Section 5. Finally we observe that we have chosen to integrate out explicitly the supermoduli parameters in (1.5) and (1.6). A superspace formulation where moduli and supermoduli are on the same footing will be given in Chapter III.

# II. DETERMINANTS AND MODULI GEOMETRY

## 3. Constant Curvature geometries

One way to analyze the structure of the string partition function and scattering amplitudes is to choose a convenient slice S, and express (1.4)(1.5)(1.6) as integrals on S of scalar quantities. A natural choice is the slice S of metrics of constant curvature, and the measure $d\mu_S$ will reduce to the Weil-Petersson measure. Recall that the Weil-Petersson metric on moduli space is obtained by representing tangent vectors to moduli by quadratic differentials, and their inner products by pairing and integrating them over M using constant curvature metrics. Determinants of Laplacians should then be taken with respect to constant curvature metrics. We begin with a discussion of the one-loop case, and will generalize next to multiloops.

The moduli space for the torus can be identified with $\{\tau=\tau_1+i\tau_2;\ \tau_2>0\}/SL(2,Z)$, where $SL(2,Z)$ is the mapping class group. With proper normalization, the Weil-Petersson measure is just the $SL(2,Z)$ invariant measure $2d\tau d\bar{\tau}\ \tau_2^{-2}$, and all relevant determinants can be expressed in terms of modular forms. In the bosonic case

$$\tau_2^{-1}\det{}'\Delta = \tau_2|\eta(\tau)|^4$$

$$\det{}'P_1^+P_1 = \frac{1}{2}\det{}'\Delta \tag{2.1}$$

while in the fermionic case, we can represent spin structures by boundary conditions

$$\phi(\xi_1,\varepsilon_2+\tau) = e^{2\pi i\nu_2}\phi(\xi_1,\xi_2),$$

$$\phi(\xi_1+1,\xi_2) = e^{2\pi i\nu_1}\phi(\xi_1,\xi_2)$$

and chiral Dirac operators become $\bar{\partial}$ operators:

$$\det{}_\nu 2\bar{\partial} = \frac{\theta_1(\nu_2-\nu_1\tau,\tau)e^{\tau\pi i\nu_1/2}}{\eta(\tau)} \qquad (\nu_1,\nu_2) \neq (0,0)$$

$$\det{}'_\nu 2\bar{\partial} = (\tau_2/2\pi)^{1/2}\theta_1'(0,\tau)/\eta(\tau) \qquad (\nu_1,\nu_2) = (0,0) \qquad (2.2)$$

Here $\eta(\tau)$ and $\theta_1(z,\tau)$ denote the Dedekind eta function and the usual theta function. In particular the partition function for the bosonic string can be expressed as

$$Z = \int d\tau d\bar{\tau}\,(4\pi\tau_2)^{-2}(2\pi\tau_2)^{-12}|\eta(\tau)|^{-48}$$

$$|\tau| \geq 1, |\mathrm{Re}\tau| \leq 1/2$$

As for fermionic strings, it is easy to see that the spin structure $(\nu_1,\nu_2) = (0,0)$ does not contribute because of a zero mode of the Dirac operator that decouples from supermoduli modes. The other spin structures do not have supermoduli, and modular invariance requires the following relative phases for the products of all the chiral determinants involved:

$$\eta(\tau)^{-4}[\theta_2{}^4(0,\tau) + \theta_4{}^4(0,\tau) - \theta_3{}^4(0,\tau)]$$

Thus the Jacobi identity implies that $Z_{NSR} = Z_{HET} = 0$, which can be viewed as a manifestation of space-time supersymmetry.

Surfaces of constant negative curvature (in the case of multiloops) can be represented as the upper half space with the metric $ds^2 = y^{-2}dzd\bar{z}$, moded out by cocompact discrete, fixed point free subgroups of SL(2,R). We shall refer to these as Fuchsian groups. To such groups we can associate Selberg zeta functions

$$Z_\nu(s) = \prod_{\gamma \text{ primitive}} \prod_{p=1}^{\infty} (1-\nu(\gamma)^{2n} e^{-(s+p)\ell_\gamma}) \quad ,n = 0,1 \tag{2.3}$$

where $\ell_\gamma = 2$ Arccosh (trace $\gamma/2$) is the length of the closed geodesic in the homotopy class of $\gamma$, $\nu(\gamma)\varepsilon\{\pm 1\}$ determines the spin structure. Note the evident analogy with the eta function of the one loop case. The absolute values of the relevant functional determinants are then given by [9][18] [19] and [20]

$$\det{}'\Delta = \exp(-c_0\chi)\ Z_\nu'(1)$$
$$\det P_1^+P_1 = \exp(-c_1\chi)Z_\nu(2)$$
$$\det P_{1/2}^+P_{1/2} = \exp(-c_{1/2}\chi)Z_\nu(3/2)$$

$$\det{}'\bar{D}^+D = \exp(-c_{1/2}\chi)Z_\nu^{(2N)}(1/2)/(2N)! \tag{2.4}$$

Here n in (2.3) should be taken to be either 0 or 1 depending on whether the relevant fields are tensors or spinors, N is the number of harmonic spinors, and $c_n$ are numerical constants:

$$c_n = \frac{1}{2} \sum_{o \le m < 1/2} (2|n|-m-1)\log(2|n|-m) - ([n]+1/2)^2 + (n-[n])^2 + (n+\frac{1}{2})\log 2\pi + 2\zeta'(-1)$$

An advantage of Selberg zeta function type formulas is that the asymptotic behavior of determinants near the boundary of moduli space can be easily read off. In the Deligne-Mumford compactification, the boundary of moduli is obtained by pinching to 0 the length of a closed geodesic $\ell_o$ (at most 3h-3 geodesics can be pinched simultaneously.) As $\ell_o \to 0$ the main contributions to $Z_\nu(s)$ for $n = 0$ come from the product

$$\prod_{p=1}^{\infty} (1-e^{-(s+p)\ell_o}) \sim \ell_o^{-s+1/2} \exp(-\pi^2/6\ell_o) \qquad \mathrm{Re}(s)>1 \tag{2.5}$$

while the remaining factors' contributions are bounded. In fact a collar around $\ell_o$ of length $\sim \log 1/\ell_o$ will stretch out, and the lengths of geodesics crossing $\ell_o$ will tend to $\infty$; as for the noncrossing ones they will form a well behaved Selberg zeta function for the punctured surfaces obtained afterwards. For $\mathrm{Res} \le 1$ analytic continuation of (2.3) is needed and the heuristic principle is that the same asymptotics coming from (2.5) should hold, with a possible uncertainty of $O(\ell_o^{-\varepsilon})$. When Res=1 this has been substantiated with incisive arguments by Gava et al., and Wolpert [21-22] who gave the precise formula:

$$Z'(1) \sim \ell_o^{-1} \exp(-\pi^2/3\ell_o) \prod_{0<\lambda_n<1/4} \lambda_n \tag{2.6}$$

Here $\lambda_n$ are the eigenvalues of the Laplacian on scalars, and the powers have been doubled to account for contributions from both $\gamma$ and $\gamma^{-1}$. Note that there are at most 4h-2 eigenvalues <1/4, and that the lowest eigenvalue is of the size $\ell_0$ when the pinching geodesic is as in Fig 1. More precisely the first eigenvalues $\lambda_n$, n <2h-2 can be determined in terms of lengths of curves separating the surface into n+1 components [23]. This together with asymptotics for the Weil-Petersson measure completely determine the divergence structure of the bosonic string in flat spacetime. To see this, it is instructive to introduce complex coordinates. The model for degenerations is then the plumbing fixture $\{zw=t;\ |z|,|w|,|t|<1\}$, for which t is a "complex coordinate for moduli", and a hyperbolic metric on each fiber $t\neq 0, \{|t|<|z|<1\}$ can be written down explicitly:

$$ds_t^2 = (\pi/\log|t|)\ \mathrm{cosec}\ (\pi\log|z|/\log|t|)\,|dz/z|^2$$

Note that the geodesic in each fibre is $\{|z|=|t|^{1/2}\}$, its length is

$$\ell = 2\pi^2/(\log 1/|t|) \qquad (2.7)$$

the hyperbolic metric is complete, and the collar discussed earlier can be viewed as the subset

$\{|t|\exp(2\pi)<|z|<\exp(-2\pi)|t|\}$. The Weil-Petersson measure on the t space is readily calculated to be

$$ds^2_{WP} = 2\pi^3(\log 1/|t|)^3|dt/t|^2 \qquad (2.8)$$

These exact correspondences between hyperbolic and complex structures for the plumbing fixture model have been shown to be asymptotic in general after careful analysis [22] [24], and the bosonic integrand becomes

$$Z'(1)^{-13}Z(2)d\mu_{WP} \sim \prod_j |ds_j^2||s_j|^{-4} \tag{2.9}$$

where we have ignored smaller factors, and accounted for simultaneous pinching of several geodesics. This is essentially the double pole behavior established by Belavin and Knizhnik for the trivializing section of the line bundle $K \otimes \lambda^{-13}$ (c.f. (2.10) below). Note that a full comparison including lower order terms would require a careful study of the boundary behavior of the period matrix. This double pole can be attributed to the presence of tachyons in the particle spectrum. Indeed a vertex for such particles is the $V_{-1}$ of (1.3).

Turning now to fermionic strings, we assume that the behavior of Z(1/2) obeys the same heuristic rules, although this has not been rigorously established. If this is indeed the case, for spin structures with $\nu(\gamma) = -1$, the asymptotic behavior of the pinching geodesic becomes instead

$$\prod_{p=1}^{\infty} (1+e^{-(s+p)\ell_0}) \sim e^{\pi^2/12\ell_0}$$

which worsens the behavior of the combined bosonic and fermionic determinants. Thus finiteness (if true) of superstring amplitudes is contingent upon delicate cancellations between various spin structures. Recently phases of chiral determinants such as $P_{1/2}$ have been shown to behave like theta functions [25]. This is actually a bosonization statement, as the theta factors can be viewed as contributions from the instanton sector of the bosonic theory. In practice, even with this key additional information it is difficult to see the cancellation mechanism, especially with the corrections from supermoduli modes. For the heterotic string, the authors of [26] have succeeded in exploiting the absence of supermoduli from the left movers and factorization assumptions to obtain a proof of the vanishing of the cosmological constant. Other authors [27] have produced Ansätze for integrations over supermoduli to obtain the same result. These are difficult issues which require further investigation.

Finally we would like to mention that constant curvature geometries also admit an elegant parametrization in terms of Fenchel-Nielsen coordinates, which are symplectic coordinates for the Weil-Petersson Kähler form [31]. A surface is decomposed with a given combinatorial pattern into "pairs of pants". This decomposition is well adapted to Feynman path integrals and may lead to an explicit construction of an off-shell three string vertex. Steps in this direction were taken on [28].

## 4. Complex geometry of bosonic strings

With Weyl invariance and the necessity of separating left from right movers for fermionic strings, it is natural to try to express string amplitudes in terms of Cauchy-Riemann operators and the holomorphic structure of moduli space [29][30]. Indeed we may consider the operators $\bar\partial_j$ on (j,0) tensors given by $\bar\partial_j(\phi dz^j) = (\partial\phi/\partial\bar z)dz^j d\bar z$. The Laplacian $\Delta$ is then just $\bar\partial_0^+\bar\partial_0$, while the operator $P_1^+$ can be identified with the symmetrized version $\bar\partial_2 + \partial_2$ of $\bar\partial_2$, acting on tensors of the form $\phi dz^2+\bar\phi d\bar z^2$. Now if we deform the complex structure by a Beltrami differential $\eta$, the basic operator $\bar\partial_0$ will become $\partial/\partial\bar z-\eta\partial/\partial z$ which depends holomorphically on $\eta$. Analogous statements hold for $\bar\partial_j$. The key issue is whether this holomorphic dependence can be maintained upon quantization, e.g., whether the regularized determinant of $\bar\partial_j^+\bar\partial_j$ is the square of a holomorphic function. The answer is provided by the following formula of Belavin-Knizhnik which gives the second variation of the determinants

$$\delta_\eta \delta_{\bar\eta} \log \frac{\det \bar\partial_j^+ \bar\partial_j}{\det\langle\phi_\alpha|\bar\phi_\beta\rangle \det\langle\psi_\alpha|\bar\psi_\beta\rangle} = - \frac{6j^2-6j+1}{6\pi} \int d^2z\, \rho^{-2}(\partial(\rho\bar\eta)\bar\partial(\rho\eta)$$

$$+ \rho^2|\eta|^2 \times \partial\bar\partial \log \rho) \qquad (2.10)$$

when the zweibein undergoes the deformation $\rho dz \to \rho(dz+\bar\eta d\bar z)$. Here, $\phi_\alpha$ and $\psi_\beta$ are bases for the kernels of $\bar\partial_j$ and $\bar\partial_{1-j}$ which vary holomorphically. Note that there is a harmless arbitrariness in the choices of conformal factors. Thus there is a "holomorphic anomaly" which will cancel in the critical dimension in exactly the same way as the conformal anomaly. With this, the bosonic partition function (1.4) can now be rephrased in purely complex analytic terms: Let K be the canonical bundle of moduli space and $\lambda$ be the Hodge bundle, i.e., maximum wedge powers of the spaces of quadratic and abelian differentials respectively. Then there is a unique (up to a constant) holomorphic non-vanishing section F of $K \otimes \lambda^{-13}$, and Z is the integral over moduli of $|F\wedge\bar F|_{\lambda^{-13}}$, where $|\cdot|_{\lambda^{-13}}$ is the norm coming from the natural pairing $\langle\phi_i|\phi_j\rangle = \int_M \phi_i\wedge\bar\phi_j$ of abelian differentials. The poles encountered in Section 3 (see equation (2.9)) can now be obtained by characteristic class arguments: $K \otimes \lambda^{-13}$ is trivial over $\mathcal{M}_h$, but over the compactified moduli space $\overline{\mathcal{M}}_h$ its natural extension (by dualizing differentials) is not, and in fact admits $[-2\Delta]$ as divisor. Since the components of $\Delta$ are independent [31], F must have a second order pole along $\Delta$. It is remarkable that the triviality of $K \otimes \lambda^{-13}$ appeared in earlier work of Mumford [32].

The complex structure of strings has led to nice formulas in terms of period matrices and theta functions [33].

5. Determinant bundle interpretations

The fact that functional determinants only transform well when accompanied as in (2.10) by a finite dimensional determinant of bases of zero modes is very suggestive of a line bundle structure. This is actually a familiar idea from the index theory interpretation of chiral anomalies [34]: the Dirac operator interchanges the $S_+, S_-$ spaces of positive and negative chirality spinors so that its determinant requires a choice of bases within each space, i.e., it is an element of $( \overset{max}{\Lambda} S_+)^* \times ( \overset{max}{\Lambda} S_-)$. When coupled to vector potentials, these form a line bundle over the potentials. Chiral anomalies correspond to the non-triviality of this bundle.

More recently Witten has suggested a more precise picture, namely that determinant bundles carry an intrinsic connection whose non-vanishing curvature would correspond to perturbative anomalies, and whose non trivial holonomy would indicate the presence of global anomalies. It is remarkable that such a geometric study of determinant bundles had been undertaken by Quillen [35] in a number theory context. Since the number of zero modes may jump, the determinant bundle of a family of $\bar{\partial}$ (or Dirac) operators depending on parameters in a base space B, is rather defined by choosing patches $B_\alpha$ in B where $\alpha>0$ is not anywhere an eigenvalue of $\bar{\partial}^+\bar{\partial}$ or $\bar{\partial}\,\bar{\partial}^+$ and setting

$$L_\alpha = ( \overset{max}{\Lambda} E_\alpha^+)^* \times ( \overset{max}{\Lambda} E_\alpha^- )$$

Here $E_\alpha^\pm$ are the eigenspaces with eigenvalues $\lambda$ less than $\alpha$. The $L_\alpha$'s are algebraically isomorphic to $( \overset{max}{\Lambda} \mathrm{Ker}\bar{\partial})^* \otimes ( \overset{max}{\Lambda} \mathrm{Ker}\bar{\partial}^+)$, and can be patched smoothly into a line bundle $(\mathrm{DET}\bar{\partial})$ with the transition functions

$t_{\alpha\beta} = \prod_{\alpha<\lambda<\beta} \lambda$ . With metrics on the surface we can construct a natural norm $\|\cdot\|_\alpha^2$ on $L_\alpha$ by setting

$$\|(\phi_1\wedge..)^{-1}\otimes(\psi_1\wedge..)\|_\alpha^2 = \det\langle\phi_j|\phi_k\rangle^{-1}\det\langle\psi_j|\psi_k\rangle$$

for $\phi_i\psi_j$ bases of $E_\alpha^+$ and $E_\alpha^-$ respectively. The presence of the transition functions $t_{\alpha\beta}$ requires however a correction term, and we are led to the metric

$$\|(\phi_1\wedge..\phi_k)^{-1}\otimes(\psi_1\wedge.\ \psi_\ell)\|_Q^2 = \|(\phi_1\wedge..\phi_k)^{-1}\otimes(\psi_1\wedge..\psi_\ell)\|_\alpha^2 \prod_{\lambda>\alpha}\lambda$$

introduced by Quillen. The infinite product on the RHS is to be regularized by zeta functions.

Returning now to the $\bar\partial_j$ operators of the Belavin-Knizhnik treatment, duality identifies $\overline{\mathrm{DET}\partial_j}$ with $(\overset{max}{\wedge}\mathrm{Ker}\bar\partial_j)^*\otimes(\overset{max}{\wedge}\mathrm{Ker}\bar\partial_{1-j})^*$, so that the Quillen norm becomes

$$\|(\phi_1\wedge..\ \phi_k)^{-1}\times(\psi_1\wedge..\ \psi_\ell)^{-1}\|_Q^2 = \frac{\det'\bar\partial_j^+\bar\partial_j}{\det\langle\phi_i|\phi_j\rangle\det\langle\psi_i|\psi_j\rangle} \qquad (2.11)$$

Since the determinant bundle in this setting is well-known to carry a holomorphic structure, we recognize the Belavin-Knizhnik formula (2.10) as giving precisely the curvature of $(\mathrm{DET}\bar\partial_j)$ with respect to the unique connection determined by the holomorphic structure and the Quillen metric.

Finally we should mention that a general construction of connections on determinant bundles has been given by Bismut and Freed [36]. Their constructions apply in the absence of holomorphic structure, and depend on a choice of horizontal spaces in the "Teichmüller curve", i.e. the fiber bundle over B whose fiber at each point $\tau$ is the Riemannian manifold indexed by $\tau$. When there is a holomorphic structure their connection will coincide with the one arising from the Quillen metric. The curvature of the Bismut-Freed connection is given by the index density formula

$$[2\pi \int_{\mathrm{Fibre}} A(R^{\mathrm{Vert}}/2\pi)\mathrm{Tr}\ \exp(-L/2\pi i)]_{(2)} \qquad (2.12)$$

Here (2) indicates keeping only 2-forms, and L is the curvature of external gauge couplings. $R^{\mathrm{Vert}}$ is the curvature of the connection $\omega^{\mathrm{Vert}}$ obtained by projecting onto the tangent to the fiber a Levi-Civita connection on the full Teichmuller curve constructed out of the metric in the fiber, the horizontal subspace, and an arbitrary metric on the base. For $\bar{\partial}_j$ operators and B = Teichmuller space there is a natural horizontal subspace arising from the Bers imbedding, and $\omega^{\mathrm{Vert}}$ will coincide with the natural holomorphic connection on the Teichmuller curve. Formula (2.12) and curvature computations for the Teichmuller curve [37] will then exhibit the curvature of $\mathrm{DET}(\bar{\partial}_j)$ as $(6j^2-6j+1)$(Weil-Petersson Kahler form)/$6\pi$. This is another way of stating (2.10). Bismut and Freed also provide expressions for the holonomy in terms of eta invariants, generalizing earlier work of Witten.

# III. SUPERHOLOMORPHIC STRUCTURES AND ANOMALIES

## 6. N=1 two dimensional supergravity

The preceding sections have revealed a rich complex structure underlying string amplitudes. It is natural to investigate whether there is a similar structure for superstrings, and to this issue we now turn. For the Neveu-Schwarz-Ramond model, the natural setting is that of N=1 two dimensional supergravity, where the basic fields are a superzweiben $dz^M E_M{}^A$ and a superconnection $dz^M \Omega_M$. Covariant derivatives on tensors of weight $\ell$ are given by

$$\mathcal{D}_\pm = E_\pm{}^M(\partial_M + i\ell\Omega_M)$$

$$[\mathcal{D}_A, \mathcal{D}_B] = -T_{AB}{}^C \mathcal{D}_C + i\ell R_{AB}$$

$$\{\mathcal{D}_+, \mathcal{D}_-\} = i\ell R_{+-}$$

and the connection must satisfy the torsion constraints

$$T_{\beta\gamma}{}^\alpha = T^a_{bc} = 0 \qquad T^a_{\beta\gamma} = 2i(\gamma^a)_{\beta\gamma} \tag{3.1}$$

Note that $E_M{}^A$ and $\Omega_M$ have 20 components, while (3.1) imposes 14 constraints. Thus the 8 components of $E_+{}^M$ are not independent.

Besides super reparametrizations (SR), super Lorentz (SL) transformations, there is also a notion of super Weyl (SW) scalings

$$E_M{}^a = e^\Phi E_M{}^a$$

$$E^\alpha_M = e^{\frac{1}{2}\Phi}(E_M{}^\alpha - \frac{i}{2} E_M{}^a(\gamma_a)^{\alpha\beta} D_\beta \Phi) \tag{3.2}$$

The NSR action is just

$$I = \int d^2\xi d^2\theta \,(\mathrm{sdet} E_M^A)\mathcal{D}_+ V^\mu \mathcal{D}_- V_\mu \qquad \mu=1,\ldots,d=10 \qquad (3.3)$$

where $V^\mu$ are scalar superfields. Vertex operators correspond to polynomials in $V^\mu$ and $R_{+-}$ with various polarization tensors, subjected to the requirement of super Weyl invariance. The relation with the component formalism is obtained by passing to Wess-Zumino gauge where

$$E_m^a = e_m{}^a + i\theta\gamma^a\chi_m + \frac{1}{4}\,\theta\theta e_m{}^a A$$

Thus we recapture the zweibein $e_m^a$ and the gravitino $\chi_m$, together with an auxiliary field A with no dynamics. It should now be evident that the superspace formalism (when available!) is conceptually and in practice much simpler than the component formalism.

Similarly the action for the heterotic string is just

$$I = \int d^2\xi d\theta (\mathrm{sdet} E_M{}^A)\left(-\frac{i}{\sqrt{2}}\mathcal{D}_{\bar{z}} V_\mu \mathcal{D}_+ V^\mu - \frac{1}{2}\psi^I \mathcal{D}_+ \psi^I\right)$$

with $\psi^I$ spinor superfields, I index for internal gauge group. Vertex operators can be constructed in analogy with the NSR case. For example

$$V_{IJ} = -i\int d^2\xi d\theta\, E \exp(ik_\mu V^\mu)\mathcal{D}_+ V^\mu\, \psi^I \psi^J$$

$$V_{\mu\nu} = \int d^2\xi d\theta\, E \exp(ik_\mu V^\mu)\mathcal{D}_+ V^\mu \mathcal{D}_{\bar{z}}\, V^\nu$$

are the vertices for the Yang-Mills and gravity supermultiplets. General rules for vertices can be derived after determination of all possible anomalous contractions [7].

7. Two dimensional super complex structures.

It has been shown by Howe [39] that two dimensional superspace is locally superconformally flat, i.e., after suitable SR, SL transformations, any $E_M{}^A$ can be put in the form (3.2) with $E_M{}^A$ the flat superzweibein

$$E_m^a = \delta_m^a \qquad E_m^\alpha = 0 \qquad E_\mu^\alpha = \delta_\mu{}^\alpha \quad E_\mu^a = i\theta^\lambda \gamma^a_{\lambda\mu} \tag{3.4}$$

This is not true globally however, and it is crucial for us to identify the space of all inequivalent geometries under SR, SL, and SW, since gauge-fixing will reduce superstrings amplitudes to integrals over this space. To eliminate SW scalings, we introduce the super Weyl invariant tensor

$$J_M{}^N = E_M{}^a \varepsilon_a{}^b E_b{}^N + E_M{}^\alpha (\gamma^5)_\alpha{}^\beta E_\beta{}^M \tag{3.5}$$

which is integrable thanks to the torsion constraints, and whose square is -1. Thus $J_M{}^N$ defines a super complex structure and we may define

$$\begin{aligned} \text{SUPERMODULI SPACE} &= \{\text{Supergeometries}\}/\{\text{LL,SR,SW}\} \\ &= \{\text{Supercomplex structures}\}/\text{SR} \\ &= \{\text{Supergeometries with } R_{+-} = \text{Constant}\}/\text{SR} \end{aligned}$$

The last assertion follows from solvability of the super Liouville equation. As in the bosonic case, tangent vectors to supermoduli space can be described as kernel of a gauge fixing operator. More precisely introduce $H_A{}^B = E_A{}^M \delta E_M{}^B$ and observe that $H_+{}^+, H_-{}^-$, $H_\alpha{}^b$ in Range $\mathcal{P}_1$ are the modes that can be gauged away. Here $\mathcal{P}_1$ is the superanalogue of the bosonic $P_1$ encountered earlier and is given by

$$(\mathcal{P}_1 \delta V)_\alpha{}^b = \frac{1}{2} (\gamma_c \gamma^b)_\alpha{}^\beta \mathcal{D}_\beta \delta V^c$$

With the nondegenerate inner product [40]

$$||H_A{}^B||^2 = \int E(\varepsilon^{\alpha\beta} H_\alpha{}^a H_\beta{}^a + c_1 (H_+{}^+)^2 + c_1^* (H_-{}^-)^2 + c_2 H_+{}^+ H_-{}^-] \qquad (3.6)$$

the range of $\mathcal{P}_1$ can be expected to be a non-degenerate subspace, and we may identify the remaining modes as $\mathrm{Ker}\,\mathcal{P}_1{}^+$. This is a space of dimensions (6h-6,4h-4) which is the tangent space to supermoduli. The pairing (3.6) equips then supermoduli with a metric which we shall call the super Weil-Petersson metric.

To introduce a superholomorphic structure on supermoduli space, we view the tangent space to supermoduli T (sModuli) as

$$T(\{J_M{}^N\}) = \{\delta J_M{}^N;\ J_M{}^N \delta J_N{}^P + \delta J_M{}^N J_N{}^P = 0\} \qquad (3.7)$$

on which there is a natural operator with square -1:

$$\mathcal{J}: \quad T(\mathrm{sModuli}) \to T(\mathrm{sModuli})\ \mathcal{J}(\delta J_M{}^N) = J_M{}^P \delta J_P{}^N \qquad (3.8)$$

This almost complex structure is integrable since setting

$$\Gamma_M{}^N = dJ_M{}^N - i\mathcal{J}(dJ_M{}^N)$$

yields

$$d\Gamma_M{}^N = i\ (\Gamma_M{}^P \bar{\Gamma}_P{}^N + \bar{\Gamma}_M{}^P \Gamma_P{}^N)/4 = 0 \text{ (modulo } \Gamma).$$

## 8. Variations of supergeometries to first and second orders.

Fixing a supergeometry $E_A{}^M$, our goal is to describe its deformations to second order by a traceless $H_A{}^B$. We define $\overline{H} \equiv H_+{}^{\overline{z}}$ and note that all components of $H_A{}^B$ can be written in terms of $H$ and $\overline{H}$. For example

$$H_+{}^- = -iD_-\overline{H}/2\sqrt{2} \qquad H_+{}^z = 0 \qquad H_+{}^- = 0$$

Here we have used $D$ to denote derivatives with respect to $E_A{}^M$, by opposition to $\mathcal{D}$ which will denote derivatives with respect to the deformed geometries to be defined later.

To first order a deformation $E_A{}^M$ of $E_A{}^M$ by $H_A{}^B$ is unambiguously given by

$$E_A{}^M = E_A{}^M - H_A{}^B\, E_B{}^M$$

from which it follows at once that $E_+{}^M$ depends only on $\overline{H}$. To second order on the other hand torsion constraints require

$$E_+{}^M = E_+{}^M + \frac{i}{2\sqrt{2}}(D_-\overline{H} + D_+M)\, E_-{}^M - \overline{H}E_{\overline{z}}{}^M$$

$$\mathcal{D}_+ = D_+ - HD_{\overline{z}} + \frac{i}{2\sqrt{2}}(D_-\overline{H} + D_+M) \qquad \text{(for the weight } \ell=0) \tag{3.9}$$

with

$$M = -\frac{i}{\sqrt{2}}HD_+\overline{H} + \frac{i}{2\sqrt{2}}D_+H\,\overline{H}$$

Thus we do have a dependence on $H$ (which cannot be removed by a differeent choice of superconformal gauge) which seems at first sight to spoil any superholomorphic dependence on supermoduli. Actually this is not so. This can be seen by determining the components of the $J_M{}^N$ tensor of $E_A{}^M$:

$$J_+{}^{\overline{z}} = -2\overline{H} \qquad J_+{}^- = \frac{1}{\sqrt{2}}(D_-\overline{H} + D_+M)$$

$$J_z{}^{\overline{z}} = -\sqrt{2}\,D_+\overline{H} \qquad J_z^- = \frac{i}{2}D_+J_+{}^- \tag{3.10}$$

Since $J_+^{\,-}$ , $J_z^{\,-}$ , $J_+^{\,\bar{z}}$ , $J_z^{\,\bar{z}}$ are superholomorphic on supermoduli, it is consistent to view both $\overline{H}$ and $D_-\overline{H} + D_+M$ as superholomorphic, and the geometry and basic operators of N=1 supergravity do depend superholomorphically on supermoduli.

9. Second Variations of Superdeterminants.

The main task now is to derive a superanalogue of the Belavin-Knizhnik formula. For this it is most convenient to alter the notation slightly from last section, and starting with an initial supergeometry $E_M{}^A$ deform it using $\mathcal{H}_A{}^B$ to a supergeometry $\mathcal{E}_M{}^A$ by the following process

$$\begin{array}{ccc}
E_M{}^A & \xrightarrow{\ \mathcal{H}\ } & \mathcal{E}_M{}^A \\
\Phi\downarrow & & \Phi\uparrow \\
\hat{E}_M{}^A & \xrightarrow{\ H\ } & \hat{\mathcal{E}}_M{}^A
\end{array}$$

Here $\Phi$ denotes scalings by a scalar superfield sending $E_M{}^A$ to the flat superzweibein $E_M{}^A$ of (3.4), H is the corresponding scaling of

$$\mathcal{H}_+{}^a = e^{\frac{1}{2}\Phi} H_+{}^a$$
$$\mathcal{H}_+{}^\alpha = H_+{}^\alpha - i(\gamma_+)^{\alpha\beta} D_\beta \Phi$$

and the deformation by H above is the second order one of (3.9).

Since the dependence on the superconformal factors of superdeterminants is known, we need only determine the second order variations of superdeterminants when deforming $E_M{}^A \xrightarrow{\ H\ } \mathcal{E}_M{}^A$. A superheat kernel regularization yields

$$\delta_H \delta_{\overline{H}} \log \frac{\mathrm{sdet}\,\mathcal{D}_+\mathcal{D}_-}{\mathrm{sdet}\langle\chi_j|\chi_k\rangle} = (-)^{2\ell+1}\,\frac{\ell+1/4}{4\pi}\int \partial_z H \partial_{\overline{z}}\,\overline{H}$$

with $\ell$ the Lorentz weight and otherwise obvious notations. The full variation including superWeyl scalings can then be put in covariant form

$$\delta_{\mathcal{H}} \delta_{\overline{\mathcal{H}}} \log \frac{\mathrm{sdet}\,\mathcal{D}_+\mathcal{D}_-}{\mathrm{sdet}\langle\chi_j|\chi_k\rangle} = (-)2\ell\,\frac{\ell+1/4}{4\pi}\int \mathcal{E}[-\tfrac{1}{8}\mathcal{R}_{+-}^2\,\mathcal{H}\overline{\mathcal{H}}$$

$$-\tfrac{1}{4}\mathcal{D}_+^2\,\mathcal{H}\,\mathcal{D}_-^2\overline{\mathcal{H}} + \tfrac{1}{8}\mathcal{R}_{+-}(4\mathcal{D}_+\overline{\mathcal{H}}\mathcal{D}_-\mathcal{H} + \mathcal{H}\mathcal{D}_+\mathcal{D}_-\mathcal{H} + \overline{\mathcal{H}}\mathcal{D}_-\mathcal{D}_+\mathcal{H})] \quad (3.11)$$

The right hand simplifies to $-(-)^{2\ell}(\ell+1/4)\,||\mathcal{H}||^2_{sWP}/32\pi$ for $\mathcal{H}$ in $\mathrm{Ker}\,\mathcal{P}_1^+$ so we have a complete superanalogue of the holomorphic anomaly in terms of the Weil-Petersson Kähler form.

We return now to quantization of the NSR superstring in the superfield formalism. Path integral quantization leads to

$$Z_{NSR} = \int DE_M{}^A DV_\mu\ \exp(-I)$$

$$= \int_{\mathrm{sModuli}} sd\mu_{WP}\ \ (\mathrm{sdet}\langle\chi_j|\chi_k\rangle^{1/2})(\mathrm{sdet}\langle\phi_a|\phi_b\rangle)^{-5}\ \mathcal{K}$$

where I is the action of (3.3) and $\mathcal{K}$ is the combination

$$\left(\frac{\mathrm{sdet}\,\mathcal{P}_1^+\mathcal{P}_1}{\mathrm{sdet}\langle\chi_j|\chi_k\rangle}\right)^{1/2}\left(\frac{\mathrm{sdet}\,\mathcal{D}_+\mathcal{D}_-}{\int E\ \mathrm{sdet}\langle\phi_a|\phi_b\rangle}\right)^{-5}$$

In analogy with (1.4) $\phi_a$ and $\chi_j$ are bases of superabelian and superquadratic differentials respectively. Thus the superholomorphic anomaly (3.11) cancels and we do have a superholomorphic splitting of the quantum measure. Similar arguments apply to heterotic strings. In conclusion supermoduli space has a rich and unusual structure and it is likely that a "super algebraic geometric" understanding of its aspects would be extremely helpful in superstring theory.

Details on the above can be found in [43]. Other approaches to supermoduli from the super Riemann surfaces viewpoint are in [41][42]. A discussion of the holomorphic properties of the supermoduli contributions is given in [26].

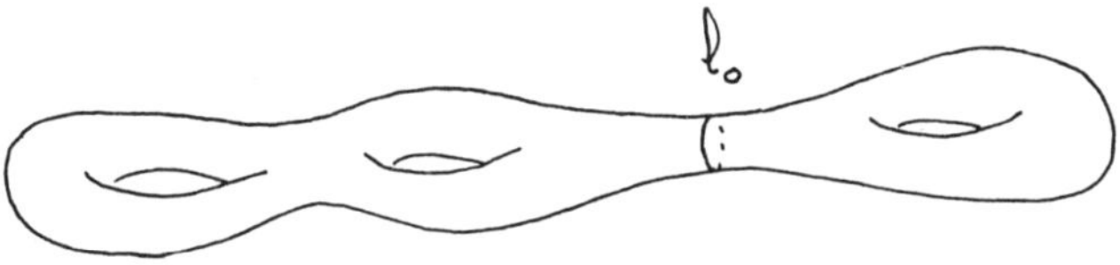


Fig. 1

## REFERENCES

[1] J.H. Schwarz, "Superstrings", Vols I & II, World Scientific, Singapore 1985.
M.B. Green and D.J. Gross, "Unified String Theories", World Scientific, Singapore (1986)

[2] A.M.Polyakov, Phys. Lett. 103B(1981) 207, 211.
D. Friedan, Les Houches 1982 Summer School Proceedings, North Holland, Amsterdam
O. Alvarez, Nucl. Phys. B216 (1983) 125

[3] S. Mandelstam, in "United String Theories" edited by M. Green and D. Green, p. 46 and p. 577, and UCB-PTH-86/16.

[4] D. Friedan, E. Martinec, and S. Shenker, Nucl Phys. B271 (1986) 93
D. Friedan and S. Shenker, Chicago Preprints EFI86-18A, 18B
D. Friedan, EFI86-22.

[5] C. Lovelace, Phys. Lett. 32B (1970) 703
A. Alessandrini, Nuovo Cimento 2A (1971) 321
A. Alessandrini and D. Amati, Nuovo Cimento 4A (1971) 790
M. Kaku and J. Scherk, Phys. Rev. D3 (1971) 430
M. Kaku and L. P. Yu, Phys. Rev. D3 (1971) 2992, 3007, 3020.

[6] S. Weinberg, Phys. Lett. 156B (1985) 309
S.P. De Alwis, Phys. Lett. 168B (1986) 59
R. Sasaki and I. Yamanaka, Phys. Lett. 165B (1985) 283

[7] E. D'Hoker and D.H. Phong, Preprint PU-PT-1030

[8] G. Moore and P. Nelson, Nucl. Phys. B266 (1986) 585

[9] E. D'Hoker and D.H. Phong, Nucl. Phys. B269 (1986) 205

[10] L. Brink, P. di Vecchia, and P. Howe, Phys. Lett. 65B (1976)471
S. Deser and B. Zumino, Phys. Lett. 65B (1976) 369

[11] E. D'Hoker and D.H. Phong, Nucl. Phys. B278 (1986)

[12] G. Moore, P. Nelson, and J. Polchinski, Phys. Lett. B169 (1986) 47

[13] S. Chaudhuri, K. Kawai, and S. H. Tye, Cornell Preprint CLNS 86/723 (1986)

[14] E. D'Hoker and D. H. Phong, The Geometry of Quantum Strings (1987) submitted to Rev. Mod. Phys.

[15] L. Alvarez Gaumé, P. Ginsparg, G. Moore, and C. Vafa, Phys. Lett. B171 (1986) 155

L. Dixon and J. Harvey, Nucl. Phys. B274 (1986) 93

N. Seiberg and E. Witten, Nucl. Phys. B276 (1986) 272

[16] J. Shapiro, Phys. Rev. D5 (1972) 1945

J. Polchinski, Comm. Math. Phys. 104 (1986) 37

[17] F. Gliozzi, J. Scherk, and D. Olive, Nucl. Phys. B122 (1977) 253.

[18] E. D'Hoker and D H. Phong, Comm. Math. Phys. 104 (1986) 537

[19] D. Fried, Invent. Math. 84 (1986) 523

[20] P. Sarnak, Stanford Mathematics Preprint (1986)

Kierlanczyk, M.I.T. Ph.D. Thesis

A. Namazie and S. Rajeev, CERN Preprint

M. Baranov and A. Schwarz, Pisma ZETF 42 (1985) 340

A. Voros, Saclay Preprint (1986)

G. Gilbert, Texas Preprint UT-TG-23-85.

[21] E. Gava, R. Iengo, T. Jayaraman, and R. Ramachandran, Phys. Lett. B168

[22] S. Wolpert, Maryland Preprint 1986

[23] R. Schoen, S. Wolpert, and S.T. Yau, Proc. Symp. Pure Math. XXXVI, 1980, 279

J. Dodziuk, T. Pignataro, B. Randol, D. Sullivan, Preprint 1986

[24] H. Masur, Duke Math. J.43 (1976) 623

[25] L. Alvarez-Gaumé, J. B. Bost, G. Moore, P. Nelson, and C. Vafa, Harvard preprint HU-TP-86/A039

J. B. Bost and P. Nelson, Phys. Rev. Lett. 57 (1986) 795

E. Verlinde and Verlinde, Utrecht preprint 1987

[26] G. Moore, J. Harris, P. Nelson, and I.M. Singer, Phys. Lett. B178 (1986) 167

J. Harris and I. Morrison, Preprint

M. Chang and Z. Ran, Preprint 1987

[27] J. Atick and A. Sen, SLAC-PUB-4083, 4131

H. Sonoda, Phys. Lett. B178 (1986) 390

[28] E. D'Hoker, D. H. Phong, Phys. Rev. Lett. 56 (1986) 912

A. Tseytlin, Phys. Lett. B168 (1986) 63

[29] A. A. Belavin and V.G. Knizhnik, Phys. Lett. B 168 (1986) 201 and Landau Institute Preprint

[30] R. Catenacci, M. Cornalba, C. Reina, and Martinelli, Phys. Lett. B174 (1986) 386

J. B. Bost and T. Jolicoeur, Phys. Lett.B174 (1986) 273

[31] S. Wolpert, Bull. AMS (N.S.) 11 (1984) 189

[32] D. Mumford, Ens. Math. 23 (1977) 39

[33] A. A. Belavin, V. Knizhnik, A. Morozov, and A. Perelomov, Landau Institute preprint, 1986, in JETP Lett. 43 (1986) 411

G. Moore, Phys. Lett. B176 (1986) 369

Yu. I. Manin, Phys. Lett. B172 (1986) 184

A. Beilinson and Yu. I. Manin, Comm. Math. Phys. 107 (1986) 359

[34] M. Atiyah and I.M. Singer, Proc. Natl. Acad. Sci. USA 81 (1984) 2597

O. Alvarez, I.M. Singer, and B. Zumino, Comm. Math. Phys. 1

[35] D. Quillen, Funk Anal. i prilozen 19 (1985) 37

[36] J. M. Bismut and D.S. Freed, Comm. Math. Phys. 106 (1986) 159, 107 (1986) 103

D. S. Freed, Comm. Math. Phys. 107 (1986) 483

[37] S. Wolpert, Invent. Math. 85 (1986) 119

A. Tromba, Preprint 1985

H. Royden, unpublished

Y.T. Siu, Harvard Preprint 1986

M. Wolf, Stanford PH.D. Thesis 1986

[38] E. Witten, in "Anomalies, Topology, and Geometry", edited by W. Bardeen and A. White, Argonne 1985, p. 81

[39] P.S. Howe, J. Phys. A12 (1979) 393

[40] E. Martinec, Phys. Rev. D28 (1983) 2604

[41] D. Friedan, in [1] M.B. Green and D. J. Gross (1986)

J. D. Cohn, EFI 86-22, preprint

J. Rabin and L. Crane, Chicago Preprint (1986)

[42] E. Martinec, Nucl. Phys. B

[43] E. D'Hoker and D. H. Phong, PU-PT-1029 Preprint (1986)

# BRST QUANTIZATION AND BRST COHOMOLOGY

Neil Marcus

*Department of Physics and Lawrence Berkeley Laboratory*
*University of California*
*Berkeley, CA 94720 U.S.A.*

Augusto Sagnotti

*Dipartimento di Fisica*
*Università di Roma II, "Tor Vergata"*
*Via Orazio Raimondo*
*00173 Roma ITALY*
*and*
*INFN-Sezione di Roma, ITALY*

Over the last year or so there has been a large interest in the construction of manifestly covariant field theory formulations of string models |1|. Light-cone formulations were obtained long ago |2|, following the original construction of the vertex by Mandelstam |3| and the understanding of light-cone first-quantization by GGRT |4|. Covariant formulations would, in principle, achieve two goals. On the one hand, one would be able to discuss the physics of string models along lines familiar from ordinary field theory. On the other hand, the huge symmetries of these models would somehow be made transparent.

Whereas the program is not yet completed, a large progress undoubtedly followed Siegel's observation |5| that the BRST operator of the first-quantized string theory enters crucially the construction of a <u>gauge-fixed</u> version of the corresponding free field theories. Following this work, several groups arrived, by different routes, at free <u>gauge-invariant</u> kinetic terms for strings |1|. What appeared more surprising for a while is that there exists, at least for the open string, a gauge-invariant formulation of free field theory based, again, on the BRST operator of the first-quantized model. This was obtained in refs. |6,7|. It is this apparent coincidence that we intend to discuss in the following. The source is a recent paper of ours |8|. The conclusion will be that the appearance of the BRST operator is not at all coincidental. Rather, it can be well understood by tracing the steps of the transition from first to second quantization in the more familiar case of non-relativistic point particles. The result is rather general. If $\Phi$ denotes a string field, Q the corresponding BRST (Becchi-Rouet-Stora-Tyutin) |9,10| charge, and g the ghost number operator (an operator that counts the number of first-quantized ghosts minus the number of corresponding antighosts in the field),

$$Q\,\Phi = 0 \quad , \tag{1}$$

$$g\,\Phi = 0 \tag{2}$$

are gauge-invariant equations of motion for all string theories. Since the BRST operator is nilpotent, just like exterior derivatives so familiar to mathematicians, the solutions of eqs. (1) and (2) are in correspondence with its nontrivial cohomology classes. Remarkably, it is rather simple to study such cohomology classes. The basic points were discussed by Kato and Ogawa |8| a few years ago.

To understand matters, let us therefore start back from the familiar example of non-relativistic particles. In this case one derives the Schrödinger equation from the particle Lagrangian as follows. One starts from

$$L = \frac{1}{2}\, m\, (\dot{q})^2 \tag{3}$$

and obtains the Hamiltonian by the standard Legendre transformation

$$H = p\,\dot{q} - L \quad . \tag{4}$$

Then one promotes the fundamental Poisson bracket

$$\{q \;,\; p\} = 1 \tag{5}$$

to a commutation relation between quantum operators, realized in the coordinate basis by

$$p = \frac{1}{i} \frac{\partial}{\partial q} \quad . \tag{6}$$

The Hamiltonian becomes a differential operator, and one arrives at the Schrödinger equation for the wavefunction

$$-\frac{1}{2m} \frac{\partial^2}{\partial q^2} \Phi = i \frac{\partial}{\partial t} \Phi \quad . \tag{7}$$

All of this, of course, is well known. One further point worth emphasizing is that eq. (7) is the free part of the classical field equation of the corresponding second-quantized theory. In other words, the Schrödinger equation is a bridge between the two formulations, and it is rather natural to go through the Hamiltonian formalism in this case (the corresponding steps in the Lagrangian formalism lead to Feynman path integrals, and are thus somewhat more difficult to follow).

So much for the non-relativistic particle. In more complicated cases, things become more difficult because of the Legendre transformation in eq. (4). The difficulty is associated with the appearance of constraints. This follows from the fact that the equations defining the canonical momenta

$$p = \frac{\partial L}{\partial \dot{q}} \quad . \tag{8}$$

may not be invertible. It is not easy to deal with such constraints in the general case. Dirac first discussed a way to treat them in the 1950's |12|. Basically, his prescription is as follows. The constraints, including those coming from enforcing conservation at different times, are split into two different kinds, called "first-class" and "second-class" respectively. The first class constraints form a closed Poisson-bracket algebra. The second-class constraints can be disposed of by a suitable redefinition of the brackets (the brackets one obtains in the end are usually called Dirac brackets). Basically, in this way one is reduced to dealing with first class constraint alone, which are imposed individually on the states (or wavefunctions). The result is a set of equations replacing the non-relativistic Schrödinger equation. This is the way one arrives at the Klein-Gordon equation

$$( \partial_\mu \partial^\mu - m^2 ) \Phi = 0 \tag{9}$$

from the relativistic particle Lagrangian.

The problem of a general set-up of a Hamiltonian formalism for constrained systems was reconsidered in the mid seventies, after the work of Becchi-Rouet-Stora and Tyutin |9,10|. The main idea in the BRST method, as expounded mostly by the Russian school |13| and the Kyoto group |14|, can be explained as follows. Basically, given a system with gauge invariances (and thus first-class constraints), one can construct an equivalent system where the constraints disappear in favor of extra variables of wrong statistics (Faddeev-Popov ghosts), and of a global symmetry associated with a nilpotent charge (the BRST charge). The BRST charge is thus a well-defined derivation which finds its place already at the classical level. The gain is twofold. On the one hand, the system looses its gauge invariances, and so the Legendre transformation is straightforward. On the other hand, the many conditions of the Dirac method coalesce into those of eqs. (1) and (2). Furthermore, the nilpotent BRST charge makes eq. (1) invariant under the gauge transformation

$$\delta \Phi = Q \Lambda \tag{10}$$

By construction, the BRST charge has ghost number one, so the gauge parameter $\Lambda$ has ghost number lower by one unit than that of the field $\Phi$. It should be stressed that, in this procedure, the arguments of the wavefunction include ghost coordinates. In this sense, eq. (2) is rather natural. It expresses somehow the fact that physical states should only know about the variables of proper statistics. Actually, one can do better than this, expanding the non-semisimple algebra of Q and g to a larger simple algebra (these results are quite recent, and are discussed in detail in ref. |15|).

Thus, one is led somehow to deal with an exterior derivation, and with a cohomology problem, something whose importance is being increasingly felt in Physics. The appearance of the BRST charge in such a fundamental way in string field equations is just a manifestation of the general procedures of refs. |13| and |14|, and thus of the BRST approach to the quantization of constrained systems.

The next question then becomes solving eqs. (1) and (2) for all strings. This is not very difficult, if one makes proper use of the work of Kato and Ogawa |11|. There is one surprise |8|. Whereas for the open string one finds a nontrivial cohomology class corresponding to a field of ghost number -1/2, so that one can write a Lagrangian of the form

$$L = \Phi \; Q \; \Phi \quad , \tag{11}$$

which is ghost number conserving, and thus non vanishing, for all other cases in which this is not possible one finds <u>two</u> (rather than one) nontrivial cohomology classes. These always correspond to ghost numbers j and (-1-j), so that a "doubled" Lagrangian exists:

$$L = \Phi_j \; Q \; \Phi_{(-1-j)} \quad . \tag{12}$$

Whereas this is not completely satisfying, it is intriguing that it can be done at all.

At last, we would like to mention the latest developments concerning a point that we raised (perhaps too pessimistically) in ref. |8|. Witten has shown |6| that, at least for the bosonic string and for the superstring |16|, interactions can be accomodated very elegantly by completing the term in eq. (11) into a Chern-Simons form

$$L = \int (\Phi * Q \Phi + \frac{2}{3} g \, \Phi * \Phi * \Phi) \quad , \tag{13}$$

where multiplication and integration correspond to suitably defined non-local operations in the space of strings. As far as this discussion is concerned, we need the field equation from eq. (13), which is

$$Q \; \Phi + g \; \Phi * \Phi = 0 \quad . \tag{14}$$

This equation has the gauge invariance

$$\delta \Phi = Q \Phi + g(\Phi * \Lambda - \Lambda * \Phi) \tag{15}$$

as a consequence of the nilpotency of the BRST operator. The gauge invariance, in its turn, has a further gauge invariance, and so on, only if one enforces the field equation, eq. (14). This is to be compared with the free case (set g=0 above), where there is no need to impose the field equation. Thus, whereas in the free case one expects an infinite chain of ghosts-for-ghosts, there appears to be a problem in the interacting case. The resolution of the difficulty has been recently achieved in refs. |17,|18|. It is based upon the work of ref. |19|, which deals with general methods for quantizing systems with linearly dependent constraints (i.e. gauge invariances of gauge invariances). Basically the non-closure of the algebra off shell, manifested in the lack of nilpotency of the "interacting" BRST charge (the object in the gauge transformations), turns out to be similar to what happens in certain supergravity theories, where the gauge algebra lacks the proper auxiliary fields for closure (see, for example, ref. |20|). The proper quantization is non trivial, in that it requires cubic ghost couplings. The result is remarkably simple. The gauge-fixed action is basically identical to the gauge-invariant action, but with a field of unspecified ghost number. This includes all the (second-quantized) ghost variables.

The conclusion of this discussion is that the most basic steps in these constructions lie in the proper first quantization. Strictly speaking, one has a straight path to field equations, not Lagrangian. Indeed, the BRST approach does not seem to give Lagrangians in general, although the more general methods of ref. |15| do. More basic is the question of what the content of off-shell formulations for strings can be. Open string are in this sense somewhat exceptional, since they couple to external matter in the form of closed strings, and so are naturally expected to involve off-shell gauge invariant currents to this end. An alternative bootstrap-like approach centered on on-shell quantities has been recently advocated in ref. |21|.

We are grateful to the Miller Institute, the University of

California at Berkeley and the University of California at San Diego for financial support during our participation at this Conference.

## REFERENCES


|1| W. Siegel and B. Zwiebach, *Nucl. Phys.* **B263** (1986) 105;
T. Banks and M. Peskin, *Nucl. Phys.* **B264** (1986) 515;
K. Itoh, T. Kugo, H. Kunitomo and H. Ooguri, *Prog. Theor. Phys.* **75** (1986) 162;
A. Neveu and P.C. West, *Nucl. Phys.* **B268** (1986) 125;
an extensive list of references can be found in:
H. Hata, K. Itoh, T. Kugo, H. Kunitomo and K. Ogawa, *Phys. Rev.* **D34** (1986) 2360.

|2| M. Kaku and K. Kikkawa, *Phys. Rev.* **D10** (1974) 1110, 1823 ;
E. Cremmer and J.-L. Gervais, *Nucl. Phys.* **B76** (1974) 209, **B90** (1975) 410.

|3| S. Mandelstam, *Nucl. Phys.* **B64** (1973) 205, **B69** (1974) 77.

|4| P. Goddard, J. Goldstone, C. Rebbi and C.B. Thorn, *Nucl. Phys.* **B56** (1973) 109.

|5| W. Siegel, *Phys. Lett.* **151B** (1985) 391, 396.

|6| E. Witten, *Nucl. Phys.* **B268** (1986) 253.

|7| A. Neveu, H. Nicolai and P.C. West, *Phys. Lett.* **167B** (1986) 307.

|8| N. Marcus and A. Sagnotti, *Phys. Lett.* **178B** (1986) 343.

|9| C. Becchi, A. Rouet and R. Stora, *Phys. Lett.* **52B** (1974) 344; *Ann. Phys. (N.Y.)* **98** (1976) 29.

|10| I.V. Tyutin, *Lebedeev Preprint* **FIAN** n. 39 (1975).

|11| M. Kato and K. Ogawa, *Nucl. Phys.* **B212** (1983) 443;
S. Hwang, *Phys. Rev.* **D28** (1983) 2614.

|12| P.A.M. Dirac, *Lectures on Quantum Mechanics* (New York, Academic Press, 1967).

|13| I.A. Batalin and and G.A. Vilkovisky, *Phys. Lett.* **69B** (1977) 309;
E.S. Fradkin and T.E. Fradkina, *Phys. Lett.* **72B** (1978) 343.
For a review see: M. Henneaux, *Phys. Reports* **126** (1985) 1.

|14| T. Kugo and I. Ojima, *Phys. Lett.* **73B** (1978) 459.

|15| W. Siegel and B. Zwiebach, *Maryland Preprints* **86-105, 87-97** (1986);
W. Siegel, *Maryland Preprint* **87-60** (1986);
L. Baulieu, W. Siegel and B. Zwiebach, *Maryland Preprint* **87-71** (1986).

|16| E. Witten, *Nucl. Phys.* **B276** (1986) 291.

|17| M. Bochicchio, *University of Rome "La Sapienza" Preprint* (1986); *Princeton Preprint* (1986).

|18| C.B. Thorn, *Princeton IAS Preprint* (1986).

|19| I.A. Batalin and G.A. Vilkovisky, *Phys. Rev.* **D28** (1983) 2567.

|20| For a review see: P. van Nieuwenhuizen, *Phys. Reports* **68** (1981) 189.

|21| D. Friedan and S. Shenker, *Phys. Lett.* **175B** (1986) 287.

# ANALYTIC STRUCTURE OF TWO-DIMENSIONAL QUANTUM FIELD THEO...

Philip Nelson*

Lyman Laboratory of Physics
Harvard University
Cambridge, Massachusetts 02138

There has been considerable progress in the past year in understanding the structure of two-dimensional quantum field theories. Some of this progress has come from an improved understanding of how algebraic geometry enters into the theory of determinants of differential operators on Riemann surfaces. In these notes I will sketch the connection briefly, en route to a description of some recent work done with L. Alvarez-Gaumé, J.-B. Bost, G. Moore, and C. Vafa.[1] Further details appear in Ref. 2. The motivation for these investigations is of course the fact that 2d field theory is an essential ingredient in the formulation of string theories, as discussed by many other speakers here. In particular the Fermi-Bose equivalence to be described below is a key tool used in many constructions.

It has been known for many years that quantum field theory in two dimensions has remarkable special features. For example, in many cases such theories are exactly solvable. The study of such solvable models led to the discovery that in some instances theories with commuting fields were exactly equivalent to other theories with anticommuting fields. In the following paragraphs I will outline some features of this equivalence and sketch how it can be demonstrated. A key feature of the development is that it applies to theories formulated on compact orientable

*Work supported in part by the Harvard Society of Fellows and by the National Science Foundation under Grant Number PHY82-15249.

surfaces of arbitrary genus, as it must in order to be of use for string theory.

A simple example of a theory with commuting (Bose) fields is the following. Consider a compact Riemann surface X and the "field space" $\mathscr{X}$ of all real functions $\phi: X \to \mathbb{R}$ satisfying some sort of Sobolev condition, which we will not specify in detail. Define the functional $S_b[\phi] = \|\partial\phi\|^2 \equiv \int_X \partial\phi\,\bar{\partial}\phi$. $S_b$ is called the action functional of a free massless boson; it depends on $\phi$ and also on the conformal structure chosen for X. It is possible to define a probability measure $[d\phi]$ on $\mathscr{X}$ in such a way that

$$Z_b = \int_{\mathscr{X}} [d\phi] e^{-S_b[\phi]} \tag{1}$$

is well-defined. There are two complications, however. First, to define the measure $[d\phi]$ we must "regulate" it, i.e. suppress the contributions from rapidly-varying maps $\phi$. We can only do this if we are given not just a conformal structure on X, but a *metric* g in the given conformal class. Thus $Z_b = Z_b[g]$. Secondly, as it stands $Z_b$ includes an integral over constant maps $\phi$, all of which have $S_b = 0$. To remedy this trivial infinity we must eliminate the zero mode of $\partial$ from $\mathscr{X}$ and, it turns out, divide the measure by the metric norm of this zero mode, i.e. by $\|1\|^2 = \int_X \mathrm{vol}_g$, the area of X. It turns out that we can define (1) by

$$Z_b[g] = \left(\frac{\det{}'\bar{\partial}^\dagger\bar{\partial}}{\|1\|^2}\right)^{-\frac{1}{2}}, \tag{2}$$

which is just the naive answer to (1) obtained by Gaussian integration, but made meaningful by removing the zero mode and defining the determinant via zeta-function regularization.

Equation (2) is called the partition function of the commuting field theory. It does not characterize the measure $[d\phi]$, since it is just one number for each g. Rather it is the simplest of a collection of moments of $[d\phi]$, which are called correlation functions. We will restrict attention to (2), but our conclusions generalize to the more complicated functions as well.

Next we can define an anticommuting (or Fermi) field theory. Again let X be a compact Riemann surface, but this time equipped with an even theta characteristic. This amounts to choosing a spin structure on X, or a holomorphic line bundle L such that $L \otimes L \cong K$, the canonical bundle on X. Analogous to (1) one can define

$$Z_f = \int [db][dc][d\bar{b}][d\bar{c}]e^{-S_f} \qquad (3)$$

The functional measures in (3) are more formal than the one in (1). For the present purposes it suffices to define (3) by a formula analogous to (2)

$$Z_f[g;L] = \det \bar{\partial}_L^\dagger \bar{\partial}_L \ . \qquad (4)$$

Here $\bar{\partial}_L$ is the Cauchy-Riemann operator on the holomorphic line bundle L. Again a metric is needed to define the "measures" in (3); it enters (4) via the adjoint. Since generically $H^0(X;L) = 0$, $\bar{\partial}_L$ has no zero mode and no prime or normalizing factor is needed in (4).

The expression (4) is the partition function of an anticommuting field theory. It too admits generalizations to correlation functions which we will not discuss here.

Suppose that X is the Riemann sphere. Then the statement

of Bose-Fermi equivalence begins with the claim that

$$Z_b[g] = c\, Z_f[g] \tag{5}$$

where c is some constant independent of g. In fact (5) is true, but in a fairly uninteresting way. Both sides of (5) are unchanged if g is replaced by $f^*g$ for some diffeomorphism f. Moreover, both sides change in the same simple way if we change g in its conformal class, i.e. if we replace g by $e^{\sigma}g$ for a function $\sigma$. (See e.g. Ref. 3.) Since up to diffeomorphism there is just one conformal structure on $S^2$, (5) follows at once.

Equation (5) is bound to be more interesting for surfaces of genus $\gamma > 0$, since here there is a nontrivial space of conformal classes, the *moduli space* of genus $\gamma$. On such an X (5) cannot be literally true, however, since one side depends on a $\theta$-characteristic while the other does not. Thus (5) should be replaced by

$$Z_b[g] = c \sum_L Z_f[g;L] \ . \tag{5'}$$

Moreover from earlier work involving the case where X is an infinite cylinder, we know that $\phi$ in (1) must be taken to be not a function to $\mathbb{R}$ but rather a circle-valued function. The distinction matters when X is not simply-connected.

When X has genus $\gamma > 0$ we can classify maps $\phi$ into topological sectors as follows. Choose a canonical homology basis for X $\{a_i, b_i\}$ (see e.g. Refs. 4, 5) and let $\phi$ wind $n_i, m_i$ times around $S^1$ as one walks around $a_i, b_i$ respectively. Viewing $S^1$ as a group, we can realize any $\phi$ as $\phi_{n,m} \cdot \tilde{\phi}$ where $\tilde{\phi}$ has zero winding and $\omega_{n,m} \equiv \phi_{n,m}^{-1}\, d\, \phi_{n,m}$ is harmonic, by the Hodge theorem. Then (2) becomes

$$Z_b[g] = \left(\sum_{n,m} e^{-\|\omega_{n,m}\|^2}\right)\left(\frac{\det' \bar\partial^\dagger \bar\partial}{\|1\|^2}\right)^{-\frac{1}{2}} . \tag{6}$$

The foregoing analysis was performed by the authors of Ref. 5, who then summed (6) to get

$$Z_b[g] = \left(\sum_{\varepsilon} |\vartheta[\varepsilon](\tau)|^2\right)\left(\frac{\det' \bar\partial^\dagger \bar\partial}{\det(\omega^i,\omega^j)\cdot\|1\|^2}\right)^{-\frac{1}{2}} . \tag{7}$$

Here $\vartheta$ is the Riemann theta function with characteristics and we sum over the half-points of the Jacobian of X; this is because the physically appropriate prescription for (6) sums over half-integers $n_i, m_i$. $\tau$ is the period matrix of X in the given homology basis, and we have rewritten $\det \operatorname{Im}\tau$ as the determinant of the matrix of inner products of the Abelian differentials $\omega^i$ determined by the homology basis.[4]

In Ref. 5 Fermi-Bose equivalence and Quillen's theorem[6] were used to deduce

$$\det \bar\partial_L^\dagger \bar\partial_L = c\left(\frac{\det' \bar\partial^\dagger \bar\partial}{\det(\omega^i,\omega^j)\cdot\|1\|^2}\right)^{-\frac{1}{2}} |\vartheta[\varepsilon](\tau)|^2 \tag{8}$$

where on the right $\varepsilon$ is the point in the Jacobian corresponding to the spin bundle L via the chosen homology basis and Riemann's theorem.[4,5] We will sketch how (8) can be derived directly.

Both sides of (8) are functions of g with the same behavior under conformal rescaling. Thus their quotient F is a function on the moduli space $\mathcal{M}$ of curves of genus $\gamma$ (or more precisely on its cover $\hat{\mathcal{M}}$, curves with a $\theta$-characteristic). We need two key facts about moduli space:

(a) $\hat{\mathcal{M}}$ is a complex space, and $\bar\partial, \bar\partial_L$ vary holomorphically.

(b) $\hat{\mathcal{M}}$ has only one holomorphic function in genus $\gamma > 2$;

> indeed $\hat{\mathcal{M}}$ has only one pluriharmonic function. This follows from the existence of the Satake compactification of $\mathcal{M}$, which has codimension two.

Thus to show that F is a constant, and hence prove (8), one has only to show that

$$\delta_i \bar{\delta}_j \log F = 0 \qquad (9)$$

where $\delta$, $\bar{\delta}$ are holomorphic and antiholomorphic derivatives on $\hat{\mathcal{M}}$.

To show (9), we note that the determinants in (8) are closely related to Quillen's construction of norms on determinant line bundles.[6] Indeed over moduli space we have determinant bundles $D_0 = \mathrm{DET}\ \bar{\partial}$, $D_{\frac{1}{2}} = \mathrm{DET}\ \bar{\partial}_L$, and by the Riemann-Roch-Grothendieck theorem we have

$$D \equiv D_{\frac{1}{2}} \otimes D_{\frac{1}{2}} \otimes D_0 \cong \text{trivial}.$$

Moreover the trivializing section is $\sigma \equiv \sigma_{\frac{1}{2}} \otimes \sigma_{\frac{1}{2}} \otimes \sigma_0$, where $\sigma_{\frac{1}{2}}$ is the holomorphic section of $D_{\frac{1}{2}}$ vanishing whenever $H^0(X;L) \neq 0$, and

$$\sigma_0 = (\omega^1 \wedge \ldots \wedge \omega^\gamma)^{-1} \otimes (1)^{-1} \cdot \vartheta[\varepsilon](\tau)^{-2}$$

(We are being careless about torsion since we are going to take the absolute square of $\sigma$.) $\sigma$ never vanishes or blows up, since $\varepsilon$ is chosen to make $\vartheta$ vanish whenever $\bar{\partial}_L$ has a zero mode.

The quotient of the two sides of (8) is then just $F = \|\sigma_{\frac{1}{2}} \otimes \sigma_{\frac{1}{2}} \otimes \sigma_0\|$, and so $\delta\bar{\delta} \log F$ is the Hermitian curvature of Quillen's norm on D. But remarkably, Quillen's norm has the property of being *flat* on any trivial combination of determinant bundles.[7] Hence $\delta\bar{\delta} \log F = 0$, F is a constant, and (8) is established for genus $\gamma > 2$. (The cases $\gamma = 1,2$ can be verified directly by other means.[8])

Arguments like these based on the near-compactness of $\mathcal{M}$ also let one prove Fermi-Bose equivalence for the correlation functions mentioned above. They have other applications to string theory as well, as discussed for instance in Ref. 2.

References


1. L. Alvarez-Gaumé, J. Bost, G. Moore, P. Nelson, and C. Vafa, Phys. Lett. B178, 41 (1986); manuscript to appear; J. Bost and P. Nelson, Phys. Rev. Lett. 57, 795 (1986).
2. P. Nelson, "Lectures on Strings and Moduli Space", to appear in Phys. Rep.; L. Alvarez-Gaumé and P. Nelson, "Riemann Surfaces and String Theories", CERN preprint to appear in *Supersymmetry, Supergravity, and Superstrings '86* (World Scientific, 1986). These articles contain many references to the literature.
3. O. Alvarez, Nucl. Phys. 216, 125 (1983).
4. D. Mumford, *Theta Lectures on Theta* (Birkhaüser, 1983).
5. L. Alvarez-Gaumé, G. Moore, and C. Vafa, "Theta Functions, Modular Invariance, and Strings", Comm. Math. Phys. 106, 1 (1986).
6. D. Quillen, Funk. Anal. i Prilozen 19, 37 (1985) [=Funct. Anal. Appl. 19, 31 (1986).
7. A. Belavin and V. Knizhnik, Phys. Lett. B168, 201 (1986); "Complex Geometry and Theory of Quantum Strings", Landau Inst. preprint submitted to ZETF; J. Bost and J. Jolicoeur, Phys. Lett. B174, 273 (1986).
8. G. Moore, talk presented at the VIIIth International Congress on Mathematical Physics, Marseille, August 1986, and references therein.

# The Geometrical Meaning of Currents in String Theory

**Orlando Alvarez**
**Paul Windey**

*Department of Physics*
*and*
*Lawrence Berkeley Laboratory*
*University of California*
*Berkeley, CA 94720*

## ABSTRACT

A new geometrical interpretation of the partition function of a string theory is given. It is seen as a the norm, with respect to a *physicist metric*, of a section of a line bundle on the moduli space of Riemann surfaces. The flat connection compatible with it is the energy-momentum tensor. The latter is seen as the fundamental object of the theory from which one can reconstruct the partition function. Any anomaly is an obstruction to the flatness integrability condition which can be seen as a fundamental equation of motion to be satisfied by any string theory.

# 1 Introduction

The main result presented here is a new geometrical interpretation of the energy-momentum tensor of string theories as a flat connection of the partition function line bundle on the moduli space of Riemann surfaces. The flatness condition is proposed as a fundamental dynamical equation that any string theory must satisfy. In particular this condition implies the vanishing of the conformal or gravitational anomalies, This result follows from a simple question: how to reconstruct the partition function of a string theory from physical data specified in the form of a collection of energy-momentum tensors. The functional approach of Polyakov [1,2] provides the necessary framework. It is by now well known that it gives an expression for the partition function $Z(y^i, \bar{y}^j)$ of the string at order $g$ as a "function" of the $3g-3$ complex parameters of the moduli space $\mathcal{M}_g$ [3,4]. More precisely $Z$ is not a *bona fide* function but the section of a potentially non-trivial line bundle $\mathcal{L}$ over $\mathcal{M}_g$ (see for example [5,6,7]). We will refer to this bundle as the *partition line bundle*. Only when certain consistency conditions are satisfied such as the cancellation of anomalies can one interpret $Z$ as an honest density which can be unambiguously integrated. We shall rephrase in a geometrical language the conditions under which $Z_g$ is in a form suitable to do physics, *i.e.* to extract a number from the integration over $\mathcal{M}_g$. To wit: the energy-momentum tensor provides a connection on the partition line bundle over moduli space. This connection is naturally associated with a particular metric called the *physicist's metric*. The energy-momentum is precisely the flat connection compatible with this metric. However the partition line bundle $\mathcal{L}$ is not flat in general and therefore one has global topological obstructions to imposing the physicist's metric. In such a situation the above flat connection has only a local meaning and the partition function will not have a physical interpretation . We shall consider this flatness condition as providing a fundamental equation of motion for any string theory. We also analyze in some detail the meaning of the integrability condition and link it with properties of some two points correlation functions[1]. Most of our results were obtained by analyzing the case of a string in a flat background space-time geometry. It is known that in that case the determinant line bundles are the germane objects of interest and all our constructions can be made explicit. It is

[1]The idea of that there are natural connections associated with determinants has been around for a while [8]. The emphasis of these authors was in a different direction.

however possible to see that our conclusions do not depend upon these particular premises and must also pertain to the generic case of arbitrary background[2] In Section 2 we discuss the geometry of line bundles over the parametrization space for the partition function. We show how the notion of connections on this space is intimately related to the definition of *generalized currents.* Section 3 we review briefly the minimal understanding of determinant line bundles which is necessary to understand the point of this note. Determinant line bundles arise in the simple case where the background metric on space-time is flat. This section is based on the discussion given in [10]. We explicitly construct connections, currents, metrics and flat sections whose general properties are analyzed in Section 2 and Section 4. In Section 4 we discuss a new interpretation of anomalies based on the potentially dangerous introduction of a special metric on the space of field configurations. This interpretation is geometrical in nature as opposed to purely topological. It covers all possible anomalies.

## 2 Currents as connections

In theories of physical interest the partition function $Z$ must admit an interpretation as the generating functional for the moments of a probability distribution. If we are interested in studying the partition function in the presence of background fields (collectively denoted by $A$) then the *generalized currents* $j[A]$ associated with the background fields are defined by the variational equation

$$dZ[A] + \left(\int_X j[A]dA\right) Z[A] = 0\,. \tag{2.1}$$

In the above one can interpret $d$ as the exterior derivative in the space of background fields. We will denote the space of fields by $Y$. The fields $A$ may be scalars, vectors or even higher rank tensors. One should think of $dA$ as being a generic variation. The associated *generalized currents* will have the appropriate tensorial properties. Remember that the currents will not be *bona fide* tensors but object with anomalous transformation properties! For example, if $A$ is the metric tensor $g_{ab}$ then the associated current will be the energy-momentum tensor density $T^{ab}$. If $A$ is a vector field then the current will be a vector. We are interested in the cases where there is an explicit background field or where one

[2]The treatment of this general case can be found in Dan Friedan contribution to these Proceedings and in [9].

integrates over all $A$ field configurations. In the latter the term partition function always refers to the non-integrated density over the parameter space. In string theories, one could conceivably couple the string to abelian or nonabelian background fields. For example, in the abelian situation the partition function will depend on a larger space such as the product of moduli space with the Jacobian variety.

At first glance, $\int j\, dA$ resembles a one form $\omega$ on the space of background fields:

$$\omega = \int_X j[A]\, dA\,. \tag{2.2}$$

If $Y$ is the space of all vector potentials then one should think of the above as a one form on this infinite dimensional space. If $Y$ is a finite dimensional subspace with coordinates $\{y^i\}$ then the above should be interpreted as

$$\omega = \left(\int_X j[A]\, \frac{\partial A}{\partial y^i}\right)\, dy^i \tag{2.3}$$

In fact, equation (2.1) looks just like the equation for the parallel transport of a section of a line bundle. The object $\omega$ appears to be a connection on the space of background fields. The integrability condition for equation (2.1) follows from $d^2 = 0$:

$$\begin{aligned} 0 &= (d\omega + \omega \wedge \omega) Z \\ &= (d\omega) Z = \Omega\, Z\,. \end{aligned} \tag{2.4}$$

Notice that since $\omega$ is a one form one has that $\omega \wedge \omega = 0$. In the above one can interpret $d\omega$ as the curvature $\Omega$ of a line bundle. In fact, one has a flat line bundle since the curvature vanishes as demonstrated by (2.4). The equation $\Omega = 0$ may be seen as a dynamical equation.

All these mathematical remarks have a very important physical interpretation. Given the physical currents $j[A]$ one would like to reconstruct the partition function $Z[A]$. This requires solving equation (2.1). The latter cannot be solved unless the integrability conditions $\Omega = 0$ are satisfied. An elementary calculation shows that the curvature is given by

$$\Omega = \int_1 \int_2 \left(\langle j(1)\, j(2)\rangle - \langle j(1)\rangle\, \langle j(2)\rangle\right) dA(1) \wedge dA(2)\,. \tag{2.5}$$

This is the antisymmetric part of the connected two point current correlation function. It should vanish by Bose statistics. In fact, the antisymmetric part of the $\langle j \rangle \langle j \rangle$ term vanishes by itself and therefore the antisymmetric part of $\langle j(1)j(2) \rangle$ is automatically zero. All this generalizes to the fermionic case.

In addressing the issues raised by the above one has to ask the following questions:

1. Is the partition function an ***honest*** function or is it a section of some line bundle over the parameter space?
2. What are the gauge transformation properties of the partition function?
3. Does the generalized current actually transform as a connection?
4. Are the integrability conditions compatible with additional constraints one might impose such as gauge invariance under some Yang-Mills group or under diffeomorphisms?

In the above we are considering a general variation $dA$ and there is not much content in the above equations. The equations become non-trivial as soon a one starts restricting the allowed class of variations. It is important to bear in mind that the equations above have very different meaning according to the specific variation one chooses to do. Some interesting cases will be analyzed below. It suffices to mention two broad classes of variations: gauge transformations and variations with respect to the residual parameters of the theory after elimination of the gauge group. Both of them can be treated using line bundle concepts. The integrability condition will be the vanishing of the curvature of the appropriate line bundle. This will be discussed at different places throughout this paper. In anticipation of the results of the ensuing sections we will assume that the generalized currents are genuine connections.

Of particular interest is the case where the exterior derivative $d$ on $Y$ may be invariantly split into the form $d = d_1 + d_2$, where $d_i^2 = 0$. As an example, consider the case where the field space $Y$ is a complex manifold. In this case the exterior derivative splits off into the holomorphic derivative $\partial$ and the anti-holomorphic derivative $\bar{\partial}$, *i.e.* $d = \partial + \bar{\partial}$. One has integrability condition with respect to the operators $d$, $\partial$ and $\bar{\partial}$ which are analogous to the conditions found in (2.4). One also has integrability conditions associated with the mixed derivative constraint $\partial\bar{\partial} + \bar{\partial}\partial = 0$. This condition may be written as

$$\bar{\partial}\omega + \partial\bar{\omega} = 0 \tag{2.6}$$

where $\omega = -\partial \log Z$ and $\bar{\omega} = -\bar{\partial} \log Z$. To be more explicit one can see that

$$\bar{\partial}\omega = \int_1 \int_2 \left( \langle \bar{\jmath}(1)\, \jmath(2) \rangle - \langle \bar{\jmath}(1) \rangle \langle \jmath(2) \rangle \right) d\bar{A}(1) \wedge dA(2) \tag{2.7}$$

In the above we are considering the situation where $A$ depends holomorphically on the complex coordinates $\{y^i\}$ of $Y$ and $\bar{A}$ depends holomorphically on the complex conjugate coordinates $\bar{y}^i$. The $\langle \bar{\jmath}\jmath \rangle$ correlation function does not have any particular symmetry properties since $\bar{\jmath}$ and $\jmath$ are distinct objects. The above really gives the connected part of the two point correlation function $\langle \bar{\jmath}\jmath \rangle$. The total curvature has parts which are forms of degree $(2,0)$, $(1,1)$ and $(0,2)$. The $(1,1)$ part of the curvature is precisely $\bar{\partial}\omega + \partial\bar{\omega}$. If $\bar{\partial}\omega$ vanishes then $\partial\omega$ also vanishes because of constraint (2.6).

The interesting situation occurs when subcurvatures associated with the existence of a holomorphic structure vanish by themselves. For example, $\bar{\partial}\omega$ might vanish by itself. In such a situation [5], the partition function will factorize into a holomorphic and antiholomorphic part. The reason is that one then has independent integrability in both the holomorphic and antiholomorphic directions. We would like to stress that such a strong integrability condition will not be the general case. It is interesting seeing what $\bar{\partial}\omega$ is in terms of correlation functions.

The integrability conditions are geometric constraints on the quantum theory. They contain more information than anomaly cancellations which are often purely topological in nature.

Another example is the case where $Y = Y_1 \times Y_2$. Similarly one has $d = d_1 + d_2$ in an obvious notation. The discussion is analogous to the complex manifold case. If the "mixed curvatures" $d_1\omega_2$ and $d_2\omega_1$ each vanish then the partition function will factorize into a product of a $Y_1$ and a $Y_2$ part.

As a concrete example we discuss the energy momentum tensor in a two dimensional conformal field theory. The standard definition of the energy momentum tensor density is

$$d \log Z[g] = \frac{1}{2} \int T^{ab}\, dg_{ab}\,. \tag{2.8}$$

Comparing with definition (2.1) we see that $T^{ab}$ is a connection on the space of all metrics on the Riemann surface. If the two point energy-momentum correlation function is symmetric then the integrability conditions will be satisfied. We would like stronger integrability conditions. In fact we would like the partition function to be invariant under the action of the diffeomorphism group. This requires that the connection vanish in the direction of the infinitesimal diffeomorphism action:

$$\int T^{ab}\, \pounds_\xi\, g_{ab} = 0\ . \tag{2.9}$$

In the above $\xi$ is the vector field which generates the infinitesimal diffeomorphism. The above is just the statement that the gravitational anomaly vanish. We will come back to the more interesting case of the variation of the complex structure in another section. Briefly we mention that in the notation of complex tensors, the connection in the direction of the Beltrami differential $\mu^z{}_{\bar z}$ is given by $\int T_{zz}\, \mu^z{}_{\bar z}\, dz \wedge d\bar z$.

# 3 The Determinant Line Bundle

In this section we attempt to explain the main ideas which form the basis for a proper definition of the determinant line bundle of a family of elliptic operators. These ideas will be used later in this work. The most important thing to remember is that the word *determinant* is used in several different ways. To avoid confusion we will use different fonts for the distinct usages.

Given a complex vector space of complex dimension $m$, one can associate a one dimensional complex vector space $\mathsf{DET}(V) = \bigwedge^m V$, where $\bigwedge^m$ denotes the $m$–th exterior power of $V$. This one dimensional complex vector space is called the *determinant* of the vector space $V$.

If $L : V \to W$ is a linear transformation then there is induced a linear transformation $\det L$, called the *determinant* of $L$:

$$\det L : \mathsf{DET}(V) \longrightarrow \mathsf{DET}(W)\ . \tag{3.1}$$

This transformation is defined in a very natural way. If $\{e_i\}$ ($\{f_a\}$) is a basis for $V$ (respectively $W$), then $e_1 \wedge \cdots \wedge e_m$ is a basis for $\mathsf{DET}(V)$. The transformation $\det L$ is defined by

$$\det L(e_1 \wedge \cdots \wedge e_m) = Le_1 \wedge Le_2 \wedge \cdots \wedge Le_m = \Delta_L f_1 \wedge \wedge \cdots \wedge f_m\ . \tag{3.2}$$

This is a perfectly sensible definition except that $\Delta_L$ cannot be interpreted as a number since it transforms under changes of basis. Except of course if $V = W$, in which case one recovers the fact that the determinant is a number. Let $V^*$ be the dual vector space to $V$. $V^*$ are the linear transformation from $V$ to $\mathbf{C}$. The dual basis is denoted by $\{\theta^i\}$ and defined by $\theta^i(e_j) = \delta^i{}_j$. The linear transformation $L$ may be viewed as an element of $V^* \otimes W$:

$$L = L^a{}_i \, f_a \otimes \theta^i \ . \tag{3.3}$$

The space $V^* \otimes W$ is the space of linear transformations between $V$ and $W$ which is also called the homomorphisms between $V$ and $W$, Hom($V$,$W$). Similarly, $\det L$ may be viewed as an element of $\mathsf{DET}(V)^* \otimes \mathsf{DET}(W)$:

$$\det L = \Delta_L \Big( f_1 \wedge \cdots \wedge f_m \Big) \otimes \Big( \theta^1 \wedge \cdots \wedge \theta^m \Big) \ . \tag{3.4}$$

Note that $\mathsf{DET}(V)^* \otimes \mathsf{DET}(W)$ may be viewed as the fiber of a complex line bundle over the vector space $V^* \otimes W$. The line bundle in question is the trivial bundle $\mathcal{L} = (\mathsf{DET}(V)^* \otimes \mathsf{DET}(W)) \times (V^* \otimes W)$. The object det is a section of $\mathcal{L}$ since to every $L$ in $V^* \otimes W$ it assigns an element $\det L$ of $\mathsf{DET}(V)^* \otimes \mathsf{DET}(W)$. The transition functions for this bundle are precisely the ones for $\Delta_L$. This line bundle is called the *determinant line bundle* of the family of all linear transformations. We will use the abbreviation $\mathbf{DET}(L)$ for the determinant line bundle associated with all linear transformations $L : V \to W$.

We will now briefly explain how these considerations are modified in the infinite dimensional case. Let $Y$ be a compact manifold parametrizing a family $\{D_y\}_{y \in Y}$ of elliptic operators mapping sections of the vector bundle $E_y^+$ into sections of the vector bundle $E_y^-$:

$$D_y : \Gamma(E_y^+) \longrightarrow \Gamma(E_y^-) \ . \tag{3.5}$$

The capital gamma denotes the sections of the respective vector bundle. In the case of conformal field theory one will require that $Y$ be a complex manifold and that $D_y$ depends on $y$ holomorphically. on $y$. The vector bundles live over some manifold $X$ and are endowed with hermitian metrics. Thus one can define adjoint operators $D_y^\dagger : \Gamma(E_y^-) \to \Gamma(E_y^+)$. The example we will be interested in involve a Riemann surface $X$ and a family of $\bar\partial$-operators on $X$ depending on some parameter space $Y$. The space of primary interest will be the moduli space of $X$. For example for the free bosons which describe the embedding space of the string, the object of interest will be the determinant of the operator $\bar\partial$ acting on functions, while for the Faddeev-Popov ghosts it will be $\bar\partial$ acting on the tangent bundle of $X$. Note that in the finite dimensional case, the parametrizing space was

$V^* \otimes W$. First one has to be careful about defining $\mathsf{DET}(\Gamma(E_y^{\pm}))$ since the spaces of sections are infinite dimensional. Following [11] and [10] we will construct the determinant line bundle over $Y$ by actually constructing the transition functions and the associated open cover of $Y$.

Firstly we define two Laplacians $\Delta^+$ and $\Delta^-$:

$$\Delta^+ \; = D^\dagger D \; : \Gamma(E^+) \longrightarrow \Gamma(E^+)\,, \tag{3.6}$$

$$\Delta^- \; = D D^\dagger \; : \Gamma(E^-) \longrightarrow \Gamma(E^-)\,. \tag{3.7}$$

Both of these are non-negative self-adjoint elliptic operators. Let $\{\phi_n\}$ be a basis of eigensections of $\Delta^+$:

$$\Delta^+ \phi_n = \lambda_n \, \phi_n\,. \tag{3.8}$$

At fixed $y \in Y$, there is an isomorphism of $(\ker D_y)^\perp$, the orthogonal complement to the kernel of $D_y$, and $(\ker D_y^\dagger)^\perp$ given by:

$$\psi = D_y \phi \in (\ker D_y^\dagger)^\perp\,. \tag{3.9}$$

where $\phi \in (\ker D_y)^\perp$ . Therefore, to every eigensection $\phi_n$ of $\Delta^+$ with eigenvalue $\lambda_n > 0$ there corresponds and eigensection $\psi_n$ of $\Delta^-$. One has the direct sum decompositions

$$\Gamma(E_y^+) \;\; = \;\; \ker D_y \oplus (\ker D_y)^\perp\,, \tag{3.10}$$

$$\Gamma(E_y^-) \;\; = \;\; \ker D_y^\dagger \oplus (\ker D_y^\dagger)^\perp\,. \tag{3.11}$$

It is well known that the *index* of $D$ is a topological invariant and defined by

$$\mathcal{I}(D) = \dim\ker D - \dim\ker D^\dagger. \tag{3.12}$$

Thus for fixed $y$, the space of sections differs by the kernels of the respective operators. The difficulty in making an identification between the sections of the two different vector bundles is that the kernels jump in dimension as one varies $y$. For example, if $D$ has index one at a point $y_1$ in $Y$, it may have two zero modes in the kernel and one in the cokernel. At a nearby point $y_2$ it may have only one zero mode in the kernel. The reason is that generically, perturbations will lift zero modes.

We now construct the determinant line bundle [11,10]. The construction involves two steps. Firstly, a useful open cover of $Y$ is given. Secondly, information equivalent to specifying the transition functions is provided. The idea is remarkably simple and consists essentially in introducing a cutoff $a$ which does not belong to the spectrum of the operator. For $a > 0$, let $U_a$ be the set of points in $Y$ where $a$ is not an eigenvalue of $\Delta^+$. This is an open set. Clearly since the operator $D_y$ depends smoothly on $y$, the number of eigenvalues of $\Delta_y^+$ less than $a$ is constant in $U_a$ . Note that it is important that $a > 0$. The reason is that the spectrum is bounded below by zero. A similar statement about the kernels and cokernels, *i.e.*, "$a = 0$", is not possible.

Let $(E_y^\pm)_a$ be the space spanned by eigensections of $\Delta_y^\pm$ with eigenvalues less than $a$. The determinants of these spaces will be pieced together over $Y$ in order to construct the determinant line bundle. Over $U_a$ consider the space $\mathsf{DET}((E_y^+)_a)^* \otimes \mathsf{DET}((E_y^-)_a)$. This is a one dimensional vector space. Let $s_a$ be a section:

$$s_a = \sigma_a \frac{\psi_1 \wedge \cdots \wedge \psi_\ell}{\phi_1 \wedge \cdots \wedge \phi_k}\,, \tag{3.13}$$

where $\phi_1, \cdots, \phi_k$ is an arbitrary basis for $(E^+)_a$ and $\psi_1, \cdots, \psi_\ell$ is an arbitrary basis for $(E^-)_a$. Note that the index of $D$ is given by $k - \ell$. Similarly one considers $U_b$ and $s_b$ for $b > a$. On $U_a \cap U_b \neq \emptyset$, let $(E_y^\pm)_{(a,b)}$ be the space spanned by eigensections with eigenvalues in the open interval $(a, b)$. Then for fixed $y$

$$(E_y^\pm)_b = (E_y^\pm)_a \oplus (E_y^\pm)_{(a,b)}\,. \tag{3.14}$$

On the intersection $U_a \cap U_b \neq \emptyset$ one can make the following canonical identification of sections:

$$s_b = s_a \otimes \frac{D\phi_{a+} \wedge \cdots \wedge D\phi_{b-}}{\phi_{a+} \wedge \cdots \wedge \phi_{b-}}\,. \tag{3.15}$$

In the above we have used $a+$ and $b-$ symbolically to delimit the product of the eigensections of $(E_y^\pm)_{(a,b)}$. The number of eigensections in the numerator and denominator in the last term is the same because of equation (3.9). The above only depends on the operator $D_y$ and not on the choice of basis for $E^+_{(a,b)}$. In fact this is just like a finite dimensional determinant. It is easy to verify that the analogue of the cocycle condition $g_{ca} = g_{cb} g_{ba}$ for $c > b > a$ is satisfied on a non-empty overlap $U_a \cap U_b \cap U_c$:

$$\begin{aligned} s_c &= s_b \otimes \frac{D\phi_{b+} \wedge \cdots \wedge D\phi_{c-}}{\phi_{b+} \wedge \cdots \wedge \phi_{c-}} \\ &= s_a \otimes \frac{D\phi_{a+} \wedge \cdots \wedge D\phi_{b-}}{\phi_{a+} \wedge \cdots \wedge \phi_{b-}} \otimes \frac{D\phi_{b+} \wedge \cdots \wedge D\phi_{c-}}{\phi_{b+} \wedge \cdots \wedge \phi_{c-}} \\ &= s_a \otimes \frac{D\phi_{a+} \wedge \cdots \wedge D\phi_{c-}}{\phi_{a+} \wedge \cdots \wedge \phi_{c-}} \end{aligned} \tag{3.16}$$

where in the last line we used the multiplication properties of finite dimensional determinants. The line bundle thus constructed is called the *determinant line bundle* and will be denoted by $\mathcal{L} = \mathbf{DET}(D)$. Note that the infinite wedge product $\mathsf{DET}(\Gamma(E^{\pm}))$ was in a sense constructed using:

$$\lim_{a\to\infty}(E^{\pm})_a = \Gamma(E^{\pm}) . \tag{3.17}$$

The canonical identifications allows us to push the 'cutoff' $a$ towards zero. Consequently, at fixed $y \in Y$, one has that the fiber $\mathcal{L}_y$ of $\mathcal{L}$ over the point $y$ may be identified with the finite dimensional space

$$\mathbf{DET}(\ker D_y)^* \otimes \mathbf{DET}(\ker D_y^{\dagger}) . \tag{3.18}$$

The next step is to define a hermitian metric on $\mathcal{L}$. There is an induced metric on each fiber given by the metrics on $\Gamma(E^{\pm})$. To be more explicit, let us denote the inner product on the space of sections by $\langle\cdot,\cdot\rangle$. In fact the induced norm of sections given in $U_a$ by (3.13) is

$$||s_a||^2_{ind} = |\sigma_a|^2 \frac{\det(\langle\psi_\mu,\psi_\nu\rangle)}{\det(\langle\phi_\alpha,\phi_\beta\rangle)} \tag{3.19}$$

The induced norm of the same section in patch $U_b$ may be computed using (3.15):

$$||s_b||^2_{ind} = ||s_a||^2_{ind} \prod_{a<\lambda_n<b} \lambda_n . \tag{3.20}$$

Note that the product term is independent of the choice of section for $(E^+)_{(a,b)}$. The actual norm of a section is given in terms of a hermitian metric $h$ such that

$$\begin{aligned} ||s||^2_Q &= h_a\,||s_a||^2_{ind} && (3.21)\\ &= h_b\,||s_b||^2_{ind} && (3.22)\end{aligned}$$

such that

$$h_a = \prod_{\lambda_n>a} \lambda_n; . \tag{3.23}$$

For simplicity let us temporarily assume that the operator has index zero. The determinant line bundle has a canonical section called $\det D$ and given over $U_a$ by

$$(\det D)_a = \frac{D\phi_1\wedge\cdots\wedge D\phi_k}{\phi_1\wedge\cdots\wedge\phi_k} . \tag{3.24}$$

This is a well defined section which corresponds to the naive section

$$(\det D)_\infty = \frac{D\phi_1 \wedge D\phi_2 \wedge \cdots}{\phi_1 \wedge \phi_2 \cdots} \tag{3.25}$$

The induced naive norm of this section is

$$\|\det D\|_\infty^2 = \prod \lambda = \det(D^\dagger D) \ . \tag{3.26}$$

One would like to define the hermitian metric on $\mathcal{L}$ in such a way that the norm of the determinant section is the naive norm. It is clear that we have to choose (3.23). This is Quillen's choice. The most important observation is that Quillen's metric corresponds to the naive induced metric on $\mathsf{DET}(\Gamma(E^+))^* \otimes \mathsf{DET}(\Gamma(E^-))$. One can go in the infrared direction and study the metric on the fiber $\mathcal{L}_y$ over $y$ where we identify the fiber with (3.18). The norm of a section

$$s = \sigma \frac{\psi_1 \wedge \cdots \wedge \psi_\ell}{\phi_1 \wedge \cdots \wedge \phi_k} \tag{3.27}$$

at the point $y$ is given by

$$\|s(y)\|_Q^2 = |\sigma(y)|^2 \ H_0 \ , \tag{3.28}$$

where

$$H_0 = \frac{\det \langle \psi_\mu, \psi_\nu \rangle}{\det \langle \phi_\alpha, \phi_\beta \rangle} \ \det{}'(D_y^\dagger D_y) \ , \tag{3.29}$$

and the prime denotes the omission of zero modes. In the above, $\{\phi_1, \cdots, \phi_k\}$ is a basis for $\ker D_y$ and $\{\psi_1, \cdots, \psi_\ell\}$ is a basis for $\operatorname{cok} D_y$. Note finally that for a holomorphic family of operators, is possible to verify that the association given by (3.15) actually leads to a holomorphic line bundle. Given a hermitian metric on a holomorphic line bundle there is a unique metric compatible connection of type $(1,0)$. To simplify the equations we assume that the kernels and cokernels are empty. For such operators, we can work all the way down at the bottom of the spectrum and use equation (3.28). The natural $(1,0)$ connection is given by

$$\omega_0 = H_0^{-1} \, \partial_Y H_0 = \operatorname{Tr}\left(\frac{1}{D}\,(\partial_Y D)\right) \ . \tag{3.30}$$

More explicitly, if FP denotes the finite part then the metric in proper time regularization is given by

$$\log H_0 = -\mathrm{FP} \int_\epsilon^\infty \frac{dt}{t} \ \operatorname{Tr}\exp(-t\Delta^+) \ ; \tag{3.31}$$

while

$$\omega_0 = \text{FP}\,\text{Tr}\left(\frac{1}{D}(\partial_Y D)\exp(-\epsilon\Delta^+)\right) . \tag{3.32}$$

In the above $\epsilon$ is a short distance cutoff used to extract the finite part, It is clear that this is a current. From its form we see that it is covariant under changes of coordinates of the manifold $X$. The associated curvature is given by

$$\Omega_0 = \bar{\partial}_Y \omega_0 = -\text{FP}\int_0^\epsilon dt\,\text{Tr}\left[\exp\Big(-(\epsilon - t)\Delta^+\Big)(\bar{\partial}_Y D^\dagger)\exp\Big(-t\Delta^-\Big)(\partial_Y D)\right] . \tag{3.33}$$

We can now state an important result valid regardless of the dimension of the manifold $X$. The curvature $\Omega_0$ is a local $(1,1)$–form on $X$. The reason is that in the integrand we have factors of $\exp(-s\Delta^\pm)$ for $0 < s < \epsilon$. These are just smeared out Dirac delta functions. If the variations of $D$ and $D^\dagger$ are local then expression (3.33) will be local. This leads to the Belavin-Knizhnik result [5,12]. If $\mu = \mu^z{}_{\bar{z}}\, d\bar{z}/dz$ is the Beltrami differential which generates the change in the complex structure and $q = q_{zz}(dz)^2$ is the Serre dual holomorphic quadratic differential. There are only two only possible dimension two operators:

$$\int_X |\nabla_z q_{zz}|^2 \,(g^{z\bar{z}})^2\, dz \wedge d\bar{z} , \tag{3.34}$$

$$\int_X R\, |q_{zz}|^2\, g^{z\bar{z}}\, dz \wedge d\bar{z} , \tag{3.35}$$

where $R$ is the curvature scalar of $X$. A term such as $\nabla^z q_{zz}$ does not appear because of the holomorphicity of the quadratic differentials. In addition there is an integrability condition that must be satisfied. Differentiating (3.30) one finds $\partial_Y \omega_0 = 0$.

Since Quillen's construction is completely rigorous, the Quillen metric is a well defined structure on the holomorphic line bundle $\mathcal{L}$. This metric has an associated metric compatible connection of type $(1,0)$ and curvature form $\Omega_Q$. In particular, if $Y$ is compact then the real first Chern class $c_1(\mathcal{L}, \mathbf{R})$ may be represented by the differential form

$$\frac{i}{2\pi}\Omega_Q ; . \tag{3.36}$$

The Chern class of the line bundle $\mathcal{L}$ is proportional to the central charge $c$ of the Virasoro algebra when the parameter space is moduli space. If $\mathcal{Z}$ is a two cycle on $Y$ then the evaluation of $c_1(\mathcal{L}, \mathbf{R})$ on $\mathcal{Z}$ is given by integration

$$c_1(\mathcal{L}, \mathbf{R})(\mathcal{Z}) = \int_{\mathcal{Z}} \frac{i}{2\pi}\Omega_Q . \tag{3.37}$$

In particular, is $c_1(\mathcal{L}, \mathbf{R}) \neq 0$ then the holomorphic line bundle $\mathcal{L}$ is non-trivial[3]. If $\mathcal{L}$ is non-trivial then the determinant of the operators $D_y$ will not be globally defined [13,14].

It is here that contact may be made most directly with the computation of anomalies, the problem of holomorphic factorization and ultimately the Mumford theorem [15]. There are several viable ways of calculating the curvature. One elegant way is to use the Riemann-Roch-Grothendieck theorem:

$$c_1(\mathcal{L}, \mathbf{R}) = \int_{\mathcal{X}} \mathrm{Todd}(TX)\mathrm{ch}(\mathcal{L}) \ , \tag{3.38}$$

where $\mathrm{Todd}(TX)$ is the Todd class of the tangent bundle $TX$ of $X$ and $\mathrm{ch}(\mathcal{L})$ is the Chern character of $\mathcal{L}$. This is the most direct way of calculating anomalies [16][10][7]

# 4 The Physicist's Metric

The Quillen metric is a very useful object but it is not the main object of interest in physics. In chiral theories the object of interest is not $\det(D^\dagger D)$ since this involves both left and right movers. One is interested in the chiral determinant $\det D$. We advocate a more physical approach than Quillen's where one directly starts from the chiral determinant and asks whether it can be sensibly defined. We defined it with the aid of an auxiliary fiducial operator $D_0$ [4] :

$$\det D \equiv \det(D_0^\dagger D) \ . \tag{4.1}$$

The energy momentum tensor $T_{zz}$ will be determined by the change in the chiral determinant as one changes the complex structure. Assume the operator $D$ is the $\bar\partial$ operator acting on tensors of type $K^n$ where $K$ is the canonical bundle on $X$. As one changes the complex structure not only does the operator $\bar\partial$ change but also the space of complex tensors. In general, the equation (4.1) is the object whose variation gives the physical current $J$:

$$J = \mathrm{Tr}\left[\exp(-\epsilon D_0^\dagger D)\ D^{-1}\,\partial_Y D\right] \tag{4.2}$$

[3]The vanishing of the total first Chern class of a *holomorphic* line bundle is a necessary but not sufficient condition for triviality.

[4]In reality one has to be very careful using the following prescription. For more details see [17] [18, page 30]. The above provides a definition which will be good enough for our purposes as the example demonstrates.

The above is essentially the diagonal element of the Green function which is well known to correspond to the vacuum expectation value of the associated current.

In this section we show that one can heuristically define a metric on $\mathcal{L}$ which has the physical current $J$ as the associated holomorphic connection. This metric is well defined locally but it cannot always be imposed due to topological obstructions. The conclusions can be rephrased as answering the question of when does 4.1 make sense.

We now give a description of a different metric that one may *attempt* to impose on $\mathcal{L}$. The local description of the metric seems to be fine but there are global obstructions to imposing this metric. For simplicity we discuss the index zero case where the physicist metric is restricted to a small neighborhood of an invertible operator $D_0$. In such a neighborhood, the operator $D$ will also be invertible. The physicist norm is given by

$$\|s\|_P^2 = \frac{\left|\det(D_0^\dagger D)\right|^2}{\det(D_0^\dagger D_0)} |\sigma|^2 \qquad (4.3)$$

$$= h_P |\sigma|^2 . \qquad (4.4)$$

The normalization factor is chosen such that the above coincides with the Quillen metric at $D_0$. Locally, this is a fine metric to impose on the determinant line bundle $\mathcal{L}$. In fact one can compute the associated holomorphic connection $J$

$$J = h_P^{-1} \partial_Y h_P \qquad (4.5)$$

$$= \mathrm{Tr}[\exp(-\epsilon D_0^\dagger D)\, D^{-1}\, \partial_Y D] . \qquad (4.6)$$

One verifies that $J$ [5] satisfies the consistency condition $\partial_Y J = 0$ This is just the statement that the two point function $\langle J\, J \rangle$ satisfies Bose statistics.

The curvature associated with the physicist metric will be denoted by $\Omega_P = \bar{\partial}_Y J = 0$.It vanishes since $J$ depends holomorphically on $y$ (cf 4.6. The previous equation generally signals a disaster. If $h_P$ can be extended smoothly to all of $Y$ then one would have an everywhere flat connection. We already know that $\Omega_P$ vanishes where $D$ is invertible. Generically, the places where $\ker D \neq 0$ are of codimension greater that zero in $Y$. By continuity a smooth extension of $h_P$ would lead to an everywhere vanishing curvature $\Omega_P$. The discussion shows

[5] The letter $J$ was used since the above coincides with the physical current $J$.

that it is easy to describe $h_P$ where $D$ is invertible. The difficulties arise with the zero modes just as in the discussion of anomalies [13,14,19]. If $h_P$ is an admissible metric on $\mathcal{L}$ then $\mathcal{L}$ would have to be a flat bundle thus the real Chern classes of $\mathcal{L}$ would have to vanish. This leads to a topological obstruction for the admissibility of $h_P$. We know that the Quillen metric rigorously exists on $\mathcal{L}$ therefore one can compute $c_1(\mathcal{L}, \mathbf{R})$. If this is non-zero then we immediately know that one cannot impose the physicist's metric $h_P$.

The above is a new interpretation of the current. The physical current defines a potential metric on the determinant line bundle with an associated flat connection. It may happen that such a connection and the associated metric are not admissible. In such a situation one cannot define a physically sensible theory due to the presence of anomalies [6][13,14,19].

There is a very important interplay between between what goes on in the Riemann surface $X$ and what goes on in the parameter space $Y$. This is best illustrated by considering an example. Consider a gauged **bc** first order system [20] in a background $A = A_{\bar{z}}\, d\bar{z}$ field with defining Lagrangian

$$\mathbf{b}\,(\partial_{\bar{z}} + A_{\bar{z}})\,\mathbf{c}\ . \tag{4.7}$$

The anomaly may be computed by considering what happens to the vacuum functional when one performs a gauge transformation $A_{\bar{z}} \to A_{\bar{z}} + \partial_{\bar{z}}\lambda(z, \bar{z})$. Alternatively, the determinant line bundle will have the Quillen curvature form $\Omega_Q$ which may be seen by the discussion following (3.33) to be both a two-form on $X$ and a two-form on $Y$. This is the "four-form" that enters into the index theorem calculations of the anomaly [13,14,10]. The geometrical situation in this case should be spelled out in more detail. The space of all $(0,1)$-forms will be denoted by $\mathcal{V}$ and the group of gauge transformations[7] will be denoted by $\mathcal{G}$. Notice that $A_{\bar{z}}$ is left invariant by holomorphic gauge transformations. The family of $\bar{\partial} + A$ operators is a family over $Y = \mathcal{V}/\mathcal{G}$. The two form $\Omega_Q$ on $Y$ can be transgressed [13] to a one form on the fiber $\mathcal{G}$. This *local* one form is the anomaly! There is a very delicate connection between the cohomological information in the Riemann surface $X$ which we used to determine the anomaly and the geometrical information which was used to construct the determinant line bundle over the parameter space $Y$. If the anomaly vanishes then the associated Chern class on

[6]There are also some global questions which should be discussed. For a discussion see [17].

[7]We will not worry about technicalities concerning the exact definition of $\mathcal{G}$, [13].

the line bundle is trivial. This is equivalent to knowing that one can impose a flat connection on $\mathcal{L}$. Hence we are back to the issue of whether one can impose the physicist's metric or not.

The conclusion that the reader should draw from this section is that anomalies now have an interpretation in terms of the physicist's metric. A theory will be anomaly free if one can impose the physicist's metric. We now see why it was necessary to discuss the relation between transformation laws on $X$ and anomalies, and also why the currents are connections on $Y$. It appears as if one is working on two completely different objects but they are related by the transgression operation which relates the cohomology of the fiber $\mathcal{G}$ of a principal bundle $\mathcal{V}$ with the cohomology of the base $Y = \mathcal{V}/\mathcal{G}$.

This same discussion can be reproduced for case of the energy momentum tensor. For conclusions and outlook, we refer the reader to the discussion in the introduction.

## Acknowledgments

It is a pleasure to acknowledge many conversations with Michael Dugan at the early stages of this work. We would like to thank the organizers of the conference for an opportunity to present this work in a stimulating and pleasant atmosphere and Daniel Friedan for many conversations about the relation of the above with his and Steve Shenker's work. We would like to thank our colleagues for many discussions especially Dan Freed, Costas Kounnas, Neil Marcus, Augusto Sagnotti, Hidenori Sonoda and Bruno Zumino. O.A. would also like to thank the Alfred P. Sloan Foundation for its support.

# References

[1] A.M. Polyakov. Quantum geometry of bosonic strings. *Phys. Lett.* **103B**, 207 (1981).

[2] A.M. Polyakov. Quantum geometry of fermionic strings. *Phys. Lett.* **103B**, 211 (1981).

[3] D. Friedan. Introduction to Polyakov's string theory. In R. Stora and J-B. Zuber, editors, *Recent Advances in Field Theory and Statistical Mechanics*, 1982 Les Houches Summer School, North-Holland, 1983.

[4] O. Alvarez. Theory of strings with boundaries: fluctuations, topology and quantum geometry. *Nucl. Phys.* **216**, 125 (1983).

[5] A.A. Belavin and V.G. Knizhnik. Algebraic geometry and the geometry of quantum strings. *Phys. Lett.* **B168**, 201 (1986).

[6] R. Catenacci, M. Cornalba, M. Martellini, and C. Reina. Algebraic geometry and path integral for closed strings. *Phys. Lett.* **B172**, 328 (1986).

[7] J.B. Bost and T. Jolicœur. A holomorphy property and critical dimension in string theory from an index theorem. *Phys. Lett.* **B174**, 273 (1986).

[8] A. Niemi and G.W. Semenoff. Quantum holonomy and the chiral gauge anomaly. *Phys. Rev. Lett.* **55**, 927 (1985).

[9] D. Friedan and S. Shenker. The analytic geometry of two-dimensional conformal field theory. Chicago preprint EFI–86–18A.

[10] J.M. Bismut and D. Freed. The analysis of elliptic families I, II. *Comm. Math. Phys.* **106**, 159 (1986).

[11] D. Quillen. Determinants of Cauchy-Riemann operators over a Riemann surface. *Funk. Anal. i Prilozen* **19**, 37 (1985).

[12] A.A. Belavin and V.G. Knizhnik. Complex geometry and theory of quantum strings. to appear in *ZETF*.

[13] M. Atiyah and I. Singer. Dirac operators coupled to vector potentials. *Proc. Nat. Acad. Sci. (USA)* **81**, 2597 (1984).

[14] L. Alvarez-Gaumé and P. Ginsparg. The topological meaning of non-abelian anomalies. *Nucl. Phys.* **B243**, 449 (1984).

[15] D. Mumford. Stable curves over moduli space. *L'Ens. Math.* **23**, 39 (1977).

[16] O. Alvarez. Conformal anomalies and the index theorem. to appear in *Nucl. Phys.*

[17] O. Alvarez and P. Windey. The energy momentum tensor as geometrical datum. Preprint UCB-PTH-86/36.

[18] P. Nelson. Lectures on strings and moduli space. Harvard preprint, HUTP–86/A047.

[19] G. Moore and P. Nelson. The ætiology of sigma model anomalies. *Comm. Math. Phys.* **100**, 83 (1985).

[20] D. Friedan, E. Martinec, and S. Shenker. Conformal invariance, supersymmetry, and string theory. *Nucl. Phys.* **B271**, 93 (1986).

# STRING FIELD THEORY AND THE GEOMETRY OF MODULI SPACE*

Steven B. Giddings†
*Joseph Henry Laboratories*
*Princeton University, Princeton N.J. 08544*

## ABSTRACT

Some recent work on string field theory and on its relation to the moduli space of Riemann surfaces is reviewed. A few motivational remarks and a review of some basics of string field theory are presented, followed by discussions of the cell decomposition of moduli space obtained from the open string field theory and of the the connection between string field theory and the Polyakov measure on moduli space.

## 1. INTRODUCTION

Much recent work on the foundations of string theory has been focused on some of the elegant features of the Polyakov approach[1] to strings, in which the perturbation series appears as a sum over Riemann surfaces. Here I wish to discuss a different but related approach of string theory, namely string field theory. Before beginning a discussion of string field theory, however, it is worthwhile to suggest why physicists are even interested in such a thing. In particular, it is known that the string perturbation theory treated as a sum over Riemann surfaces is an exceptionally beautiful and rich structure that can be explored by a

---

⋆ Invited talk at the conference "Mathematical Aspects of String Theory," University of California, San Diego, July 21 - August 2, 1986.

† Supported in part by a National Science Foundation Graduate Fellowship.

variety of mathematical methods which have been discussed at this conference and elsewhere. The question is why we might hope for anything more in string theory.

There are at least two answers to this question. The first is that the coupling constant may likely not be small, so that the perturbation series does not converge. If this is the case, it isn't clear that we could find sensible answers even for perturbative physics. Secondly, even if we could *sum* this series we would still miss non-perturbative effects, in which there is expected to be important physics. Specifically, a string field theory appears to be needed for a complete study of problems such as cosmology and compactification, supersymmetry and gauge group breaking and the determination of the masses of the fermions, and for a resolution of the question of singularities in the theory of General Relativity. In other words, it seems that a workable string field theory would be very useful, if not necessary, to study a large subset of the physical questions that physicists hope string theory will be able to answer.

One can add a more abstract motivation for studying string field theory: physicists expect that there should be some deep principle underlying string theory much as the principle of equivalence and Riemannian geometry underly the theory of General Relativity. When discussing the sum over Riemann surfaces we are only scratching the surface. Work on string field theory is directed toward unearthing this deeper structure.

We might illustrate one of the troubles with perturbation theory by way of analogy. Instead of exploring string theory through its perturbation theory, we could imagine exploring the theory of gravity through its perturbation theory (if it had a well defined perturbation theory). This would involve studying scattering processes of spin-two massless particles. If we expected to describe the whole theory of gravity by this approach we would in particular expect to be able to find Schwarzschild's solution corresponding to a black hole. Studying the perturbation theory for spin-two massless particles, however, is clearly *not* the way to find black hole solutions – they are manifestly not perturbative phenomena. (Similarly in Yang-Mills theory, to find interesting things like instanton solutions we must leave the realm of perturbation theory.) Of course, to find black holes we must go beyond perturbation theory to invent General Relativity, the field theory for the gravitational field, in the process recognizing the deep relation between gravity and Riemannian geometry. Then we may proceed to study solutions of non-trivial geometry such as the Schwarzschild solution. Similarly, the expectation is that to mine the many riches of the string we must invent string field theory; since string theory is a generalization of the theory of gravity, we expect string field theory to somehow involve some deeper generalization of Riemannian geometry.

In this talk I will principally focus on the field theory for open strings that has recently been invented by Witten[2] in a generalization of the BRST approach to string field theory initially proposed by Siegel.[3] From this theory we will see several general features arising that should possibly also be features of the closed string field theory, once such a theory is developed. One of these features is the cell decomposition of moduli space* in which the cells correspond to string Feynman diagrams. A second feature is provided by the reparametrization ghosts that must be used to define the string field theory; these ghosts give us the correct measure for the integration of the string amplitude over moduli space. This measure can be taken to be provided by the Weil-Petersson measure on moduli space, although this is only one way to get the correct measure. In this talk I will restrict myself to the case of the bosonic string. Witten has also constructed a field theory for the open superstring,[4] and similar features should appear for that theory in the context of super moduli space.[5] I will begin by giving a brief sketch of some basics of string field theory, and in particular of Witten's string field theory and the Feynman rules for that theory. This will be followed by a discussion of the triangulation of moduli space obtained from these Feynman rules, after which I will sketch how the reparametrization ghosts give the string

---

⋆ Recall that the moduli space of a surface is the space of metrics on that surface modulo diffeomorphisms and conformal (Weyl) rescalings.

integration measure. I have tried to give some useful references, but they by no means should be taken to be a complete set.

## 2. BASIC STRING FIELD THEORY

Before launching into the construction of string field field theory we should give a picture of roughly what we mean by a string field theory. It helps to first think about the field theory of the massless particle. For the non-interacting particle we could begin with the action

$$S = \int d\tau \sqrt{g} g^{-1} \frac{dX^\mu}{d\tau}\frac{dX^\nu}{d\tau}\eta_{\mu\nu}$$

corresponding to the length of the particle's trajectory. Here $X^\mu(\tau)$ gives the position of the particle, $g$ is the one-dimensional metric tensor, and $\eta_{\mu\nu}$ is the flat background metric for a space-time of dimension $D$. To calculate amplitudes for this particle to propogate from an initial to a final state we then could do the path integral using this action. This corresponds roughly to the Polyakov approach to string theory. (Of course for the particle this picture gives no natural way to treat interactions.) We can alternately switch to a modified description of the particle. This is given in terms of a quantum field $\phi(X)$, for which we have the action

$$I = \int d^D X (\tfrac{1}{2}\partial_\mu \phi \partial^\mu \phi + \frac{\lambda}{3!}\phi^3)$$

The variable $X^\mu$ labels the position of the particle, so we see that $\phi$ is a function of the possible configurations of the particle. The first term in the action is the kinetic piece, which gives a prescription for the propagation of the particle; the second term is the interaction piece which we couldn't naturally include in the previous description of the particle. Here for illustration I have taken a cubic interaction, which is shown pictorally in Fig. 1(a); in Fig. 1(b) I show an example of a Feynman diagram that would come from this action in the calculation of an amplitude for two particles to scatter off one another.

Now string field theory should just be a generalization of this. Since in string theory we are trying to picture elementary particles as tiny strings, we know that Fig. 1(b) should *really* look like Fig. 1(c); If we had a good enough microscope we would see that the particle is in reality a string and that the particle trajectory is in reality a two-dimensional surface. (Of course propagation of different modes of vibration of the string corresponds to propagation of what we would call different "elementary particles.") So as before we can calculate the amplitude for a process by doing a functional integral over all different intermediate configurations using the Polyakov action

$$S = \frac{1}{2\pi}\int d^2\sigma\sqrt{g}g^{ab}\partial_a X^\mu \partial_b X^\nu \eta_{\mu\nu} \tag{2.1}$$

(Here interactions *are* well defined since they don't change the fact that we have a manifold.) But instead of taking this approach we would like to introduce a string field theory. As before, we define a quantum field $A[X(\sigma)]$; since we are now dealing with a string we have a field functional of string configurations. Once again we want to introduce an action with a kinetic term and an interaction term; because of the observation that there is really only one fundamental interaction in string theory, namely that in which two strings meet to form a third string, we expect that we should be able to find an action with a single cubic interaction term.

In constructing the string field theory it is useful to consider a string field that is not just a functional of the physical string configurations $X^\mu(\sigma)$, but also of the reparametrization ghosts[6] $c(\sigma), b(\sigma)$; these can be expanded in normal modes with coefficients $c_n$, $b_n$ that obey canonical anticommutation relations $\{c_m, b_n\} = \delta_{m+n,0}$. In the Polyakov approach to string theory the reparametrization ghosts are just local Grassman fields on the string world-sheet; the path integral over these fields provides the Fadde'ev Popov determinant from the gauge fixing of the reparametrization invariance.[7]* The string field $A[X(\sigma), c(\sigma), b(\sigma)]$ can itself

---

★ This will be further discussed in Section 4.

then be expanded in terms of the normal mode coefficients $X_n, c_n, b_n$, which are to be treated as operators acting on a vacuum state in the usual type of Fock space construction. Notice that the presence of the zero modes $c_0, b_0$, with $\{c_0, b_0\} = 1$, indicate that the ground state is doubly degenerate; we call the two ground states $|\uparrow\rangle$ and $|\downarrow\rangle$ with relations

$$b_0 \,|\downarrow\rangle = 0, \;\; c_0 \,|\downarrow\rangle = |\uparrow\rangle \,, \;\; c_0 \,|\uparrow\rangle = 0, \;\; b_0 \,|\uparrow\rangle = |\downarrow\rangle \,, \;\; \langle\uparrow \mid \downarrow\rangle = 1 \,.$$

Once the reparametrization gauge invariance has been fixed, the reparametrization invariance is in a sense recaptured by BRST invariance.[8] The BRST invariance of a free string is generated by the BRST operator[6,9] $Q$ which takes the form (: : denotes normal ordering)

$$Q = \sum_n (c_n L^X_{-n} + \tfrac{1}{2} : c_n L^{gh}_{-n} :) \,.$$

Here $L^X_n$ are the generators of the ordinary Virasoro algebra

$$[L^X_m, L^X_n] = L^X_{m+n} + \tfrac{D}{12} m(m^2 - 1)\delta_{m+n,0} \,,$$

and $L^{gh}_n = \sum_m (n-m) : b_{n+m} c_{-m} :$ are the Virasoro generators for the ghost system. The operator $Q$ has the important property $Q^2 = 0$; this is true only in the critical space-time dimension[6] $D = 26$. Another important operator to keep in mind is the ghost number operator $N_g$; this simply counts the number of ghost excitations minus the number of antighost excitations of a state. $N_g$ is given by

$$N_g = \sum_{n>0} (c_{-n} b_n - b_{-n} c_n) + \tfrac{1}{2}(c_0 b_0 - b_0 c_0) \,.$$

To write the open string action, Witten introduced for string fields a multiplication law $*$, an integration law $\int$, and a grading of the algebra of the fields.

This construction, together with the BRST operator which acts as a derivation of the $*$-algebra, is a concrete realization of the axioms of Connes non-commutative geometry.[10] The multiplication law maps two string fields to a third; it is defined as the limit of a path integral on the region shown in Fig. 2(a). Notice that there is a curvature singularity of strength $R = -\pi\delta^2(\sigma - \sigma')$ at the interaction point. Pictorally, the expression $A = B * C$ can be thought of as an overlap integral setting the right half of string $B$ equal to the left half of string $C$; the product $A$ then has left half equal to the left half of string $B$ and right half equal to the right half of string $C$, as shown in Fig. 2(b). All of this can be made much more precise; see Ref. 2 for more details. The integration $\int$ maps string fields into the complex numbers. The integration $\int A$ can be thought of as an overlap integral setting the left half of string $A$ equal to the right half. The string field $A$ is also given a given the $Z_2$ grading $(-1)^A = (-1)^{N_g(A)-\frac{1}{2}}$ depending on ghost number; the physical states have ghost number $-\frac{1}{2}$ and are odd elements of the algebra. The $*$-algebra is associative but non-commutative; however, under the integral sign we have

$$\int A * B = (-)^{AB} \int B * A \ .$$

Crucial in the construction of a gauge-invariant action is the fact that the BRST operator acts as a derivation of the $*$-algebra:

$$Q(A * B) = QA * B + (-)^A A * QB \ .$$

Further confirming the analogy with the $d$ and $\int$ operators acting on ordinary differential forms on a manifold is the identity

$$\int QA = 0$$

which is true for any field $A$. The action itself is written

$$I = \int (A * QA + \tfrac{2}{3} A * A * A) \ . \tag{2.2}$$

Using the above identities one easily checks the invariance of this action under the gauge transformation

$$\delta A = Q\epsilon + A * \epsilon - \epsilon * A. \tag{2.3}$$

To calculate scattering amplitudes with the action (2.2) one must first fix a gauge; otherwise the kinetic operator cannot be inverted to define a propagator and a set of Feynman rules. The gauge fixing for the full non-linear invariance (2.3) appears to be non-trivial.[11] However, there is an ansatz for the non-linear gauge fixing motivated by the gauge fixing in the non-interacting case; this ansatz should give the correct answer for the tree diagrams* and more general arguments[14] that will be sketched in the sequel indicate that this ansatz should also be correct for loop diagrams. We begin with the gauge fixing for the the linear invariance of the non-interacting theory,

$$\delta A = Q\epsilon \ .$$

A suitable gauge choice proves to be the Siegel gauge choice[3] $b_0 A = 0$ which implies $A = \tilde{A} b_0 \left|\uparrow\right\rangle$, where $\tilde{A}$ is a state excluding the ghost zero modes. On such a state the kinetic part of the action becomes

$$\int A * QA = \left\langle \tilde{A} \right| \left\langle \uparrow \right| b_0 Q b_0 \left| \uparrow \right\rangle \left| \tilde{A} \right\rangle \ .$$

We then use the relation $b_0 Q b_0 = b_0 \Delta$, where $\Delta = L_0^X + L_0^{gh}$ is the ordinary world-sheet Hamiltonian, to reduce the kinetic term to $\left\langle \tilde{A} \middle| \Delta \middle| \tilde{A} \right\rangle$. Now the kinetic operator does have an inverse; it is given by $\frac{1}{\Delta} = \int_0^\infty d\tau e^{-\tau\Delta}$. Since $\Delta$ is the world-sheet Hamiltonian, the operator $e^{-\tau\Delta}$ just tells us to propagate a string freely for a time $\tau$; this builds up a flat rectangular world-sheet of width $\pi$

---

* This has been explicitly checked for the four point tachyon scattering amplitude in Ref. 12. It has also been shown[13] that the vertex in this formulation of field theory reproduces the usual three-point couplings between states.

and length $\tau$. The integral over $\tau$ tells us to sum over all such world-sheets. Of course, instead of completely dropping the zero mode part of the string field we could have left it in; then the $b_0$ insertion would be represented as a line integral $\int d\sigma b(\sigma)$ across the rectangle.

When we introduce interactions, they just tells us to tie together such flat rectangles three at a time using vertices like that shown in Fig. 2(a). Let me once again emphasize that for such a picture the entire surface is flat except at the interaction point where there is a curvature delta function, and the boundaries are all straight lines in this flat metric. Thus in the Feynman rules of Witten's string field theory we encounter all Riemann surfaces obtained by gluing together such flat strips of any given length in the manner prescribed by the vertex.

## 3. STRING FIELD THEORY AND THE TRIANGULATION OF MODULI SPACE

With the Feynman rules for the string field theory in hand we can now tackle the issue of modular invariance. For simplicity, we will consider only the vacuum diagrams of the theory, *i.e.* those diagrams without external states. The discussion easily extends to the case of the more general diagrams.

Suppose we consider a diagram like that shown in Fig. 3(a). It is not hard to convince oneself that this diagram is topologically a torus with a disk removed. (To see this note that the surface has one boundary component and is orientable; one can then look at the homology basis, or better yet, count the curvature singularities and use Gauss-Bonnet, to conclude that the surface has genus one.) Furthermore, we see that we have a three parameter family of such surfaces, corresponding to the different values of the three propagator variables $\tau_1, \tau_2, \tau_3$. But notice that three is precisely the dimension of the moduli space for the torus with a disk removed. Therefore as the three parameters $\tau_i$ vary over the interval $(0, \infty)$ we sweep out some open set in the moduli space. If we are interested in obtaining a modular invariant description of string theory, then the question is whether this open set covered by the field theory Feynman diagrams is in fact a one-to-one cover of moduli space for the torus with one hole.

More generally, one can consider diagrams such as the one shown in Fig. 3(b). Such a diagram is topologically a surface of genus $g$ with one boundary component. Of course there will be many other Feynman diagrams from our Feynman rules that will give surfaces of genus $g$ with one boundary; distinct diagrams just correspond to different combinatorical gluing patterns for the propagators. We can in a similar manner also obtain surfaces of arbitrary genus with more than one boundary component. Therefore we would like to check whether from such diagrams we get a one-to-one cover for the moduli space of a surface of genus $g$ with one or more boundary components.

To explore the question of whether or not we have a cover of moduli space we will first put the Feynman diagrams in a simpler form. This is done by cutting each propagator in half along its midline; since the cuts never touch the boundary the Feynman diagram then falls apart into a cylinder (or, in the case where there are several boundary components, into several cylinders.) Such a cylinder is pictured in Fig. 4, with the gluing pattern necessary to return to the original surface indicated; we will call this a *canonical presentation* of the Riemann surface. To each Feynman diagram we have a unique canonical presentation, and conversely, to each canonical presentation we have a unique Feynman diagram. (Notice that we have to be careful to forbid gluing patterns such as those shown in Fig. 5(a),(b); the former give curvature singularities that cannot come from our Feynman rules, and the latter are redundant specifications.) The fact that we have an injective cover of moduli space then comes from the theorem that given a Riemann surface with one or more boundary components, there exists a unique (up to combinatorical factors) canonical presentation of that surface.[15] Therefore for each Riemann surface there is a unique string field theory Feynman diagram.[16] This may strike one as surprising – modular invariance was not an ingredient of Witten's string field theory. Rather, gauge invariance served as an input and the single cover of moduli space later appeared in a somewhat mysterious fashion.

Of course in asserting that we have a single cover of moduli space we must be mindful of the Feynman symmetry factors for the graphs; for example, in the case of the torus with a hole we really get a six-fold cover of the moduli space when we allow $\tau_1, \tau_2, \tau_3$ to each range over the interval $(0, \infty)$. This is compensated by the Feynman symmetry factor 3! for the string field theory graph. A relation between the problem of computing the symmetry factors for the cell decomposition of moduli space and the combinatorics of field theory was in fact obtained by Penner before the correspondence between the string field theory Feynman rules and the cell decomposition of moduli space was known;[17] the symmetry factors were found by considering an interacting field theory of Hermetian matrices that can be thought of as roughly corresponding to the Chan-Paton factors used to introduce non-abelian gauge symmetries into the theory of open strings.

For completeness I will give a brief sketch of the proof of the above theorem. For simplicity I will treat here only the case of a single boundary component, although the results extend to the case of multiple boundary components.[18] Consider an arbitrary Riemann surface with one boundary; we will embed in this Riemann surface a cylinder such that one boundary of the cylinder is identified with the boundary of the surface. This cylinder will be taken to cover the Riemann surface, so along the other boundary of the cylinder identifications of boundary arcs are made. Other than these identifications we take the embedding to be one-to-one. This situation is presented for the genus one case in Fig. 6. The embedded torus is of course conformally equivalent to a "standard" right circular flat cylinder of height $b$, circumference $a$, and modulus $m = \frac{b}{a}$. However, in general the conformal rescalings necessary to make the metric of the embedded cylinder flat can be different as we approach the identified boundary arcs from different sides. Therefore in general we cannot glue together the boundary segments on the conformally equivalent flat cylinder; generically the boundary segments to be identified will have different lengths in the new Euclidean metric.

To avoid this problem we use our freedom to adjust the embedding of the cylinder. It is not hard to convince oneself on grounds of compactness that there ought to be at least one embedded cylinder for which the modulus $m$ attains a local maximum. Consider this cylinder and its conformally equivalent flat cylinder. Suppose that on the flat cylinder the lengths don't match up. Then the conformal rescaling factors must be different as we approach the boundary of the cylinder from opposite sides ; call these factors $\rho, \rho'$. One can imagine varying the identified boundary segments of the embedded cylinder such that an amount of area $\delta A$ is moved from one side of the boundary to the other; see Fig. 7. Then in the conformally equivalent flat cylinder we have added an amount of area $\rho\delta A$ to one part of the cylinder and subtracted an amount of area $\rho'\delta A$ from another part, for a total change $\Delta A = (\rho - \rho')\delta A$ in the area of the cylinder. Therefore if we use a conformal mapping to return us once again to a standard right circular flat cylinder, we will make a change in the modulus that is first order in $\delta A$. This is not possible if we are located at a local maximum in modulus, so $\rho$ must equal $\rho'$ and consequently identified boundary segments must have equal lengths. This shows that if we take an embedded cylinder of maximal modulus we obtain a canonical presentation of the surface.

For the second part of the proof we must show that such a canonical presentation is unique. Suppose we had two such canonical presentations, which we will represent as rectangles with opposite sides identified as in Fig. 8. Of course if we have different presentations of the Riemann surface then the top boundaries of the rectangles don't correspond. We can by an overall scaling set the circumferences of both $R$ and $R'$ equal to the common value $a$; the heights will then be $b$ and $b'$. Consider a horizontal curve $C$ on $R$ and its image $C'$ on $R'$. Let $\rho(x, y)$ be the conformal scaling necessary to pass from the metric of $R$ to that of $R'$; here $(x, y)$ are the coordinates on $R$. Clearly curve $C'$ must have length greater than $a$; in the coordinates on $R$ this is written $a \leq \int_C dx \sqrt{\rho(x, y)}$. Applying the Cauchy-Schwarz inequality, we see that $a \leq \int_C dx \rho(x, y)$. We then integrate this inequality over $y$, finding $ab \leq \int_R dx dy \rho(x, y)$; however, this latter expression is just the area in the metric of $R'$. Therefore $ab \leq ab'$ or $b \leq b'$. But we could have just as well reversed the roles of $R, R'$ to conclude $b' \leq b$, so $b$ must equal $b'$. This means we must have obtained equality in the above inequalities, and this in turn implies that the two canonical presentations are one and the same.

This argument completes the demonstration that conformal structures on a given topological surface are in one-to-one correspondence with string field theory Feynman diagrams. We see that we have a construction that is manifestly invariant under the action of the mapping class group; at no point in the argument have we made reference to the choice of marking on the surface. Therefore the Feynman rules give a cell decomposition of the moduli space for a given Riemann surface with at least one boundary component; the cells correspond to the different combinatorical gluing patterns, *i.e.* to the different Feynman diagrams that one can construct to represent the given topological surface. When we add up the different Feynman diagrams with their appropriate Feynman symmetry factors we therefore obtain a one-to-one cover of moduli space. For more details the reader is referred to Ref. 16.

It is worth mentioning that there has been work done on a different approach to string field theory[19] inspired by the light-cone gauge or interacting string picture approach to the construction of string Feynman diagrams.[20] It can be shown that the Feynman diagrams in this formalism also provide a cell decomposition of moduli space;[21] this cell decomposition is distinct from the one discussed above and is apparently previously unknown to mathematicians.

# 4. STRING FIELD THEORY AND THE POLYAKOV MEASURE

Another feature of Witten's string field theory for the open bosonic string is that it should provide the Polyakov measure for integration over moduli space.* This has been checked explicitly in the case of the tree amplitude with four external tachyons,[12] and is expected to extend without difficulty to arbitrary tree amplitudes. For the loop amplitudes the problem is more difficult. To see this note that in fixing the linear gauge invariance of the free theory we encounter ghosts-for-ghosts.†[22] The reason for this is that in the linear case we have invariance under changes $\delta A = Q\epsilon$, but once this is used to fix a particular gauge for $A$ we still have an invariance under the transformation $\delta\epsilon = Q\lambda$, necessitating another gauge-fixing determinant. This process continues indefinitely, leading to an infinite number of ghosts-for-ghosts. In the interacting theory the same thing also happens on-shell, *i.e.* when we are working with a field configuration that satisfies the equations of motion $QA + A * A = 0$, but off-shell things are more subtle. It is of course necessary to have a gauge-fixing procedure that works even off-shell since in doing the path integral for string fields we include off-shell string fields; there will be ghosts propagating in loops. There certainly appear to be redundant gauge invariances, but they are complicated. In tree diagrams the ghosts decouple so this is not much of a problem, but for loops a satisfactory resolution of this question is necessary to get the proper integration measure on moduli space.

---

⋆ The Polyakov measure can of course be thought of as arising from the Weil-Petersson measure on moduli space, but since the Weil-Petersson measure is not invariant under Weyl rescalings, it is more natural to treat the string measure as a whole without reference to the Weil-Petersson measure. The full Polyakov measure is, of course, invariant under Weyl rescalings. I would like to thank P. Nelson and G. Moore for calling these facts to my attention.

† The ghosts we consider here are distinct from the reparametrization ghosts $c(\sigma), b(\sigma)$; they are perhaps more aptly referred to as "ghost strings" since they are entire string fields that play the role of Fadde'ev-Popov ghosts in fixing the gauge invariance of the classical action for the string field.

It turns out that the problem can be attacked from a different direction: we can start from the Polyakov measure and try to work back to something that looks like it should come from the full gauge-fixed string field theory. Remarkably, by this route it looks like one can come rather close to what should be the string field theory result.[14] We begin with the familiar Polyakov path integral for an $n$ point correlation function ($n$ could of course be zero, in which case we would be computing the partition function), on a surface of genus $g$ with $s$ boundary components.‡ This takes the form

$$A(1,\ldots,n) = \int_{\mathcal{G}^g_{n,s}} \frac{\mathcal{D}g_{ab}}{Vol(Conf)\cdot Vol(Diff)} \langle \mathcal{V}_1,\ldots,\mathcal{V}_n\rangle \ .$$

Here we integrate over the space of metrics $\mathcal{G}^g_{n,s}$ on a surface $\Sigma^g_{n,s}$ of topology dictated by the scattering process in question. The $\mathcal{V}_i$ are vertex operators corresponding to the external states, and the expectation value

$$\langle(\cdots)\rangle = \int_{\mathcal{E}^g_{n,s}} \mathcal{D}X^\mu(\cdots)e^{-S}$$

is computed by doing the path integral over the space $\mathcal{E}^g_{n,s}$ of maps of the surface $\Sigma^g_{n,s}$ into the target manifold, where we use the Polyakov action (2.1). As usual we must divide out by the volumes of the groups of Weyl rescalings and diffeomorphisms to make the path integral well defined; this accounts for the volume factors in the denominator. The measure $\mathcal{D}g_{ab}$ for integration over metrics can be taken to be provided by the metric on $\mathcal{G}^g_{n,s}$,

$$\|\delta g_{ab}\|^2 = \int_{\Sigma^g_{n,s}} d^2\sigma\sqrt{g}g^{ab}g^{cd}\delta g_{ac}\delta g_{bd} \ .$$

In order to factor out the volumes of the symmetry groups we first pick a gauge slice $S$ through the space of metrics $\mathcal{G}^g_{n,s}$ such that every metric in $\mathcal{G}^g_{n,s}$

‡ Here I follow Refs. 7,23,24.

can be reached from a unique metric $\hat{g} \in S$ by a combined diffeomorphism and Weyl rescaling. This situation is pictured schematically in Fig. 9. The slice $S$ corresponds to a family $\hat{g}_{ab}(\tau^j)$ of metrics parametrized by the $6g-6+2n+3s$ modular parameters[§] $\tau^j$. Suppose we pick a point $\tau_0^j$; then the metric $\hat{g}_{ab}(\tau_0^j)$ defines a complex structure $z$ on the surface $\Sigma^g_{n,s}$. In these coordinates the metric for a neighboring complex structure $z' = z + \delta z$, $\delta z \equiv \delta\tau^j \tilde{V}_j^z$, (here we take $|\delta\tau^j| \ll 1$) on $S$ is given by $ds^2(\tau_0^j + \delta\tau^j) = g_{z\bar{z}}|dz + \delta\tau^j \nabla_{\bar{z}} \tilde{V}_j^z d\bar{z}|^2 + \delta g_{z\bar{z}}|dz|^2$. Therefore we see that the tangent vectors to the slice $S$ are, up to Weyl rescalings, the *Beltrami differentials* $\mu^z_{(j)\bar{z}} = \nabla_{\bar{z}} \tilde{V}_j^z$. The parameters $\tau^j$ labelling the slice $S$ will provide the coordinates in which we will compute the measure for integration over the moduli space $\mathcal{M}^g_{n,s}$.

To compute this measure, we work in the local complex coordinates $z, \bar{z}$ on the Riemann surface. In these coordinates the orthogonal decomposition of an arbitrary variation of the metric is written

$$\delta g_{zz} = \nabla_z \delta V_z + \sum_j \phi^j_{zz} \delta\tau_j \ , \ \delta g_{\bar{z}\bar{z}} = \nabla_{\bar{z}} \delta V_{\bar{z}} + \sum_j \phi^j_{\bar{z}\bar{z}} \delta\bar{\tau}_j \ , \ \delta g_{z\bar{z}} = \delta\Omega g_{z\bar{z}} \ . \quad (4.1)$$

Here $\delta V_z, \delta V_{\bar{z}}$ are continuous vector fields that generate diffeomorphisms of the surface; $\phi^j_{zz}, \phi^j_{\bar{z}\bar{z}}$ are holomorphic quadratic differentials that serve as the basis for the cotangent space to moduli space;[¶] $\delta\tau_i, \delta\bar{\tau}_i$ parametrize the modular deformations; and $\delta\Omega$ represents the change in conformal factor. Using the orthogonal decomposition (4.1) we can decompose the measure,

$$\mathcal{D}g_{ab} = \mathcal{D}\Omega \mathcal{D}g_{ab}{}^{(diff)} \mathcal{D}g_{ab}{}^{(mod)} \ . \quad (4.2)$$

§ If the surface has conformal Killing vectors, then the number of modular parameters is $6g-6+2n+3s+\rho$, where $\rho$ is the real dimension of the continuous group of conformal Killing vectors on the surface.

¶ In the case where the surface has boundary components we take the Riemann double along these components to find a surface of genus $2g+s-1$ with $2n$ specified points. Thus we have a $6g-6+2n+3s$ complex dimensional space of holomorphic quadratic differentials. The number of parameters in the general variation is then halved, to $6g-6+2n+3s$ real parameters, by the restriction of symmetry of the doubled surface about the former boundary components.

The integrand in the Polyakov path integral is invariant under diffeomorphisms and Weyl rescalings,[1,7,23] so we can pull this measure back to the slice $S$. The integral over the factor $\mathcal{D}\Omega$ cancels the factor $Vol(Conf)$. To cancel $Vol(Diff)$ we change variables of integration, in the process picking up a Jacobian determinant:

$$\mathcal{D}g_{ab}{}^{(diff)} = det'(\nabla_z)\mathcal{D}V_z \cdot det'(\nabla_{\bar{z}})\mathcal{D}V_{\bar{z}} \ .$$

Of course we have to be careful in defining the determinants in this expression; for details see Ref. 25. Notice that we have used $det'$, since we must omit the zero modes which correspond to the deformations of moduli. The factor $Vol(Diff)$ is cancelled by the integral over $\mathcal{D}V_z, \mathcal{D}V_{\bar{z}}$, leaving us with the Jacobian determinants.

To properly treat the last factor in (4.2), notice that the tangent vectors $\mu^z_{(j)\bar{z}}$ to the slice $S$ in general have components both in the direction of diffeomorphisms and in the direction of pure modular deformations; see Fig. 9. To get the correct volume element for integration over the $\tau^j$ we must project the tangent vectors onto the basis covectors for modular deformations, namely the holomorphic quadratic differentials. The natural inner product between the Beltrami differentials and the quadratic differentials is given by

$$\langle \mu^z_{\bar{z}}, \phi_{zz} \rangle = \int\limits_{\Sigma^g_{n,s}} d^2\sigma \sqrt{g} g^{z\bar{z}} \mu^z_{\bar{z}} \phi_{zz} \quad .$$

Therefore for the last factor in (4.2) we obtain

$$\mathcal{D}g_{ab}{}^{(mod)} = det\langle \mu_j, \phi_k \rangle \prod_{i=1}^{6g-6+2n+3s} d\tau^i \ , \tag{4.3}$$

where we must be careful to normalize the $\phi_k$ properly.

It is useful to represent the determinant $det'(\nabla_z)$ and its anti-holomorphic counterpart in terms of path integrals[7] over the local anticommuting ghost and anti-ghost fields $c'^z, b'_{zz}$:

$$det'(\nabla_z) = \int \mathcal{D}c'\mathcal{D}b' e^{-S'_G} \tag{4.4}$$

where

$$S'_G = \frac{1}{2\pi} \int_{\Sigma^g_{n,s}} d^2\sigma \sqrt{g} b'_{zz} \nabla_{\bar{z}} c'^z g^{z\bar{z}} = \frac{1}{2\pi} \langle b'_{zz}, \nabla_{\bar{z}} c'^z \rangle \quad .$$

In this integral we of course omit the zero modes of $b_{zz}$, which are the holomorphic quadratic differentials. This is indicated by the prime; since each such non-zero mode of $b_{zz}$ is in one-to-one correspondence with a continuous vector field on the surface, we also put a prime on the ghost field $c$ to indicate the correspondence.

As it turns out, we can also represent the finite dimensional determinant in (4.3) as a path integral by a slight extension of the ghost and anti-ghost fields. Of course it is easy to extend the anti-ghosts; we just include Grassman fields $b^j_{zz}$ corresponding to the holomorphic quadratic differentials. To extend the ghost fields, recall that a tangent vector to $S$ is given by $\nabla_{\bar{z}} \tilde{V}^z_j$. If $\tilde{V}^z_j$ were continuous on the Riemann surface it would generate a pure diffeomorphism rather than a deformation of complex structure. To obtain a change in the modulus of the surface we take $\tilde{V}^z_j$ to be a *discontinuous* vector field on the surface. Correspondingly, we introduce the extra ghost fields $c^z_j$ which are discontinuous on the surface. Then the finite dimensional determinant in (4.3) can be written

$$det \langle \mu_j, \phi_k \rangle = \int \prod_j \mathcal{D}c_j \mathcal{D}\bar{c}_j \mathcal{D}b^j \mathcal{D}\bar{b}^j exp\{-\tfrac{1}{2\pi}(\langle b^j_{zz}, \nabla_{\bar{z}} c^z_j \rangle + \langle \bar{b}^j_{\bar{z}\bar{z}}, \nabla_z \bar{c}^{\bar{z}}_j \rangle)\} \quad . \tag{4.5}$$

The two path integrals (4.4), (4.5) can be combined into a single path integral with action

$$S_G = \frac{1}{2\pi}\left\langle b'_{zz}, \nabla_{\bar{z}} c'^{z}\right\rangle + \frac{1}{2\pi}\sum_j \left\langle b^j_{zz}, \nabla_{\bar{z}} c^z_j\right\rangle + \text{c.c.}$$

where *c.c.* indicates that we add also the action for the complex conjugate fields. We can lump the zero modes and the non-zero modes together into a single field by defining $c = c' + \sum_j c_j$, $b = b' + \sum_j b^j$. Then the action can be rewritten

$$S_G = \langle b_{zz}, \nabla_{\bar{z}} c^z\rangle - \sum_j \left\langle b'_{zz}, \nabla_{\bar{z}} c^z_j\right\rangle \; + \; \text{c.c.} \; ,$$

where we use the fact that $\left\langle b^j_{zz}, \nabla_{\bar{z}} c'^{z}\right\rangle = 0$, as can easily be seen by integration by parts. The cross term involving $b'$ and $c_j$ does not contribute to the path integral, so after assembling all the pieces, we find for the Polyakov path integral

$$A(1,\cdots,n) = \int_{\mathcal{M}^g_{n,s}} \prod_{k=1}^{6g-6+2n+3s} d\tau_k \int \mathcal{D}c\mathcal{D}b\mathcal{D}\bar{c}\mathcal{D}\bar{b}$$
$$exp\{-\tfrac{1}{2\pi}\int d^2\sigma\sqrt{g}g^{z\bar{z}}(b_{zz}\nabla_{\bar{z}}c^z + \bar{b}_{\bar{z}\bar{z}}\nabla_z\bar{c}^{\bar{z}})\}\,\langle \mathcal{V}_1\cdots\mathcal{V}_n\rangle \; . \tag{4.6}$$

Here the coordinates $\tau_j$ provide the coordinates for the moduli space $\mathcal{M}^g_{n,s}$.

We should not end our discussion of the ghost contributions to the Polyakov path integral without mentioning the ghost insertions due to the curvature of the Riemann surface and at the external states. When we actually perform the ghost path integral in (4.6) we will find the anomaly in the conservation of ghost number.[26] This is most easily explored by bosonizing the ghosts,[27] $c^z = e^{i\phi(z)}$ , $b_{zz} = e^{-i\phi(z)}$. In such expressions the coefficient of $i\phi$ in the exponent corresponds to the ghost number $N_G$ of the operator. The action for the bosonized ghost $\phi(z)$ couples to the Euler character density on the Riemann surface[28]

$$S_\phi = \frac{1}{2\pi} \int\limits_{\Sigma^g_{n,s}} d^2\sigma \sqrt{g} g^{ab} \partial_a \phi \partial_b \phi$$
$$- \frac{3i}{2\pi} \Big[ \int\limits_{\Sigma^g_{n,s}} d^2\sigma \sqrt{g} R\phi + \int\limits_{\partial\Sigma^g_{n,s}} ds k\phi + \sum_j (\pi - \theta_j)\phi(\sigma_j) \Big] \quad . \tag{4.7}$$

(Here $k$ is the extrinsic curvature of $\partial\Sigma^g_{n,s}$ and $\theta_j$ are the interior angles of the corners on $\partial\Sigma^g_{n,s}$.) This action then gives a term $e^{i\frac{3}{2}\phi(z)}$ (corresponding to a ghost number $\frac{3}{2}$) at each of the curvature singularities at the interaction points in the Feynman diagrams. Furthermore, in the computation of the ghost correlation function the external states must also serve as sources for the ghost fields. For the external states considered as strips coming in from large time, the ghost number of is just the ghost number for physical states, $N_G = -\frac{1}{2}$. In actually computing the ghost correlation function it proves useful to conformally map the semi-infinite strips to finite points; for open string diagrams the conformal mapping is usually taken so that these finite points lie on a straight boundary of the new world sheet. From the action (4.7) it is clear that this conformal mapping changes the ghost number by $+\frac{3}{2}$ since we are mapping the end of a long rectangle into a straight boundary, in the process flattening out the two right angles. Therefore the vertex operator that should be used to represent the external state on a straight boundary has ghost number $+1$. The natural (BRST invariant) vertex operators to use are thus operators of the form $c^z V_X$, where $V_X$ is one of the usual vertex operators in the Koba-Nielsen formulation.[29]

We are now quite close to having the Feynman rules that would come from gauge fixing the string field theory. To see this, notice that the moduli for the field theory Feynman diagrams are the Schwinger parameters $\tau_i$. A deformation of a particular $\tau_i$ just corresponds to a deformation lenthening the rectangular strip that represents the propagator by an amount $\delta\tau$. The discontinuous vector field that accomplishes this can easily be written in terms of the natural complex parameter $z$ on the strip. The parameter $z = x + iy$ is chosen such that $x$ ranges from 0 to $\pi$ and $y$ ranges from 0 to $\tau$ on the propagator. Then the desired vector field is

$$\delta z = \delta\tau \tilde{V}^z = \delta\tau \frac{z - \bar{z}}{2a} \; if \; y < a$$
$$\delta z = \delta\tau \tilde{V}^z = 0 \; if \; y > a \; .$$

This vector field has its discontinuity along the line $y = a$. Notice that the vector field is taken to have support only on the propagator whose modulus we are changing; it is zero elsewhere on the Feynman diagram. The ghost field corresponding to this deformation vector field is simply $c^z_j = \theta_j \tilde{V}^z_j$ where $\theta_j$ is some constant basis element of the Grassman algebra. In the path integral we can do the Grassman integral over the $c_j$, which pulls down into the correlation function a product of terms $\int d^2\sigma \sqrt{g} g^{z\bar{z}} b_{zz} \nabla_{\bar{z}} \tilde{V}^z_j$. We can now integrate by parts; although $b_{zz}$ is not purely holomorphic, the full correlation function with $b_{zz}$ inserted is, as a function of $z$, holomorphic away from sources in the ghost field. All such sources are outside the support of $\tilde{V}^z_j$, so upon integration by parts we get zero for the term with $\nabla_{\bar{z}}$ acting on $b_{zz}$. However, this is not the end of the story, since the field $\tilde{V}^z_j$ is discontinuous we get from the integration by parts a "surface term" that is proportional to the discontinuity in $\tilde{V}^z_j$. This term takes the form $\int_{C_j} dz b_{zz}$, where $C_j$ is the contour along which $\tilde{V}_j$ is discontinuous. Therefore we find precisely the ghost line integral insertions that appear in the gauge fixing ansatz motivated by the non-interacting theory, namely $b_0 = \int_{C_j} dz b_{zz}$.

For the Feynman rules of the gauge-fixed nonlinear action we expect to find all values of ghost number circulating in loops (we have in the Polyakov approach nothing restricting the ghost number in loops to that of physical states, $N_g = -\frac{1}{2}$). We have for the propagation the ordinary kinetic operator $\Delta$ corresponding to free propagation of the string, together with a $b_0$ insertion. The gauge-fixed action for the string field theory can then be written schematically as

$$S_{gf} = \int (\mathcal{A} * b_0 \Delta \mathcal{A} + \tfrac{2}{3} \mathcal{A} * \mathcal{A} * \mathcal{A}) \tag{4.8}$$

where $\mathcal{A}$ is a string field with no restriction on its ghost number. Recent work indicates that such an action has a second quantized BRST invariance and in fact does reproduce the Feynman rules deduced from the Polyakov approach.[30] Furthermore, a general gauge fixing procedure has been applied to the action (2.2) and from this procedure it appears possible to derive a gauge fixed action of the form (4.8) by passing to the Siegel gauge.[11]

## 5. CONCLUSION

I have described how Witten's open string field theory captures two aspects of the geometry of the moduli space of Riemann surfaces. The first is the topology of moduli space that is described by the cell decomposition of moduli space in which each distinct string field theory diagram corresponds to a cell. The second feature is the expression of an integration measure over moduli space provided by the reparametrization ghosts. This measure can optionally be thought of as having its origen in the Weil-Petersson measure on moduli space.

A big open problem is that of inventing a field theory structure that describes closed strings. It seems that the local measure for integration in such a theory is also provided by the ghosts in the manner discussed above, but it is not known how to invent a gauge invariant structure for the closed string that gives the expected ghost insertions. Further, a closed string field theory might be presumed to give a natural cell decomposition of moduli space. One of the possible problems with this is the appearance of closed Riemann surfaces in the closed string Feynman diagrams.* It is presently not known how to triangulate the moduli space of a Riemann surface without boundaries or punctures. Furthermore, closed string structures appear to be closely tied to the complex structure of moduli space, so we would hope to have a cell decomposition of moduli space in which the complex structure appears in a more transparent guise.

* This is potentially not a problem if the Polyakov path integrals over the closed surfaces all vanish; this has been shown to happen in certain cases (see Ref. 31.)

# ACKNOWLEDGEMENTS

I would like to thank my collaborators, E. Martinec and E. Witten, who have greatly influenced my thinking on string field theory. Thanks also to the organizers of the conference for the invitation to speak.

## REFERENCES


1. A.M. Polyakov, *Phys. Lett.* **103B** (1981), 207, 211.
2. E. Witten, *Nucl. Phys* **B268** (1986), 253.
3. W. Siegel, *Phys. Lett.* **149B** (1984), 157,162; **151B** (1985) 391,396; W. Siegel and B. Zweibach, *Nucl. Phys* **B263** (1986), 105.
4. E. Witten, *Nucl. Phys* **B276** (1986), 291.
5. See *e.g.* D. Friedan, in *Unified String Theories*, proceedings of the Santa Barbara workshop on Unified String Theories, M. Green and D. Gross (eds.), World Scientific (1986); E. Martinec, *Conformal Field Theory on a (super) Riemann Surface*, Princeton preprint (April 1986); L. Crane and J.M. Rabin, *Super Riemann Surfaces: Uniformization and Teichmüller Theory*, Enrico Fermi preprint EFI 86-25 (June 1986).
6. M. Kato and K. Ogawa, *Nucl. Phys.* **B212** (1983), 443.
7. D. Friedan, in 1982 Les Houches Summer School, *Recent Advances in Field Theory and Statistical Mechanics*, J.-B. Zuber and R. Stora (eds.), North Holland (1984).
8. C. Becchi, A. Rouet, and R. Stora, *Phys. Lett.* **52B** (1975), 344; I.V. Tyutin, Lebedev preprint FIAN no. 39 (1975), in Russian, unpublished.
9. See also I.B. Frenkel, H. Garland, and G.J. Zuckerman, *Semi-Infinite Cohomology and String Theory*, Yale Math. preprint.
10. A. Connes, in *Algébres D'Operateurs*, P. de la Harpe (ed.) Springer (1979); in *C.R. Acad. Sci. Paris* **290** (1980); and IHES preprints (1982-1984).
11. M. Bochicchio, *Gauge Fixing for the Field Theory of the Bosonic String*, Rome preprint No. 527, and Princeton preprint (in preparation).
12. S. Giddings, *The Veneziano Amplitude from Interacting String Field Theory*, Princeton preprint (January 1986), to appear in *Nucl. Phys.* **B278** (1986) 242.

13. D. Gross and A. Jevicki, *Operator Formulation of Interacting String Field Theory*, Princeton preprint (May 1986); E. Cremmer, A. Schwimmer, and C. Thorn *The Vertex Function in Witten's Formulation of String Field Theory*, Univ. of Florida preprint UFTP 86-8, to appear in *Phys. Lett.*

14. S. Giddings and E. Martinec, *Conformal Geometry and String Field Theory*, Princeton preprint (April 1986), to appear in *Nucl. Phys.*

15. This theorem is based on an idea due to W. Thurston (unpublished), with details provided by Bowditch and Epstein; the first complete proof was by D. Mumford (unpublished) and J. Harer. See J. Harer, *The Virtual Cohomological Dimension of the Mapping Class Group of an Orientable Surface*, Maryland Math. preprint (July, 1984) or J. Harer, *The Cohomology of the Moduli Space of Curves*, C.I.M.E. lecture notes.

16. S. Giddings, E. Martinec, and E. Witten, *Phys. Lett.* **176B** (1986), 362.

17. R.C. Penner, *Perturbative Series and Moduli Space*, preprint (1986).

18. J. Harer, *op. cit.*, J. Harer and D. Zagier, *The Euler Characteristic of the Moduli Space of Curves*, Maryland Math. preprint (April, 1986).

19. W. Siegel, in Ref. 3; H. Hata, K. Itoh, T. Kugo, H. Kunitomo, and K. Ogawa, Kyoto University preprints (December 1985, March 1986); M.A. Awada, *Phys. Lett.* **172B** (1986), 32, and Cambridge preprints (January, February, and September 1986); A. Neveu and P. West, *Phys. Lett.* **168B** (1986), 192, and CERN preprints (1986).

20. S. Mandelstam, *Nucl Phys.* **B64** (1973), 205, *Phys. Rep.* **13C** (1974), 259; see also S. Mandelstam in *Unified String Theories, op. cit.*

21. S. Giddings and S. Wolpert, *A Triangulation of Moduli Space From Light-Cone String Theory*, Princeton preprint PUPT-1025 (October 1986).

22. This fact has been noticed by several physicists.

23. O. Alvarez, *Nucl. Phys.* **B216** (1983), 125.

24. E. d'Hoker and D.H. Phong, *Nucl. Phys.* **B269** (1986), 205; G. Gilbert, Texas preprint UTTG-23-85.

25. D. Quillen, *Funct. Anal. and Appl.* **19** (1985), 31; A. Belavin and V.G. Knizhnik, *Phys. Lett.* **168B** (1986), 201; L. Alvarez-Gaumé, G. Moore, and C. Vafa, *Comm. Math. Phys.* **106** (1986), 40.

26. K. Fujikawa, *Phys. Rev* **D25** (1982), 2584.

27. *Cf.* Siegel and Zweibach, Ref. 3, and references therein; see also D. Friedan, E. Martinec, and S. Shenker, *Phys. Lett* **160B** (1985), 55, *Nucl. Phys.* **B271** (1986), 93; L. Alvarez-Gaumé, G. Moore, P. Nelson, C. Vafa, and J.B. Bost, *Phys. Lett.* **178B** (1986), 41, and J. Bost and P. Nelson, Harvard preprint HUTP-86/A044.

28. For the bulk curvature term see Marnelius, *Nucl. Phys.* **B211** (1983), 14.

29. Z. Koba and H.B. Nielsen, *Nucl. Phys.* **B12** (1969), 517; S. Mandelstam, *Nucl. Phys.* **B64** (1973), 205.

30. C.B. Thorn, *Perturbation Theory for Quantized String Fields*, IAS preprint (October, 1986).

31. G. Moore, J. Harris, P. Nelson, and I. Singer, *Phys. Lett.* **178B** (1986), 167; M.C. Chang and Z. Ran, *Divisors on $\bar{M}_g$ and the Cosmological Constant*, preprint (1986).

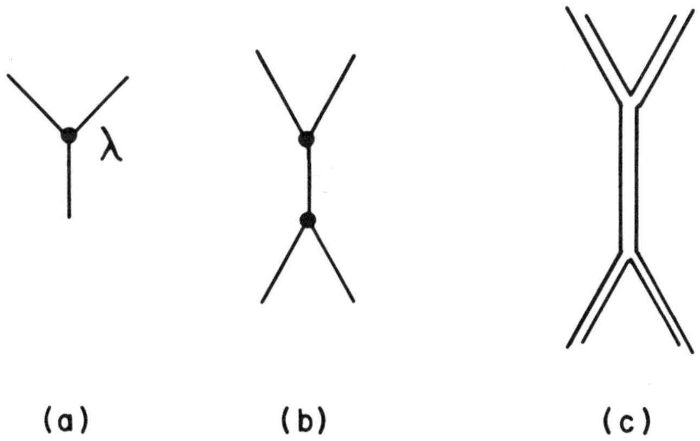


Fig. I

In (a) is shown the basic cubic interaction for a particle; (b) shows the Feynman diagram corresponding to a scattering process. In (c) we see the Feynman diagram at extreme magnification, where we find that the particle is really a string.

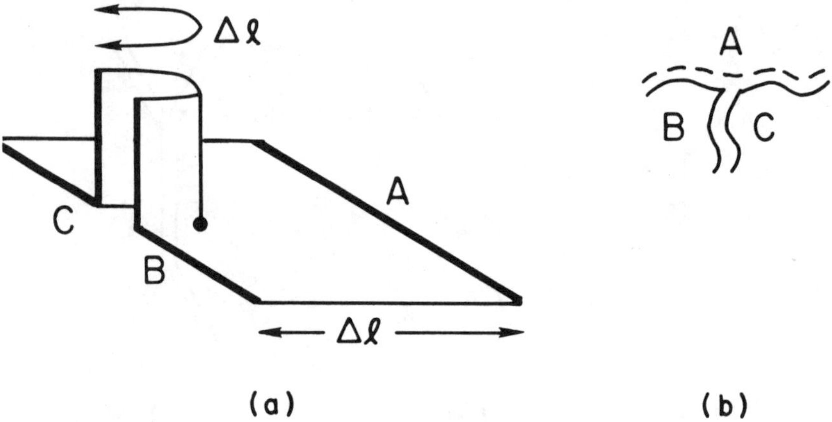


Fig. 2

The open string multiplication law. In (a) is the world-sheet picture, in which we do the path integral with boundary conditions imposed by string fields $B$ and $C$ to find the string field functional $A(X(\sigma))$. In this path integral we take the limit as $\Delta l \to 0$. In (b) we see the interpretation of this multiplication law in terms of strings; the right half of $B$ is equated to the left half of $C$ and the leftover halves make up the string $A$.

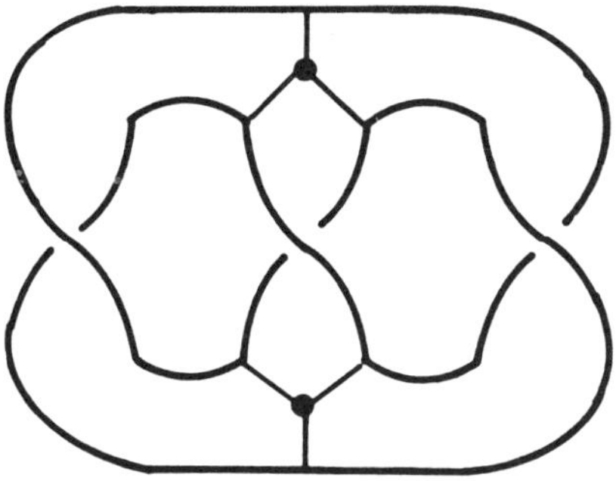

(a)

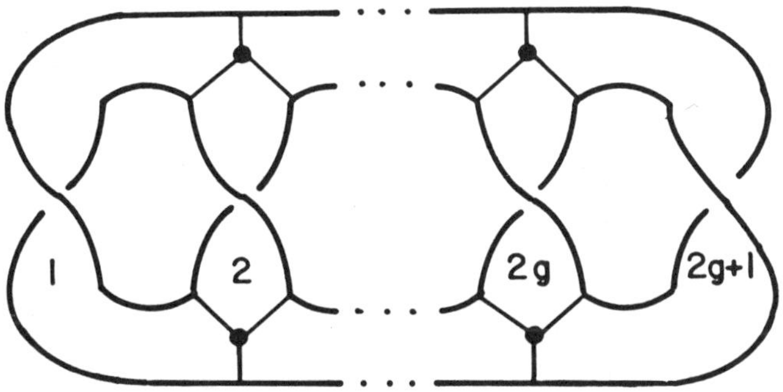


(b)

Fig. 3

In (a) we have an open string vacuum diagram with three twisted propagators; topologically this is equivalent to a torus with one hole. In (b) is a vacuum diagram with $2g+1$ twisted propagators; topologically this surface is a surface of genus $g$ with a disk removed.

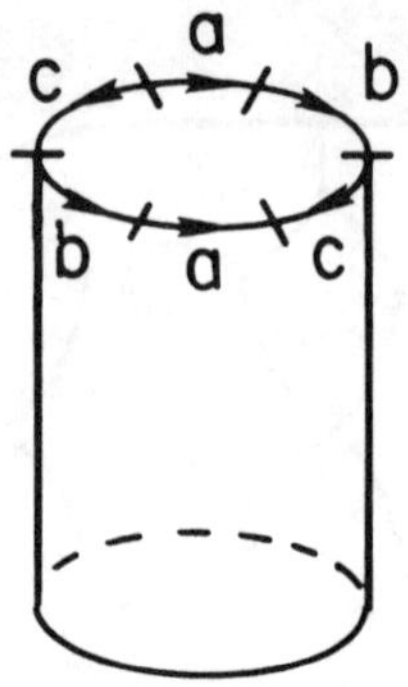


Fig. 4

The cylinder obtained by cutting a Feynman diagram apart along the midlines of the propagators; the lower boundary of the cylinder corresponds to the original boundary of the surface, and along the upper boundary we have indicated the gluing pattern necessary to restore the original surface. (Shown is the case of the torus, Fig. 3(a))

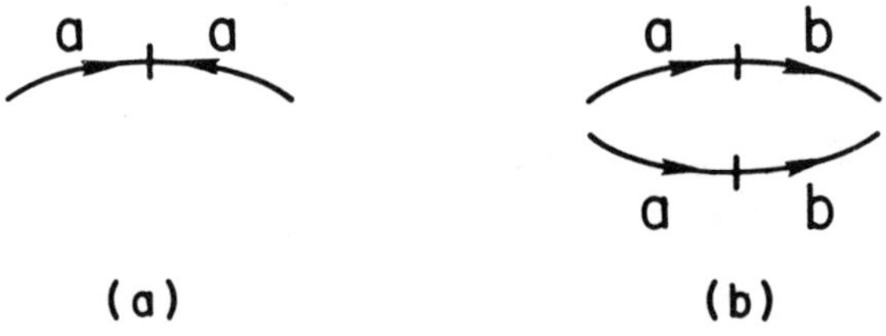


Fig. 5

Gluing patterns that are forbidden in considering canonical presentations. In (a) is shown a pattern that does not contribute to the type of Feynman diagram we are considering, and in (b) is shown a redundant pattern.

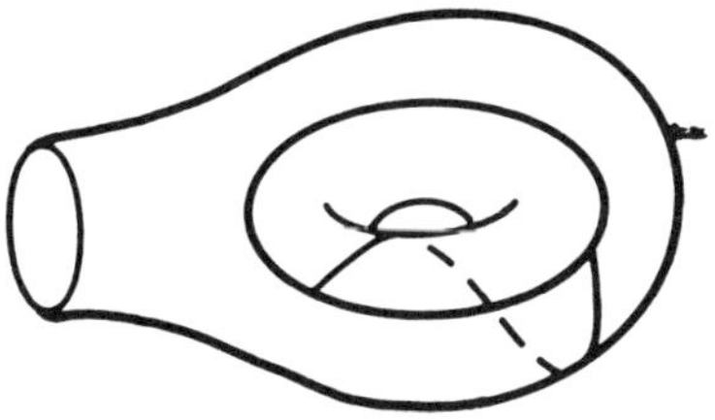

Fig. 6

The embedding of a cylinder in the torus with one hole; one boundary of the cylinder is identified with the boundary of the torus, and the other boundary just touches itself so that we get a one-to-one cover of the torus except along the identified segments.

Fig. 7

A small variation of the boundary of the embedded cylinder such that an area $\delta A$ is moved from one side of the boundary to the other; on the conformally equivalent cylinder $\rho\delta A$ is added to one part and $\rho'\delta A$ is subtracted from another part.

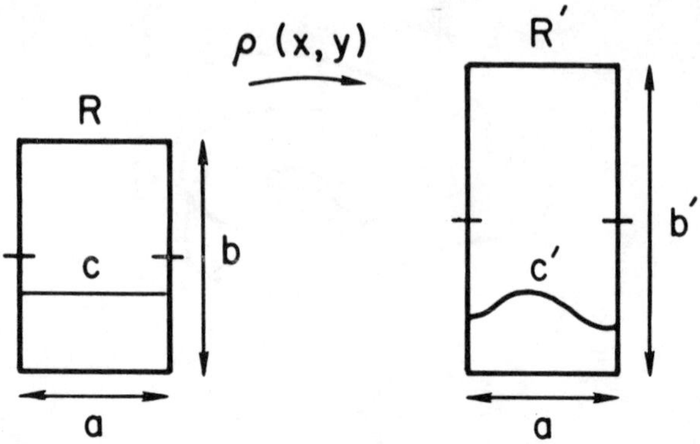


Fig. 8

Two canonical presentations of a given Riemann surface, represented as two rectangles $R, R'$ with vertical sides identified. The lower boundaries corresponds to the boundary of the Riemann surface, and the identifications of arcs are made on the upper boundaries. Also shown is a curve $C$ and its image $C'$.

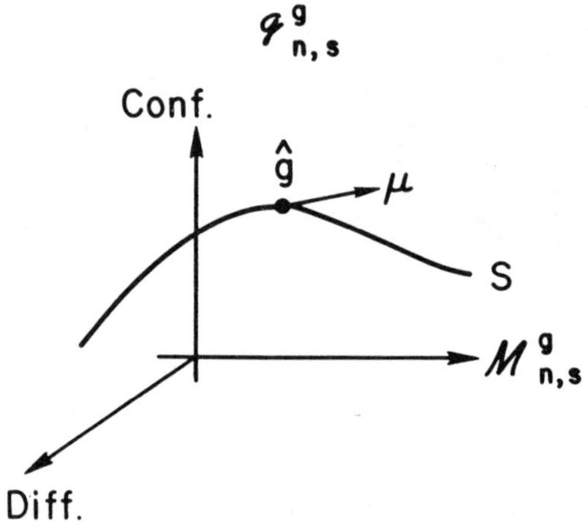


Fig. 9

A representative slice through the space of metrics. This slice has the same dimension as the moduli space and is transverse to the orbits of the diffeomorphism group and group of Weyl rescalings. Also shown is a tangent vector to the slice, a Beltrami differential.

# String Theory Without a Background Spacetime Geometry*

Gary T. Horowitz**

*Department of Physics*
*University of California*
*Santa Barbara CA 93106*

ABSTRACT: An approach to (bosonic) string field theory is described which contains no explicit dependence on a background spacetime metric. The spacetime metric arises from classical solutions to the field equation of the theory.

* To appear in the Proceedings of the conference on "Mathematical Aspects of String Theory," San Diego (July 1986)
** Alfred P. Sloan Research Fellow

# I. INTRODUCTION

String theory is believed to be a consistent quantum theory unifying gravity with the other known forces and matter fields. As such, the theory should be formulated in a way that the spacetime geometry (which according to general relativity represents the classical gravitational field) should be determined by particular solutions to the classical field equation of the theory. One should not have to specify a preferred background metric in order to write down the field equation. This is certainly the case for general relativity itself where the field equation - the vanishing of the Ricci tensor - does not require that one split the metric into a background plus a perturbation.

Unfortunately such a formulation of string theory is not yet available. The difficulty inherent in finding such a formulation can be understood as follows. In most theories of physics one considers a finite number of tensor fields (including a spacetime metric) on a manifold $M$. The differential equations these fields must satisfy always involve the metric (which may or may not be dynamical). String theory can be viewed as a theory with an infinite number of tensor fields on $M$. As we will see, this infinity of fields is unified into a single field $\Phi$ on a much larger space - essentially the loop space of $M$. But string theory does not provide a metric on this larger space. In the absence of a metric, it is difficult to see how one can write down any non-trivial equation for $\Phi$.

In the face of this problem, string theory has progressed mainly by focusing on perturbation theory about particular backgrounds. The background spacetimes are selected by the requirement that the quantum mechanical treatment of a single string propagating in this background is invariant under conformal rescalings of the two dimensional metric. There is evidence[1] that such backgrounds correspond to solutions of the (still unknown) fundamental string equations of motion. One background which satisfies this condition is flat spacetime of dimension 26 or 10 depending on whether one is considering the bosonic or supersymmetric string. Another background, for the superstring, is $M_4 \times K$ where $M_4$ is four dimensional Minkowski spacetime and $K$ is a Calabi-Yau manifold[2] i.e. a complex three manifold with vanishing first Chern class. Of course a background for string theory will in general contain other fields in addition to a spacetime metric. Including these fields (in particular a massless scalar $\phi$ called the dilaton and a massless antisymmetric tensor $B_{\mu\nu}$) has led to the discovery of more general backgrounds.[3]

Perturbation theory about one of these backgrounds involves functional integrals over maps from arbitrary genus Riemann surfaces into spacetime. As discussed elsewhere in these proceedings much progress has been made recently in this area - especially for the case of flat spacetime.[4] However in order to gain a better understanding of non-perturbative effects as well as a more fundamental description of strings in general, it is necessary to find an equation for the string field $\Phi$ which does not refer to a particular spacetime metric. I would like to describe one attempt in this direction. For more details see [5]. Since this problem arises even for the bosonic string, I will restrict attention to this case.

We begin in section (2) by reviewing the theory of a single string in flat spacetime, with emphasis on the treatment of the constraints. This theory depends explicitly on the spacetime metric - and is expected to. The problem of metric independence arises only in the larger context of string field theory. This is discussed in section (3) and a possible solution is proposed.

## 2. QUANTIZATION OF A SINGLE STRING

To begin, consider a single string propagating in a flat 26 dimensional spacetime. This is described classically the the action

$$S_1 = \frac{-1}{2\pi} \int d^2\sigma \sqrt{-q}\, q^{ab} \partial_a X^\mu \partial_b X^\nu \eta_{\mu\nu} \tag{1}$$

where the integral is over a two dimensional manifold - the worldsheet - with Lorentzian metric $q_{ab}$, and $X^\mu$ is the embedding of the worldsheet into flat spacetime. There are two types of strings depending on the topology of the worldsheet. For open strings, the worldsheet is $R^2$ and $X^\mu$ must have vanishing normal derivative along the timelike boundaries. For closed strings, the worldsheet is $S^1 \times R^1$. The action (1) is clearly invariant under conformal rescalings of $q_{ab}$ and diffeomorphisms of the worldsheet. Classically, this gauge invariance implies that there are constraints on initial data i.e. one cannot specify $X^\mu$ and its time derivative arbitrarily on a spacelike curve and obtain a solution. Quantum mechanically the gauge invariance implies that there are constraints on the quantum states. There are several ways to deal with these constraints. One approach (the BRST approach) introduces anti-commuting ghost fields associated with fixing part of the gauge invariance. Another approach, which is essentially equivalent to the BRST approach, is more geometrical. I will now discuss the main ideas of this second approach. For more details see [6].

One begins by adding to $S_1$ the action*

$$S_2 = -\frac{9}{32\pi} \int d^2\sigma d^2\sigma' \sqrt{-q(\sigma)} \sqrt{-q(\sigma')} R(\sigma) R(\sigma') G(\sigma, \sigma') \tag{2}$$

where $R$ is the scalar curvature of the metric $q_{ab}$ and $G$ is the Green's function for the two dimensional wave equation. Although this action is non-local, the "non-locality is pure

* Even though $S_2$ is not invariant under rescalings of $q_{ab}$ (except for constant rescalings), the combined action $S_1 + S_2$ when treated quantum mechanically is conformally invariant.

gauge" in the sense that one can fix part of the diffeomorphism invariance in such a way that the action becomes local. The most convenient choice is the conformal gauge

$$ds^2 = -e^{2\rho} du dv \tag{3}$$

where $u$ and $v$ are null coordinates on the worldsheet. In fact one can show that $S = S_1 + S_2$ is the most general action for a string in flat spacetime which is diffeomorphism invariant, renormalizable, globally scale invariant, and local in conformal gauge. Varying $S$ with respect to $q_{ab}$ and taking the trace yields $R = 0$. So the worldsheet is flat.

Consider first the open string. The boundary conditions on $\rho$ and $G$ are the same as $X^\mu$: their normal derivatives must vanish along the timelike boundaries. Geometrically this implies that the boundary curves are geodesics of $q_{ab}$ and have zero extrinsic curvature. The general solution for $\rho$ is thus

$$\rho(u,v) = \frac{1}{2}[\rho(u) + \rho(v)] \tag{4}$$

The worldsheet has a unique time translation symmetry which preserves the boundaries of the string. This is generated by the killing field:

$$\xi = e^{-\rho(u)} \frac{\partial}{\partial u} + e^{-\rho(v)} \frac{\partial}{\partial v} \tag{5}$$

The conserved quantity associated with this symmetry is the Hamiltonian

$$H = \int_\Sigma \xi^a T_{ab} d\Sigma^b \tag{6}$$

where $T_{ab} = 2\frac{\delta S}{\delta q^{ab}}$ and $\Sigma$ is any spacelike curve stretching from one boundary to the other.

Now consider the quantum theory. In any quantum theory, the Hilbert space of states can be represented as wave functions on the space of configurations of the system. In the present case, a configuration of the string at $u + v$ = constant consists of the pair $X^\mu(\sigma), \rho(\sigma)$ where $\sigma \equiv v - u$. So quantum states are complex-valued functionals $\psi[X^\mu(\sigma), \rho(\sigma)]$.

The collection of all such functionals clearly form a vector space and can be turned into a Hilbert space by introducing the inner product

$$< \psi_1|\psi_2 >= \int DX(\sigma)D\rho(\sigma)\bar{\psi_1}\psi_2 \tag{7}$$

Of course the path integral on the right hand side must be suitably defined. In the present case of a string propagating in a flat spacetime, one defines the inner product as follows. First one expands $X^\mu(\sigma)$ and $\rho(\sigma)$ in a Fourier series and notices that classically, each Fourier mode acts like a simple harmonic oscillator. One then constructs the Hilbert space of each oscillator and takes the tensor product over all the modes.

The question naturally arises as to what class of functions $X^\mu(\sigma), \rho(\sigma)$ form the domain of the wave function. This can be answered as follows. For simplicity, we will focus on just $X^\mu(\sigma)$. For each harmonic oscillator one can introduce creation and annihilation operators $a_n^{\mu\dagger}, a_n^\mu$ satisfying $[a_m^\mu, a_n^{\nu\dagger}] = \delta_{m,n}\eta^{\mu\nu}$. (Strictly speaking, one should analytically continue to a Euclidean metric on spacetime.) Define the ground state (or vacuum state) by $a_m^\mu|0 >= 0$ for all positive $m$. The Hilbert space is spanned by a finite number of $a_m^{\mu\dagger}$ acting on $|0 >$. In terms of the Fourier modes $X_m^\mu$ this corresponds to finite degree polynomials multiplied by the ground state wave function:

$$|0 >= e^{-\Sigma_{n=1}^{\infty} nX_n^2} \tag{8}$$

The states therefore have support only if $\Sigma_{n=1}^{\infty} nX_n^2 < \infty$. This corresponds to functions $X^\mu(\sigma)$ which lie in a subspace of $L^2$, but are not necessarily continuous and bounded.

We must now impose the constraints. As usual, a classical observable such as the Hamiltonian (6) becomes an operator on quantum states by expressing it in terms of the variables $X^\mu(\sigma), \rho(\sigma)$ and their conjugate momentum $P_\mu(\sigma), \pi(\sigma)$ (which are essentially their time derivatives) and then replacing $P_\mu(\sigma) = -i\frac{\delta}{\delta X^\mu}(\sigma)$and $\pi(\sigma) = -i\frac{\delta}{\delta\rho(\sigma)}$. Requiring that quantum states be invariant under the diffeomorphism generated by the killing field $\xi$ requires

$$H|\psi>=0 \tag{9}$$

The operator $H$ turns out to have two remarkable properties. First let $|\psi_1>$ and $|\psi_2>$ be any two solutions of (9). Then one can show that

$$<\psi_1|O|\psi_2>=0 \tag{10}$$

where $O$ is any operator corresponding to the generator of a conformal diffeomorphism on the worldsheet. This is established by showing that one can always express $O$ as an (anti) commutator of $H$ with another operator in the theory. In this sense the single equation (9) incorporates all the constraints. Solutions of (9) are called ***physical states.*** Second, $H$ is nilpotent: $H^2 = 0$. This results from the fact that in two dimensions certain exponentials of bosonic fields anticommute, so $H$ (which involves $e^{-\rho}$) becomes an anticommuting operator. A simple consequence of this is that if $|\psi>$ is a physical state, so is

$$|\tilde{\psi}>=|\psi>+H|\varepsilon> \tag{11}$$

for any state $|\varepsilon>$ .

The transformation (11) has a suprising interpretation. Different states of the string correspond to different pointlike particles in spacetime. Many of these are invariant under certain spacetime gauge transformations. For example, the state $A_\mu(X_0)a_1^{\mu\dagger}|0>$ corresponds to a massless vector particle in the sense that the physical state condition (9) is equivalent to Maxwell's equations for $A_\mu$. It turns out that when restricted to massless states, (11) corresponds to $\tilde{A}_\mu = A_\mu + \nabla_\mu\lambda$ which of course is the usual Maxwell gauge invariance. Similarly, when restricted to massive states, (11) again corresponds to the appropriate spacetime gauge transformations. This leads to two remarkable conclusions. First, the generator of time translations on the worldsheet automatically generates the apparently unrelated spacetime gauge transformations. Second, gauge inequivalent physical states of the string are equivalence classes of states in the kernel of $H$, where two are equivalent if they are related by (11). In other words, inequivalent physical states are just the cohomology classes of $H$.*

The above discussion was restricted to the open string. For the closed string the analysis is similar.[6] However now there exist a one parameter family of timelike killing fields on the cylindrical worldsheet. If $H$ is the Hamiltonian associated with one of them, $\xi$, it is still true that $H|\psi>= 0$ implies that matrix elements of all the constraints are satisfied. However $H^2 \neq 0$. Instead, $HP = 0$ where $P$ is the operator generating spacelike translations orthogonal to $\xi$. So spacetime gauge transformations are naturally associated with spacelike translations on the worldsheet.

* Mathematically, $H$ is equivalent to the open string BRST operator Q with $\rho$ playing the role of the bosonized ghost.

## 3. STRING FIELD THEORY

In the previous section we considered the quantum mechanics of a single string. To consider many strings or dynamical background fields one must pass to the string field theory. The transition is summarized below

| *Single String Theory* | | *String Field Theory* |
|---|---|---|
| Wave functions $\psi[X(\sigma), \rho(\sigma)]$ | $\rightarrow$ | Operators $A[X(\sigma), \rho(\sigma)]$ which create or destroy entire strings. |
| Describes an infinite number of point particles on spacetime | $\rightarrow$ | Describes an infinite number of fields on spacetime |
| Physical state condition $H\psi = 0$ | $\rightarrow$ | Linearized field equation $HA = 0$ |

A familiar illustration of the last step is given by the free relativistic particle. The first quantized physical state condition $(p^2 + m^2)|\psi>= 0$ becomes the Klein-Gordon field equation $(\nabla^2 - m^2)\phi = 0$.

Notice that the passage from the single string theory to string field theory only yields the *linearized* field equation. This is to be expected: The non-linear terms describe how several strings interact. This is a new physical input which is not present in the theory of a single string. Recall that the basic assumption is that strings interact by splitting and joining so that the worldsheet remains a smooth two-manifold. (See Fig. 1) This assumption is the basis behind the rules for perturbation theory which involve summing over different smooth two-manifolds. One motivation for this comes from the fact that in order to recover local equations for the infinite number of spacetime fields described by the string one needs an interaction which is local in spacetime. Since the interactions of many strings can be obtained by iterating the one in Fig. 1 involving three strings, we expect just a $A^3$ term in the action and hence a $A^2$ term in the equations of motion. The simplest possibility would thus be

$$HA + A^2 = 0 \tag{12}$$

where the second term is the usual pointwise multiplication of fields. Unfortunately, this is wrong. The reason is that $A^2[X(\sigma), \rho(\sigma)]$ creates two strings in exactly the same configuration so it would correspond to the interaction shown in Fig. 2. To recover the simple picture of strings splitting in spacetime we need a *non-local* product on loop space.

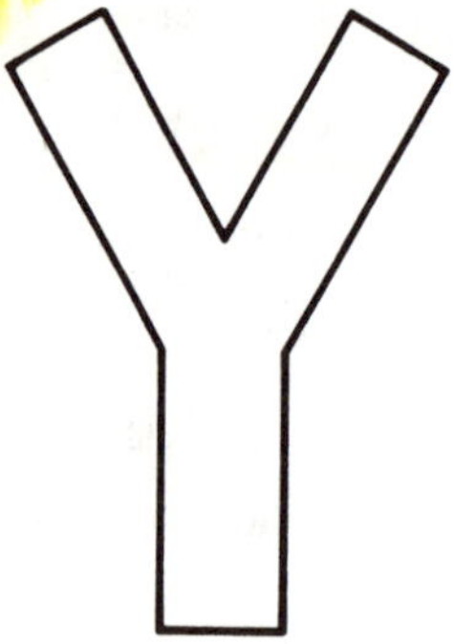

Fig. 1

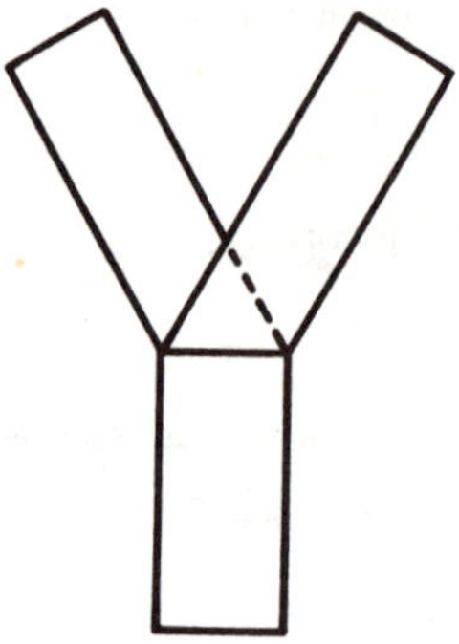

Fig. 2

Fig 1. The worldsheet of one string splitting into two.

Fig 2. Pointwise multiplication of string fields correspond to interactions in which the worldsheet is not a smooth manifold.

Witten[7] has proposed a product of exactly this type for the open string.* (In the interest of clarity we will suppress the dependence on the conformal factor $\rho(\sigma)$ in the following discussion.) Since $A$ is a function on the space of parametrized strings, we can consider each string as consisting of a left half $X_L(\sigma)$ $\sigma\varepsilon[0,\frac{\pi}{2}]$ and a right half $X_R(\sigma)$ $\sigma\varepsilon[\frac{\pi}{2},\pi]$. Define

$$(A*B)(X_L,X_R)\equiv\int DY\,A[X_L,Y]B[\bar{Y},X_R] \tag{13}$$

where the integral is a functional integral over all half strings $Y(\sigma),\sigma\varepsilon[\frac{\pi}{2},\pi]$ with boundary conditions $Y(\frac{\pi}{2})=X_L(\frac{\pi}{2})$ and $Y'(\pi)=0$ and $\bar{Y}(\sigma)\equiv Y(\pi-\sigma)$. The $*$ product resembles matrix multiplication and is not commutative. One has thus turned the space of string fields into a non-commutative algebra. ** Our candidate field equation is now[7]

$$HA+A*A=0. \tag{14}$$

One can show that this equation reproduces the standard (tree-level) scattering of open strings in a flat background.[8]

We now come to the main point: $H$ involves a spacetime metric. Indeed it was the Hamiltonian for a string propagating in 26 dimensional flat spacetime. We would like to find an equivalent equation which does not have this explicit metric dependence. The main idea for how to proceed is the following.[5,9] (This idea has also been investigated in the context of a different string field theory in [10].) The metric dependence in $H$ should come from a particular string solution $\Phi_0$. The field $A$ should not be regarded as the fundamental string field, but rather the difference between the fundamental string field $\Phi$ and the particular solution $\Phi_0$. With this in mind set $\Phi=\Phi_0+A$ and consider the

* A theory of non-interacting open strings does not describe gravity. However once interactions are included, open strings can form closed strings which do describe gravity. It is thus believed that Witten's theory (or some slight modification if it) includes both open and closed strings, although the details of how this comes about are not yet well understood.

** Actually, the algebra is $Z$ - graded because of "ghost number" i.e. dependence on the zero mode of $\rho$.

equation:

$$\Phi * \Phi = 0. \tag{15}$$

It is immediate that if $\Phi_0$ satisfies

$$\Phi_0 * \Phi_0 = 0 \tag{16a}$$

and

$$\Phi_0 * A + A * \Phi_0 = HA \tag{16b}$$

then (15) reduces to (14). Thus, if there is a solution to (16a) and (16b) then $\Phi * \Phi = 0$ is equivalent to (14). Note that $\Phi * \Phi = 0$ does not contain explicit reference to a spacetime metric.

In fact using formal properties of the star product one can show[5] that a solution exists to (16a) and (16b). Let $I$ denote the identity element of the $*$ algebra: $\Phi * I = I * \Phi = \Phi$. (Roughly speaking $I$ is a $\delta$ - function with support when $X_L(\sigma) = X_R(\pi - \sigma)$.) Define $H_L \equiv \int_0^{\frac{\pi}{2}} \xi^a T_{ao} d\sigma$ i.e. $H_L$ is the integral over half the string of the current which yields the Hamiltonian $H$. Then using the fact that $H^2 = 0$, one can show that $\Phi_0 = H_L I$ satisfies (16a) and (16b).

The first condition on $\Phi_0$ (16a) just says that $\Phi_0$ is a classical solution to the field equation (15). The second condition (16b) can be interpreted as follows. String theory is a *unified* theory of gravity with an infinite number of other fields. One does not expect to be able to unambiguously extract a spacetime metric from an arbitrary solution of string theory. Only for particularly simple solutions (presumably corresponding to ground states) is this possible. Equation (16b) can be viewed as the condition for the solution $\Phi_0$ to correspond to a set of background fields on spacetime. The situation is quite analogous to a Kaluza-Klein theory which is another attempt to unify gravity with gauge fields in the context of higher dimensional spacetimes. In that case the basic field is a metric in $N$ dimensions satisfying $R_{\mu\nu} = 0$. In general, there is no way to extract a four dimensional metric. However, for certain solutions e.g. those of the product form $M^4 \times M^{N-4}$ a four dimensional metric is defined.

Thus $\Phi * \Phi = 0$ is a candidate for a fundamental field equation for string theory. Note that it is the non-local nature of the $*$ product which has enabled us to "shift away" the derivative term in (14). Although no explicit reference to a spacetime metric appears, one might wonder whether there is background dependence implicit in $*$, since this product involves an integral and one needs a metric or at least a measure to define the integral. This implicit dependence can perhaps be eliminated by requiring that $\Phi$ be a density of appropriate weight.

Recall that spacetimes corresponding to conformally invariant quantum theories of a single string are believed to be classical solutions of the string field theory. The present approach allows a simple and direct proof of this result. One can show that the condition for conformal invariance is equivalent to the condition for $H^2$ to vanish where $H$ is the Hamiltonian of the string in the background spacetime.[11] Thus, given a conformally invariant background, one can construct $H_L I$ and show that this is a classical solution to (15).

We now discuss the gauge invariance of the string field theory. The equation $\Phi * \Phi = 0$ is clearly invariant under the infinitesimal transformations $\delta\Phi = \Phi * \Lambda - \Lambda * \Phi$ where $\Lambda$ is an arbitrary string field. These are of course just the inner derivations of the $*$ algebra. In the formulation proposed by Witten,[7] the gauge transformations pick up an inhomogeneous term

$$\delta A = H\Lambda + A * \Lambda - \Lambda * A. \tag{17}$$

As a consequence, the usual ground state solution $A = 0$ is not invariant under all the symmetries of the theory, but only those satisfying $H\Lambda = 0$. One can show that these include the usual Yang-Mills gauge invariance. The rest of the symmetries are "spontaneously broken." By contrast, in the formulation (15), the natural ground state $\Phi = 0$ is invariant under all inner derivations and hence corresponds to a state of unbroken symmetry. Since physical fields should be real, $\Phi$ is required to be hermitean in the sense that $\Phi(X_L, X_R) = \Phi^*(\bar{X_R}, \bar{X_L})$ where, as before, $\bar{X}_R(\sigma) = X_R(\pi - \sigma)$ and $*$ denotes complex conjugation. Similarly, $\Lambda$ is required to be antihermitean $\Lambda(X_L, X_R) = -\Lambda^*(\bar{X_R}, \bar{X_L})$. One can thus (formally) exponentiate the infinitesimal transformation to find the finite transformation law:

$$\Phi' = U^{-1} * \Phi * U \tag{18}$$

where $U = e^{\Lambda} \equiv I + \Lambda + \frac{1}{2}\Lambda * \Lambda + \cdots$ Thus the theory is invariant under an infinite dimensional generalization of $U(N)$.

If a string field theory based on $\Phi * \Phi = 0$ is to be viable, it is clearly necessary to go beyond relying on formal properties of the star product to show the equivalence of this equation to (14). One needs a more precise (regularized) definition of the star product to check whether the formal properties have anomalies. In fact, it turns out that there *is* an anomaly[12] which can be seen even in the special case when one is given a flat metric on spacetime. In this case one can express classical string fields $A$ as elements of the first quantized state space. The star product can then be described in terms of a state $|V_3>$ which is an element of the tensor product of the three string state spaces:[13]

$$|A * B> = <A| <B|V_3> \tag{19}$$

Due to the infinite number of modes of the string, the star product involves infinite sums. In particular one can obtain double sums which are not absolutely convergent but given an order, all one dimensional sums are. This gives rise to anomalies in associativity.[12] These anomalies allow spacetime translations to be described in terms of inner derivations of the star algebra despite the fact that naively the star product does not move the string midpoint. One can show that this anomaly does not affect the equivalence of (15) and (14) for describing the scattering of open string states in a flat space background.[14] But there are strong indications that it will give rise to subtle differences in the treatment of closed string states.[15]

Ultimately, one would like a formulation of string theory where not just the spacetime metric, but even the manifold topology is determined by the field equation. One way that this might happen[16] is the following. For each manifold one can construct the space of maps $\{X(\sigma), \rho(\sigma)\}$, define string fields $A[X(\sigma), \rho(\sigma)]$, and the star algebra. It may turn out that the resulting star algebras are all isomorphic. In other words, the star algebra may be independent of the manifold one starts with. One might recover a manifold in an appropriate limit from the fact that a particular string solution might be more naturally described in terms of a collection of tensor fields on one manifold than another. As an analogy, consider the Hilbert space H of square integrable functions on a manifold M with metric $g_{ab}$. Since H is separable and all separable Hilbert spaces are isomorphic, H is in fact independent of M and $g_{ab}$. However, the action of a given operator may be simpler for one choice of M, $g_{ab}$ than another.

The formulation we have arrived at looks more algebraic than geometrical: The solutions of the field equation $\Phi * \Phi = 0$ are just the nilpotent elements of a non-commutative star algebra, and the gauge transformations are the inner derivations of this algebra. Clearly a lot of work remains to be done, but we may be seeing the beginnings of a background independent formulation of string theory.

ACKNOWLEDGEMENTS

It is a pleasure to thank my collaborators on the work described here: J. Lykken, R. Rohm, and especially A. Strominger. This work was supported in part by the National Science Foundation under Grant No. PHY85-06686.

REFERENCES

1) C. Callan, E. Martinec, M. Perry, and D. Friedan, *Nucl. Phys.* **B262**, 593, (1985).

A. Sen, *Phys. Rev. Lett.* **55**, 1846, (1985).

C. Callan, I. Klebanov, and M. Perry, *Nucl. Phys.* **B278**, 78, (1986).

2) P. Candelas, G. Horowitz, A. Strominger, and E. Witten, *Nucl. Phys.* **B258**, 46, (1985);

D. Nemeschansky and A. Sen, *Phys. Lett.* **B178**, 365, (1986).

3) E. Witten, *Nucl. Phys.* **B268**, 79, (1986).

A. Strominger, *Nucl. Phys.* **B274**, 253, (1986).

C. Hull, *Phys. Lett.* **B178**, 357, (1986).

4) A. Belavin and V. Knizhnik, *Phys. Lett.* **B168**, 201, (1986).

Yu. Manin, *Phys. Lett.* **B172**, 184, (1986).

5) G. Horowitz, J. Lykken, R. Rohm, and A. Strominger, *Phys. Rev. Lett.* **57**, 283, (1986).

6) G. Horowitz and A. Strominger, *Phys. Rev. Lett.* **57**, 519, (1986); *Phys. Rev. D* (to appear).

7) E. Witten, *Nucl. Phys.* **B268**, 253, (1986).

8) S. Giddings, *Nucl. Phys.* **B278**, 242, (1986).

S. Giddings and E. Martinec, *Nucl. Phys.* **B278**, 91, (1986).

9) Similar ideas have been discussed by:

D. Friedan, *Nucl. Phys.* **B271**, 540, (1986).

E. Witten, *Nucl. Phys.* **B276**, 291, (1986).

T. Yoneya, in Proceedings of the Seventh Workshop on Grand Unificiation, Toyama, Japan (to be published).

10) H. Hata, K. Itoh, T. Kugo, H. Kunitomo, and K. Ogawa, *Phys. Lett.* **B175**, 138, (1986).

11) T. Banks, D. Nemeschansky, and A. Sen, *Nucl. Phys.* **B277**, 67, (1986).

12) G. Horowitz and A. Strominger, *Phys. Lett. B* (to appear).

13) D. Gross and A. Jevicki, *Nucl. Phys. B* (to appear); ITP preprint (1986).

S. Samuel, CERN preprints TH 4365-Rev and TH-4495.

E. Cremmer, A. Schwimmer, and C. Thorn, *Phys. Lett.* **B179**, 57, (1986).

14) A. Strominger, IAS preprint (1986).

15) A. Strominger, in preparation; G. Horowitz and A. Strominger, in progress.

16) E. Witten, to appear in the Proceedings of the Second Nobel Symposium on Elementary Particle Physics, ed. L. Brink et. al.

# HOLOMORPHIC CURVES ON MANIFOLDS OF SU(3) HOLONOMY

EDWARD WITTEN*

*Joseph Henry Laboratories*
*Princeton University*
*Princeton, New Jersey 08544*

* Supported in part by NSF grants PHY80-19754 and PHY86-16129

The purpose of this note is to draw attention to a mathematical problem that is of some significance in connection with compactification of ten dimensional supersymmetric string theories. Before stating the problem, I will try to give a flavor of how it arises in physics; unfortunately, the discussion will not be self-contained in that regard.

Let $K$ be a complex manifold of complex dimension three with vanishing first Chern class. Then, from Yau's proof of the Calabi conjecture, we know that $K$ admits Ricci-flat Kahler metrics, i.e., metrics of $SU(3)$ holonomy; the moduli in these metrics correspond to a choice of the Kahler class and the complex structure of $K$.

Let $M_t$ be the moduli space of metrics of $SU(3)$ holonomy and volume $t$. The spaces $M_t$, for various $t$, are naturally isomorphic, since the corresponding metrics just differ by scaling. The moduli space for all metrics of $SU(3)$ holonomy is $M \simeq M_1 \times R^+$, $R^+$ denoting the positive real numbers ($M_1$ is $M_{t=1}$).

In string theory, complex threefolds of $c_1 = 0$ are of interest because of their relation to unbroken supersymmetry.[1] However, in string theory, the equation that one wishes to solve on a threefold of $c_1 = 0$ is not simply the Einstein-Kahler equation. Schematically, the appropriate equation is

$$R_{i\bar{j}} + \frac{1}{t}\,\partial_i\,\partial_{\bar{j}}\,[R^3] + \cdots \;=\; 0 \qquad (1)$$

where by $[R^3]$ I mean the cubic polynomial in the curvature tensor whose integral over $K$ gives its Euler characteristic. An infinite series of higher corrections are indicated by the $(\cdots)$.

It is possible to show[2] that order by order in an expansion in $t^{-1}$, equation (1) has solutions depending on the same moduli that enter in solving the equation $R_{i\bar{j}} = 0$. This is a non-trivial result that depends on the specific form of the correction terms in (1). Moreover, there is strong evidence[3] that for large enough $t$, the equation (1) has exact solutions, not just order by order in $t^{-1}$, depending on the same moduli that entered in the Ricci-flat equation at $t^{-1} = 0$. If we denote as $N_t$ the moduli space of solutions of (1), with volume $t$, one expects that $N_t$ for large $t$ is isomorphic to $M_t$. For smaller $t$ this need not be so; it is very plausible that $N_t$ might be empty for small enough $t$. The moduli space, $N$, of "string theory vacuum states," i.e. solutions of (1), looks for large $t$ like the moduli space $M$ of Ricci flat Kahler metrics on $K$, but once $t$ is not large there is no reason to believe that $M$ and $N$ have the same topology. It may be that $N$ is compact except for one "end" at $t = \infty$. If so, this would be reminiscent of Donaldson's work on four manifolds; we would have $\partial N \simeq M_\infty \simeq M_1$, i.e. $N$ would be a naturally occurring space with moduli space of Ricci flat metrics of fixed volume as its boundary.

It is possible to generalize all this in the following way. Let $V$ be a stable holomorphic vector bundle over $K$ with the same second Chern class as the tangent bundle $T$. Then one can try to find a "string theory vacuum state" depending on the choice of $V$ as well as a choice of Kahler class and complex structure on K. This possibility is extremely interesting for physical applications of manifolds of $SU(3)$ holonomy.

If one tries to introduce in this way a vector bundle V, the equation (1) is generalized to an even more complex equation. It is still possible to argue[2] that (1) has a formal solution in an expansion in powers of $1/t$, but convergence can be obstructed by the existence of rational curves, i.e. holomorphic curves of genus zero, in $K$. (By such a curve I mean simply a holomorphic map $\phi : CP^1 \to K$; the image of $CP^1$ in $K$ may have self-intersections.) If $K$ does not contain rational curves, then it is believed that the series solution of the generalization of (1) converges. If $K$ does contain rational curves, and they generate $H_2(K; R)$ (second homology of $K$ with *real* coefficients), it is believed that the series solution of (1) does not converge (and that solutions do not exist). If $K$ has rational curves which do not generate $H_2(K; R)$, then (1) may have solutions, depending on additional properties of $V$.

It is easy to see that many easily constructed examples of threefolds of $c_1 = 0$ contain rational curves. For example, there are 375 rational curves in the generic quintic hypersurface in $CP^4$.[3] The question I want to pose in this note is whether every threefold of $c_1 = 0$ contains rational curves, and if not, to describe some counterexamples. Also of interest would be threefolds of $c_1 = 0$ in which the rational curves do not generate $H_2(K; R)$. It is interesting that rational curves play a role in the mathematical theory of threefolds of $c_1 > 0$.[4]

Finally, I want to remark that equation (1) is of a fundamental nature to which I have not tried to do justice in the above. It is the specialization to Kahler geometry of the string theoretical generalization of the Einstein equations. Riemannian geometry is not the proper category for posing this equation. Its proper setting is in a new category that has not yet been correctly formulated, and whose correct formulation would be of utmost interest. I have tried to convey a taste of the latter problem elsewhere.[5]

# REFERENCES


1. P. Candelas, G. Horowitz, A. Strominger, and E. Witten, *Nucl. Phys.* **B258** (1986), 47.
2. E. Witten, *Nucl. Phys.* **B268** (1986), 79.
3. M. Dine, N. Seiberg, X.-G. Wen, and E. Witten, "*Non-Perturbative Effects on the String World Sheet.*"
4. S. Mori and S. Mukai, *Manuscripta Math.* **36** (1981), 147;
   S.-T. Yau, private communication.
5. E. Witten, *Physics and Geometry*, to appear in the *Proceedings of the International Congress of Mathematicians* (Berkeley, 1986).

# VERTEX OPERATOR CALCULUS

I.B. Frenkel[1]

Department of Mathematics
Yale University
New Haven, CT 06520

J. Lepowsky[2]

Department of Mathematics
Rutgers University
New Brunswick, NJ 08903

A. Meurman[3]

Department of Mathematics
University of Stockholm
S-113 85 Stockholm, Sweden

## 1. INTRODUCTION

We present a general treatment of the vertex operators of bosonic string theory as generators of infinite-dimensional Lie algebras.

First we review the construction of certain level 1 standard modules (basic modules) for the untwisted affine Lie algebras of types

[1]Partially supported by the Sloan Foundation and NSF Grant DMS86-02091.

[2]Partially supported by NSF Grant DMS86-03151.

[3]Partially supported by the Swedish Natural Sciences Research Council.

$\hat{A}_n$, $\hat{D}_n$, $\hat{E}_n$ by untwisted vertex operators, following [FK, S]; see also [Fr2, GO1] and [BHN, Hal]. Then we show how to calculate the commutators of untwisted vertex operators associated with arbitrary lattice elements by means of a "formal calculus" related to operator product expansion theory and developed in [FLM5]. This leads to a result of [B] on the Lie algebra spanned by the components of "generalized vertex operators". Such operators are also familiar in string theory (see e.g. [ADDF]; cf. [DNPS, SY, We]), and their possible use in Lie theory was conjectured at the end of the article [Mde]. They are in fact related to non-affine Kac-Moody algebras (see [B]; cf. [Fr2, GO1]). See also [TK].

Next, following the treatment in [L1], we explain the twisted vertex operators (twisted by a lattice automorphism) which have been introduced to construct basic modules for twisted affine algebras of types A, D, E [Fr1, FLM1, KKLW, KP3, L1, LW1]; cf. [C1,2, CF, Fi]. Such operators and modules are closely related to orbifold compactification (a variant of Calabi-Yau compactification [CHSW]) of the heterotic string [GHMR1,2]; see [DHVW1,2] and [FV, Har, NSV, V]; cf. [ABLR, My]. Finally "generalized twisted vertex operators" are introduced and a general result on the structure of the Lie algebra spanned by their components is announced [FLM5, L2]. For convenience of exposition, we restrict our attention to even lattices. The case in which the twisting automorphism is $-1$ is carried out in [FLM5], and together with the result of [B] helps illuminate the construction [FLM2] of the Monster [Gr]. The main point is that the generalized vertex operator algebra is essentially independent of the twisting automorphism. This fact is related to the existence of a Monster element which mixes spaces with different twists, and it should be related to the study of the analogous phenomenon for orbifold compactification discussed in [DFMS, HV]. The construction of the Monster in [FLM2] can be viewed in essence as a string theory on a $\mathbb{Z}_2$-orbifold. The general result in Section 3.3 [L2] may also play a role in new constructions of the Monster and constructions of other sporadic groups.

We shall not discuss the construction of non-basic standard modules for affine algebras (cf. [K]) in this paper, except to say a few words here. A general approach to the explicit construction of such modules -- the method of "Z-algebras" -- was introduced in [LW2,3], and has been used to construct a number of cases [LP1,2, LW4, Mdi, MP, Mis1-3, Wa]; see also [AMW, D, GOS]. While Z-algebras are studied by methods of vertex operator calculus (or operator product expansions), one obtains not commutation relations but "generalized commutation relations", which correspond to "nonlocal" or "parafermionic" current algebras. Such ideas and the modular transformation properties of characters of standard modules (see [KP1,2]) are connected with certain aspects of two-dimensional conformal field theory models [Ge, GQ, JMO, Miw, NY, ZF].

Some recent expositions relating to the present subject are [Fr3, FLM4, Go, GO2, GSW, PS, Wi].

This paper was prepared for the Proceedings of the Conference on Mathematical Aspects of String Theory, La Jolla, July 21 - August 2, 1986. Section 2 is an elaboration of the talk presented there by one of us, J.L., who would like to thank the organizers of the conference for their hospitality and for a stimulating occasion. The results announced in Sections 3.2 and 3.3 were completed after the conference. We are very grateful to many people, especially O. Alvarez, L. Dixon, J.A. Harvey, G. Moore, J. Shapiro, C. Vafa and E. Witten for numerous discussions on vertex operators and orbifolds.

## 2. THEORY OF UNTWISTED VERTEX OPERATORS

### 2.1 The Operators $Y(\alpha,z)$

We shall present a definition of the (untwisted) vertex operators of bosonic string theory. See e.g. [Mde] for an exposition of their origin.

Let $\underline{h}$ be a finite-dimensional complex vector space equipped with a symmetric nondegenerate bilinear form $\langle\cdot,\cdot\rangle$, which need not be the complexification of a real bilinear form of any special type (such as positive definite or Minkowskian). Consider the infinite-dimensional Lie algebra

$$\hat{\underline{h}} = \underline{h} \otimes_{\mathbb{C}} \mathbb{C}[t,t^{-1}] \oplus \mathbb{C}c \tag{1}$$

with structure defined by:

$$\begin{gathered}[x\otimes t^m, y\otimes t^n] = \langle x,y\rangle m\delta_{m+n,0}c \quad \text{for} \quad x,y \in \underline{h}, \quad m,n \in \mathbb{Z}, \\ [c,\hat{\underline{h}}] = 0.\end{gathered} \tag{2}$$

The algebra $\hat{\underline{h}}$ is the (<u>untwisted</u>) <u>affine</u> <u>Lie</u> <u>algebra</u> associated with $\underline{h}$ (which is viewed as an abelian Lie algebra) and with the form $\langle\cdot,\cdot\rangle$. It is $\mathbb{Z}$-graded, with $x\otimes t^n$ having degree $n$ and $c$ having degree $0$.

More generally, if $\underline{g}$ is a Lie algebra with a symmetric bilinear form $(\cdot,\cdot)$ which is <u>invariant</u> in the sense that

$$([x,y],z) = (x,[y,z]) \quad \text{for} \quad x,y,z \in \underline{g}, \tag{3}$$

then the corresponding (<u>untwisted</u>) <u>affine</u> <u>Lie</u> <u>algebra</u> is the $\mathbb{Z}$-graded algebra

$$\hat{\underline{g}} = \underline{g} \otimes_{\mathbb{C}} \mathbb{C}[t,t^{-1}] \oplus \mathbb{C}c \tag{4}$$

with

$$\begin{gathered}[x\otimes t, y\otimes t^n] = [x,y]\otimes t^{m+n} + \langle x,y\rangle m\delta_{m+n,0}c \quad \text{for} \quad x,y \in \underline{g}, \quad m,n \in \mathbb{Z}, \\ [c,\hat{\underline{g}}] = 0.\end{gathered} \tag{5}$$

If $\underline{g}$ is a finite-dimensional simple Lie algebra of type $A_n$, $B_n$, $\cdots$ or $G_2$, then $\hat{\underline{g}}$ is the corresponding (untwisted) affine Kac-Moody algebra $\hat{A}_n$, $\hat{B}_n$, $\cdots$ or $\hat{G}_2$ (cf. [K]).

The subalgebra

$$\hat{\underline{h}}_* = \bigoplus_{n\neq 0} \underline{h} \otimes t^n \oplus \mathbb{C}c \tag{6}$$

of $\hat{\underline{h}}$ is a <u>Heisenberg</u> <u>algebra</u>, that is, it has a basis consisting of homogeneous elements $p_i$, $q_i$ $(i > 0)$ and $c$ satisfying the relations

$$[p_i, p_j] = [q_i, q_j] = 0, \quad [p_i, q_j] = \delta_{ij} c, \quad [c, \hat{\underline{h}}_*] = 0. \tag{7}$$

No connection is assumed between the subscript $i$ and $\deg p_i$ or $\deg q_i$, but we do assume that $\deg p_i > 0$ and $\deg q_i < 0$.

The "canonical realization" of these "Heisenberg commutation relations" may be described as follows: Set

$$\hat{\underline{h}}^{\pm} = \bigoplus_{\pm n > 0} \underline{h} \otimes t^n,$$

$$\underline{b} = \hat{\underline{h}}^{+} \oplus \underline{h} \; (\otimes \; t^0) \oplus \mathbb{C}c$$

and view $\mathbb{C}$ as the 1-dimensional $\underline{b}$-module such that

$$c \cdot 1 = 1 \quad \text{and} \quad (\hat{\underline{h}}^{+} \oplus \underline{h}) \cdot 1 = 0.$$

Form the ($\mathbb{Z}$-graded) induced $\hat{\underline{h}}$-module

$$S = U(\hat{\underline{h}}) \otimes_{U(\underline{b})} \mathbb{C}, \tag{8}$$

$U(\cdot)$ denoting universal enveloping algebra. Then $S$ is isomorphic to $S(\hat{\underline{h}}^{-})$, the symmetric algebra of $\hat{\underline{h}}^{-}$ (the algebra of polynomials on a basis of $\hat{\underline{h}}^{-}$). Note that the grading of $S$ is bounded above:

$$S = \bigoplus_{n \leq 0} S_n.$$

For $x \in \underline{h}$ and $n \in \mathbb{Z}$ we shall write $x(n)$ for the operator (of degree $n$) on $S$ associated with $x \otimes t^n \in \hat{\underline{h}}$:

$$x \otimes t^n \mapsto x(n). \tag{9}$$

The $\hat{\underline{h}}$-module $S$ is irreducible and is essentially a "Fock space" in the sense of quantum field theory. The operators corresponding to the $q_i$ are "creation operators" (multiplication operators) and those corresponding to the $p_i$ are "annihilation operators" (partial differentiation operators with respect to the basis $\{q_i\}$ of $\hat{\underline{h}}^{-}$). For $n < 0$ the $x(n)$ are linear combinations of the $q_i$, for $n > 0$ the $x(n)$ are linear combinations of the $p_i$, and for $n = 0$ the operators $x(n)$ are zero.

The graded dimension (sometimes called the character) of the graded module $S$ is the generating function of the dimensions of the homogeneous component subspaces of $S$:

$$\dim_* S = \sum_{n\geq 0} (\dim S_{-n}) q^n = \prod_{j\geq 1} (1-q^j)^{-\dim \underline{h}}.$$

In order to give $\dim_* S$ modular transformations properties, we shift the grading of $S$ by adding $(1/24)\dim \underline{h}$ to the degree of every homogeneous element of $S$, thereby changing the graded dimension to

$$\dim_* S = \eta(q)^{-\dim \underline{h}}, \tag{10}$$

where $\eta(q)$ is Dedekind's $\eta$-function,

$$\eta(q) = q^{1/24}\prod_{j\geq 1} (1-q^j),$$

a certain kind of modular form when the formal variable $q$ is replaced by $e^{2\pi i\tau}$, $\tau$ in the complex upper half plane. The degree of the constant polynomial $1 \in S$, for instance, has been shifted to $(1/24)\dim \underline{h}$ (essentially the "zero-point energy").

The abelian Lie subalgebra $\underline{h}$ of $\hat{\underline{h}}$ commutes with all of $\hat{\underline{h}}$, which is why only $\hat{\underline{h}}_*$ (see (6)) and not all of $\hat{\underline{h}}$ has a basis satisfying the Heisenberg commutation relations (7). Thus we enlarge $\hat{\underline{h}}$ by adjoining the dual $\underline{h}^*$ of $\underline{h}$, which we identify with $\underline{h}$ by means of the form $\langle\cdot,\cdot\rangle$. For $\alpha \in \underline{h}$ we write $\alpha^*$ for the corresponding element of $\underline{h}^*$. Then

$$\underline{\ell} = \hat{\underline{h}} \oplus \underline{h}^*$$

becomes a Heisenberg algebra when we define

$$[x,y^*] = -[y^*,x] = \langle x,y\rangle c \quad \text{for } x,y \in \underline{h}$$

$$[\underline{h}^*,\underline{h}^*] = [\underline{h}^*,\hat{\underline{h}}_*] = [c,\underline{\ell}] = 0.$$

Now the Heisenberg subalgebra $\underline{h} \oplus \underline{h}^* \oplus \mathbb{C}c$ of $\underline{\ell}$ acts on the space $F$ of analytic functions on $\underline{h}$ by the following prescription:

$c$ acts as the identity operator, $\alpha \in \underline{h}$ acts as the corresponding directional differentiation operator, which we denote by $\alpha(0)$, and $\alpha^* \in \underline{h}^*$ acts as the corresponding multiplication operator. We view the "zero-mode" operators $\alpha(0)$ as annihilation operators and the corresponding operators $\alpha^*$ as creation operators. The algebra $\underline{\ell}$ acts in a natural way on the space $S \otimes F$, with $c$ still acting as 1, the operators $\alpha(n)$ for $n \neq 0$ acting only on $S$, and the operators $\alpha(0)$ and $\alpha^*$ acting only on $F$. The space $\underline{h}$ corresponds to momentum space in bosonic string theory, and $\underline{h}^*$ to position space.

For $\alpha \in \underline{h}$, define

$$\alpha(z) = \sum_{n \in \mathbb{Z}} \alpha(n) z^{-n-1}, \tag{11}$$

a generating function of the operators $\alpha(n)$ (recall (9)). In string theory, $z$ is understood as a nonzero complex variable. Here we take an algebraic approach instead and understand $z$ to be a formal Laurent series variable.

We shall need the concept of normal ordering: When an expression is enclosed between open colons, all the creation operators are to be placed to the left of all the annihilation operators before the expression is to be interpreted as an operator. Normal ordering permits "infinite" expressions to act as meaningful operators.

Before defining the vertex operator rigorously, we introduce it in a suggestive unrigorous way: For $\alpha \in \underline{h}$, take the corresponding vertex operator, denoted $Y(\alpha,z)$, to be

$$Y(\alpha,z) = {}^{\circ}_{\circ} e^{\int \alpha(z)dz} {}^{\circ}_{\circ}. \tag{12}$$

Noting (still unrigorously) that

$$\int \alpha(z)dz = \sum_{n \neq 0} \frac{\alpha(n)}{-n} z^{-n} + \alpha(0)\log z + C,$$

with $C$ a constant of integration which we take to be $\alpha^*$, we find that formally,

$$Y(\alpha,z) = {}^{\circ}_{\circ}e^{-\Sigma_{n\neq 0}(\alpha(n)/n)z^{-n}}\,{}^{\circ}_{\circ}e^{\alpha^*}e^{\alpha(0)\log z}$$

$$= e^{-\Sigma_{n<0}(\alpha(n)/n)z^{-n}}\,e^{-\Sigma_{n>0}(\alpha(n)/n)z^{-n}}\,e^{\alpha^*}z^{\alpha(0)}, \qquad (13)$$

where the first two occurrences of exponentiation in (13) refer to the formal exponential series, and the third occurrence refers to the operation of multiplication by the function $e^{\alpha^*}$, which we shall designate simply as $e^{\alpha}$, on F. The symbol $z^{\alpha(0)}$, which we write briefly as $z^{\alpha}$, denotes an operator which, when acting on an eigenvector for $\alpha(0)$ in F, multiplies it by $z^{x}$, x being the eigenvalue. Examples of eigenvectors are the functions $e^{\beta}$, $\beta \in \underline{h}$. Thus

$$z^{\alpha}\cdot e^{\beta} = z^{\langle\alpha,\beta\rangle}e^{\beta}. \qquad (14)$$

Returning to a rigorous viewpoint, we denote by $\mathbb{C}[\underline{h}]$ the group algebra of the abelian group $\underline{h}$. Then $\mathbb{C}[\underline{h}]$ has a basis consisting of elements $e^{\alpha}$ for $\alpha \in \underline{h}$ and with multiplication given by

$$e^{\alpha}e^{\beta} = e^{\alpha+\beta} \quad \text{for} \quad \alpha,\beta \in \underline{h}. \qquad (15)$$

Identifying $e^{\alpha}$ with the analytic function $e^{\alpha}$ on $\underline{h}$ (and also with the corresponding multiplication operator), we identify $\mathbb{C}[\underline{h}]$ with a subalgebra of the commutative algebra F. Then the expression (13) acts rigorously on the space $S \otimes \mathbb{C}[\underline{h}]$, and for $\alpha \in \underline{h}$ we define the (<u>untwisted</u>) <u>vertex</u> <u>operator</u> $Y(\alpha,z)$ acting on $S \otimes \mathbb{C}[\underline{h}]$ to be

$$Y(\alpha,z) = e^{-\Sigma_{n<0}(\alpha(n)/n)z^{-n}}\,e^{-\Sigma_{n>0}(\alpha(n)/n)z^{-n}}\,e^{\alpha}z^{\alpha}, \qquad (16)$$

with $z^{\alpha}$ defined as in (14). Then $Y(\alpha,z)$ maps $S \otimes \mathbb{C}[\underline{h}]$ to the larger vector space consisting of all generalized formal Laurent series in the formal variable z (allowing infinite linear combinations of arbitrary <u>complex</u>, not just integral, powers of z) with coefficients in $S \otimes \mathbb{C}[\underline{h}]$. Note that exponential notation has been used in several "compatible" ways.

We may expand (16) in powers of z to define operators $x_{\alpha}(n)$

(for $n \in \mathbf{C}$) taking the space $S \otimes \mathbf{C}[\underline{h}]$ to itself:

$$Y(\alpha,z) = \sum_{n\in\mathbf{C}} x_\alpha(n) z^{-n-\langle\bar{a},\bar{a}\rangle/2}, \tag{17}$$

the shift $-\langle\bar{a},\bar{a}\rangle/2$ being analogous to the shift $-1$ in the exponent in (11); we shall motivate it later. The operators $x_\alpha(n)$ would be very cumbersome to define and compute with directly, without the generating function (16).

## 2.2 The Operators $Y_1(a,z)$

Here we give an exposition of the main result of [FK,S].

In order to perform subtle operations on vertex operators, especially, the computation of commutators, we must refine the notion. Let $L$ be a lattice -- the $\mathbf{Z}$-span of a basis of a real finite-dimensional vector space equipped with a $\mathbf{Z}$-valued symmetric nondegenerate bilinear form $\langle\cdot,\cdot\rangle$. We do not assume the form to be positive definite. Suppose that $L$ is even, that is, that $\langle\alpha,\alpha\rangle \in 2\mathbf{Z}$ for all $\alpha \in L$. An even lattice is integral, that is, $\langle\alpha,\beta\rangle \in \mathbf{Z}$ for all $\alpha,\beta \in L$. Examples of positive definite even lattices are the root lattices of the simple Lie algebras of types $A_n$, $D_n$, $E_n$ (always normalized so that the roots constitute the set of lattice elements $\alpha$ such that $\langle\alpha,\alpha\rangle = 2$) and the Leech lattice, which is involved in the construction of the Monster.

Form the complex vector space $\underline{h} = L \otimes_{\mathbf{Z}} \mathbf{C}$ (the complexification of $L$). Then all the considerations of Section 2.1 apply to this choice of $\underline{h}$. But we shall need to replace the space $\mathbf{C}[\underline{h}]$ by something more delicate.

There is an essentially unique (nonabelian) group $\hat{L}$ containing as a central subgroup the two-element multiplicatively written group $\langle\pm1\rangle$, such that $\hat{L}/\langle\pm1\rangle \simeq L$ and

$$aba^{-1}b^{-1} = (-1)^{\langle\bar{a},\bar{b}\rangle} \quad \text{for } a,b \in \hat{L}, \tag{18}$$

where $a \mapsto \bar{a}$ denotes the canonical "double covering map" from $\hat{L}$ to $L$. This central extension $\hat{L}$ of $L$ may be constructed by first constructing a $\mathbb{Z}$-bilinear map

$$\epsilon_0 : L \times L \to \mathbb{Z}/2\mathbb{Z}$$

such that

$$\epsilon_0(\alpha,\beta) - \epsilon_0(\beta,\alpha) = \langle\alpha,\beta\rangle \bmod 2\mathbb{Z} \quad \text{for} \quad \alpha,\beta \in L$$

and then taking $\hat{L}$ to be the set $\{\pm e_\alpha | \alpha \in L\}$ with multiplication determined by the formula

$$e_\alpha e_\beta = (-1)^{\epsilon_0(\alpha,\beta)} e_{\alpha+\beta} \quad \text{for} \quad \alpha,\beta \in L.$$

It is sometimes (but not always!) convenient to use this concrete realization of $\hat{L}$. The map $\epsilon_0$ is a 2-cocycle.

Now we make the group algebra $\mathbb{C}[L]$ an $\hat{L}$-module by defining

$$e_\alpha \cdot e^\beta = \epsilon_0(\alpha,\beta) e^{\alpha+\beta} \quad \text{for} \quad \alpha,\beta \in L,$$

where we use exponential notation as in (15) for elements of $\mathbb{C}[L]$. We shall use the notation $\mathbb{C}\{L\}$ for the space $\mathbb{C}[L]$ when it is given this $\hat{L}$-module structure. Invariantly (without using the cocycle $\epsilon_0$), $\mathbb{C}\{L\}$ is the induced $\hat{L}$-module

$$\mathbb{C}\{L\} = \mathbb{C}[\hat{L}] \otimes_{\mathbb{C}[\langle\pm 1\rangle]} \mathbb{C} \tag{19}$$

(brackets as usual denoting group algebras), where $\mathbb{C}$ is viewed as the natural module for the group $\langle\pm 1\rangle$. Here $\hat{L}$ acts by left multiplication on the first tensor factor $\mathbb{C}[\hat{L}]$. For $a \in \hat{L}$, we shall write $\iota(a)$ for the image $a\otimes 1$ of $a$ in $\mathbb{C}\{L\}$.

Now set

$$V_L = S \otimes \mathbb{C}\{L\}. \tag{20}$$

We shall define vertex operators which act on the space $V_L$ instead of the space $S \otimes \mathbb{C}[\underline{h}]$ considered above. Interpreting $\underline{h}$ as momentum space and $\underline{h}^*$ as position space as mentioned above, we see that we have constrained momenta to lie on a lattice and thus we have effectively compactified position space to a torus. But we have also introduced the $\pm$ signs associated with $\hat{L}$. (These correspond to

"Klein transformations" in physics.)

As above, we make $\hat{\underline{h}}_*$ act on $V_L$ by acting on the first factor $S$ alone, we make $\hat{L}$ act on the second factor $\mathbb{C}\{L\}$ alone, and we make $\underline{h}$ act on the second factor alone by

$$\alpha(0)\cdot\iota(a) = \langle\alpha,\bar{a}\rangle\iota(a) \quad \text{for} \quad \alpha\in\underline{h}, \quad a\in\hat{L}. \tag{21}$$

We also define an action of $z^{\alpha}$ on the second factor by

$$z^{\alpha}\cdot\iota(a) = z^{\langle\alpha,\bar{a}\rangle}\iota(a). \tag{22}$$

We give $\mathbb{C}\{L\}$ the (well-defined) $\mathbb{Z}$-gradation determined by

$$\deg\iota(a) = -\frac{1}{2}\langle\bar{a},\bar{a}\rangle \quad \text{for} \quad a\in\hat{L} \tag{23}$$

and we give $V_L$ the tensor product gradation. Then the graded dimension of $V_L$ becomes the modular function

$$\dim_* V_L = \frac{\theta_L(q)}{\eta(q)^{\dim\underline{h}}}, \tag{24}$$

where $\theta_L(q)$, the $\theta$-function of $L$, is defined by

$$\theta_L(q) = \sum_{\alpha\in L} q^{\langle\bar{a},\bar{a}\rangle/2}$$

(see (10)).

For $a\in\hat{L}$ we define the (untwisted) vertex operator $Y(a,z)$ (also denoted $Y_1(a,z)$, or, as in [FLM1-5], $Y_{\mathbb{Z}}(a,z)$, to distinguish it from the twisted operators defined below) by:

$$Y(a,z) = Y_1(a,z) = {}^{\circ}_{\circ}e^{-\Sigma_{n\neq0}(\alpha(n)/n)z^{-n}}az^{\bar{a}}{}^{\circ}_{\circ} \tag{25}$$

(cf. (16)), a formal Laurent series in $z$ with operator coefficients, involving only integral powers of $z$ by the assumed integrality of the lattice $L$. As in (17), we define operators $x_a(n)$ on $V_L$ for $n\in\mathbb{Z}$ by the expansion

$$Y(a,z) = \sum_{n\in\mathbb{Z}} x_a(n)z^{-n-\langle\bar{a},\bar{a}\rangle/2}. \tag{26}$$

(n ranging through $\mathbb{Z}$ because L is even). Then one finds that

$$\deg x_a(n) = n, \tag{27}$$

that is, the operator $x_a(n)$ shifts the degrees of homogeneous elements of $V_L$ by adding n. One of several ways of motivating both the shift $-\langle\bar{a},\bar{a}\rangle/2$ in the exponent in (26) and the definition of the gradation of $\mathbb{C}\{L\}$ (see (23)) is that these lead to the natural formula (27) while giving modular transformation properties to the graded dimension of $V_L$ (see (24)).

<u>Theorem 1</u> [FK, S]. Let L be the root lattice of a Lie algebra $\underline{g}$ which is a product of simple algebras of types $A_n$, $D_n$ and $E_n$. Then the operators $x_a(n)$ for $\langle\bar{a},\bar{a}\rangle = 2$ and $n \in \mathbb{Z}$ generate a copy of the affine Kac-Moody algebra $\hat{\underline{g}}$, and the $\hat{\underline{g}}$-module $V_L$ is irreducible.

See also [BHN, Hal]. For $\underline{g} = E_8 \times E_8$, this construction appears in the heterotic string [GHMR1,2].

2.3 Calculation of Commutators

We shall sketch a proof of Theorem 1 using an algebraic "formal calculus", motivated by operator product expansion theory, complex analysis and distribution theory (cf. [Fr2, FK, GO1, S]). The proof here is essentially the algebraic one we learned from H. Garland. Later we shall sketch an extension of this method.

It would be virtually impossible to compute $[x_a(m), x_b(n)]$ directly. Naturally, we use the generating functions and instead compute $[Y(a,z_1), Y(b,z_2)]$, reading off the coefficients of monomials in $z_1$ and $z_2$ afterward. Here we take $z_1$ and $z_2$ to be distinct commuting formal Laurent series variables rather than the nonzero complex variables customary in string theory.

For arbitrary $a, b \in \hat{L}$,

$$Y(a,z_1)Y(b,z_2) = {}^{\circ}_{\circ}Y(a,z_1)Y(b,z_2){}^{\circ}_{\circ}(z_1-z_2)^{\langle\bar{a},\bar{b}\rangle}, \tag{28}$$

where the normal ordered product is defined in an obvious way, and $(z_1-z_2)^{\langle\bar{a},\bar{b}\rangle}$ is understood as a formal Laurent series in $z_1$ and $z_2$ which is the formal binomial expansion in *nonnegative integral powers of* $z_2$. To justify (28), we apply the formal rule

$$e^A e^B = e^B e^A e^{[A,B]} \quad \text{if } [A,B] \text{ commutes with } A \text{ and } B$$

to

$$A = -\sum_{n>0} (\bar{a}(n)/n) z_1^{-n}, \quad B = -\sum_{n<0} (\bar{b}(n)/n) z_2^{-n},$$

keeping in mind the commutation relations (2) and the fact that $\exp(\log X) = X$, provided the formal series are well-defined. We also use the formula

$$z_1^{\bar{a}} b = b z_1^{\bar{a}+\langle\bar{a},\bar{b}\rangle},$$

which may be checked by applying both sides to suitable elements of $V_L$.

Similarly,

$$Y(b,z_2)Y(a,z_1) = {}^{\circ}_{\circ}Y(b,z_2)Y(a,z_1){}^{\circ}_{\circ}(z_2-z_1)^{\langle\bar{a},\bar{b}\rangle} \tag{29}$$

where $(z_2-z_1)^{\langle\bar{a},\bar{b}\rangle}$ is to be expanded in *nonnegative integral powers of* $z_1$, *not* $z_2$ (cf. (28)). But

$${}^{\circ}_{\circ}Y(b,z_2)Y(a,z_1){}^{\circ}_{\circ} = (-1)^{\langle\bar{a},\bar{b}\rangle}\,{}^{\circ}_{\circ}Y(a,z_1)Y(b,z_2){}^{\circ}_{\circ} \tag{30}$$

from (18), and this formula motivates the introduction of the central extension $\hat{L}$. In fact, (28), (29) and (30) lead to

$$[Y(a,z_1),Y(b,z_2)] = {}^{\circ}_{\circ}Y(a,z_1)Y(b,z_2){}^{\circ}_{\circ}D \tag{31}$$

where $D$ may be written in the strange-looking way

$$D = (z_1-z_2)^{\langle\bar{a},\bar{b}\rangle} - (z_1-z_2)^{\langle\bar{a},\bar{b}\rangle}. \tag{32}$$

Of course, $D$ is not zero but rather a formal Laurent series in $z_1$

and $z_2$ which is the difference of a formal power series in $z_2$ and a formal power series in $z_1$. We might call D an "expansion of zero". We may think of D as the expansion of a distribution supported at one point (cf. [S]; see below). If we want to work instead with nonzero complex variables $z_1$ and $z_2$, then (28) makes sense only if $|z_1| > |z_2|$ and (29) makes sense only if $|z_1| < |z_2|$. This incompatibility requires the use of suitable contour integration, as in [Fr2, FK, GO1].

Let us examine a few special cases. If $\langle\bar{a},\bar{b}\rangle \geq 0$, then $(z_1-z_2)^{\langle\bar{a},\bar{b}\rangle}$ is a polynomial in $z_1$ and $z_2$ and so $D = 0$ and

$$[Y(a,z_1),Y(b,z_2)] = 0. \tag{33}$$

Suppose $\langle\bar{a},\bar{b}\rangle = -1$. Then

$$\begin{aligned} D &= z_2^{-1}\left[(1-z_1/z_2)^{-1}+(z_2/z_1)(1-z_2/z_1)^{-1}\right] \\ &= z_2^{-1}\delta(z_1/z_2), \end{aligned} \tag{34}$$

where we use the indicated binomial expansions, and

$$\delta(z) = \sum_{n\in\mathbb{Z}} z^n. \tag{35}$$

Formally, this is the Fourier expansion of the $\delta$-function at $z = 1$, and this observation motivates the fundamental principle

$$f(z)\delta(z) = f(1)\delta(z) \tag{36}$$

for a Laurent <u>polynomial</u> $f(z)$ (the expressions being undefined if $f(z)$ were an infinite series). Actually, (36) is trivial: First observe its truth for $f(z) = z^n$ and then use linearity. Similarly,

$$f(z_1,z_2)\delta(z_1/z_2) = f(z_2,z_2)\delta(z_1/z_2) \tag{37}$$

for any formal Laurent series $f(z_1,z_2)$ such that

$$\lim_{z_1\to z_2} f(z_1,z_2) \text{ exists.}$$

Here the limit is understood in an algebraic sense: Only finite sums are to occur in each coefficient when the variables are set equal; ordinary convergence of infinite sums is not allowed. If $f(z_1, z_2)$ is operator-valued, then it is to be applied to a typical vector before the algebraic limit is taken. Hence we find that

$$\begin{aligned}[Y(a,z_1),Y(b,z_2)] &= z_2^{-1}\,{}^{\circ}_{\circ}Y(a,z_1)Y(b,z_2){}^{\circ}_{\circ}\,\delta(z_1/z_2) \\ &= z_2^{-1}Y(ab,z_2)\delta(z_1/z_2);\end{aligned} \tag{38}$$

note that

$$\lim_{z_1 \mapsto z_2} {}^{\circ}_{\circ}Y(a,z_1)Y(b,z_2){}^{\circ}_{\circ} \quad \text{exists}$$

because the expression is normal ordered. It is easy to extract components from (38): For $m,n \in \mathbb{Z}$,

$$[x_a(m),x_b(n)] = x_{ab}(m+n).$$

For $\langle \bar{a},\bar{a}\rangle = \langle \bar{b},\bar{b}\rangle = 2$, this constitutes a part of the bracket relations for $\hat{\mathfrak{g}}$ in Theorem 1 (cf. (5)).

We have computed the commutator of the two vertex operators when $\langle \bar{a},\bar{b}\rangle \geq -1$. The remaining case needed for Theorem 1 is the case $\langle \bar{a},\bar{b}\rangle = -2$. One can proceed as above, obtaining the derivative of $\delta(z)$ in place of $\delta(z)$. The analogue of (36) involves the derivative of $f(z)$, and $[Y(a,z_1),Y(b,z_2)]$ is easily found (cf. [Fr2, FK, GO1, S]).

In the construction of the Monster in [FLM1-5], one needs the commutator for $\langle \bar{a},\bar{b}\rangle$ as low as $-4$ (since the shortest nonzero vectors $\alpha$ in the Leech lattice have $\langle \alpha,\alpha\rangle = 4$), and the computation starts to become long and the result starts to look complicated, third derivatives being required. There is a very general method, however, which handles all cases of $\langle \bar{a},\bar{b}\rangle$ (and all derivatives) at once. This method, presented in [FLM5], is also related to operator product expansion ideas. We sketch this next. We also comment that [FLM5] develops the method of formal calculus in systematic detail.

## 2.4 General Commutators of Untwisted Vertex Operators

To compute $[Y(a,z_1),Y(b,z_2)]$ for <u>any</u> (integral) value of $\langle\bar{a},\bar{b}\rangle$, we begin with (31), (32). As in (34), we have

$$D = -(-z_2)^{\langle\bar{a},\bar{b}\rangle}\left[(1-z_1/z_2)^{\langle\bar{a},\bar{b}\rangle}-(-1)^{\langle\bar{a},\bar{b}\rangle}(z_1/z_2)^{\langle\bar{a},\bar{b}\rangle}(1-z_2/z_1)^{\langle\bar{a},\bar{b}\rangle}\right],$$

where the first term in the parentheses on the right comes from the <u>second</u> term in (32) and vice versa. But then using (34), (35) we find that $D$ is the coefficient of $z_0^{-\langle\bar{a},\bar{b}\rangle-1}$ in

$$z_2^{-1}e^{-z_0\frac{\partial}{\partial z_1}}\delta(z_1/z_2)$$

($z_0$ being a new variable and the exponential indicating the exponential series), since

$$\frac{1}{n!}\left(\frac{d}{dz}\right)^n\frac{1}{1-z} = \frac{1}{(1-z)^{n+1}} \quad \text{for all} \quad n \geq 0.$$

Now $e^{-z_0(\partial/\partial z_1)}$ acts as a "one-parameter group of translations", by "Taylor's theorem": For any formal Laurent series $f(z_1,z_2)$,

$$e^{-z_0(\partial/\partial z_1)}f(z_1,z_2) = f(z_1-z_0,z_2),$$

where the right-hand side is to be expanded in nonnegative integral powers of $z_0$. Hence $[Y(a,z_1),Y(b,z_2)]$ is the coefficient of $z_0^{-\langle\bar{a},\bar{b}\rangle-1}$ in

$$z_2^{-1}\,{\circ\atop\circ}Y(a,z_1)Y(b,z_2){\circ\atop\circ}\,e^{-z_0(\partial/\partial z_1)}\delta(z_1/z_2) =$$

$$= z_2^{-1}\,{\circ\atop\circ}Y(a,z_1)Y(b,z_2){\circ\atop\circ}\,\delta((z_1-z_0)/z_2)$$

$$= z_2^{-1}\,{\circ\atop\circ}Y(a,z_2+z_0)Y(b,z_2){\circ\atop\circ}\,\delta((z_1-z_0)/z_2) \qquad (39)$$

$$= z_2^{-1} {}^\circ_\circ Y(a,z_2+z_0)Y(b,z_2){}^\circ_\circ e^{-z_0(\partial/\partial z_1)}\delta(z_1/z_2). \quad (40)$$

The calculation of $[Y(a,z_1),Y(b,z_2)]$ is essentially concentrated in the one step (39), which follows formally from the basic principle (37). This expression for the commutator is a reformulation of the result announced at the end of [L], for the special case of untwisted vertex operators, and it is also a special case of a result of [B] (see below).

On the other hand, ${}^\circ_\circ Y(a,z_2+z_0)Y(b,z_2){}^\circ_\circ$ can be expanded by a rigorous refinement of the following steps, motivated by (12):

$$\begin{aligned}
{}^\circ_\circ Y(a,z_2+z_0)Y(b,z_2){}^\circ_\circ &= {}^\circ_\circ e^{\int\bar{a}(z_2+z_0)+\int\bar{b}(z_2)}{}^\circ_\circ \\
&= {}^\circ_\circ e^{e^{z_0(d/dz_2)}\int\bar{a}(z_2)+\int\bar{b}(z_2)}{}^\circ_\circ \\
&= {}^\circ_\circ e^{\Sigma_{n\geq 1}(1/n!)(d/dz_2)^{n-1}\bar{a}(z_2)z_0^n+\int\bar{a}(z_2)+\int\bar{b}(z_2)}{}^\circ_\circ \\
&= {}^\circ_\circ e^{\Sigma_{n\geq 1}(1/n!)(d/dz_2)^{n-1}\bar{a}(z_2)z_0^n}Y(ab,z_2){}^\circ_\circ. \quad (41)
\end{aligned}$$

The final step (41) is in fact rigorously true. Observe that (40) and (41) give an expression for the commutator of the two vertex operators generalizing the simple cases (33), (38).

An interesting philosophical point arises here: A vertex operator is a generating function of component operators such as $x_a(n)$ which would be virtually impossible to work with directly (although it was via such components that the twisted vertex operator was built up and discovered in [LW1]!). On the other hand, as we mentioned in the last section, the direct computation of the bracket $[Y(a,z_1),Y(b,z_2)]$ becomes increasingly complicated as $\langle\bar{a},\bar{b}\rangle$ becomes a larger and larger negative integer. But these bracket computations become easy, as we have just explained, when we do them all at once by what is essentially a "generating function" method. The use of this method should be thought of as analogous to the use of vertex operators themselves to study the operators $x_a(n)$. But there is more than just analogy here. The exponential factor in the normal ordered expression (41) is reminiscent of the definition of vertex operator. This suggests the definition (43) below of generalized vertex operator and thus the elegant reformulation of the commutator in the form (46), which generalizes dramatically to (48).

## 2.5 Generalized Vertex Operators and their Commutators

We next give an exposition of a result announced in [B]. First we define <u>generalized</u> (<u>untwisted</u>) <u>vertex</u> <u>operators</u> $Y(v,z)$ $(v \in V_L)$, which are parametrized by the space $V_L$ in such a way that $Y(a,z)$ corresponds to $\iota(a) = a\otimes 1 \in V_L$ for $a \in \hat{L}$: Let

$$a \in \hat{L}, \quad \alpha_1,\cdots,\alpha_k \in \underline{h}, \quad n_1,\cdots,n_k > 0,$$

set

$$v = \alpha_1(-n_1)\cdots\alpha_k(-n_k) \otimes \iota(a) \in V_L, \tag{42}$$

define

$$Y(v,z) = Y_1(v,z) =$$

$$= {}^{\circ}_{\circ}\left[\frac{1}{(n_1-1)!}\left[\frac{d}{dz}\right]^{n_1-1}\alpha_1(z)\right]\cdots\left[\frac{1}{(n_k-1)!}\left[\frac{d}{dz}\right]^{n_k-1}\alpha_k(z)\right]Y(a,z)\,{}^{\circ}_{\circ}, \tag{43}$$

and extend to all $v \in V_L$ by linearity. Even more generally, we shall sometimes allow $v$ to be a formal Laurent series with coefficients in $V_L$; for instance, it is clear what we mean by $Y(Y(a,z_0)\iota(b),z)$ for $a,b \in \hat{L}$. Also define component operators $x_v(n)$ for $v$ as in (42) and $n \in \mathbb{Z}$ by

$$Y(v,z) = \sum_{n\in\mathbb{Z}} x_v(n)z^{-n+\deg v-(\dim \underline{h})/24}, \tag{44}$$

and again use linearity to define $x_v(n)$ for all $v \in V_L$. Then

$x_{\iota(a)}(n) = x_a(n)$ for $a \in \hat{L}$ (see (26)) and

$$\deg x_v(n) = n \quad \text{for} \quad v \in V_L \quad \text{and} \quad n \in \mathbb{Z}, \tag{45}$$

generalizing (27). Note that the term $-(\dim \underline{h})/24$ in (44) comes from the gradation shift which makes the graded dimension of $V_L$ a modular function (recall (10), (24)). Generalized vertex operators are also familiar in string theory (see e.g. [ADDF]; cf. [DNPS, SY, We]).

It is not hard to check that the result in the last section can be expressed as the formula

$$\begin{aligned}&[Y(\iota(a),z_1),Y(\iota(b),z_2)] = \\ &= \operatorname{Res}_{z_0} z_2^{-1} Y(Y(\iota(a),z_0)\iota(b),z_2) e^{-z_0(\partial/\partial z_1)} \delta(z_1/z_2)\end{aligned} \tag{46}$$

for $a,b \in \hat{L}$, where $\operatorname{Res}_{z_0}$ refers to the residue -- the coefficient of $z_0^{-1}$. Or we can extract the coefficients of powers of the variables and express the result in the form

$$\begin{aligned}&[x_{\iota(a)}(m),x_{\iota(b)}(n)] = \\ &= \sum_{i\in\mathbb{Z}} \begin{bmatrix} m-1+(\dim \underline{h})/24-\deg \iota(a) \\ i-1+(\dim \underline{h})/24-\deg \iota(a) \end{bmatrix} x_{x_{\iota(a)}(i)\cdot\iota(b)}(m+n)\end{aligned} \tag{47}$$

for $a,b \in \hat{L}$ and $m,n \in \mathbb{Z}$; the sum on the right-hand side is in fact finite.

It is a remarkable fact that (46) and (47) remain valid when $\iota(a)$ and $\iota(b)$ are replaced by arbitrary elements of $V_L$, so that in particular, the operators $x_v(n)$ span a Lie algebra with precisely determined structure:

Theorem 2 [B]. We have

$$[Y(u,z_1),Y(v,z_2)] = \tag{48}$$
$$= \mathrm{Res}_{z_0} z_2^{-1} Y(Y(u,z_0)v,z_2) e^{-z_0(\partial/\partial z_1)} \delta(z_1/z_2)$$

for $u,v \in V_L$. If $u$ is homogeneous,

$$[x_u(m),x_v(n)] = \sum_{i\in\mathbb{Z}} \binom{m-1+(\dim \underline{h})/24-\deg u}{i-1+(\dim \underline{h})/24-\deg u} x_{x_u(i)\cdot v}(m+n) \tag{49}$$

for $m,n \in \mathbb{Z}$ (finite sum).

The announcement [B] actually contains a variant of (49), and does not include a formulation of the result in terms of generating functions, as in (48). This theorem may be proved by an extension of the method of Sections 2.3 and 2.4.

## 3. THEORY OF TWISTED VERTEX OPERATORS

### 3.1 The Operators $Y_\nu(a,z)$

In this section, we systematically construct a setting which generalizes that of Sections 2.1 and 2.2 to incorporate an arbitrary isometry $\nu$ of our lattice $L$, in such a way that the earlier setting corresponds to the identity automorphism $\nu = 1$. The reader should observe how the various steps here generalize those above. The important and suggestive case $\nu = -1$ is treated in [FLM1-5], and the reader should also notice how the theory looks in this case. The general case is treated in [L1], which we follow here; see also [KP3] for an independent approach which does not generalize in the directions we are interested in (see below). Although some of the constructions in this section may seem unintuitive, they are in fact quite canonical. One may think of Theorem 3 below and the results of Sections 3.2 and 3.3 as motivation.

First we define the notion of twisted affine Lie algebra (cf. [K, KKLW]). With the space $\underline{h}$ and its symmetric nondegenerate bilinear form $\langle\cdot,\cdot\rangle$ as in Section 2.1, let

$$\nu:\underline{h} \to \underline{h}$$

be a linear automorphism which is also an isometry, that is,

$$\langle\nu\alpha,\nu\beta\rangle = \langle\alpha,\beta\rangle \quad \text{for} \quad \alpha,\beta \in \underline{h},$$

and assume that $\nu$ has finite order:

$$\nu^M = 1 \quad \text{for some} \quad M > 0. \tag{50}$$

It will be convenient to allow $M$ to be larger than the exact order of $\nu$; for instance, if $\nu$ has order 2 we are free to choose $M$ to be any positive even number. For $n \in \mathbf{Z}$ set

$$\underline{h}_{(n)} = \{x \in \underline{h} \mid \nu x = \omega^n x\}, \quad \text{where} \quad \omega = e^{2\pi i/M} \tag{51}$$

and define the $\nu$-twisted affine Lie algebra associated with $\underline{h}$ (regarded as an abelian Lie algebra) and $\langle\cdot,\cdot\rangle$ to be

$$\hat{\underline{h}}[\nu] = \bigoplus_{n\in\mathbf{Z}} \underline{h}_{(n)} \otimes t^{n/M} \oplus \mathbf{C}c$$

with

$$[x\otimes t^m, y\otimes t^n] = \langle x,y\rangle m\delta_{m+n,0}c \quad \text{for} \quad m,n \in \tfrac{1}{M}\mathbf{Z}, \quad x \in \underline{h}_{(mM)}, \quad y \in \underline{h}_{(nM)}$$

$$[c,\hat{\underline{h}}[\nu]] = 0. \tag{52}$$

This algebra is $(1/M)\mathbf{Z}$-graded, with $x\otimes t^n$ of degree $n$ and $c$ of degree 0. Note that the untwisted algebra $\hat{\underline{h}}$ of (2) is the same as the twisted algebra $\hat{\underline{h}}[\nu]$ for $\nu$ the identity automorphism.

More generally, if $\underline{g}$ is a Lie algebra with a symmetric bilinear form $(\cdot,\cdot)$ which is invariant as in (3) and if $\nu$ is a Lie algebra automorphism and an isometry satisfying (50), we define the corresponding $\nu$-twisted affine Lie algebra to be the $(1/M)\mathbf{Z}$-graded algebra

$$\hat{\underline{g}}[\nu] = \bigoplus_{n\in\mathbf{Z}} \underline{g}_{(n)} \otimes t^{n/M} \oplus \mathbf{C}c, \tag{53}$$

with $\underline{g}_{(n)}$ defined as in (51) and with brackets given by

$$[x\otimes t^m, y\otimes t^n] = [x,y]\otimes t^{m+n} + \langle x,y\rangle m\delta_{m+n,0}c,$$
$$[c,\hat{\underline{g}}[\nu]] = 0 \tag{54}$$

for $m,n \in \frac{1}{M}\mathbb{Z}$, $x \in \underline{g}_{(mM)}$, $y \in \underline{g}_{(nM)}$. If $\underline{g}$ is a finite-dimensional simple Lie algebra of type $X_n$, then $\hat{\underline{g}}$ is a twisted affine Kac-Moody algebra of type $\hat{X}_n = X_n^{(1)}$ when $\nu$ is an inner automorphism and of type $X_n^{(2)}$ or $X_n^{(3)}$ when $\nu$ is an outer automorphism (in which case X must be A, D or E).

Now the subalgebra

$$\hat{\underline{h}}[\nu]_* = \bigoplus_{n\neq 0} \underline{h}_{(n)} \otimes t^{n/M} \oplus \mathbb{C}c$$

is a Heisenberg algebra (cf. (6)). Set

$$\hat{\underline{h}}[\nu]^{\pm} = \bigoplus_{\pm n>0} \underline{h}_{(n)} \otimes t^{n/M},$$

$$\underline{b} = \hat{\underline{h}}[\nu]^{+} \oplus \underline{h}_{(0)} \;(\otimes t^0) \oplus \mathbb{C}c$$

and view $\mathbb{C}$ as the 1-dimensional $\underline{b}$-module such that

$$c\cdot 1 = 1 \quad\text{and}\quad (\hat{\underline{h}}[\nu]^{+} \oplus \underline{h}_{(0)})\cdot 1 = 0.$$

As in (8), form the $(1/M)\mathbb{Z}$-graded induced $\hat{\underline{h}}$-module

$$S[\nu] = U(\hat{\underline{h}}[\nu]) \otimes_{U(\underline{b})} \mathbb{C}, \tag{55}$$

which is isomorphic to the symmetric algebra $S(\hat{\underline{h}}[\nu]^{-})$ and whose grading is bounded above:

$$S[\nu] = \bigoplus_{n\leq 0} S[\nu]_n. \tag{56}$$

Also, $S[\nu]$ is irreducible under $\hat{\underline{h}}[\nu]$. We shall comment later on the appropriate grading shift (cf. (10)) for this "twisted Fock space"; for now, we are not concerned with the degree of the constant polynomial $1 \in S[\nu]$. For $x \in \underline{h}$ and $n \in (1/M)\mathbb{Z}$, we denote by $x_{(nM)}$ the projection of $x$ to $\underline{h}_{(nM)}$ with respect to the $\omega$-eigenspace decomposition of $\underline{h}$ (recall (51)) and we write $x(n)$

for the operator of degree $n$ on $S$ associated with $x_{(nM)}\otimes t^n \in \hat{\underline{h}}[\nu]$:

$$x_{(nM)}\otimes t^n \mapsto x(n).$$

As in (11), set

$$\alpha(z) = \sum_{n\in(1/M)\mathbf{Z}} \alpha(n)z^{-n-1},$$

where we now allow fractional powers of $z$. Then

$$\int(\alpha(z)-\alpha(0)z^{-1})dz = \sum_{n\neq 0}\frac{\alpha(n)}{-n}z^{-n}. \tag{57}$$

Now we start with an even lattice $L$ (still not assumed to have any particular signature) and an arbitrary isometry of $L$ (necessarily of finite order). We take our space $\underline{h}$ to be $L\otimes_{\mathbf{Z}}\mathbf{C}$ as in Section 2.2, and our isometry $\nu$ to be the extension of the lattice automorphism to $\underline{h}$. So far, the positive integer $M$ has been chosen to satisfy only the condition (50). For our deeper considerations we now also assume the following:

$$\text{If } M \text{ is even, then } \langle\nu^{M/2}\alpha,\alpha\rangle \in 2\mathbf{Z} \text{ for all } \alpha\in L. \tag{58}$$

In fact, this condition can always be arranged by doubling $M$ if necessary; recall that $M$ need not be the exact order of $\nu$. Set

$$\omega_0 = (-1)^M\omega. \tag{59}$$

Now using (58) it can be shown that that there is an essentially unique (nonabelian) group $\hat{L}_\nu$ containing as a central subgroup the multiplicative group $\langle\omega_0\rangle$ of roots of unity generated by $\omega_0$ (this group having order $M$ or $2M$ according as $M$ is even or odd), such that $\hat{L}_\nu/\langle\omega_0\rangle \simeq L$ and

$$aba^{-1}b^{-1} = \prod_{p=1}^{M}(-\omega^p)^{\langle\nu^p\bar{a},\bar{b}\rangle}, \tag{60}$$

where as in (18), $a \mapsto \bar{a}$ denotes the canonical map from $\hat{L}_\nu$ to $L$. It can also be shown that $\nu$ can be "lifted" to $\hat{L}_\nu$ in the following sense: There is an automorphism $\hat{\nu}$ of $\hat{L}_\nu$ such that

$$(\hat{\nu}a)^- = \nu\bar{a} \quad \text{for all} \quad a \in \hat{L}_\nu \tag{61}$$

and such that

$$\hat{\nu}a = a \quad \text{whenever} \quad \nu\bar{a} = \bar{a}. \tag{62}$$

Then it turns out that

$$\hat{\nu}^M = 1$$

(a nontrivial fact!).

Denote by $N$ the orthogonal complement of $\underline{h}_{(0)}$ in $L$:

$$N = \{\alpha \in L \mid \langle \alpha, \underline{h}_{(0)} \rangle = 0\}.$$

By $\hat{N}$ we shall mean the pullback of $N$ to $\hat{L}_\nu$: the subgroup of $\hat{L}_\nu$ consisting of those $a \in \hat{L}_\nu$ such that $\bar{a} \in N$. Then there exist irreducible (necessarily finite-dimensional) $\hat{N}$-modules $T$ on which $\omega_0 \in \hat{N}$ acts as multiplication by $\omega_0$ on $T$ and on which for every $a \in \hat{L}_\nu$, the element $a\hat{\nu}a^{-1}$ of $\hat{L}_\nu$, which always lies in $\hat{N}$, acts as the following scalar multiplication on $T$:

$$a\hat{\nu}a^{-1} \mapsto \omega^{-M\langle \bar{a}_{(0)}, \bar{a}_{(0)} \rangle / 2}$$

(the exponent on the right being an integer because of the assumption (58)). In fact, in [L] there is a classification and explicit construction of <u>all</u> such modules $T$, which turn out to have the same (explicitly computed) dimension and to be finite-group-theoretic analogues of the "Fock spaces" $S$ and $S[\nu]$ (recall (8), (55)). Modules like $T$ are studied in [FLM5].

Consider the induced $\hat{L}_\nu$-module

$$U = \mathbb{C}[\hat{L}_\nu] \otimes_{\mathbb{C}[\hat{N}]} T$$

and set

$$V_L^T = S[\nu] \otimes U,$$

generalizing (19) and (20). We require $\hat{\underline{h}}[\nu]_*$ to act on this space by acting only on the first factor, and we make $\hat{L}_\nu$ act only on the second factor. We also define an action of $\underline{h}_{(0)}$ on the second factor by

$$\alpha(0)\cdot a\otimes\tau = \langle\alpha,\bar{a}\rangle a\otimes\tau \quad \text{for} \quad \alpha \in \underline{h}_{(0)}, \quad a \in \hat{L}_\nu, \quad \tau \in T$$

and we set

$$z^{\alpha_{(0)}}\cdot a\otimes\tau = z^{\langle\alpha_{(0)},\bar{a}\rangle} a\otimes\tau \quad \text{for} \quad \alpha \in L, \quad a \in \hat{L}_\nu, \quad \tau \in T$$

(cf. (21), (22)).

We provide the space U with the (well-defined) $(1/M)\mathbb{Z}$-gradation determined by

$$\deg\ (a\otimes\tau) = -\frac{1}{2}\langle\bar{a}_{(0)},\bar{a}_{(0)}\rangle \quad \text{for} \quad a \in \hat{L}_\nu, \quad \tau \in T, \tag{63}$$

and we give $V_L^T$ the corresponding tensor product gradation (for which, we recall, the grading shift is left undetermined).

We now define the $\nu$-twisted vertex operator $Y_\nu(a,z)$ as follows, for $a \in \hat{L}_\nu$:

$$Y_\nu(a,z) =$$

$$= M^{-\langle\bar{a},\bar{a}\rangle/2}\sigma(\bar{a})\, {}^{\circ}_{\circ} e^{-\Sigma_{n\neq 0}(\alpha(n)/n)z^{-n}}\, a z^{\bar{a}_{(0)}+\langle\bar{a}_{(0)},\bar{a}_{(0)}\rangle/2-\langle\bar{a},\bar{a}\rangle/2} {}^{\circ}_{\circ}. \tag{64}$$

Here the normalization factor $\sigma$ is defined by

$$\sigma(\alpha) = \begin{cases} \sigma'(\alpha)2^{\langle\nu^{M/2}\alpha,\alpha\rangle/2} & \text{if} \quad M \in 2\mathbb{Z} \\ \sigma'(\alpha) & \text{if} \quad M \in 2\mathbb{Z}+1 \end{cases}$$

for $\alpha \in L$, where

$$\sigma'(\alpha) = \prod_{0<p<M/2} (1-\omega^{-p})^{\langle \nu^p\alpha,\alpha\rangle}.$$

Observe that $Y_\nu(a,z)$ indeed generalizes the untwisted operator $Y_1(a,z)$, the case $\nu = 1$. For the case $\nu = -1$ introduced in [FLM1-5], we see that the group $\hat{L}_{-1}$ is the same as the group $\hat{L} = \hat{L}_1$ of Section 2.2, the module U is just the finite-dimensional module T, and the vertex operator becomes

$$Y_{-1}(a,z) = 2^{-\langle\bar{a},\bar{a}\rangle} {}^{\circ}_{\circ} e^{-\Sigma_{n\neq 0}(\alpha(n)/n)z^{-n}} a z^{-\langle\bar{a},\bar{a}\rangle/2} {}^{\circ}_{\circ}. \tag{65}$$

(In [FLM1-5], the subscript $\mathbb{Z}+1/2$ is used instead of the subscript $-1$ to distinguish this twisted case from the untwisted case $Y_{\mathbb{Z}} = Y_1$.) For the case $\dim \underline{h} = 1$, this is the twisted vertex operator of [LW1] (except for the factor $z^{-\langle\bar{a},\bar{a}\rangle/2}$ and the use of the variable z rather than its inverse). The operator (65) was also introduced in [CF], without the factor a. Note that we are considering a version of the twisted vertex operator which contains the factor $z^{-\langle\bar{a},\bar{a}\rangle/2}$, a choice which is natural from the viewpoint of calculations (see for example the treatment of the case $\nu = 1$ above) and from the viewpoint of string theory (especially, conformal transformation properties of vertex operators). The numerical factors at the front of (64) and (65) (see [FLM1-5] for $\nu = -1$ and [L1] in general) lead to just the right form for the general results below.

Defining component operators $x_a^\nu(n)$ for $n \in (1/M)\mathbb{Z}$ by the expansion

$$Y_\nu(a,z) = \sum_{n\in(1/M)\mathbb{Z}} x_a^\nu(n) z^{-n-\langle\bar{a},\bar{a}\rangle/2}$$

we obtain

$$\deg x_a^\nu(n) = n \tag{66}$$

as in (26), (27). The following result generalizes Theorem 1:

<u>Theorem 3</u>. Suppose that L is the root lattice of a Lie algebra $\underline{g}$ which is a product of simple algebras of types $A_n$, $D_n$ and $E_n$. Then the operators $x_a^{\nu}(n)$ for $\langle\bar{a},\bar{a}\rangle = 2$ and $n \in (1/M)\mathbb{Z}$ generate a copy of the twisted affine Kac-Moody algebra $\hat{\underline{g}}[\nu_1]$, where $\nu_1$ is a suitable automorphism of $\underline{g}$. The space $\underline{h}$ is realized as a Cartan subalgebra of $\underline{g}$ and $\nu_1$ is an extension to $\underline{g}$ of the automorphism $\nu$ of L. Moreover, the $\hat{\underline{g}}$-module $V_L^T$ is irreducible.

The case $\dim \underline{h} = 1$, $\nu = -1$ was proved in [LW1]. This was generalized in [KKLW] to $\nu$ a Coxeter element of the Weyl group, or more generally, a twisted Coxeter automorphism of the root lattice. (Now L is arbitrary again.) The result of [KKLW] has been recovered by vertex operator calculus methods in [Fi]. As we have mentioned, the untwisted case $\nu = 1$ was done in [FK,S]. Involutions of the root lattice associated with Dynkin diagram involutions were studied in [Fr1]. The case $\nu = -1$ was carried out in [FLM1]. The general theorem is treated in [L1] and independently in [KP3] (which does not consider the direct commutation of vertex operators). (The hypothesis that L be a root lattice is missing from the irreducibility result, Proposition 9.2, of [L1].) The method in [L1], based on vertex operator calculus, includes the precise determination of the Lie algebra structure constants and generalizes to Theorem 5 below.

3.2 Generalized Twisted Vertex Operators and their Commutators: the Case $\nu = -1$

Theorems 1 and 3 exhibit a uniformity among vertex operators of different twistings, at least with reference to those commutators $[Y_{\nu}(a,z_1),Y_{\nu}(b,z_2)]$ for which $|\langle\bar{a},\bar{b}\rangle| \leq 2$. For the cases $\nu = 1$ and $\nu = -1$, a more subtle uniformity in the commutator structure also emerged more generally for $|\langle\bar{a},\bar{b}\rangle| \leq 4$, in [FLM1]. We shall present a result which generalizes Theorems 1, 2 and 3 and clarifies all this. In this section, we restrict to the case $\nu = -1$, treated in [FLM5] and important in the construction of the Monster. We recall that the groups $\hat{L}$ and $\hat{L}_{-1}$ are the same.

Extensions and refinements of the method of Section 2.4 enable us to compute $[Y_{-1}(a,z_1),Y_{-1}(b,z_2)]$ for any $a,b \in \hat{L}$. The result is more complicated than the one given by (40) and (41), and suggests that one should look for a definition of generalized twisted vertex operator which somehow absorbs the complications.

It turns out that generalized twisted vertex operators are to be parametrized by the untwisted Fock space $V_L$, not by the twisted space $V_L^T$. For $v \in V_L$ of the form (42), first set

$$W_{-1}(v,z) = \tag{67}$$
$$= {\circ \atop \circ}\left[\frac{1}{(n_1-1)!}\left(\frac{d}{dz}\right)^{n_1-1}\alpha_1(z)\right]\cdots\left[\frac{1}{(n_k-1)!}\left(\frac{d}{dz}\right)^{n_k-1}\alpha_k(z)\right]Y_{-1}(a,z){\circ \atop \circ},$$

where the right-hand side is an operator on $V_L^T$, and extend to all $v \in V_L$ by linearity (cf. (43)). (Normal ordering for operators on $V_L^T$ is defined just as in Section 2.1 for operators on $V_L$ -- all creation operators are placed to the left of all annihilation operators before the expression is interpreted as an operator; as for the "zero-mode" operators, $a \in \hat{L}$ is understood as a creation operator and $\alpha \in \underline{h}_{(0)}$ and $z^{\alpha}$ as annihilation operators.)

The correct generalized twisted vertex operator $Y_{-1}(v,z)$ is composed of $W_{-1}(v,z)$ and an operator of the form $e^{\Delta_z}$, where $e$ as usual denotes the exponential series and $\Delta_z$ is a "quadratic differential operator".

Let $\{\beta_1,\cdots,\beta_\ell\}$ be an orthonormal basis of $\underline{h}$. We define constants $c_{mn} \in \mathbb{C}$ for $m,n \geq 0$ by the formula

$$\sum_{m,n\geq 0} c_{mn}x^m y^n = -\log\left[\frac{(1+x)^{1/2}+(1+y)^{1/2}}{2}\right], \tag{68}$$

the log being well-defined since the expression in parentheses on the right is the form

$$1 + \tfrac{1}{4}x + \tfrac{1}{4}y + \cdots,$$

and we set

$$\Delta_z = \sum_{m,n\geq 0} \sum_{i=1}^{\ell} c_{mn}\beta_i(m)\beta_i(n)z^{-m-n}. \tag{69}$$

Then since $c_{00} = 0$, $e^{\Delta_z}$ is defined in our usual algebraic sense, and for $v \in V_L$, $e^{\Delta_z}v$ is a <u>polynomial</u> in $z^{-1}$ with coefficients in $V_L$. Thus the following definition of <u>generalized twisted vertex operator</u> $Y_{-1}(v,z)$ makes sense: We take

$$Y_{-1}(v,z) = W_{-1}(e^{\Delta_z}v,z) \quad \text{for} \quad v \in V_L. \tag{70}$$

Expressions similar to $e^{\Delta_z}$ appear in [CF].

By analogy with (44), we define operators $x_v^{-1}(n)$ on $V_L^T$ for homogeneous elements $v \in V_L$ and $n \in (1/2)\mathbb{Z}$ by

$$Y_{-1}(v,z) = \sum_{n\in(1/2)\mathbb{Z}} x_v^{-1}(n)z^{-n+\deg v-(\dim \underline{h})/24}, \tag{71}$$

and we use linearity to define $x_v^{-1}(n)$ for all $v \in V_L$. Then $x_{\iota(a)}^{-1}(n) = x_a^{-1}(n)$ for $a \in \hat{L}$ and

$$\deg x_v^{-1}(n) = n \quad \text{for} \quad v \in V_L, \tag{72}$$

(cf. (45), (66)).

We remark that the automorphism $\hat{\nu}$ $(= (-1)^{\wedge})$ of $\hat{L}$ acts in a natural way on the space $V_L$.

As we noted above, the definition (70) absorbs the complications appearing in the commutator $[Y_{-1}(a,z_1),Y_{-1}(b,z_2)]$, and we can express this commutator in a form analogous to that of (46). (This serves to motivate the expressions (68), (69), (70).) More generally, a somewhat delicate argument gives the analogue of Theorem 2 as well:

Theorem 4 [FLM5]. Let $\nu$ be the automorphism $-1$ of $L$. Then

$$[Y_{-1}(u,z_1),Y_{-1}(v,z_2)] =$$
$$= \frac{1}{2}\sum_{p=0,1} \mathrm{Res}_{z_0} z_2^{-1} Y_{-1}(Y_1(\hat{\nu}^p u,z_0)v,z_2) e^{-z_0(\partial/\partial z_1)} \delta((-1)^p z_1^{1/2}/z_2^{1/2}) \tag{73}$$

for $u,v \in V_L$. (Note the different roles of $Y_{-1}$ and $Y_1$.) If $u$ is homogeneous, then as operators on $V_L^T$,

$$[x_u^{-1}(m),x_v^{-1}(n)] =$$
$$= \frac{1}{2}\sum_{p=0,1} (-1)^{2pm} \sum_{i\in\mathbb{Z}} \begin{bmatrix} m-1+(\dim \underline{h})/24-\deg u \\ i-1+(\dim \underline{h})/24-\deg u \end{bmatrix} x^{-1}_{x_{\hat{\nu}^p u}(i)\cdot v}(m+n) \tag{74}$$

for $m,n \in (1/2)\mathbb{Z}$ (finite sum). (On the right-hand side, $x_{\hat{\nu}^p u}(i)$ is an operator on $V_L$.)

The similarity with Theorem 2 (the untwisted case) is very significant. Moreover, this similarity increases when we restrict $u$ to the subspace of $V_L$ fixed by $\hat{\nu}$: For such $u$ and for arbitrary $v \in V_L$,

$$[Y_{-1}(u,z_1),Y_{-1}(v,z_2)] =$$
$$= \mathrm{Res}_{z_0} z_2^{-1} Y_{-1}(Y_1(u,z_0)v,z_2) e^{-z_0(\partial/\partial z_1)} \delta(z_1/z_2) \tag{75}$$

and if $u$ is homogeneous, then

$$[x_u^{-1}(m),x_v^{-1}(n)] = \sum_{i\in\mathbb{Z}} \begin{bmatrix} m-1+(\dim \underline{h})/24-\deg u \\ i-1+(\dim \underline{h})/24-\deg u \end{bmatrix} x^{-1}_{x_u(i)\cdot v}(m+n) \qquad (76)$$

for $m,n \in \mathbb{Z}$; on the right-hand side, $x_u(i)$ is an operator on $V_L$.

## 3.3 The General Theorem

Now we return to a the case of a general lattice isometry $\nu$. Here we announce a very general result [L2] which includes Theorems 1-4 as special cases.

The first step is to modify the group $\hat{L}$ slightly, since $\hat{L}$ is not in general the same as the group $\hat{L}_\nu$. There is an essentially unique group $\hat{L}_{[\nu]}$ containing as a central subgroup the group $\langle\omega_0\rangle$ (recall (59)) such that $\hat{L}_{[\nu]}/\langle\omega_0\rangle \simeq L$ and

$$aba^{-1}b^{-1} = (-1)^{\langle\bar{a},\bar{b}\rangle},$$

where $a \mapsto \bar{a}$ denotes the canonical map from $\hat{L}_{[\nu]}$ to $L$ (cf. (18), (60)). Next, there is a natural <u>set</u>-<u>theoretic</u> identification (which is usually not an isomorphism of groups!) between the groups $\hat{L}_{[\nu]}$ and $\hat{L}_\nu$ such that the respective group multiplications (here distinguished by subscripts) are related by

$$a \times_{[\nu]} b = \prod_{-M/2<p<0} (-\omega^{-p})^{\langle\nu^p\bar{a},\bar{b}\rangle} a \times_\nu b \quad \text{for} \quad a,b \in \hat{L}_{[\nu]}$$

(cf. (60) and formula [4.6] in [L1]). (Note that for the cases $M = 1,2$, this identification is also an identification of groups, as we expect from the discussion above.) Moreover, the automorphism $\hat{\nu}$ of $\hat{L}_\nu$ (recall (61), (62)) is also an automorphism of $\hat{L}_{[\nu]}$.

Natural modifications of all the constructions for the untwisted theory ((19), (20), and so on) allow us to imitate precisely that whole theory using the group $\hat{L}_{[\nu]}$ in place of $\hat{L}$. We shall use the notations $V_L^{[\nu]}$, $Y_{\mathbb{Z}}^{[\nu]}$, and $x_v^{[\nu]}$ $(v \in V_L^{[\nu]})$ for the indicated objects in this modified untwisted theory.

Generalized $\nu$-twisted vertex operators are to be parametrized by the untwisted Fock space $V_L^{[\nu]}$. For $v \in V_L^{[\nu]}$ of the form (42), set

$$W_\nu(v,z) = \tag{77}$$
$$= {}^\circ_\circ\left[\frac{1}{(n_1-1)!}\left(\frac{d}{dz}\right)^{n_1-1}\alpha_1(z)\right]\cdots\left[\frac{1}{(n_k-1)!}\left(\frac{d}{dz}\right)^{n_k-1}\alpha_k(z)\right]Y_\nu(a,z){}^\circ_\circ,$$

where the right-hand side is an operator on $V_L^T$. (Cf. (67).) Note that we are using the set-theoretic identification between $\hat{L}_{[\nu]}$ and $\hat{L}_\nu$. Now extend to all $v \in V_L^{[\nu]}$ by linearity. Let $\{\beta_1,\cdots,\beta_\ell\}$ be an orthonormal basis of $\underline{h}$ as above, and define constants $c_{mnp} \in \mathbb{C}$ for $m,n \geq 0$ and $p = 0,\cdots,M-1$ by the formulas

$$\sum_{m,n\geq 0} c_{mn0}x^m y^n = -\frac{1}{2}\sum_{p\neq 0}\log\left[\frac{(1+x)^{1/M}-\omega^{-p}(1+y)^{1/M}}{1-\omega^{-p}}\right],$$

$$\sum_{m,n\geq 0} c_{mnp}x^m y^n = \frac{1}{2}\log\left[\frac{(1+x)^{1/M}-\omega^{-p}(1+y)^{1/M}}{1-\omega^{-p}}\right] \quad \text{for} \quad p \neq 0.$$

Set

$$\Delta_z = \sum_{m,n\geq 0}\sum_{p=0}^{M-1}\sum_{i=1}^{\ell} c_{mnp}(\nu^{-p}\beta_i)(m)\beta_i(n)z^{-m-n}, \tag{78}$$

a generalization of (69). Then $e^{\Delta_z}$ is defined since $c_{00p} = 0$ for all $p$, and for $v \in V_L^{[\nu]}$, $e^{\Delta_z}v$ is a polynomial in $z^{-1}$. This allows us to define <u>generalized</u> $\nu$-<u>twisted</u> <u>vertex</u> <u>operators</u> $Y_\nu(v,z)$ as follows: We take

$$Y_\nu(v,z) = W_\nu(e^{\Delta_z}v,z) \quad \text{for} \quad v \in V_L^{[\nu]}. \tag{79}$$

We also define operators $x_v^\nu(n)$ on $V_L^T$ for homogeneous elements

$v \in V_L^{[\nu]}$ and $n \in (1/M)\mathbf{Z}$ by

$$Y_\nu(v,z) = \sum_{n\in(1/M)\mathbf{Z}} x_v^\nu(n) z^{-n+\deg v-(\dim \underline{h})/24}, \tag{80}$$

and we use linearity to define $x_v^\nu(n)$ for all $v \in V_L^{[\nu]}$. Then $x_{\iota(a)}^\nu(n) = x_a^\nu(n)$ for $a \in \hat{L}_{[\nu]}$ and

$$\deg x_v^\nu(n) = n \quad \text{for} \quad v \in V_L^{[\nu]}. \tag{81}$$

The automorphism $\hat{\nu}$ of $\hat{L}_{[\nu]}$ acts in a natural way on the space $V_L^{[\nu]}$.

We can now state the result which generalizes Theorems 1-4:

<u>Theorem 5</u> [L2]. We have

$$[Y_\nu(u,z_1),Y_\nu(v,z_2)] =$$
$$= \frac{1}{M}\sum_{p=0}^{M-1} \mathrm{Res}_{z_0} z_2^{-1} Y_\nu(Y_1^{[\nu]}(\hat{\nu}^p u,z_0)v,z_2) e^{-z_0(\partial/\partial z_1)} \delta(\omega^p z_1^{1/M}/z_2^{1/M}) \tag{82}$$

for $u,v \in V_L^{[\nu]}$. In particular, if $\hat{\nu}u = u$, then

$$[Y_\nu(u,z_1),Y_\nu(v,z_2)] =$$
$$\mathrm{Res}_{z_0} z_2^{-1} Y_\nu(Y_1^{[\nu]}(u,z_0)v,z_2) e^{-z_0(\partial/\partial z_1)} \delta(z_1/z_2), \tag{83}$$

and if in addition $u$ is homogeneous, then

$$[x_u^\nu(m),x_v^\nu(n)] = \sum_{i\in\mathbf{Z}} \binom{m-1+(\dim \underline{h})/24-\deg u}{i-1+(\dim \underline{h})/24-\deg u} x^\nu_{x_u^{[\nu]}(i)\cdot v}(m+n) \tag{84}$$

for $m,n \in \mathbf{Z}$.

Note the resemblance between (48), (49) on the one hand and (83), (84) on the other hand. Of course, formula (82) includes a precise version of Theorem 3.

We close with a few comments about the Virasoro algebra. Let $\gamma$

be the canonical quadratic element of $V_L^{[\nu]}$ given by

$$\gamma = \frac{1}{2}\sum_{i=1}^{\ell} \beta_i(-1)^2 \otimes \iota(1),$$

$\{\beta_i\}$ being the orthonormal basis of $\underline{h}$, as above, and set

$$L_\nu(z) = Y_\nu(\gamma, z),$$

$$L^\nu(n) = x_\gamma^\nu(n) \quad \text{for} \quad n \in \mathbb{Z},$$

so that

$$L_\nu(z) = \sum_{n\in\mathbb{Z}} L^\nu(n) z^{-n-2}$$

and

$$\deg L_\nu(n) = n \quad \text{for} \quad n \in \mathbb{Z}$$

by (80) and (81). Then by Theorem 5, the operators $L_\nu(n)$ and the identity operator span a copy of the Virasoro algebra with the standard commutation relations

$$[L_\nu(m), L_\nu(n)] = (m-n)L_\nu(m+n) + \frac{1}{12}m(m^2-1)(\dim \underline{h})\delta_{m+n,0} \quad \text{for} \quad m,n \in \mathbb{Z}.$$

The cases $\nu = 1$ and $\nu = -1$ play significant roles in the construction of the Monster [FLM1-5]. Note that the general objects (78) and (79) tell us what the twisted Virasoro operators should be, _including the correct scalar which must be incorporated into the operator of degree_ 0. Finally, this scalar determines the standard "correct" grading shift ("zero-point energy") for the twisted Fock space $V_L^T$; as we recall (cf. (55), (63)), this was left undetermined. For instance, in the case $\nu = -1$, the right value for the degree of the constant polynomial $1 \in S[\nu]$ turns out to be $-\frac{1}{48}\dim \underline{h}$.

## REFERENCES

[ABLR] Altschüler, D., Béran, Ph., Lacki, J. and Roditi, I., String models with twisted vertex operators, preprint.

[ADDF] Ademollo, M., Del Giudice, E., DiVecchia, P. and Fubini, S., Couplings of three excited particles in the dual resonance model, Nuovo Cimento 19A, 181-203 (1974).

[AMW] Alvarez, O., Mangano, M. and Windey, P., Vertex operator construction of the SO(2n+1) Kac-Moody algebra and its spinor representation, Nucl. Phys. B277, 317-331 (1986).

[BHN] Banks, T., Horn, D. and Neuberger, H., Bosonization of the SU(N) Thirring models, Nucl. Phys. B108, 119 (1976).

[B] Borcherds, R.E., Vertex algebras, Kac-Moody algebras, and the Monster, Proc. Natl. Acad. Sci. U.S.A. 83, 3068-3071 (1986).

[CHSW] Candelas, P., Horowitz, G., Strominger, A. and Witten, E., Vacuum configurations for superstrings, Nucl Phys. B258, 46-74 (1985).

[C1] Corrigan, E., Twisted vertex operators and representations of the Virasoro algebra, Phys. Lett. 169B, 259-263 (1986).

[C2] Corrigan, E., Representations of the Virasoro algebra, talk at Intern. Symp. Elementary Particles, Ahrenshoop, GDR, 1985.

[CF] Corrigan, E. and Fairlie, D.B., Off-shell states in dual resonance theory, Nuc. Phys. B91, 527-545 (1975).

[DNPS] DiVecchia, P., Nakayama, R., Petersen, J.L. and Sciuto, S., Properties of the three-Reggeon vertex in string theories, preprint.

[DFMS] Dixon, L., Friedan, D., Martinec, E. and Shenker, S., The conformal field theory of orbifolds, Nucl. Phys. B282, 13-73 (1987).

[DHVW1] Dixon, L, Harvey, J.A., Vafa, C. and Witten, E., Strings on orbifolds, Nucl. Phys. B261, 651 (1985).

[DHVW2] Dixon, L, Harvey, J.A., Vafa, C. and Witten, E., Strings on orbifolds II, Nucl. Phys. B274, 285 (1986).

[D] Dong, C., Structure of some nonstandard modules for $C_n^{(1)}$, J. Algebra, to appear.

[Fi] Figueiredo, L., Calculus of principally twisted vertex operators, Ph.D. thesis, Rutgers University (1985) and Contemporary Math. (1987).

[FV] Freed, D.S. and Vafa, C., Global anomalies on orbifolds, preprint.

[Fr1] Frenkel, I.B., Two constructions of affine Lie algebra representations and boson-fermion correspondence in quantum field theory, J. Funct. Anal. 44, 259-327 (1981).

[Fr2] Frenkel, I.B., Representations of Kac-Moody algebras and dual resonance models, in: Applications of Group Theory in Physics and Mathematical Physics, Chicago Summer Seminar, 1982, Proceedings, Lectures in Applied Math. 21, American Math. Soc., 325-353 (1985).

[Fr3] Frenkel, I.B., Beyond affine Lie algebras, Proceedings of the International Congress of Mathematicians, 1986, Berkeley, to appear.

[FK] Frenkel, I.B. and Kac, V.G., Basic representations of affine Lie algebras and dual resonance models, Invent. Math. 62, 23-66 (1980).

[FLM1] Frenkel, I.B., Lepowsky, J. and Meurman, A., An $E_8$-approach to $F_1$, in: Finite Groups - Coming of Age, Proc. 1982 Montreal Conference, ed. by J. McKay, Contemporary Math. 45, 99-120 (1985).

[FLM2] Frenkel, I.B., Lepowsky, J. and Meurman, A., A natural representation of the Fischer-Griess Monster with the modular function J as character, Proc. Natl. Acad. Sci. U.S.A. 81, 3256-3260 (1984).

[FLM3] Frenkel, I.B., Lepowsky, J. and Meurman, A., A Moonshine Module for the Monster, in: Vertex Operators in Mathematics and Physics, Proc. 1983 M.S.R.I. Conference, ed. by J. Lepowsky, S. Mandelstam and I.M. Singer, Publ. Math. Sciences Res. Inst. #3, Springer-Verlag, New York, 231-273 (1985).

[FLM4] Frenkel, I.B., Lepowsky, J. and Meurman, A., An introduction to the Monster, in: Workshop on Unified String Theories, 1985, Proceedings, Inst. for Theoretical Physics, ed. by M. Green and D. Gross, World Scientific, Singapore, 533-546 (1986).

[FLM5] Frenkel, I.B., Lepowsky, J. and Meurman, A., Vertex operators and the Monster, monograph, in preparation.

[Ge] Gepner, D., New conformal field theories associated with Lie algebras and their partition functions, preprint.

[GQ] Gepner, D. and Qiu, Z., Modular invariant partition functions for parafermionic field theories, preprint.

[Go] Goddard, P., Vertex operators and algebras, Lectures at Summer Workshop on High Energy Physics and Cosmology, Trieste, 1985.

[GO1] Goddard, P. and Olive, D., Algebras, lattices and strings, in: Vertex Operators in Mathematics and Physics, Proc. 1983 M.S.R.I. Conference, ed. by J. Lepowsky, S. Mandelstam and I.M. Singer, Publ. Math. Sciences Res. Inst. #3, Springer-Verlag, New York, 51-96 (1985).

[GO2] Goddard, P. and Olive, D., Kac-Moody and Virasoro algebras in relation to quantum physics, preprint, DAMTP 86-5.

[GOS] Goddard, P., Olive, D. and Schwimmer, A., Vertex operators for non-simply-laced algebras, preprint.

[GSW] Green, M.B., Schwarz, J.H. and Witten, E., Superstring Theory, Volumes 1 and 2, Cambridge University Press, Cambridge (1987).

[Gr] Griess, R.L., Jr., The Friendly Giant, Invent. Math. 69, 1-102 (1982).

[GHMR1] Gross, D.J., Harvey, J.A., Martinec, E. and Rohm, R., Heterotic string theory (I). The free heterotic string, Nucl. Phys. B256, 253-284 (1985).

[GHMR2] Gross, D.J., Harvey, J.A., Martinec, E. and Rohm, R., Heterotic string theory (II). The interacting heterotic string, Nucl. Phys. B267, 75-124 (1986).

[Hal] Halpern, M.B., Quantum "solitons" which are SU(N) fermions, Phys. Rev. D12, 1684-1699 (1975).

[HV] Hamidi, S. and Vafa, C., Interactions on orbifolds, Nucl. Phys. B279 465-513 (1987).

[Har] Harvey, J.A., Twisting the heterotic string, in: Workshop on Unified String Theories, 1985, Proceedings, Inst. for Theoretical Physics, ed. by M. Green and D. Gross, World Scientific, Singapore, 704-718 (1986).

[JMO] Jimbo, M., Miwa, T. and Okado, M., Solvable lattice models with broken $Z_N$ symmetry and Hecke's indefinite modular forms, preprint.

[K] Kac, V.G., Infinite-dimensional Lie Algebras, Second edition, Cambridge University Press, Cambridge (1985).

[KKLW] Kac, V.G., Kazhdan, D.A., Lepowsky, J. and Wilson, R.L., Realization of the basic representations of the Euclidean Lie algebras, Advances in Math. 42, 83-112 (1981).

[KP1] Kac, V.G. and Peterson, D.H., Affine Lie algebras and Hecke modular forms, Bull. Amer. Math. Soc. (New Series) 3, 1057-1061 (1980).

[KP2] Kac, V.G. and Peterson, D.H., Infinite-dimensional Lie algebras, theta functions and modular forms, Advances in Math. 53, 125-264 (1984).

[KP3] Kac, V.G. and Peterson, D.H., 112 constructions of the basic representation of the loop group of $E_8$, in: Symposium on Anomalies, Geometry, Topology, 1985, Proceedings, ed. by W.A. Bardeen and A.R. White, World Scientific, Singapore, 276-298 (1985).

[L1] Lepowsky, J., Calculus of twisted vertex operators, Proc. Natl. Acad. Sci. U.S.A. 82, 8295-8299 (1985).

[L2] Lepowsky, J., The Lie algebra of generalized twisted vertex operators, to appear.

[LP1] Lepowsky, J. and Primc, M., Standard modules for type one affine Lie algebras, Number Theory, New York, 1982, Springer Lecture Notes in Mathematics 1052, 194-251 (1984).

[LP2] Lepowsky, J. and Primc, M., Structure of the standard modules for the affine Lie algebra $A_1^{(1)}$, Contemporary Math. 46 (1985).

[LW1] Lepowsky, J. and Wilson, R.L., Construction of the affine Lie algebra $A_1^{(1)}$, Comm. Math. Phys. 62, 43-53 (1978).

[LW2] Lepowsky, J. and Wilson, R.L., A new family of algebras underlying the Rogers-Ramanujan identities and generalizations, Proc. Natl. Acad. Sci. U.S.A. 78, 7254-7258 (1981).

[LW3] Lepowsky, J. and Wilson, R.L., The structure of standard modules, I: Universal algebras and the Rogers-Ramanujan identities, Invent. Math. 77, 199-290 (1984).

[LW4] Lepowsky, J. and Wilson, R.L., The structure of standard modules, II: The case $A_1^{(1)}$, principal gradation, Invent. Math. 79, 417-442 (1985).

[Mde] Mandelstam, S., Introduction to string models and vertex operators, in: Vertex Operators in Mathematics and Physics, Proc. 1983 M.S.R.I. Conference, ed. by J. Lepowsky, S. Mandelstam and I.M. Singer, Publ. Math. Sciences Res. Inst. #3, Springer-Verlag, New York, 15-35 (1985).

[Mdi] Mandia, M., Structure of the level one standard modules for the affine Lie algebras $B_\ell^{(1)}$, $F_4^{(1)}$ and $G_2^{(1)}$, Ph.D. thesis, Rutgers University (1985) and Contemporary Math. (1987).

[MP] Meurman, A. and Primc, M., Annihilating ideals of standard modules of $s\ell(2,\mathbb{C})^\sim$ and combinatorial identities, Advances in Math., to appear.

[Mis1] Misra, K.C., Structure of the standard modules for $A_n^{(1)}$ and $C_n^{(1)}$, Ph.D. thesis, Rutgers University (1982).

[Mis2] Misra, K.C., Structure of certain standard modules for $A_n^{(1)}$ and the Rogers-Ramanujan identities, J. Algebra 88, 196-227 (1984).

[Mis3] Misra, K.C., Structure of some standard modules for $C_n^{(1)}$, J. Algebra 90, 385-409 (1984).

[Miw] Miwa, T., Integrable lattice models and branching coefficients, Proceedings of the International Congress of Mathematicians, 1986, Berkeley, to appear.

[My] Myhill, R.G., Automorphisms of Lie algebra root systems which leave only the origin fixed, preprint.

[NSV] Narain, K.S., Sarmadi, M.H. and Vafa, C., Asymmetric orbifolds, preprint.

[NY] Ninomiya, M. and Yamagishi, K., Nonlocal SU(3) current algebra, preprint.

[PS] Pressley, A. and Segal, G., Loop Groups, Oxford Mathematical Monographs, Clarendon Press, Oxford (1986).

[SY] Sasaki, R. and Yamanaka, I., Vertex operators for a bosonic string, Phys. Lett. 165B, 283-288 (1985).

[S] Segal, G., Unitary representations of some infinite-dimensional groups, Comm. Math. Phys. 80, 301-342 (1981).

[TK] Tsuchiya, A. and Kanie, Y., Vertex operators and conformal field theory on $\mathbb{P}^1$ and monodromy representations of braid groups, in: Conformal Field Theory and Solvable Lattice Models, Advanced Studies in Pure Mathematics, Kinokuniya, to appear.

[V] Vafa, C., Modular invariance and discrete torsion on orbifolds, Nucl. Phys B273, 592 (1986).

[Wa] Wang, X., Structure of some level two standard modules for $A_{n-1}^{(1)}$, preprint, Harbin Normal University, P.R.C.

[We] Weinberg, S., Coupling constants and vertex functions in string theories, Phys. Lett. 156B, 309-314 (1985).

[Wi] Witten, E., Physics and geometry, Proceedings of the International Congress of Mathematicians, 1986, Berkeley, to appear.

[ZF] Zamolodchikov, A.B. and Fateev, V.A., Nonlocal (parafermion) currents in two-dimensional conformal quantum field theory and self-dual critical points in $Z_N$-symmetric statistical systems, Sov. Phys., JETP 62(2), 215-225 (1985).

# On Determinant Line Bundles

DANIEL S. FREED

Department of Mathematics
Massachusetts Institute of Technology

Determinant line bundles entered differential geometry in a remarkable paper of Quillen [**Q**]. He attached a holomorphic line bundle $\mathcal{L}$ to a particular family of Cauchy-Riemann operators over a Riemann surface, constructed a Hermitian metric on $\mathcal{L}$, and calculated its curvature. At about the same time Atiyah and Singer [**AS2**] made the connection between determinant line bundles and anomalies in physics. Somewhat later, Witten [**W1**] gave a formula for "global anomalies" in terms of $\eta$-invariants. He suggested that it could be interpreted as the holonomy of a connection on the determinant line bundle. These ideas have been developed by workers in both mathematics and physics. Our goal here is to survey some of this work.

We consider arbitrary families of Dirac operators $D$ on a smooth compact manifold $X$. The associated Laplacian has discrete spectrum, which leads to a patching construction for the determinant line bundle $\mathcal{L}$. The determinant $\det D$ is a section of $\mathcal{L}$. Quillen uses the *analytic torsion* of Ray and Singer [**RS1**] to define a metric on $\mathcal{L}$. An extension of these ideas produces a unitary connection whose curvature and holonomy can be computed explicitly. The holonomy formula reproduces Witten's global anomaly. Section 1 represents joint work with Jean-Michel Bismut, whose proof of the index theorem for families [**B**] is a crucial ingredient in the curvature formula (1.30).

These basic themes allow many variations, two of which we play out in §2 and §3. Suppose $X$ is a complex manifold and the family of Dirac operators (or Cauchy-Riemann operators) varies holomorphically. Then $\mathcal{L}$ carries a natural complex structure, and under appropriate restrictions on the geometry the canonical connection is compatible with the holomorphic structure. The proper geometric hypothesis, that the total space swept out by $X$ be Kähler, at least locally in the parameter space, also ensures that the operators vary holomorphically. Many special cases of this result can be found in the literature; the version we prove is due to Bismut, Gillet, and Soulé [**BGS**].

---

The author is partially supported by an NSF Postdoctoral Research Fellowship

One novelty here is the observation that $\det D$ has a natural square root[1] if the dimension of $X$ is congruent to 2 modulo 8. On topological grounds one can argue the existence of $\mathcal{L}^{1/2}$ using Rohlin's theorem, which is linked to real $K$-theory. However, one needs the differential geometry to see that $\det D$ also admits a square root. There is an extension of the holonomy theorem to $\mathcal{L}^{1/2}$.

In §4 we study Riemann surfaces. This is the case originally considered by Quillen. Faltings [**Fa**] considered determinants on Riemann surfaces in an arithmetic context. These determinants also form the cornerstone of Polyakov's formulation of string theory [**P**]. For elliptic curves there are explicit formulas for the determinant, which illustrate the general theory beautifully. A recent paper of Atiyah [**A2**] studies the transformation law for Dedekind's $\eta$-function in this context. On higher genus surfaces the determinants are related to $\vartheta$-functions. Precise formulas have been derived in the physics literature, where they are applied to "bosonization." In a mathematical vein they clarify the relationship between Quillen's metric and the metric constructed by Faltings. Here and elsewhere in this section we have benefited from Bost [**Bo**].

The determinant line bundle carries topological information. In §5 we explore a particular example related to Witten's global anomalies (in string theory). One could speculate that this anomaly is related to orientation in some generalized cohomology theory. Of more technical interest is our use of Sullivan's $\mathbb{Z}/k$-manifolds to pass from $K$-theory to integral cohomology.

Although this paper is rather lengthy, by necessity we have omitted many important topics. For example, the determinant of the Dirac operator can be studied as an energy function on appropriate configuration spaces. Donaldson [**D**] illustrates this idea in his construction of Hermitian-Einstein metrics on stable bundles over projective varieties. Osgood, Phillips, and Sarnak [**OPS**] treat the uniformization theorem in Riemann surface theory from a similar point of view. We have not mentioned the approach to determinants using Selberg $\zeta$-functions. Cheeger [**C**] and Singer [**S**] give other geometric interpretations of Witten's global anomaly that we leave untouched. We have not given justice to work of Beilinson and Manin, nor to work of Deligne [**De**], among others. Such omissions are inevitable, though regrettable.

My understanding of determinants owes much to my collaborators Jean-Michel Bis-

[1] Lee and Miller also construct this square root.

mut and Cumrun Vafa. I have also benefited greatly from the insight of Michael Atiyah, Raoul Bott, Gunnar Carlsson, Jeff Cheeger, Pierre Deligne, Simon Donaldson, Richard Melrose, Ed Miller, Haynes Miller, Greg Moore, John Morgan, Phil Nelson, Dan Quillen, Isadore Singer, and Ed Witten on the various topics covered here, and I take this opportunity to thank them all. I am grateful to Peter Landweber for his careful reading of an earlier version of this paper.

## §1 The Determinant Line Bundle

Suppose $X$ is a compact spin manifold. Then the chiral Dirac operator maps positive (right-handed) spinor fields to negative (left-handed) spinor fields. Symbolically, we write $D\colon \mathcal{H}_+ \to \mathcal{H}_-$. If $\mathcal{H}_\pm$ were finite dimensional and of equal dimension, then on the highest exterior power there would be an induced map $\det D\colon \det \mathcal{H}_+ \to \det \mathcal{H}_-$. The determinant of $D$ would be regarded as an element of the complex line $(\det \mathcal{H}_+)^* \otimes (\det \mathcal{H}_-)$. As the Dirac operator acts on infinite dimensional spaces, some care must be exercised to define its determinant. In this section we recall Quillen's construction of the metrized determinant line associated to a Dirac operator [**Q**]. The $\varsigma$-functions of Ray and Singer [**RS1**] play a crucial role. For a family of Dirac operators a patching construction then produces a metrized line bundle over the parameter space. We review in detail the construction of a natural connection on the determinant line bundle [**BF1**]. At the end of this section we briefly recall the formulas for its curvature and holonomy [**BF2**]. The holonomy formula is one possible interpretation of Witten's global anomaly [**W1**]. As stated in the introduction, this section is based on joint work with J.-M. Bismut. The reader may wish to refer to [**F1**,§1] for related expository material.

The basic setup for our work is given by the

Geometric Data 1.1.

(1) *A smooth fibration of manifolds* $\pi\colon Z \xrightarrow{X} Y$. *To define a family of Dirac operators we need to add the topological hypothesis that the tangent bundle along the fibers* $T(Z/Y) \to Z$ *has a fixed Spin (or Spin*$^c$*) structure.*

(2) *A metric along the fibers, that is, a metric* $g^{(Z/Y)}$ *on* $T(Z/Y)$.

(3) *A projection* $P\colon TZ \to T(Z/Y)$.

(4) *A complex representation $\rho$ of* Spin$(n)$.

(5) *A complex vector bundle $E \longrightarrow Z$ with a hermitian metric $g^{(E)}$ and compatible connection $\nabla^{(E)}$.*

REMARKS:

(1) To say that $\pi$ is a smooth fibration of manifolds simply means that $Z$ is a smooth manifold, $\pi$ is a smooth map, and for small open sets $U \subset Y$ the inverse image $\pi^{-1}(U)$ is diffeomorphic to $U \times X$. Here we glue local products $U \times X$ using diffeomorphisms of X. The identification of $X$ with the fiber $X_y$ at $y \in Y$ is only up to an element of Diff$(X)$. Equivalently, $Z \longrightarrow Y$ is associated to a smooth principal Diff$(X)$-bundle $\mathcal{P} \longrightarrow Y$. A point in the fiber of $\mathcal{P}$ at $y$ is an identification $X_y \cong X$. There are technical difficulties in constructing a smooth Lie group out of the space of smooth diffeomorphisms of $X$. Here we understand that the transition function $\varphi \colon U \longrightarrow \mathrm{Diff}(X)$ is smooth if and only if $\langle y, x\rangle \mapsto \varphi(y)x$ is smooth. The topological hypothesis amounts to a reduction of this bundle to the subgroup of diffeomorphisms fixing a given orientation and spin structure on $X$.

(2) The kernel of the projection $P$ defines a distribution $\ker P$ of horizontal subspaces on $Z$. Then $\ker P \cong \pi^*TY$ and $TZ = T(Z/Y) \oplus \ker P$ splits as a direct sum. Therefore, we can lift a tangent vector $\xi \in T_yY$ to a horizontal vector field $\tilde{\xi}$ along the fiber $Z_y$. The projection $P$ is equivalent to a connection on the bundle $\mathcal{P} \longrightarrow Y$.

(3) We may, of course, choose $\rho$ to be the trivial representation. Also, we allow virtual representations, that is, differences of representations. The data in (1) and (2) determine a Spin$(n)$ bundle of frames Spin$(Z) \longrightarrow Z$ consisting of "spin frames" of the vertical tangent space, and the representation $\rho$ gives an associated vector bundle $V_\rho \longrightarrow Z$. For example, if $\rho$ is the n-dimensional representation of Spin$(n)$, then $V_\rho$ is simply $T(Z/Y)$. The Dirac operator we construct is coupled to $V_\rho$; this is an intrinsic coupling in that it only involves bundles associated to the tangent bundle. In case $V_\rho = T(Z/Y)$ we obtain the Rarita-Schwinger operator. If $\rho$ is the total spin representation $\sigma_+ \oplus \sigma_-$ we obtain the signature operator.

(4) The vector bundle $E$ determines the extrinsic coupling of the Dirac operator.

The family of Dirac operators $\{D_y\}$ corresponding to the data in (1.1) is constructed fiberwise. For each $y \in Y$ the fiber $Z_y$ is a smooth spin manifold and so has a Levi-Civita connection. There is an induced connection on $V_\rho|_{Z_y}$. The connection $\nabla^{(E)}$ restricts to a connection on $E|_{Z_y}$. Let $S_\pm \to Z$ be the bundles associated to Spin($Z$) via the half-spin representations. Then the ordinary chiral Dirac operator on $Z_y$, which maps positive spinor fields to negative spinor fields, couples to the bundles $V_\rho|_{Z_y}$ and $E|_{Z_y}$ via the connections. Thus we obtain an operator

$$D_y : C^\infty\big((S_+ \otimes V_\rho \otimes E)\big|_{Z_y}\big) \to C^\infty\big((S_- \otimes V_\rho \otimes E)\big|_{Z_y}\big). \tag{1.2}$$

The metrics on $S_\pm$, $V_\rho$, and $E$, together with the volume form on $Z_y$, induce $L^2$ completions $(\mathcal{H}_\pm)_y$ of the $C^\infty$ spaces in (1.2). Then $D_y$ extends to an unbounded operator on these $L^2$ spaces. Furthermore, as $y$ varies over $Y$ these spaces fit together to form *continuous* Hilbert bundles $\mathcal{H}_\pm \to Y$. (These bundles are not smooth since the composition $L^2 \times C^\infty \to L^2$ is not differentiable.) Thus we can view $D_y$ as a bundle map $\mathcal{H}_+ \xrightarrow{D} \mathcal{H}_-$. Notice that the Hilbert bundles $\mathcal{H}_\pm$ carry $L^2$ metrics by definition. Our constructions below only use finite dimensional subbundles of the Hilbert bundles, and so the technicalities associated with infinite dimensional bundles are not a problem here.

The geometric data (1.1) determine a connection $\nabla^{(Z/Y)}$ on $T(Z/Y)$ as follows. Fix an arbitrary Riemannian metric $g^{(Y)}$ on $Y$. The metric $g^{(Z/Y)}$ along the fibers together with the lift of $g^{(Y)}$ to the horizontal subspaces (the kernel of the projection $P$) combine to form a metric $g^{(Z)}$ on $Z$. Let $\nabla^{(Z)}$ denote its Levi-Civita connection.

LEMMA 1.3. *The projection of $\nabla^{(Z)}$ to a connection $\nabla^{(Z/Y)}$ on $T(Z/Y)$ is independent of the choice of metric on $Y$.*

We omit the easy proof. The connection $\nabla^{(Z/Y)}$ restricts to the Levi-Civita connection on each fiber $Z_y$.

We use this connection to induce connections on $\mathcal{H}_\pm$. Since these Hilbert bundles are only continuous, the "connections" we define will operate only on smooth sections. To obtain *unitary* connections we must modify $\nabla^{(Z/Y)}$ slightly, since the volume form vol of $g^{(Z/Y)}$ changes from fiber to fiber. Now although this volume originally comes as a section of $\Lambda^n(T(Z/Y))^*$, the projection $P$ allows us to regard **vol** as an $n$-form on $Z$. Fix a tangent vector $\eta$ to $Z$, and use the splitting $TZ \cong T(Z/Y) \oplus \ker P$ to define a 1-form $\gamma \in \Omega^1(Z)$:

$$\iota(\eta)d(\mathbf{vol}) = 2\gamma(\eta)\,\mathbf{vol} + \text{nonvertical terms}. \tag{1.4}$$

Finally, set

$$\tilde{\nabla}^{(Z/Y)} = \nabla^{(Z/Y)} + \gamma. \tag{1.5}$$

This defines a new connection on $T(Z/Y)$ and its associated bundles. The tilde serves as a reminder of the correction term. Hence we obtain connections

$$\tilde{\nabla}^{(\pm)} = \dot{\sigma}_{\pm}\big(\tilde{\nabla}^{(Z/Y)}\big) \oplus \dot{\rho}\big(\nabla^{(Z/Y)}\big) \oplus \nabla^{(E)} \tag{1.6}$$

on the bundles $S_{\pm} \otimes V_{\rho} \otimes E$. (Here $\dot{\rho}$ denotes the Lie algebra representation induced by $\rho$, and $\sigma_{\pm}$ are the half spin representations.) Of course, the metrics $g^{(Z/Y)}$ and $g^{(E)}$ induce metrics $(\cdot\,,\cdot)$ on these bundles. The modification (1.5) was introduced so that if $s_1, s_2$ are smooth sections of one of these bundles (over $Z$), then

$$d\{(s_1,s_2)\,\mathbf{vol}\} = (\tilde{\nabla}^{(\pm)}s_1,s_2)\,\mathbf{vol} + (s_1,\tilde{\nabla}^{(\pm)}s_2)\,\mathbf{vol} + \text{nonvertical terms}. \tag{1.7}$$

We can consider $s_i$ as sections of the Hilbert bundle $\mathcal{H}_{\pm} \to Y$. The $L^2$ metric on $\mathcal{H}_{\pm}$ is defined by

$$(s_1,s_2)_{\mathcal{H}_{\pm}} = \int_{Z/Y} (s_1,s_2)\,\mathbf{vol}. \tag{1.8}$$

Let $\xi$ be a tangent vector to $Y$ and $\tilde{\xi}$ its horizontal lift to a fiber in $Z$. Then for a section $s$ of $\mathcal{H}_{\pm}$ we define

$$\tilde{\nabla}_{\xi}^{(\mathcal{H}_{\pm})}s = \tilde{\nabla}_{\tilde{\xi}}^{(\pm)}s \tag{1.9}$$

acting pointwise. We view $\tilde{\nabla}^{(\mathcal{H}_{\pm})}$ as "connections" on $\mathcal{H}_{\pm}$, which act only on the dense subspace of smooth sections $C^{\infty}(S_{\pm} \otimes V_{\rho} \otimes E) \subset L^2(S_{\pm} \otimes V_{\rho} \otimes E)$. Since integration along the fiber commutes with $d$, and nonvertical terms integrate to zero, equation (1.7) implies that $\tilde{\nabla}^{(\mathcal{H}_\pm)}$ preserves the $L^2$ metric (1.8).

Let $\xi_1, \xi_2$ be vector fields on $Y$, denote by $\tilde{\xi}_1, \tilde{\xi}_2$ their horizontal lifts to $Z$, and set

$$T(\xi_1, \xi_2) = [\tilde{\xi}_1, \tilde{\xi}_2] - \widetilde{[\xi_1, \xi_2]}. \tag{1.10}$$

It is easy to check that $T(\xi_1, \xi_2)$ is a vertical vector field on $Z$ which is tensorial in $\xi_i$. In fact, $T$ is essentially the curvature of the connection on the $\mathrm{Diff}_0(X)$-bundle $\mathcal{P} \to Y$ (c.f., remarks (1) and (2) above). The next result will enter our discussion of holomorphic families in §2.

PROPOSITION 1.11. *The curvature $\Omega^{(\mathcal{H}_\pm)}(\xi_1, \xi_2)$ is the first order differential operator*

$$\Omega^{(\mathcal{H}_\pm)}(\xi_1, \xi_2) = \tilde{\nabla}^{(\pm)}_{T(\xi_1, \xi_2)} + \tilde{\Omega}^{\pm}(\tilde{\xi}_1, \tilde{\xi}_2),$$

*where $\tilde{\Omega}^{\pm}$ is the curvature of $\tilde{\nabla}^{(\pm)}$.*

The determinant line of the single Dirac operator $D_y$ is

$$\mathcal{L}_y = (\det \ker D_y)^* \otimes (\det \operatorname{coker} D_y).$$

(Recall that $\det(V)$ is the highest exterior power of the vector space V.) Since $\dim \ker D_y$ may jump as $y$ varies, we need a construction to patch these lines together.

THEOREM 1.12 [**BF1**]. *The Geometric Data (1.1) determine a smooth line bundle $\mathcal{L} \to Y$ along with a natural Hermitian metric $g^{(\mathcal{L})}$ (the Quillen metric) and unitary connection $\nabla^{(\mathcal{L})}$. There is a canonical section $\det D$ of $\mathcal{L}$ over components of $Y$ where $D$ has numerical index zero.*

Let us first recall some basic consequences of ellipticity. Each fiber $\mathcal{H}_y$ of the Hilbert bundle $\mathcal{H}_+$ (resp. $\mathcal{H}_-$) decomposes into a direct sum of eigenspaces of the nonnegative elliptic operator $D^*D$ (resp. $D\,D^*$). We delete $y$ from the notation for convenience. The spectra of these operators are discrete, the nonzero eigenvalues $\{\lambda\}$ of $D^*D$ and $DD^*$ agree, and $D$ is an isomorphism between the corresponding eigenspaces. For $a > 0$ not in the spectrum of $D^*D$, let $\mathcal{H}_\pm^{(a)}$ be the sum of the eigenspaces for eigenvalues less than $a$. Then $\mathcal{H}_\pm^{(a)}$ is finite dimensional and consists of smooth fields. There is an exact sequence

$$0 \to \ker D^*D \to \mathcal{H}_+^{(a)} \xrightarrow{D} \mathcal{H}_-^{(a)} \to \ker DD^* \to 0. \tag{1.13}$$

If $b > a$ we set $\mathcal{H}_\pm^{(a,b)} = \mathcal{H}_\pm^{(b)} \ominus \mathcal{H}_\pm^{(a)}$; then

$$D\big|_{\mathcal{H}_+^{(a,b)}} = D^{(a,b)} : \mathcal{H}_+^{(a,b)} \to \mathcal{H}_-^{(a,b)}$$

is an isomorphism. (Here we could use the quotient $\mathcal{H}_\pm^{(b)}/\mathcal{H}_\pm^{(a)}$ instead of the Hilbert space difference.) We represent the situation schematically in Figure 1.

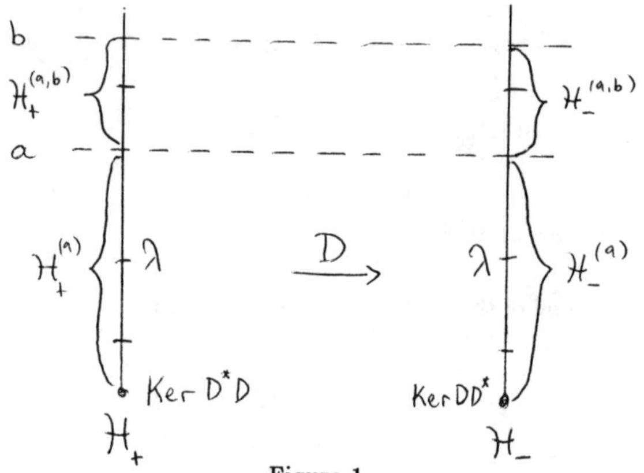


Figure 1

The spaces $\mathcal{H}_\pm^{(a)}$ fit together to form *smooth* finite dimensional vector bundles (of locally constant rank) over the open subset $U^{(a)} \subset Y$ in which $a \notin \operatorname{spec}(D^*D)$. Smoothness follows since $\mathcal{H}_\pm^{(a)}$ consists of $C^\infty$ fields, which transform smoothly under diffeomorphisms of $X$. Therefore, the bundles $\mathcal{H}_\pm^{(a)}$ can be smoothly patched together using the patching in the fibration of manifolds $Z \to Y$. Since the spectrum of an elliptic operator is discrete, it follows that $\{U^{(a)}\}$ forms an open cover of $Y$. Define a line bundle $\mathcal{L}^{(a)} \to U^{(a)}$ by

$$\mathcal{L}^{(a)} = \left(\det \mathcal{H}_+^{(a)}\right)^* \otimes \left(\det \mathcal{H}_-^{(a)}\right). \tag{1.14}$$

The sequence (1.13) implies that for each $y \in Y$ there is a canonical isomorphism

$$\mathcal{L}_y^{(a)} \cong (\det \ker D_y)^* \otimes (\det \ker D_y^*).$$

We emphasize that the expression on the right hand side does not define a line bundle, even locally, because of the possibly jumping dimension of the kernels. However, the line bundle $\mathcal{L}^{(a)}$ is well-defined and smooth since $\mathcal{H}_{\pm}^{(a)}$ are smooth bundles over $U^{(a)}$. Now we must see how to patch together $\mathcal{L}^{(a)}$ and $\mathcal{L}^{(b)}$ over $U^{(a)} \cap U^{(b)}$. On that intersection

$$\mathcal{L}^{(b)} \cong \mathcal{L}^{(a)} \otimes \mathcal{L}^{(a,b)} \tag{1.15}$$

where

$$\mathcal{L}^{(a,b)} = \left(\det \mathcal{H}_{+}^{(a,b)}\right)^{*} \otimes \left(\det \mathcal{H}_{-}^{(a,b)}\right).$$

Since $D^{(a,b)}\colon \mathcal{H}_{+}^{(a,b)} \to \mathcal{H}_{-}^{(a,b)}$ is an isomorphism, it induces an isomorphism

$$\det D^{(a,b)}\colon \det \mathcal{H}_{+}^{(a,b)} \to \det \mathcal{H}_{-}^{(a,b)},$$

which we regard as a *nonzero* section of $\mathcal{L}^{(a,b)}$. This induces a canonical smooth isomorphism

$$\begin{aligned} \mathcal{L}^{(a)} &\longrightarrow \mathcal{L}^{(a)} \otimes \mathcal{L}^{(a,b)} = \mathcal{L}^{(b)} \\ s &\longmapsto s \otimes \det D^{(a,b)} \end{aligned} \tag{1.16}$$

over $U^{(a)} \cap U^{(b)}$. Finally, a differentiable line bundle $\mathcal{L} \to Y$ is defined by patching the $\mathcal{L}^{(a)}$ using the isomorphisms (1.16).

Over components of $Y$ where the numerical index of $D$ is zero, we have $\dim \mathcal{H}_{+}^{(a)} = \dim \mathcal{H}_{-}^{(a)}$. In this case each $\mathcal{L}^{(a)}$ has a canonical section $\det D^{(a)}\colon \det \mathcal{H}_{+}^{(a)} \to \det \mathcal{H}_{-}^{(a)}$, and the multiplicative property of determinants shows that $\det D^{(a)}$ and $\det D^{(b)}$ correspond under the isomorphism (1.16). Therefore, a global section $\det D$ of $\mathcal{L} \to Y$ is defined. This section is nonzero exactly where $D$ is invertible. Over components where the index is nonzero, $\det D$ is (by fiat) identically zero.

We now proceed to describe the Quillen metric on $\mathcal{L}$. Fix $a > 0$. Then the subbundles $\mathcal{H}_{\pm}^{(a)}$ inherit metrics from $\mathcal{H}_{\pm}$. By linear algebra, metrics are induced on determinants, duals, and tensor products, so that $\mathcal{L}^{(a)}$ inherits a natural metric $g^{(a)}(\cdot,\cdot)$. If $b > a$ then under the isomorphism (1.16) we have two metrics on $\mathcal{L}^{(b)}$, and their ratio is a real number $\|\det D^{(a,b)}\|^2$. Let $\psi_1, \ldots, \psi_N$ be a basis of $\mathcal{H}_{+}^{(a)}$ consisting of eigenfunctions, and $\psi_1^*, \ldots, \psi_N^*$ the dual basis; then

$$\det D^{(a,b)} = (\psi_1^* \wedge \cdots \wedge \psi_N^*) \otimes (D\psi_1 \wedge \cdots \wedge D\psi_N). \tag{1.17}$$

Thus

$$\|\det D^{(a,b)}\|^2 = \prod_{i=1}^{N} \|\psi_i^*\|^2 \|D\psi_i\|^2$$
$$= \prod_{i=1}^{N} \|\psi_i^*\|^2 (D^*D\psi_i, \psi_i)$$
$$= \prod_{a<\lambda<b} \lambda.$$

In other words, on $U^{(a)} \cap U^{(b)}$

(1.18) $$g^{(b)} = g^{(a)} \left( \prod_{a<\lambda<b} \lambda \right).$$

To correct for this discrepancy, set

(1.19) $$\bar{g}^{(a)} = g^{(a)} \cdot \det \left( D^*D\big|_{\lambda>a} \right),$$

where we define

$$\det \left( D^*D\big|_{\lambda>a} \right) = \prod_{\lambda>a} \lambda$$

using $\varsigma$-functions [**RS1**].

Recall this procedure. Let

(1.20) $$\varsigma^{(a)}(s) = \sum_{\lambda>a} \frac{1}{\lambda^s} = \operatorname{Tr} \left( \left( D^*D\big|_{\lambda>a} \right)^{-s} \right).$$

Then $\varsigma^{(a)}(s)$ is holomorphic for $\operatorname{Re} s > n/2$ and has a meromorphic continuation to $\mathbb{C}$ which is holomorphic at $s = 0$ [**Se**]. The regularized determinant is defined to be

$$\det \left( D^*D\big|_{\lambda>a} \right) = \exp \left( -\varsigma^{(a)\prime}(0) \right).$$

The crucial property of this regularization scheme is that it behaves properly with respect to a finite number of eigenvalues:

(1.21) $$\det \left( D^*D\big|_{\lambda>a} \right) = \det \left( D^*D\big|_{\lambda>b} \right) \left( \prod_{a<\lambda<b} \lambda \right)$$

on the intersection $U^{(a)} \cap U^{(b)}$. (In fact, any regularization scheme with this property is adequate.) Equation (1.21) ensures that $\bar{g}^{(a)}$ and $\bar{g}^{(b)}$ agree on the overlap, by (1.18). Thus the $\bar{g}^{(a)}$ patch together to a Hermitian metric $g^{(\mathcal{L})}$ on $\mathcal{L}$, the *Quillen metric*.

Since $\mathcal{H}_{\pm}^{(a)}$ consists of smooth fields, the "connections" $\tilde{\nabla}^{(\mathcal{H}_{\pm})}$ project to honest connections on the smooth bundles $\mathcal{H}_{\pm}^{(a)} \to U^{(a)}$. (Recall that $\tilde{\nabla}^{(\mathcal{H}_{\pm})}$ are defined by pointwise differentiation (1.9).) Furthermore, these projected connections are unitary for the restricted metrics. Connections are induced on determinants (by taking a trace), dual bundles, and tensor products, so that $\mathcal{L}^{(a)}$ inherits a natural connection $\tilde{\nabla}^{(a)}$ which is unitary for the metric $g^{(a)}$. We have two connections on $\mathcal{L}^{(b)}$ over the intersection $U^{(a)} \cap U^{(b)}$, and these differ by

$$\begin{aligned}
\tilde{\nabla}\left(\det D^{(a,b)}\right) &= \tilde{\nabla}\{(\psi_1^* \wedge \cdots \wedge \psi_N^*) \otimes (D\psi_1 \wedge \cdots \wedge D\psi_N)\} \\
&= \sum_{i=1}^{N} (\tilde{\nabla}\psi_i^*, \psi_i) \det D^{(a,b)} + \sum_{i=1}^{N} ((D\psi_i)^*, \tilde{\nabla}(D\psi_i)) \det D^{(a,b)} \\
&= \sum_{i=1}^{N} \left\{ (\tilde{\nabla}\psi_i^*, \psi_i) + (\psi_i^*, \tilde{\nabla}\psi_i) + ((D\psi_i)^*, (\tilde{\nabla}D\, D^{-1}) D\psi_i) \right\} \det D^{(a,b)} \\
&= \operatorname{Tr}\left(\tilde{\nabla}D\, D^{-1}\big|_{a<\lambda<b}\right) \det D^{(a,b)}.
\end{aligned}$$

In other words,

$$\tilde{\nabla}^{(b)} = \tilde{\nabla}^{(a)} + \operatorname{Tr}\left(\tilde{\nabla}D\, D^{-1}\big|_{a<\lambda<b}\right). \tag{1.22}$$

By analogy with the Quillen metric, set

$$\bar{\nabla}^{(a)} = \tilde{\nabla}^{(a)} + \operatorname{Tr}\left(\tilde{\nabla}D\, D^{-1}\big|_{\lambda>a}\right), \tag{1.23}$$

where the trace of the infinite operator must be regularized.

Again we use a $\zeta$-function to define this trace. Let

$$\omega^{(a)}(s) = \operatorname{Tr}\left((DD^*)^{-s}\, \tilde{\nabla}D\, D^{-1}\big|_{\lambda>a}\right). \tag{1.24}$$

The analytic properties of $\omega^{(a)}(s)$ are somewhat more complicated than those of $\zeta^{(a)}(s)$.

PROPOSITION 1.25. *The 1-form $\omega^{(a)}(s)$ is holomorphic for $\mathrm{Re}\, s > n/2$ and has a meromorphic continuation to $\mathbb{C}$. There is a simple pole at $s = 0$. The real part of the residue at $s = 0$ is $-\frac{1}{2}d(\varsigma^{(a)}(0))$.*

PROOF: Omitting '$\lambda > a$' from the notation for convenience, set

$$\tau_u^{(a)}(s) = \mathrm{Tr}\left((DD^* + u\tilde{\nabla}D\, D^*)^{-s}\right).$$

Since the operator in parentheses is elliptic for small $u$, the function $\tau_u^{(a)}$ has a well-defined meromorphic continuation to $\mathbb{C}$ which is regular at $s = 0$. By [**APSIII**,Proposition 2.9] we can differentiate in $u$ to obtain

$$\begin{aligned}\frac{d}{du}\Big|_{u=0} \tau_0^{(a)}(s) &= -s\,\mathrm{Tr}\left((DD^*)^{-(s+1)}\tilde{\nabla}D\, D^*\right) \\ &= -s\,\omega^{(a)}(s).\end{aligned}$$

Thus $\omega^{(a)}(s)$ has a meromorphic continuation to $\mathbb{C}$ with a simple pole at $s = 0$. Then differentiating (1.20), we obtain

$$\begin{aligned} d(\varsigma^{(a)}(s)) &= d\ \mathrm{Tr}\,((DD^*)^{-s}) \\ &= -s\ \mathrm{Tr}\left(\{(DD^*)^{-(s+1)}\}\{\tilde{\nabla}D\, D^* + D\,\tilde{\nabla}D^*\}\right) \\ &= -s\ \mathrm{Tr}\left(\{(DD^*)^{-s}\}\{\tilde{\nabla}D\, D^{-1} + (D^*)^{-1}\,\tilde{\nabla}D^*\}\right) \\ &= -2s\ \mathrm{Re}\,\omega^{(a)}(s), \end{aligned} \tag{1.26}$$

which proves the final assertion.

Equation (1.26) shows that $\mathrm{Re}\, s\,\omega^{(a)}(s)$ is holomorphic at $s = 0$, and

$$\mathrm{Re}\left(s\,\omega^{(a)}(s)\right)'(0) = -\frac{1}{2}d\left(\varsigma^{(a)\prime}(0)\right). \tag{1.27}$$

Define

$$\mathrm{Tr}\left(\tilde{\nabla}D\, D^{-1}\big|_{\lambda>a}\right) = \left(s\,\omega^{(a)}(s)\right)'(0). \tag{1.28}$$

This is the finite part of $\omega^{(a)}(s)$ at $s=0$. Again the $\varsigma$-function behaves well with respect to a finite number of eigenvalues: For $a<b$

$$\operatorname{Tr}\left(\tilde{\nabla} D\, D^{-1}\big|_{\lambda>a}\right)=\operatorname{Tr}\left(\tilde{\nabla} D\, D^{-1}\big|_{\lambda>b}\right)+\operatorname{Tr}\left(\tilde{\nabla} D\, D^{-1}\big|_{a<\lambda<b}\right). \tag{1.29}$$

The crucial point is that the pole in $\omega^{(a)}(s)$ is independent of $a$. Equation (1.29) ensures that $\bar{\nabla}^{(a)}$ and $\bar{\nabla}^{(b)}$ agree on the overlap $U^{(a)}\cap U^{(b)}$ (c.f. (1.22)), and so patch together to a connection $\nabla^{(\mathcal{L})}$ on $\mathcal{L}$. Furthermore, equation (1.27), which governs the real part of the correction term, guarantees that $\nabla^{(\mathcal{L})}$ is unitary for the Quillen metric.

This completes the constructions proving Theorem 1.12. The functorial reader can formulate and prove the naturality of $\mathcal{L}, g^{(\mathcal{L})}, \nabla^{(\mathcal{L})}$ under mappings of geometric families, although this should be clear.

The connection on $\mathcal{L}$ is closely related to the "Levi-Civita superconnection" constructed by Bismut in his heat equation proof of the Atiyah-Singer index theorem for families of Dirac operators [**B**]. Therefore, it is not surprising that the curvature of $\mathcal{L}$ is the 2-form in the curvature of Bismut's superconnection.

THEOREM 1.30 [**BF2**]. *The curvature of the determinant line bundle $\mathcal{L}\to Y$ is the 2-form component of*

$$\Omega^{(\mathcal{L})}=2\pi i\int_{Z/Y}\hat{A}(\Omega^{(Z/Y)})\,\mathrm{ch}(\dot{\rho}\Omega^{(Z/Y)})\,\mathrm{ch}(\Omega^{(E)}).$$

Here $\hat{A}$ and ch are the usual polynomials

$$\hat{A}(\Omega)=\sqrt{\det\left(\frac{\Omega/4\pi}{\sinh\,\Omega/4\pi}\right)}$$
$$\mathrm{ch}(\Omega)=\operatorname{Tr}\,e^{i\Omega/2\pi}$$

The integrand is a differential form of mixed degree on $Z$ which is integrated over the fibers of $Z\to Y$. The product of the first two terms in the integrand is the index polynomial for the appropriate Dirac-type operator. Thus if we consider a family of signature operators, this product is the Hirzebruch $L$ polynomial; for the $\bar{\partial}$ complex it is the Todd polynomial.

The holonomy formula is more complicated to state. Let $\gamma\colon S^1\to Y$ be a loop. By pullback we obtain a geometric family of Dirac operators parametrized by $S^1$. Thus there is an $(n+1)$-manifold $P$ fibered over $S^1$, a metric and spin structure along the fibers, etc. Introduce an arbitrary metric $g^{(S^1)}$ and endow $S^1$ with its bounding spin structure.[2] Then $P$ acquires a metric and spin structure, and so a self-adjoint Dirac operator $A$ (coupled to $V_\rho \otimes E$). Since our constructions are independent of the metric on the base, we must scale away the choice of metric on the circle. Replace $g^{(S^1)}$ in the preceding by $g^{(S^1)}/\epsilon^2$ for a parameter $\epsilon$, and let $A_\epsilon$ denote the Dirac operator for the scaled metric. Set

$$\eta_\epsilon = \eta\text{-invariant of } A_\epsilon$$
$$h_\epsilon = \dim \ker A_\epsilon$$
$$\xi_\epsilon = \frac{1}{2}(\eta_\epsilon + h_\epsilon).$$

The $\eta$-invariant is a spectral invariant defined by analytic continuation as the value of

$$\sum_{\lambda \in \operatorname{spec}(A_\epsilon)\setminus\{0\}} \frac{\operatorname{sgn}\lambda}{|\lambda|^s}$$

at $s=0$ [**APS**]. An easy argument shows $\xi_\epsilon \pmod 1$ is continuous in $\epsilon$.

THEOREM 1.31 [**W1**],[**BF2**]. *The holonomy of $\nabla^{(\mathcal{L})}$ around $\gamma$ is*

$$\operatorname{hol}_{\mathcal{L}}(\gamma) = \lim_{\epsilon\to 0} e^{-2\pi i \xi_\epsilon}.$$

## §2 Holomorphic Families

For applications of determinants to complex geometry it is crucial to understand when the purely Riemannian considerations of §1 are compatible with holomorphic structures. The prototypical result in this direction is a characterization of Kähler metrics: A Hermitian metric on a complex manifold is Kähler if and only if its Levi-Civita connection and Hermitian connection coincide. Special cases of Theorem 2.1 have appeared in the recent literature [**AW**], [**Bo**], [**BF1**], [**BJ**], [**D**], [**F1**]. Theorem 2.1 is due to Bismut, Gillet, and Soulé, who go much further than we do here. They construct a canonical smooth isomorphism of the determinant line bundle with the Knudsen-Mumford [**KM**] determinant such that the canonical connection is compatible with the induced complex structure.[3]

[2]The bundle of spin frames is the nontrivial double cover of the circle. In [**BF2**] we use the other spin structure and correspondingly obtain a formula which differs from (1.31) by a sign. The two formulas are easily reconciled by the flat index theorem of Atiyah, Patodi, and Singer [**APSIII**].
[3]I am grateful to Henri Gillet for explaining this work.

THEOREM 2.1 [**BGS**]. *Let $\pi\colon Z \xrightarrow{X} Y$ be a holomorphic fibration with smooth fibers. Suppose $Z$ admits a closed (1,1)-form $\tau$ which restricts to a Kähler form on each fiber. Let $E \to Z$ be a holomorphic Hermitian bundle with its Hermitian connection. Then the determinant line bundle $\mathcal{L} \to Y$ of the relative $\bar\partial$ complex (coupled to $E$) admits a holomorphic structure. The canonical connection on $\mathcal{L}$ is the Hermitian connection for the Quillen metric. Finally, the section $\det \bar\partial_E$ of $\mathcal{L}$ is holomorphic.*

The geometric data (1.1) used to define the canonical connection on $\mathcal{L}$ comes from $\tau$. The metric $g^{(Z/Y)}$ along the fibers is Kähler—the Kähler form is the restriction of $\tau$ to the fibers. The horizontal distribution on $Z$ is the annihilator of $T(Z/Y)$ relative to $\tau$. So $\xi \in TZ$ is horizontal if and only if $\tau(\xi,\varsigma) = 0$ for all vertical vectors $\varsigma \in T(Z/Y)$. The existence of $\tau$ implies that $Z$ is Kähler, at least over small regions in $Y$. (If $\sigma$ is a sufficiently large (local) Kähler form on $Y$, then $\pi^*\sigma + \tau$ is a Kähler form on $Z$.) Of course, Theorem 2.1 extends to other operators (e.g. Dirac operators) which are expressed in terms of the $\bar\partial$ complex. In this section we sketch a proof of Theorem 2.1. For notational simplicity we omit the bundle $E$ from our treatment.

The hypothesis of Theorem 2.1 is satisfied in many interesting cases. For example, if $Y$ is the Siegel upper half-plane, parametrizing marked abelian varieties, then the universal abelian variety $Z$ carries a 2-form $\tau$ which is the curvature of a suitable metric on the $\Theta$-line bundle. As another example consider a holomorphic fibration $\pi\colon Z \to Y$ with Kähler-Einstein metrics along the fibers. Suppose that the ratio of the Ricci curvature to the Kähler form in each fiber $Z_y$ is a positive constant $c$. Let $\Omega^{(Z/Y)}$ be the curvature of the Hermitian connection on $T(Z/Y)$ for this family of Kähler-Einstein metrics. Then we choose $\tau = \dfrac{i}{2\pi c}\operatorname{Tr}\Omega^{(Z/Y)}$. This applies to the universal curve over Teichmüller space (or moduli space) with the family of hyperbolic metrics of constant curvature $-1$. In general we can also construct suitable closed (1,1)-forms $\tau$ from relative projective embeddings of $Z \to Y$. In this case, $\tau$ is the curvature of a relatively ample line bundle over $Z$.

The following lemma states the consequences of the equation $d\tau = 0$ on the local geometry of $Z$. Recall the tensor $T$ of (1.10); it is the curvature of the horizontal distribution on $Z$.

LEMMA 2.2. *Under the hypotheses of Theorem 2.1, the connection $\nabla^{(Z/Y)}$ is the Hermitian connection on $T(Z/Y)$, the tensor $T$ is of type (1,1), and the 1-form $\gamma$ in (1.4)* *vanishes.*

PROOF: As remarked above, we can lift a Kähler metric on $Y$ to obtain a Kähler metric on $Z$. Its Levi-Civita connection is the Hermitian connection on $TZ$. Since $T(Z/Y)$ is a holomorphic subbundle of $TZ$, the projected connection $\nabla^{(Z/Y)}$ is the Hermitian connection. (Notice a difference with Lemma 1.3. There we lifted a metric on $Y$ to construct a *Riemannian submersion*. The Kähler metric on $Z$ does not lead to a Riemannian submersion in our present context, but the projected connection $\nabla^{(Z/Y)}$ is insensitive to the horizontal metric in any case.)

Suppose $\xi_1, \xi_2$ are lifts of horizontal vector fields on $Y$, and let $\varsigma$ be any vertical vector field. Then $[\xi_i, \varsigma]$ are vertical vector fields, and the 6-term formula for $d\tau$ yields $\tau([\xi_1, \xi_2], \varsigma) = \varsigma \cdot \tau(\xi_1, \xi_2)$. This implies that $[\xi_1, \xi_2]$ is horizontal if both $\xi_i$ are of type (1,0) or (0,1). Hence $T$ is of type (1,1).

Finally, the volume form $\mathbf{vol} = \tau^n$, where $n$ is the complex dimension of the fibers. Hence $d\mathbf{vol} = 0$, and $\gamma = 0$ follows from (1.4).

Suppose momentarily that the cohomology $H^i$ of the relative $\bar\partial$ complex

$$\Omega^{0,0}_{Z/Y} \xrightarrow{\bar\partial^{(Z/Y)}} \Omega^{0,1}_{Z/Y} \xrightarrow{\bar\partial^{(Z/Y)}} \cdots \xrightarrow{\bar\partial^{(Z/Y)}} \Omega^{0,n}_{Z/Y} \tag{2.3}$$

has (locally) constant rank. Then $H^i \to Y$ is a vector bundle, and it admits a natural holomorphic structure (without using the metric on $Z$). Working locally in $Y$, suppose $\alpha \in \Omega^{0,i}_{Z/Y}$ restricts to be holomorphic on each fiber. Let $\bar\xi$ be a (0,1)-vector field in $Y$. Lift to *any* (0,1)-vector field $\tilde{\bar\xi}$ along the fiber. Define $\bar\partial^{(H^i)}_{\bar\xi}\alpha = \bar\partial^{(Z)}_{\tilde{\bar\xi}}\alpha$ acting pointwise. Let $\bar\varsigma$ be a vertical vector field of type (0,1). Then (setting $\bar\partial = \bar\partial^{(Z)}$)

$$\bar\partial_{\bar\varsigma}\bar\partial_{\tilde{\bar\xi}}\alpha = \bar\partial_{\tilde{\bar\xi}}\bar\partial_{\bar\varsigma}\alpha - \bar\partial_{[\tilde{\bar\xi},\bar\varsigma]}\alpha.$$

But $[\tilde{\bar\xi}, \bar\varsigma]$ is vertical since $\tilde{\bar\xi}$ is the lift of a vector field from the base. Since $\alpha$ is holomorphic along the fiber, the right hand side vanishes. Furthermore, it is easily seen that the definition of $\bar\partial^{(H^i)}$ is independent of the lift. Hence $\bar\partial^{(H^i)}$ is well-defined. Its square is zero, since $\left(\bar\partial^{(Z)}\right)^2 = 0$, so it defines a holomorphic structure on $H^i \to Y$. Then the determinant of the cohomology

$$\bigotimes_i \left(\det H^i\right)^{(-1)^{i+1}} \tag{2.4}$$

also inherits a holomorphic structure.

In general, the cohomology does not have locally constant rank, and we use the Kähler metric to construct a holomorphic structure on the determinant line bundle. Denote the $i^{\text{th}}$ term $\Omega^{0,i}_{Z/Y}$ of (2.3) by $\mathcal{H}_i \to Y$, and denote $\bar\partial^{(Z/Y)}$ by $D$. The $\mathcal{H}_i$ are *continuous* Hilbert bundles over $Y$. Eventually we work with *smooth* finite dimensional subbundles consisting of smooth forms, so we treat the $\mathcal{H}_i$ formally as if they were smooth and finite dimensional. Let $\tilde\nabla_i$ be the unitary connection on $\mathcal{H}_i$ defined in (1.9).

PROPOSITION 2.5. *The (0,1) part of the connection $\tilde\nabla_i = \nabla_i$ defines a complex structure on $\mathcal{H}_i$. The relative $\bar\partial$ complex*

$$\mathcal{H}_0 \xrightarrow{D} \mathcal{H}_1 \xrightarrow{D} \cdots \xrightarrow{D} \mathcal{H}_n \tag{2.6}$$

*varies holomorphically over the parameter space $Y$.*

PROOF: By Lemma 2.2 and (1.5) we have $\tilde\nabla^{(Z/Y)} = \nabla^{(Z/Y)}$, hence also $\tilde\nabla_i = \nabla_i$. Furthermore, $\nabla^{(Z/Y)}$ is the Hermitian connection, so its curvature $\Omega^{(Z/Y)}$ is of type (1,1). Since $T$ is of type (1,1), it follows from Proposition 1.11 that $\Omega^{(\mathcal{H}_i)}$ is of type (1,1). Thus the (0,1) component $\bar\nabla_i$ defines a $\bar\partial$ operator on $\mathcal{H}_i$, since $(\bar\nabla_i)^2 = 0$.

To check that $D$ varies holomorphically, we introduce local frames. Let $\sigma$ be a Kähler from on $Y$ such that $\pi^*\sigma + \tau$ is a Kähler form on $Z$. Fix $\xi_\alpha$ a local unitary basis of vector fields of type (1,0) on $Y$, and lift to horizontal vector fields $\xi_\alpha$ on $Z$. Note that $\xi_\alpha$ (on $Z$) are *not* unitary. Choose vertical vector fields $\varsigma_i$ of type (1,0) which form a unitary basis on each fiber. Let $\theta^\alpha, \phi^i$ be the dual 1-forms. Then $\tau = \sum_i \phi^i \wedge \bar\phi^i$, and

$$d\Big(\sum_i \phi^i \wedge \bar\phi^i\Big) = 0. \tag{2.7}$$

As a preliminary step we show

$$\phi^k(\nabla_{\varsigma_j}\xi_\alpha) = 0. \tag{2.8}$$

For this evaluate (2.7) on vectors $\varsigma_j, \xi_\alpha, \bar\varsigma_k$. Then since the Kähler connection $\nabla^{(Z)}$ is torsionfree, we obtain

$$\bar\phi^j(\nabla_{\xi_\alpha}\bar\varsigma_k) + \phi^k(\nabla_{\xi_\alpha}\varsigma_j) - \phi^k(\nabla_{\varsigma_j}\xi_\alpha) = 0.$$

The first two terms cancel since $\nabla$ preserves the metric along the fibers:

$$\begin{aligned} 0 &= \xi_\alpha\langle \varsigma_j, \bar{\xi}_k\rangle \\ &= \langle \nabla_{\xi_\alpha}\varsigma_j, \bar{\varsigma}_k\rangle + \langle \varsigma_j, \nabla_{\xi_\alpha}\bar{\varsigma}_k\rangle \\ &= \phi^k(\nabla_{\xi_\alpha}\varsigma_j) + \bar{\phi}^j(\nabla_{\xi_\alpha}\bar{\varsigma}_k). \end{aligned}$$

The desired equation (2.8) follows.

Now

$$\begin{aligned} D &= \bar{\phi}^j \wedge \nabla_{\bar{\varsigma}_j}, \\ \overline{\nabla} &= \bar{\theta}^\alpha \otimes \nabla_{\bar{\xi}_\alpha}, \end{aligned} \tag{2.9}$$

where $\nabla = \nabla^{(Z/Y)}$ on the right hand side of (2.9), and repeated indices are summed. We compute

$$\begin{aligned} \overline{\nabla} D &= \bar{\theta}^\alpha \otimes \left(\bar{\phi}^j \wedge \nabla_{\bar{\xi}_\alpha}\nabla_{\bar{\varsigma}_j} + \nabla_{\bar{\xi}_\alpha}\bar{\phi}^j \wedge \nabla_{\bar{\varsigma}_j}\right), \\ D\overline{\nabla} &= \bar{\theta}^\alpha \otimes \left(\bar{\phi}^j \wedge \nabla_{\bar{\varsigma}_j}\nabla_{\bar{\xi}_\alpha}\right). \end{aligned} \tag{2.10}$$

Since the curvature $\Omega^{(Z/Y)}$ has type (1,1),

$$\nabla_{\bar{\xi}_\alpha}\nabla_{\bar{\varsigma}_j} - \nabla_{\bar{\varsigma}_j}\nabla_{\bar{\xi}_\alpha} = \nabla_{[\bar{\xi}_\alpha, \bar{\varsigma}_j]}. \tag{2.11}$$

Now $[\bar{\xi}_\alpha, \bar{\varsigma}_j]$ is vertical, so using (2.8) and the fact that $\nabla$ is torsionfree,

$$[\bar{\xi}_\alpha, \bar{\varsigma}_j] = \nabla_{\bar{\xi}_\alpha}\bar{\varsigma}_j = \bar{\phi}^k(\nabla_{\bar{\xi}_\alpha}\bar{\varsigma}_j)\bar{\varsigma}_k. \tag{2.12}$$

Combining (2.10)–(2.12), and writing $\nabla_{\bar{\xi}_\alpha}\bar{\phi}^j = (\nabla_{\bar{\xi}_\alpha}\bar{\phi}^j)(\bar{\varsigma}_k)\bar{\phi}^k$, we obtain

$$\overline{\nabla} D - D\overline{\nabla} = \left\{(\nabla_{\bar{\xi}_\alpha}\bar{\phi}^j)(\bar{\varsigma}_k) + \bar{\phi}^j(\nabla_{\bar{\xi}_\alpha}\bar{\varsigma}_k)\right\}\bar{\theta}^\alpha \otimes (\bar{\phi}^k \wedge \nabla_{\bar{\varsigma}_j}).$$

The expression in braces vanishes, and therefore $[\overline{\nabla}, D] = 0$.

Next, we construct the holomorphic determinant line bundle $\mathcal{L}_{\text{hol}} \to Y$ of the complex (2.6). Each fiber of $\mathcal{L}_{\text{hol}}$ is isomorphic to (2.4), but as in §1 we must use a patching argument to construct the bundle. To grasp the ideas it is best to work first in a general setting, as in Quillen [Q]. Let Fred(H) be the space of *Fredholm complexes*

$$0 \to H_0 \xrightarrow{T_1} H_1 \xrightarrow{T_2} \cdots \xrightarrow{T_n} H_n \to 0. \tag{2.13}$$

Each $T_i$ has closed range, $\operatorname{im} T_i \subset \ker T_{i+1}$, and the cohomology $\ker T_{i+1}/\operatorname{im} T_i$ is finite dimensional. Over open sets $\mathcal{U} \subset \mathrm{Fred}(\mathbf{H})$ there exist finite dimensional subcomplexes

$$0 \to V_0 \xrightarrow{T_1} V_1 \xrightarrow{T_2} \cdots \xrightarrow{T_n} V_n \to 0 \tag{2.14}$$

such that $V_i \subset \mathcal{U} \times H_i$ is a holomorphic subbundle, and the inclusion of (2.14) into (2.13) induces an isomorphism on cohomology. To construct a local model (2.14), choose a finite dimensional subspace $V_n \subset H_n$ such that $V_n$ maps onto $\operatorname{coker} T_n(y)$ at some fixed $y \in \mathrm{Fred}(\mathbf{H})$; then $V_n$ surjects onto $\operatorname{coker} T_n$ in a neighborhood of $y$. For each $i = 0, 1, \ldots, n-1$ choose an (infinite dimensional) subspace $U_i \subset H_i$ complementary to $\ker T_{i+1}(y)$; then $U_i \oplus \ker T_i = H_i$ in a neighborhood of $y$. Also, choose a finite dimensional holomorphic subbundle $W_i \subset \ker T_{i+1}$ which surjects onto the cohomology. This is possible since $\ker T_{i+1}$ is holomorphic and the dimension of the cohomology is locally bounded. Then define $V_i$ (using downward induction) via the formula

$$V_i = W_i + \left(T_{i+1}^{-1}(V_{i+1}) \cap U_i\right). \tag{2.15}$$

To each local model (2.14) corresponds a holomorphic line bundle

$$\mathcal{L}_V = \bigotimes_{i=1}^{n} (\det V_i)^{(-1)^{i+1}} \tag{2.16}$$

over $\mathcal{U}$. If the Fredholm complex (2.13) has index zero, then $\det T$ is a holomorphic section of $\mathcal{L}_{\mathrm{hol}}$. To patch with another local model $\{V_i'\}$, we may as well assume $V_i \subset V_i'$. Then in the commutative diagram

$$\begin{array}{ccccccccc}
 & & 0 & & 0 & & & 0 & \\
 & & \downarrow & & \downarrow & & & \downarrow & \\
0 & \longrightarrow & V_0 & \longrightarrow & V_1 & \longrightarrow \cdots \longrightarrow & & V_n & \longrightarrow 0 \\
 & & \downarrow & & \downarrow & & & \downarrow & \\
0 & \longrightarrow & V_0' & \longrightarrow & V_1' & \longrightarrow \cdots \longrightarrow & & V_n' & \longrightarrow 0 \\
 & & \downarrow & & \downarrow & & & \downarrow & \\
0 & \longrightarrow & V_0'/V_0 & \longrightarrow & V_1'/V_1 & \longrightarrow \cdots \longrightarrow & & V_n'/V_n & \longrightarrow 0 \\
 & & \downarrow & & \downarrow & & & \downarrow & \\
 & & 0 & & 0 & & & 0 &
\end{array}$$

the columns and the bottom row are exact. Hence there is a canonical isomorphism

$$\mathcal{L}_V \cong \mathcal{L}_V \otimes \mathcal{L}_{V'/V} = \mathcal{L}_{V'} \tag{2.17}$$

via the nonzero holomorphic section $\det T$ of $\mathcal{L}_{V'/V}$ determined by the action of $T$ on the quotient complex. This patching produces a holomorphic determinant line bundle $\mathcal{L}_{\text{hol}} \to$ Fred($\mathbf{H}$) and a canonical holomorphic section over the index zero Fredholm complexes.

For the relative $\bar\partial$ complex (2.6), the Hilbert spaces change as we vary the parameter. But since the complexes vary holomorphically, by Proposition 2.5, the preceding constructions extend easily to this case. Therefore, we obtain a holomorphic determinant bundle $\mathcal{L}_{\text{hol}} \to Y$ and a holomorphic section $\det D$ (if index $D$ vanishes).

Next[4] we define a smooth map $j\colon \mathcal{L}_{\text{hol}} \to \mathcal{L}$, where $\mathcal{L}$ is the determinant line bundle (constructed in §1) for the collapsed complex

$$D + D^*\colon \bigoplus_{i \text{ even}} \mathcal{H}_i \longrightarrow \bigoplus_{i \text{ odd}} \mathcal{H}_i. \tag{2.18}$$

As a preliminary step, consider an exact sequence of inner product spaces

$$0 \to E_0 \xrightarrow{T} E_1 \xrightarrow{T} \cdots \xrightarrow{T} E_n \to 0. \tag{2.19}$$

[4]The relationship between $\mathcal{L}_{\text{hol}}$ and $\mathcal{L}$ was elucidated in Donaldson [**D**], and we follow his arguments closely.

The determinant

$$\det T \in \bigotimes_{i=0}^{n} (\det E_i)^{(-1)^{i+1}} \tag{2.20}$$

is nonzero. Form the linear transformation

$$T + T^* : \bigoplus_{i\ \mathrm{even}} E_i \longrightarrow \bigoplus_{i\ \mathrm{odd}} E_i . \tag{2.21}$$

It is invertible, and its determinant

$$\det(T + T^*) \in \bigotimes_{i\ \mathrm{even}} (\det E_i)^{-1} \bigotimes_{i\ \mathrm{odd}} (\det E_i) \tag{2.22}$$

is nonzero. However, the determinants do *not* correspond under the natural isomorphism of the complex lines in (2.20) and (2.22). Write

$$E_i = E_i' \oplus E_i'', \qquad E_i' = \operatorname{im} T, \quad E_i'' = \ker T^* . \tag{2.23}$$

Then a short calculation shows[5]

$$\det(T + T^*) = (\det T)\Bigl(\prod_{i\ \mathrm{even}} \det\bigl(T T^*\bigl|_{E_i'}\bigr)\Bigr). \tag{2.24}$$

Fix $a > 0$. Recall the local model (1.14) for the determinant line bundle $\mathcal{L}$. Choose a local model (2.14) for $\mathcal{L}_{\mathrm{hol}}$ as follows. Fix $y \in Y$ such that $a$ is not an eigenvalue of the Laplacian at $y$, and set $V_i = \mathcal{H}_i^{(a)}$ at the point $y$. (As before, $\mathcal{H}_i^{(a)}$ is the sum of the eigenspaces for eigenvalues $< a$ of the $i^{\mathrm{th}}$ Laplacian $\Delta_i$.) Extend to a local model in a neighborhood of $y$, as described above, and denote the determinant of the resulting complex by $\mathcal{L}_{\mathrm{hol}}^{(a)}$. Then, possibly in a smaller neighborhood of $y$, the projections $P^{(a)}$ onto $\mathcal{H}_i^{(a)}$ define an isomorphism of complexes

$$\begin{array}{ccccccccccc} 0 & \longrightarrow & V_0 & \xrightarrow{D} & V_1 & \xrightarrow{D} & \cdots & \xrightarrow{D} & V_n & \longrightarrow & 0 \\ & & \big\downarrow P^{(a)} & & \big\downarrow P^{(a)} & & & & \big\downarrow P^{(a)} & & \\ 0 & \longrightarrow & \mathcal{H}_0^{(a)} & \xrightarrow{D} & \mathcal{H}_1^{(a)} & \xrightarrow{D} & \cdots & \xrightarrow{D} & \mathcal{H}_n^{(a)} & \longrightarrow & 0 \end{array} \tag{2.25}$$

[5] Observe that $\det(T + T^*) = \det T$ for complexes of length 2. So this extra factor does not enter for Riemann surfaces.

There is an induced isomorphism $\det P^{(a)}$ on the determinants. We introduce a correction factor to pass to the determinant of the collapsed complex (2.18). The Hodge decomposition of $\mathcal{H}_i$ is

$$\mathcal{H}_i = H_i \oplus \mathcal{H}_i' \oplus \mathcal{H}_i'', \qquad H_i = \ker \Delta_i, \quad \mathcal{H}_i' = \operatorname{im} D, \quad \mathcal{H}_i'' = \operatorname{im} D^*. \tag{2.26}$$

The nonzero eigenvalues of the Laplacian $\Delta_i$ split accordingly into $\{\lambda_i'\}, \{\lambda_i''\}$. Define the regularized determinant

$$\mu^{(a)} = \Bigl(\prod_{\substack{i \text{ even} \\ \lambda_i' > a}} \lambda_i'\Bigr)^{-1} \tag{2.27}$$

using $\varsigma$-functions. Then by (2.24) the maps

$$\mu^{(a)} \cdot \det P^{(a)} \colon \mathcal{L}_{\text{hol}}^{(a)} \longrightarrow \mathcal{L}^{(a)} \tag{2.28}$$

patch together into a smooth isomorphism

$$j \colon \mathcal{L}_{\text{hol}} \longrightarrow \mathcal{L}. \tag{2.29}$$

We use $j$ to pull back the metric and connection from $\mathcal{L}$.

It remains to prove that $j^*\nabla^{(\mathcal{L})}$ is the Hermitian connection for $j^*g^{(\mathcal{L})}$. Since $j^*\nabla^{(\mathcal{L})}$ is unitary, we need only check that it is compatible with the complex structure on $\mathcal{L}_{\text{hol}}$. We work in the local model (2.25) for $j$. Recall that on $\mathcal{L}^{(a)}$ the connection $\nabla^{(\mathcal{L})}$ is a sum of two terms (1.23). The first is the determinant of the projected connections $\nabla_i^{(a)} = P^{(a)}\nabla_i$ on $\mathcal{H}_i^{(a)} \subset \mathcal{H}_i$. We claim that each $P^{(a)*}\nabla_i^{(a)}$ is compatible with the complex structure on $V_i$. For if $s$ is a local section of $V_i$, then since (2.25) is a map of complexes, and $\overline{\nabla}$ commutes with $D$ (Proposition 2.5), we have (omitting the index $i$)

$$\begin{aligned} P^{(a)} D \left(P^{(a)*}\overline{\nabla}^{(a)}\right) s &= D P^{(a)} \overline{\nabla} P^{(a)} s \\ &= P^{(a)} D \overline{\nabla} P^{(a)} s \\ &= P^{(a)} \overline{\nabla} D P^{(a)} s \\ &= P^{(a)} \left(P^{(a)*}\overline{\nabla}^{(a)}\right) D s. \end{aligned} \tag{2.30}$$

Hence $j^*\overline{\nabla}^{(a)}$ commutes with $D$, as claimed.

The second contribution to the connection on $\mathcal{L}^{(a)}$ is the regularized 1-form (1.28). Relative to the Hodge decomposition (2.26) we write (c.f. (1.24))

$$\omega^{(a)}(s) = \sum_{i \text{ even}} \operatorname{Tr}\left(\Delta_i^{-s} \nabla D\, D^{-1}\big|_{\lambda' > a}\right) + \sum_{i \text{ even}} \operatorname{Tr}\left(\Delta_i^{-s} \nabla D^*(D^*)^{-1}\big|_{\lambda'' > a}\right). \tag{2.31}$$

Since $D$ varies holomorphically (Proposition 2.5), equation (2.31) is the decomposition of $\omega^{(a)}(s)$ into forms of type (1,0) and (0,1). The compatibility of (1.28) with the holomorphic structure amounts to the assertion

$$\bar{\partial}\left(\log \mu^{(a)}\right) + (0,1) \text{ part of } \left(s\,\omega^{(a)}(s)\right)'(0) = 0. \tag{2.32}$$

We verify (2.32) by a calculation similar to (1.26). Set

$$\varsigma_i^{(a)}(s) = \operatorname{Tr}\left(\Delta_i^{-s}\big|_{\lambda > a}\right). \tag{2.33}$$

(Henceforth we omit reference to '$a$' in the notation.) Choose numbers $n_i$ satisfying

$$n_{i-1} + n_i = \begin{cases} 1, & i \text{ even}, \\ 0, & i \text{ odd}. \end{cases}$$

Then $\log \mu = \sum_{i=0}^{n} n_i\, \varsigma_i'(0)$. Since $\overline{\nabla} D = 0$, differentiating (2.33) we obtain

$$\bar{\partial}\varsigma_i(s) = -s \operatorname{Tr}\left(\Delta_i^{-s}\{\overline{\nabla} D^*(D^*)^{-1} + (D^*)^{-1}\overline{\nabla} D^*\}\right).$$

Therefore,

$$\begin{aligned} \bar{\partial}\left(\sum_{i=0}^{n} n_i\, \varsigma_i(s)\right) &= -s\sum_{i=0}^{n} \operatorname{Tr}\left(\Delta_i^{-s}\{n_i\overline{\nabla} D^*(D^*)^{-1} + n_{i-1}\overline{\nabla} D^*(D^*)^{-1}\}\right) \\ &= -s \sum_{i \text{ even}} \operatorname{Tr}\left(\Delta_i^{-s}\overline{\nabla} D^*(D^*)^{-1}\right) \\ &= -s\,\omega(s)_{(0,1)}. \end{aligned} \tag{2.34}$$

The desired equation (2.32) follows by differentiating (2.34) and setting $s = 0$.

This completes the proof of Theorem 2.1. As a final note, we display the metric $j^*g^{(\mathcal{L})}$ on $\mathcal{L}_{\text{hol}}$ explicitly. At a fixed point in the parameter space we take the sequence of cohomology groups as a model for $\mathcal{L}_{\text{hol}}$. Its determinant (2.4) carries the $L^2$ metric, and the metric $j^*g^{(\mathcal{L})}$ is this $L^2$ metric multiplied by the *analytic torsion* [**RS1**]

$$\Big(\prod_{i\text{ even}} \lambda_i'\Big)^{-1}\Big(\prod_{i\text{ even}} \lambda_i''\Big).$$

## §3 A Square Root

There are several possible variations on the basic determinant line bundle construction of §1. In this section we consider families of Dirac operators on a spin manifold $X$ of dimension $8k+2$. Riemann surfaces, 10 dimensional manifolds, and 26 dimensional manifolds all fit into this framework. These arise in string theory and in its low energy field theory limits. In $8k+2$ dimensions the determinant line bundle $\mathcal{L}$ of the complex chiral Dirac operator admits a canonical square root $\mathcal{L}^{1/2}$, and $\sqrt{\det D}$ makes sense as a section of $\mathcal{L}^{1/2}$. (The Dirac operator has numerical index zero in these dimensions.) This construction extends to Dirac operators coupled to *real* vector bundles (e.g., the real Rarita-Schwinger operator). The relevance of this square root in physics is found in a discrepancy between Clifford algebras in Euclidean space and Lorentz space. Relative to the Euclidean signature the even two dimensional Clifford algebra is $\mathbb{C}$, whereas for the Lorentz signature it is $\mathbb{R}\oplus\mathbb{R}$. By periodicity, in dimension $8k+2$ these are replaced by appropriate matrix algebras. Hence in Lorentz geometry there is a real chiral Dirac operator—there exist Majarana-Weyl spinors. These are the fermions of the physical theories. But when continued to imaginary time, i.e., to Riemannian geometry, there are only *complex* chiral spinors. We view the square root of the complex Dirac determinant as the Euclidean analog of the real Lorentz determinant. It accounts for a mysterious factor of 2 that crops up in the physics literature [**AgW**], [**ASZ**], [**W1**]. This square root plays a significant role in §4 and §5. We remark that the complex Pfaffian also enters the work of Pressley and Segal [**PS**], who use it to construct the spin representation.

Topologically, the existence of this square root can be attributed to Rohlin's theorem. One generalization of Rohlin's theorem asserts that the Dirac operator on a closed spin

$(8k+4)$-manifold has even index. This follows from the fact that the even Clifford algebra in $8k+4$ dimensions is a direct sum of two quaternion algebras. The chiral Dirac operator is quaternionic, its kernel and cokernel are quaternionic vector spaces, and so have even complex dimension. Suppose $\pi: Z \to Y$ is a family of $8k+2$ dimensional spin manifolds, and $\mathcal{L} \to Y$ is the determinant line bundle for the family of complex Dirac operators. The rational Chern class of $\mathcal{L}$ is evaluated on closed 2-manifolds $\Sigma \to Y$. The fibration $\pi$ pulls back to $Q \to Y$, where $Q$ is a closed $(8k+4)$-manifold, and $c_1(\mathcal{L})[\Sigma] = \operatorname{index} D_Q$ is even (by Rohlin's theorem). To detect the integral Chern class we must also use $\mathbb{Z}/\ell$-surfaces $\bar{\Sigma}$ as in [**F2**] (c.f. §5). Then if $\bar{\Sigma} \to Y$ pulls back $\pi$ to a fibration $\bar{Q} \to \bar{\Sigma}$, we see that $c_1(\mathcal{L})[\bar{\Sigma}] = \operatorname{index} D_{\bar{Q}} \pmod{\ell}$ is the mod $\ell$ index of the Atiyah-Patodi-Singer boundary value problem on an $8k+4$ dimensional manifold with boundary [**FM**]. Again this is even because spinors are quaternionic. Thus $c_1(\mathcal{L})$ is divisible by 2, and $\mathcal{L}$ admits a topological square root. The import of Theorem 3.1 is the existence of a canonical square root of the section $\det D$, which is not guaranteed by these topological considerations.

THEOREM 3.1. *In the situation of* (1.1) *assume that* $X$ *is an* $(8k+2)$*-dimensional spin manifold,* $\rho$ *is a real representation, and* $E \to Z$ *is a real vector bundle. Then there is a canonical complex line bundle* $\mathcal{L}^{1/2} \to Y$ *together with a canonical isomorphism* $\mathcal{L}^{1/2} \otimes \mathcal{L}^{1/2} \cong \mathcal{L}$*, where* $\mathcal{L}$ *is the determinant line bundle constructed in Theorem 1.12. If* $\operatorname{index} D_E = 0$ *there is a section* $\operatorname{Pfaff}(D)$ *of* $\mathcal{L}^{1/2}$ *with* $\operatorname{Pfaff}(D)^{\otimes 2} = \det D$*. The square root* $\mathcal{L}^{1/2}$ *inherits a metric and connection. Its curvature is*

$$\Omega^{(\mathcal{L}^{1/2})} = \frac{1}{2} \cdot 2\pi i \int_{Z/Y} \hat{A}(\Omega^{(Z/Y)}) \operatorname{ch}(\dot{\rho}\Omega^{(Z/Y)}) \operatorname{ch}(\Omega^{(E)}). \tag{3.2}$$

*The holonomy around a loop* $\gamma: S^1 \to Y$ *is (c.f. Theorem 1.31 for notation)*

$$\operatorname{hol}_{\mathcal{L}^{1/2}}(\gamma) = \lim_{\epsilon \to 0} e^{-\pi i \xi_\epsilon}. \tag{3.3}$$

*For holomorphic families (as in Theorem 2.1) the bundle* $\mathcal{L}^{1/2} \to Y$ *and section* $\operatorname{Pfaff}(D)$ *are holomorphic.*

The limit in (3.3) makes sense as the self-adjoint Dirac operator in $8k+3$ dimensions is quaternionic, and so $\xi \pmod 2$ is continuous as a function of the operator. For simplicity we will only consider the pure Dirac operator ($\rho = 1$, $E$ omitted) for the rest of this section.

The construction of $\mathcal{L}^{1/2}$ relies on the *complex Pfaffian*. Let $V$ be a finite dimensional complex vector space and $T\colon V \to V^*$ a skew-symmetric linear map. Then $T$ can be identified with an element $\omega_T \in \bigwedge^2 V^*$. If $T$ is nonsingular, then $\dim V = 2r$ is even, and

$$\text{(3.4)} \qquad \mathrm{Pfaff}(T) = \frac{1}{r!}\omega_T^r \in {\textstyle\bigwedge}^{2r} V^* = \det V^*.$$

On the other hand, regarding $T \in V^* \otimes V^*$ we have

$$\text{(3.5)} \qquad \det(T) \in (\det V^*) \otimes (\det V^*).$$

An easy calculation using a normal form for $T$ shows

$$\text{(3.6)} \qquad \mathrm{Pfaff}(T)^{\otimes 2} = \det(T)$$

in $(\det V^*) \otimes (\det V^*) = (\det V^*)^{\otimes 2}$. The Pfaffian obeys the multiplicative law

$$\text{(3.7)} \qquad \mathrm{Pfaff}(T_1 \oplus T_2) = \mathrm{Pfaff}(T_1) \otimes \mathrm{Pfaff}(T_2).$$

The real Dirac operator in $n = 8k+2$ dimensions is an infinite dimensional example of a skew-symmetric complex map [**H**,§4.2] (c.f. [**AS1**]). Let $\mathrm{Spin}(X) \to X$ denote the spin bundle of frames on $X$, and set $W = \mathrm{Spin}(X) \times_{\mathrm{Spin}(n)} C_n$, where $\mathrm{Spin}(n)$ acts on the real Clifford algebra $C_n$ by left multiplication. Then $W = W_+ \oplus W_-$ according to the splitting of $C_n$ into even and odd elements, and the space $\mathcal{H}$ of sections of $W$ is a $\mathbb{Z}/2\mathbb{Z}$-graded Clifford module, since $C_n$ acts on $W$ by right multiplication. The real Dirac operator $D\colon \mathcal{H} \to \mathcal{H}$ is defined as usual by the covariant derivative and left exterior multiplication by covectors; it is an odd endomorphism of the Clifford module $\mathcal{H}$. Fix a basis $e_n, \dots, e_n$ of $\mathbb{R}^n$, and consider the operator

$$\text{(3.8)} \qquad A = e_n D\colon \mathcal{H}_+ \to \mathcal{H}_+.$$

Here $e_n$ acts by right Clifford multiplication. The operator $A$ is skew-symmetric and anticommutes with right multiplication by $J = e_1 \cdots e_n$. Now $J^2 = -1$, so that $J$ is a complex structure on $\mathcal{H}_\pm$. Since $A$ anticommutes with $J$, it is a map from $\mathcal{H}_+$ to its complex dual $\mathcal{H}_+^*$. Finally, $A\colon \mathcal{H}_+ \to \mathcal{H}_+^*$ is complex skew-adjoint. We use $e_n$ to identify $\mathcal{H}_+^*$ and $\mathcal{H}_-$, and so identify $A$ and $D$.

When $X$ is a Riemann surface ($n = 2$) the Clifford algebra $C_2 \cong \mathbb{H}$ has 4 real dimensions. Thus $W_\pm$ are one dimensional complex spaces, and $\mathcal{H}_+$ ($\mathcal{H}_-$) is the space of positive (negative) complex spinor fields on $X$. Hence $D\colon \mathcal{H}_+ \to \mathcal{H}_-$ is the usual complex Dirac operator. In terms of the holomorphic geometry on the surface, a spin structure corresponds to a holomorphic square root $K_X^{1/2}$ of the canonical bundle $K_X$. Then $\bar\partial_{K^{1/2}}: \Omega^{0,0}(K^{1/2}) \to \Omega^{0,1}(K^{1/2})$ is the Dirac operator [**H**,§2.1]. (We assume that the metric on $X$ is Kähler.) The natural pairing $\Omega^{0,0}(K^{1/2}) \otimes \Omega^{0,1}(K^{1/2}) \to \mathbb{C}$ (via integration) is the duality of the previous paragraph. The operator $\bar\partial_{K^{1/2}}$ is skew-adjoint by Leibnitz' rule and Stokes' theorem.

In higher dimensions the Clifford algebra $C_n$ contains several copies of the complex irreducible half-spin representations. The simultaneous eigenspace decomposition of the commuting operators $e_1e_2, e_3e_4, \dots, e_{n-1}e_n$ (acting by right Clifford multiplication) gives the irreducible pieces. Since $J$ commutes with these operators, each irreducible representation admits a complex structure. For our geometric constructions we replace $\mathcal{H}_\pm$ by fields in some irreducible representation. Then $D\colon \mathcal{H}_+ \to \mathcal{H}_-$ is the complex Dirac operator of §1.

PROOF OF THEOREM 3.1: For ease of notation we denote $\mathcal{L}^{1/2}$ by $\mathcal{K}$. The construction of $\mathcal{K}$ proceeds by patching, as in §1. Following the notation used there, set

$$\mathcal{K}^{(a)} = \left(\det \mathcal{H}_+^{(a)}\right)^*. \tag{3.9}$$

On the overlap $U^{(a)} \cap U^{(b)}$ we have the invertible skew-symmetric operator $D^{(a,b)}$, and

$$\begin{aligned} \mathcal{K}^{(a)} &\longrightarrow \mathcal{K}^{(a)} \otimes \mathcal{K}^{(a,b)} \cong \mathcal{K}^{(b)} \\ s &\longmapsto s \otimes \operatorname{Pfaff}\left(D^{(a,b)}\right) \end{aligned} \tag{3.10}$$

are the patching maps. The natural isomorphisms $\mathcal{K}^{(a)} \otimes \mathcal{K}^{(a)} \cong \mathcal{L}^{(a)}$ correspond on overlaps—compare (1.16) and (3.10) via (3.6)—thus giving $\mathcal{K} \otimes \mathcal{K} \cong \mathcal{L}$. Also, the sections $\operatorname{Pfaff}\left(D^{(a)}\right)$ of $\mathcal{K}^{(a)}$ fit together into a section $\operatorname{Pfaff}(D)$ of $\mathcal{K}$ by (3.7). (The Pfaffian of a singular operator is defined to be zero.) We have $\operatorname{Pfaff}(D)^{\otimes 2} = \det D$ by (3.6).

The metric and connection lift to the square root. Alternatively, they can be constructed directly as in §1, replacing determinants by Pfaffians. Note in particular that $\omega^{(a)}(s)$ in (1.24) is replaced by $\frac{1}{2}\omega^{(a)}(s)$. The curvature formula (3.2) is a direct consequence of (1.30)—the curvature of the square root of a line bundle is half the original curvature. What is not immediately apparent is the holonomy formula (3.3), which depends on the particular square root (and isomorphism $\mathcal{K} \otimes \mathcal{K} \cong \mathcal{L}$). Now the main step in the proof of Theorem 1.31 is the proof of the holonomy formula for loops $\gamma$ along which $D$ is invertible; the general case follows by perturbation [**BF2**]. But when $D$ is invertible there is a canonical trivialization of $\mathcal{L}$ given by det $D$, and the logarithm of the holonomy around $\gamma$ is $-\int_\gamma \omega^{(\delta)}$ for $\delta$ sufficiently small. The holonomy theorem for $\gamma$ is the assertion[6]

$$\lim_{\epsilon\to 0} \xi_\epsilon = \frac{1}{2\pi i}\int_\gamma \omega^{(\delta)}. \tag{3.11}$$

The important point is that (3.11) holds over the reals. Therefore, we can divide by 2 to obtain (3.3) for loops of invertible operators. The general case follows as before by perturbation (using the fact that $\frac{1}{2}\xi_\epsilon$ (mod 1) is continuous).

The topological class of $\mathcal{L}^{1/2}$ is determined in real $K$-theory. Recall the operator $A$ in (3.8). Its symbol is the $KO$ orientation along the fibers of $\pi\colon Z \to Y$, and so

$$\operatorname{index} A = \pi_!(1), \tag{3.12}$$

where

$$\pi_!\colon KO(Z) \longrightarrow KO^{-(8k+2)}(Y)$$

is the direct image map. (See [**H**,§4.2] for details.) Now $KO^{-(8k+2)} \cong KO^{-2}$ by periodicity, and there is a natural map

$$\text{Pfaff}\colon KO^{-2}(Y) \longrightarrow H^2(Y;\mathbb{Z}), \tag{3.13}$$

which we call the Pfaffian. (One can see this by doing our construction over the space of skew-adjoint complex Fredholm operators, which is a classifying space for $KO^{-2}$ [**AS1**].

[6] Equation (3.11) is the common analytic link between the various geometric interpretations of Witten's global anomaly (c.f. [**C**], [**S**]).

Another model for that classifying space is $\Omega O$, the loop space of the infinite orthogonal group, and $H^2(\Omega O; \mathbb{Z}) \cong \mathbb{Z}$.) Note that the diagram

$$\begin{array}{ccc} KO^{-2}(\Sigma) & \xrightarrow{\mathrm{Pfaff}^{\otimes 2}} & H^2(\Sigma;\mathbb{Z}) \\ \downarrow & \nearrow_{\det} & \\ K^{-2}(\Sigma) & & \end{array} \tag{3.14}$$

commutes. The vertical arrow is complexification. Summarizing,

PROPOSITION 3.15. *In the situation of Theorem 3.1 we have*

$$c_1(\mathcal{L}^{1/2}) = \mathrm{Pfaff}\,(\pi_!(1))$$

*as a topological line bundle.*

## §4 Riemann Surfaces

Determinants on Riemann surfaces arise in both mathematics and physics. Here we survey some recent literature, choosing examples which illustrate the geometric principles of §1–§3. The case of elliptic curves is most instructive, as everything can be calculated explicitly. The Pfaffian of the Dirac operator (for the distinguished spin structure) is essentially the Dedekind $\eta$-function. Atiyah [**A2**] made a very detailed study of its transformation law, relating it to Witten's formula (Theorem 1.31) on the one hand,[7] and to various topological expressions on the other. The determinant of the Dirac operator with coefficients in a flat bundle is computed by Kronecker's second limit formula [**RS2**]. The result (4.11) is ubiquitous in the string theory literature. For surfaces of arbitrary genus this determinant is essentially a $\vartheta$-function. There is a general formula (4.16) expressing the variation of the determinant under a scale change—the conformal anomaly. We give a

[7] We modify Atiyah's presentation somewhat. He uses the *signature* operator and refines Witten's formula to compute the logarithm of the holonomy (as a real number).

simple derivation due to Bost and Jolicœur. Another result due to Bost relates Quillen's norm to that occurring in Faltings' work (Proposition 4.21). The transformation law for the determinant of the Dirac operator coupled to a flat bundle of finite order can be calculated in $K$-theory, as we discuss at the end of this section.

We begin by recalling some facts about spin structures on Riemann Surfaces [**A1**]. Let $X$ be a closed Riemann surface of genus $g$. In holomorphic terms a spin structure on $X$ is a choice of a *theta characteristic* [**ACGH**,p.287], i.e., a line bundle $K^{1/2} \in \mathrm{Pic}^{g-1}(X)$ satisfying $(K^{1/2})^{\otimes 2} \cong K$. There are $2^{2g}$ such choices; the ratio of any two is a point of order two on the Jacobian $\mathrm{Pic}^0(X)$. The choice of $K^{1/2}$ produces an isomorphism $\mathrm{Pic}^0(X) \cong \mathrm{Pic}^{g-1}(X)$, by tensor product. The canonical divisor on $\mathrm{Pic}^{g-1}(X)$, consisting of bundles $L$ for which $H^0(X,L) \neq 0$, pulls back to a symmetric divisor on $\mathrm{Pic}^0(X)$, the $\Theta$-divisor. The spin structures split into two classes according as $\dim H^0(X, K^{1/2})$ is even or odd. In fact,

$$K^{1/2} \longmapsto \dim H^0(X, K^{1/2}) \pmod 2 \tag{4.1}$$

is a quadratic function on the space of spin structures. Atiyah proves that

$$\dim H^0(X, K^{1/2}) = \pi_!^X(1) \pmod 2, \tag{4.2}$$

where $\pi_!^X\colon KO(X) \to KO^{-2}(\text{point}) = \mathbb{Z}/2\mathbb{Z}$ is the direct image map in $KO$-theory determined by the spin structure. The group of orientation-preserving diffeomorphisms of $X$ acts on the set of spin structures. There are two orbits, and they are distinguished by the invariant (4.2). Spin structures correspond to quadratic forms on $H_1(X;\mathbb{Z}/2\mathbb{Z})$ which refine the intersection pairing. If $S$ is an embedded circle in $X$, then the form evaluated on the homology class of $S$ is 0 or 1 according as the spin structure restricts trivially or nontrivially to $S$. (Here 'trivial' means 'bounding.') The Arf invariant of this quadratic form is (4.2).

For an elliptic curve $X$ there are 4 spin structures. The unique odd spin structure—the trivial double cover of the frame bundle—corresponds to the identity element of $\mathrm{Pic}^0(X) \cong X$. The three even spin structures are permuted by $SL(2;\mathbb{Z})$, which acts by diffeomorphisms on $X$. (Identify $X$ with the differentiable manifold $\mathbb{R}^2/\mathbb{Z}^2$.) The odd spin structure is invariant under $SL(2;\mathbb{Z})$. The even spin structures are invariant under a subgroup isomorphic to $\Gamma_0(2)$, the congruence subgroup consisting of matrices in $SL(2;\mathbb{Z})$ whose mod 2 reduction is upper triangular. We follow the physics language by terming the odd spin structure 'Periodic-Periodic' (PP) and the even spin structure invariant under $\Gamma_0(2)$ 'Periodic-Antiperiodic' (PA). The quadratic form corresponding to PP is $s+t+st$, where $s$, $t$ are coordinates on $H_1(X;\mathbb{Z}/2\mathbb{Z})$. The Arf invariant zero forms $st$, $s+st$, $t+st$ are permuted by $SL(2;\mathbb{Z})$. The form $s+st$ corresponds to PA.

Let $H$ denote the upper half-plane with its usual complex structure. It parametrizes a set of lattices in $\mathbb{C}$; the point $\tau \in H$ corresponds to the lattice $L_\tau$ generated by 1 and $\tau$. Equivalently, $H$ parametrizes equivalence classes of marked elliptic curves $X_\tau = \mathbb{C}/L_\tau$, i.e., elliptic curves with a distinguished basis for the first homology. Let $\mathbb{Z}\times\mathbb{Z}$ act on $H\times\mathbb{C}$ by a trivial action on the first factor, and let $\langle n_1, n_2\rangle$ act as translation by $n_1+n_2\tau$ on $\{\tau\}\times\mathbb{C}$. Set $Z=(H\times\mathbb{C})/(\mathbb{Z}\times\mathbb{Z})$. Then $\pi\colon Z\to H$ is a holomorphic fibration whose fiber over $\tau$ is $X_\tau$. Endow $Z$ with the flat metric, normalized so that the fibers have unit volume. There are four possible choices of spin structure along the fibers. The hypotheses of Theorem 2.1 are satisfied, so that the determinant line bundle $\mathcal{L}\to H$ of the Dirac operator is holomorphic. By Theorem 3.1 the determinant of the Dirac operator has a natural holomorphic square root Pfaff$(D)$, a section of $\mathcal{L}^{1/2}\to H$.

The canonical connection on $\mathcal{L}^{1/2}$ is flat [**A1**,Proposition 5.14]. For according to (3.2) its curvature is

$$\Omega^{(\mathcal{L}^{1/2})} = \pi i \int_{Z/H} \hat{A}(\Omega^{(Z/H)}). \tag{4.3}$$

But the metric on $Z/H$ lifts to the constant flat metric along the fibers of $H\times\mathbb{C}\to H$, so the curvature $\Omega^{(Z/H)}$ vanishes. Since $H$ is simply connected, there is a global covariant constant nonzero section $s_0$ of $\mathcal{L}^{1/2}$, which we take to have unit norm. Then $s_0$ is determined up to an overall constant phase. The section $s_0$ is holomorphic, by Theorem 2.1, and so we can identify Pfaff$(D)$ as a holomorphic function on $H$, which depends on the underlying spin structure. For the odd (PP) spin structure it is identically zero, due to the nonzero cohomology. However, the kernel of $D$ is one dimensional for all $\tau$, and we consider instead $\text{Pfaff}'_{PP}(D)$, constructed on the orthogonal complement of the kernel; it is not holomorphic. There are explicit formulæ for these functions. Let

$$\eta(\tau) = e^{\pi i\tau/12} \prod_{n=1}^{\infty} (1 - e^{2\pi i n\tau}) \tag{4.4}$$

be the Dedekind $\eta$-function.

PROPOSITION 4.5. *We have*

$$\mathrm{Pfaff}'_{PP}(D)(\tau) = \sqrt{\operatorname{Im}\tau} \cdot \eta(\tau), \tag{4.6}$$

$$\mathrm{Pfaff}_{PA}(D)(\tau) = \sqrt{2} e^{\pi i\tau/12} \prod_{n=1}^{\infty} (1 + e^{2\pi i n\tau}). \tag{4.7}$$

Proposition 4.5 is a special case of a formula for the determinant of the Dirac operator twisted by a flat bundle (Proposition 4.10). Equation (4.7) follows immediately from the product expansion (4.11). Equation (4.6) is obtained as the limit where the twisting bundle becomes trivial, omitting the eigenvalue which tends to zero. Of course, there are similar formulas for the remaining two spin structures.

The factor of $\sqrt{\operatorname{Im}\tau}$ in (4.6) arises from the kernel of the Dirac operator. Recall that this is the cohomology $H^0(X_\tau, K^{1/2})$, which is one dimensional for all $\tau$. These fit together into a line bundle $L \to H$, and $\omega_\tau = \sqrt{dz}$ is a holomorphic section of $L$. The norm square of $\omega_\tau$ is $\operatorname{Im}\tau$.

We introduce the action of the modular group $PSL(2;\mathbb{Z})$ on $H$. It lifts to an $SL(2;\mathbb{Z})$ action on $Z$, and the metaplectic double cover $Mp(2;\mathbb{Z})$ acts on the spinors (with the odd spin structure). Hence $Mp(2;\mathbb{Z})$ also acts on the cohomology $L$ and on the Pfaffian line bundle $\mathcal{L}^{1/2}_{PP}$. There is an identification

$$L^{\otimes 4} \cong TH \tag{4.8}$$

which preserves the $SL(2;\mathbb{Z})$ action.[8] To see this we change our point of view and fix the square torus, on which $SL(2;\mathbb{Z})$ acts by diffeomorphisms. Then $\omega_\tau = \sqrt{dx + \tau dy}$ is the

[8] Here we follow some arguments of Atiyah [**A2**]. $L^{\otimes 4}$ is the cohomology bundle of the *signature* operator, which admits an action of $SL(2;\mathbb{Z})$ (in fact, of $PSL(2;\mathbb{Z})$). Our notation differs from Atiyah's; our $\omega_\tau$ is the square root of his, and our $\mu$ (below) is $-\frac{1}{8}$ times his.

holomorphic differential, $dx \wedge dy$ is the relative Kähler form, and $\begin{pmatrix} a & b \\ c & d \end{pmatrix} \in SL(2;\mathbb{Z})$ acts on the coordinates of $H \times \mathbb{R}^2$ by

$$\begin{aligned} \tau &\longmapsto \frac{a\tau + b}{c\tau + d} \\ x &\longmapsto ax - by \\ y &\longmapsto dy - cx \end{aligned}$$

and so

$$\begin{aligned} (dx + \tau dy) &\longmapsto \frac{1}{c\tau + d}(dx + \tau dy) \\ d\tau &\longmapsto \frac{1}{(c\tau + d)^2}\, d\tau. \end{aligned}$$

Thus $\omega_\tau^4/d\tau$ is invariant, which is (4.8). The holomorphic section $\omega_\tau^{-1}$ of $L^{-1}$ determines a (nonvanishing) holomorphic section $s$ of $\mathcal{L}_{PP}^{1/2}$ by patching with the Dirac operator. Under the identification $\omega_\tau^{-4} \leftrightarrow d\tau$,

$$s^4 = \eta^4(\tau)\, d\tau. \tag{4.9}$$

Hence $s$ is essentially the Dedekind $\eta$-function.

Now since the action of $SL(2;\mathbb{Z})$ on $Z$ preserves the given geometric data, the induced action of $Mp(2;\mathbb{Z})$ on $\mathcal{L}_{PP}^{1/2}$ preserves the Quillen metric and Hermitian connection. Therefore, the global flat section $s_0$ transforms according to a character $\mu\colon Mp(2;\mathbb{Z}) \to \mathbb{Z}/24\mathbb{Z}$. (The abelianization of $Mp(2;\mathbb{Z})$ is $\mathbb{Z}/24\mathbb{Z}$.) One expression for $\mu$ follows from the holonomy formula. Namely, given $\tilde{A} \in Mp(2;\mathbb{Z})$ lifting $A \in SL(2;\mathbb{Z})$ we form the torus bundle $K_A$ over the circle with monodromy $A$. The lift $\tilde{A}$ is used to glue spinors. Endow $K_A$ with the flat metric along the fibers and an arbitrary metric on the base circle. Then by (3.3) we have $\mu(\tilde{A}) = -\xi(K_A)/2 \pmod 1$, where $\xi(K_A)$ is the $\xi$-invariant of the three dimensional Dirac operator. No adiabatic limit is necessary since the curvature (4.3) is zero. Atiyah [**A2**] studies $\mu$ in great detail, equating it with many other invariants. In particular, he relates $\mu$ to Dedekind's transformation law for $\log\eta$.

Similar remarks apply to formula (4.7). Now the Dirac operator on the elliptic curve has no kernel, which simplifies matters a bit. The double cover $\widetilde{\Gamma_0(2)}$ of the congruence subgroup $\Gamma_0(2)$ acts on $\mathcal{L}_{PA}^{1/2}$. The transformation law is now a character of $\widetilde{\Gamma_0(2)}$.

As mentioned above, we can twist the Dirac operator on $X_\tau$ by a flat holomorphic bundle $L \in \mathrm{Pic}^0(X_\tau)$. The square root of the determinant is defined only when $L$ is real, i.e., when $L$ has order 2, in which case $D_L$ is the ordinary Dirac operator with respect to a different spin structure. For definiteness fix the odd spin structure (PP). There is a holomorphic fibration $q\colon Y \to H$, where the fiber over $\tau$ is $\mathrm{Pic}^0(X_\tau)$. The pullback $q^*Z \to Y$ is also a holomorphic fibration, and over $q^*Z$ we have a Poincaré line bundle $P$, whose restriction on the fiber $X_\tau$ over $L \in \mathrm{Pic}^0(X_\tau)$ is isomorphic to $L$ (c.f. [**GH**,p.328]). The union of the identity elements in $X_\tau$ forms a subvariety of $q^*Z$, and we require that $P$ restricts trivially to this subvariety. Then $P$ is determined uniquely, and there is a family of Dirac operators $D_P$ parametrized by $Y$. By Theorem 2.1 its determinant $\det D_P$ is a holomorphic section of a line bundle $\mathcal{L} \to Y$.

The line bundle $\mathcal{L}$ has curvature along the fibers of $q$. (This is exactly the situation considered in [**Q**].) Thus we cannot write $\det D_P$ as a holomorphic *function*. To display a formula we pass to the universal cover $\tilde{Y} = H \times \mathbb{C}$, where the pullback of $\mathcal{L}$ admits a (nonholomorphic) trivialization. Note that on each slice $H \times \{w\}$ the trivialization is holomorphic. Set $q = e^{2\pi i\tau}$ and let $\bar{u}, \bar{v} \in \mathbb{R}/\mathbb{Z}$ be real coordinates on $\mathbb{C}$. The point $\langle \tau, u - \tau v\rangle \in \tilde{Y}$ projects to the bundle $L \in \mathrm{Pic}^0(X_\tau)$ corresponding to the character $m\tau + n \mapsto e^{2\pi i(mu+nv)}$ on $H_1(X_\tau)$.

PROPOSITION 4.10. *We have*

$$\det D_P(q; u, v) = q^{\frac{6v(v-1)+1}{12}} \prod_{n=1}^{\infty} \left(1 - q^{n-v}e^{2\pi i u}\right)\left(1 - q^{n+v-1}e^{-2\pi i u}\right). \tag{4.11}$$

This equality makes sense only up to a constant phase. Proposition 4.10 is well-known in the physics literature (e.g. [**AMV**], [**V**])—the translation of the path integral to the operator formulation immediately gives the product expansion. The equality of *norms* in (4.11) is Kronecker's second limit formula [**Si**]; the calculation is carried out in [**RS2**,Theorem 4.1]. The curvature of $\mathcal{L}$ is $\frac{i}{2\pi}\bar{\partial}\partial$ applied to the logarithm of the prefactor in (4.11), and then the proposition follows from the equality of norms [**Q**,§4].

Notice that the right-hand side is a perfect square for $u, v \in \frac{1}{2}\mathbb{Z}$, i.e., for the real points in the Jacobian.

Now fix $L \in \mathrm{Pic}^0(X_\tau)$ of order $N$. The bundle $L_\tau$ is specified topologically by $[L] \in H^1(X;\mathbb{Z}/N\mathbb{Z})$. Suppose $A \in SL(2;\mathbb{Z})$ preserves $[L]$. This happens if $A$ belongs to the level $N$ congruence subgroup $\Gamma(N)$, for example. Since $A$ also preserves the odd spin structure, each of the lifts $\tilde{A} \in Mp(2;\mathbb{Z})$ acts on the determinant line bundle $\mathcal{L}_L \to H$ of $D_L$. Let $\mathcal{L} \to H$ denote the determinant line bundle of the ordinary Dirac operator; $\tilde{A}$ also acts on $\mathcal{L}$. The ratio of these actions is a complex number $\phi_L(\tilde{A})$ of unit norm. Formula (4.11) is related to Klein forms [**L**,Chapter XV], and the following result is related to their transformation law.

PROPOSITION 4.12. *Choose a symplectic basis $a$, $b$ of $H_1(X)$, and let $\tilde{u}, \tilde{v} \in \mathbb{Z}/N\mathbb{Z} \cong \frac{1}{N}\mathbb{Z}/\mathbb{Z} \subset \mathbb{Q}/\mathbb{Z}$ be the evaluation of $[L]$ on $a, b$. Fix rational lifts $u, v$, and suppose*

$$(1 - A^*)\begin{pmatrix} u \\ v \end{pmatrix} = \begin{pmatrix} s \\ t \end{pmatrix} \tag{4.13}$$

*for integers $s, t$. Then $\phi_L(\tilde{A})$ depends only on $A$ and is given by*

$$\phi_L(A) = (-1)^{s+t+st} e^{\pi i(sv - tu)}. \tag{4.14}$$

Note that $\phi_L(A)$ is a $(2N)^{\text{th}}$ root of unity. We discuss a more general conjecture at the end of the section. Proposition 4.12 is a special case of Proposition 4.24, which we discuss in more detail at the end of this section.

We turn now to closed Riemann surfaces $X$ of arbitrary genus $g$. Here there are many possible choices of metric, and we first determine the behavior of the determinant under conformal transformations, reproducing an argument of Bost and Jolicœur [**BJ**], [**Bo**].[9] Consider the $\bar{\partial}$ operator on $X$ coupled to $K_X^n$, the $n^{\text{th}}$ power of the canonical bundle.

PROPOSITION 4.15. *Fix a metric $g^{(X)}$ and a function $f$. Let $L$ be the determinant line associated to $\bar{\partial}_{K^n}$ and $g_0^{(L)}$, $g_f^{(L)}$ the Quillen metrics associated to $g^{(X)}$, $e^{2f} \cdot g^{(X)}$, respectively. Then*

$$\frac{g_f^{(L)}}{g_0^{(L)}} = \exp\left\{\frac{6n(n-1)+1}{6\pi i}\int_X (f\Omega^{(X)} + \partial f \wedge \bar{\partial} f)\right\}, \tag{4.16}$$

[9]Their formula differs from (4.16) by a factor of 2, as they compare *norms* rather than *metrics*.

*where $\Omega^{(X)}$ is the curvature of the holomorphic line bundle $TX$ with the metric $g^{(X)}$.*

PROOF [**Bo**]: Consider the trivial holomorphic fibration $\pi\colon \mathbb{C}\times X\to\mathbb{C}$. Fix a real function $F(y,\cdot)$ on $\mathbb{C}\times X$ which depends only on $|y|$, with $F(0,\cdot)\equiv 1$ and $F(1,\cdot)=2f$. Endow the fibers of $\pi$ with the metric $e^F\cdot g^{(X)}$. The hypotheses of Theorem 2.1 are *not* satisfied—for example, the total volume of the fibers varies with $y$, which contradicts Lemma 2.2. We sketch a separate argument here to prove the compatibility of the complex structure and the canonical connection, thereby justifying our use of the curvature formula below. (More general versions of Theorem 2.1 are clearly desirable, and will no doubt appear in [**BGS**]. There is also a version announced in [**Bo**] which covers the present situation.) For the metric $e^f\cdot g^{(X)}$ on $\mathbb{C}\times X$ a straightforward calculation shows that the connection '$\nabla^{(Z/Y)}$' on the relative tangent bundle evaluates on $\frac{\partial}{\partial\bar y}$ to be the operator $\frac{\partial}{\partial\bar y}$. Equation (1.5) calls for a correction term due to the changing volumes. Since we work with Hermitian metrics here, we choose the 1-form $\gamma$ to be the (1,0) part of $d\mathbf{vol}$ in (1.4). Thus this correction term evaluates trivially on $\frac{\partial}{\partial\bar y}$. Now since the complex structure on $X$ is constant over the family, the bundle $\mathcal{L}_{\text{hol}}$ is holomorphically trivial. Further, we can omit the computation (2.30) by choosing both complexes in (2.25) to be the cohomology (at least over compact regions in $Y=\mathbb{C}$). Finally, (2.31) is still the decomposition of $\omega$ into type, since the operator $D$ is constant over the family and so $\overline{\nabla}D=\frac{\partial}{\partial\bar y}D=0$. For Riemann surfaces the factor $\mu^{(a)}$ in (2.27) is absent, and now (2.31) is trivial. Thus the canonical connection is compatible with the complex structure, as claimed.

By Theorem 1.30 the curvature of the determinant line bundle $\mathcal{L}\to\mathbb{C}$ of $\bar\partial_{K^n}$ is

$$\Omega^{(\mathcal{L})}=2\pi i\int_X[\operatorname{Todd}(\Omega)\operatorname{ch}(-n\Omega)]_{(4)}, \tag{4.17}$$

where $\Omega$ is the relative curvature of $\pi$. The relative tangent bundle is a holomorphic line bundle over $\mathbb{C}\times X$, and we easily calculate

$$\Omega=\Omega^{(X)}+\bar\partial\partial F. \tag{4.18}$$

Plug (4.18) into (4.17) to obtain

$$\begin{aligned}\Omega^{(\mathcal{L})}&=\frac{6n(n-1)+1}{24\pi i}\int_X\left(2\bar\partial\partial F\wedge\Omega^{(X)}+\bar\partial\partial F\wedge\bar\partial\partial F\right)\\&=\bar\partial_y\partial_y\left\{\frac{6n(n-1)+1}{24\pi i}\int_X\left(2F_y\Omega^{(X)}+\partial F_y\wedge\bar\partial F_y\right)\right\},\end{aligned}$$

where $F_y = F(y,\cdot)$. But $\Omega^{(\mathcal{L})} = \bar{\partial}_y \partial_y \log \|\cdot\|^2_{\mathcal{L}}(y)$, and $\|\cdot\|_{\mathcal{L}}(y)$ depends only on $|y|$. Equation (4.16) follows immediately.

Equation (4.16) is well-known in the physics literature [**P**], [**Al1**], [**Fr**]. The relationship of this formula to the index theorem was first noticed by Alvarez [**Al2**]. The application of Proposition 4.15 to the bosonic string has been discussed extensively. For the combination of operators in the string integrand the conformal variation (4.16) vanishes. This is the famous computation which fixes the dimension of the bosonic string to be 26. Geometrically, it allows us to push the metric and connection on the string determinant bundle down to the moduli space. Then another application of Theorem 2.1 and Theorem 1.30 gives the holomorphic factorization of Belavin and Knizhnik [**BK**] (c.f. [**BJ**], [**Bo**], [**CCMR**], [**F1**]). Recently Osgood, Phillips, and Sarnak [**OPS**] used (4.16) as the starting point in a proof of the uniformization theorem for Riemann surfaces.

Determinant line bundles play a role in arithmetic. The Hermitian norm was introduced by Faltings [**Fa**] in that context, based on ideas of Arakelov [**Ar**]. They study line bundles $L$ over a Riemann surface $X$, but restrict to a class of *admissible metrics*. Let $\mu: X \to \mathrm{Pic}^0(X)$ be the Abel map; it is defined up to translations. The polarization of the Jacobian $\mathrm{Pic}^0(X)$ (coming from the intersection pairing) determines a translation invariant (1,1) form, whose pullback to $X$ we denote $\Omega^{(X)}$. It is normalized by $\frac{i}{2\pi}\int_X \Omega^{(X)} = 2-2g$, where $g$ is the genus of $X$. A metric on a holomorphic line bundle $L \to X$ is admissible if its curvature $\Omega^{(L)}$ is proportional to $\Omega^{(X)}$. Admissible metrics always exist and are unique up to a constant. Applied to the tangent bundle we obtain a metric on $X$ whose curvature is $\Omega^{(X)}$. Fix a point $Q \in X$ and consider the line bundle $\mathcal{O}(Q)$. It has a canonical section $s$ whose divisor is $Q$, and we normalize the admissible metric on $\mathcal{O}(Q)$ by requiring

$$\int_X (\log |s|^2)\, \Omega^{(X)} = 0. \tag{4.19}$$

Let $\lambda_X(L)$ denote the determinant line bundle for the $\bar{\partial}$ complex coupled to $L$. Then the section $s$ gives rise to a canonical isomorphism

$$\hat{s}: \lambda_X(L \otimes \mathcal{O}(Q)) \cong \lambda_X(L) \otimes (L \otimes \mathcal{O}(Q))[Q], \tag{4.20}$$

where the last term is the fiber of $L \otimes \mathcal{O}(Q)$ at $Q$. Faltings characterizes his metric on the determinant by requiring that $\hat{s}$ be an isometry (c.f. [**Ar**,§4]). He than proves

[**Fa**,Theorem 1] that this property, together with two other axioms, characterizes the metric up to a scale factor. The issue of whether the Faltings metric and the Quillen metric coincide was settled by Bost.

PROPOSITION 4.21 [**Bo**]. *Let $L \to X$ be a holomorphic bundle over a Riemann surface of genus $g > 2$, and suppose both $X$ and $L$ are endowed with admissible metrics.*[10] *Then the Faltings metric on $\lambda_X(L)$ agrees with the Quillen metric, up to a constant depending only on $g$.*

We omit the proof, but remark that it is similar to an argument of Donaldson [**D**,§2]. In both cases a variational formula governing the restriction of the determinant to a divisor is derived using Theorem 1.30 together with Theorem 2.1. The universal family of Riemann Surfaces with Arakelov metrics satisfies the hypothesis of Theorem 2.1, since the (1,1)-form $\tau$ can be induced from the universal family of abelian varieties via the Abel map.

As for elliptic curves, we consider the family of Dirac operators twisted by a flat holomorphic line bundle. Fix a spin structure on $X$, and let $S_g$ be the moduli space of Riemann surfaces with this spin structure.[11] Over $S_g$ lies the family of Jacobians $q\colon Y \to S_g$ with typical fiber $\mathrm{Pic}^0(X)$. The spin structure determines a $\Theta$-divisor on $Y$, consisting of bundles $L \in \mathrm{Pic}^0(X)$ with $H^0(X, K^{1/2} \otimes L) \neq 0$. Let $p\colon Z \to S_g$ be the universal family of Riemann surfaces with the fixed spin structure. We use the Arakelov metrics along the fibers. Then points of $p^*Y$ are equivalence classes of such surfaces together with a basepoint and flat bundle. Hence there is a Poincaré line bundle $P \to p^*p^*Y$, trivial over the diagonal. (The typical fiber of $p^*p^*Y$ is $\mathrm{Pic}^0(X) \times X \times X$; the first $X$ is the basepoint which defines the Poincaré bundle over $\mathrm{Pic}^0(X) \times X_{(2)}$.) The family of Dirac operators $D_P$ is defined on the fibers of $\pi\colon p^*p^*Y \to p^*Y$. Then $\det D_P$ is a section of $\mathcal{L} \to p^*Y$. The Dirac operator is blind to the basepoint, which we now discard. Finally, then, $\det D_P$ is a section of $\mathcal{L} \to Y$.

Recall[12] that to a point $Z$ in the Siegel upper half-plane $\mathcal{H}_g$ (which consists of symmetric $g \times g$ matrices $Z$ with $\mathrm{Im}\, Z > 0$) corresponds a principally polarized abelian variety

$$A_Z = \mathbb{C}^g/(\mathbb{Z}^g + Z \cdot \mathbb{Z}^g).$$

[10] The restriction to $g > 2$ can probably be circumvented.

[11] Technically, we should work on a finite cover of $S_g$, due to nontrivial automorphisms of special surfaces. The fiber of $S_g \to \mathcal{M}_g$ can be identified with the orbit of the fixed spin structure under $\mathrm{Diff}\, X$.

[12] The reader should compare this discussion with [**Fa**,§6].

The $\vartheta$-function, defined on $\mathcal{H}_g \times \mathbb{C}^g$ by

$$\vartheta(Z,z) = \sum_{n\in\mathbb{Z}^g} \exp(\pi i n \cdot Z \cdot n + 2\pi i n \cdot z), \tag{4.22}$$

is a section of a line bundle $\mathcal{O}(\Theta)_Z$ over $A_Z$. Its divisor $\Theta$ is symmetric. A metric on $\mathcal{O}(\Theta)_Z$ is determined up to a constant by requiring its curvature to be translation invariant. Fix the constant by specifying the norm square of the canonical section (integrated over $A_Z$). Identify $A$ differentiably with the square torus. Then $Sp(2g;\mathbb{Z})$ acts by diffeomorphisms on $A$. Let $\Gamma$ denote the subgroup which fixes the spin structure. The theta group $\Gamma$ also acts on $\mathcal{H}_g$, and the quotient $\mathcal{A}_g = \mathcal{H}_g/\Gamma$ parametrizes principally polarized abelian varieties with a fixed spin structure. The abelian varieties fit together into a family $\mathcal{Y} \to \mathcal{A}_g$. The metaplectic double cover $\tilde{\Gamma}$ of $\Gamma$ acts on the theta line bundle (c.f. [**JM**]), and we obtain a holomorphic line bundle $\mathcal{O}(\Theta) \to \mathcal{Y}$. The $\vartheta$-function (4.22) is a section of $\mathcal{O}(\Theta)$ and its divisor $\Theta$ sits on $\mathcal{Y}$.

Returning to the Riemann surface $X$ there is a diagram

$$\begin{array}{ccc} Y & \longrightarrow & \mathcal{Y} \\ \downarrow & & \downarrow \\ \mathcal{S}_g & \longrightarrow & \mathcal{A}_g \end{array}$$

The bottom arrow is the map which attaches the period matrix to a Riemann surface, and the lift to $Y$ sends $\mathrm{Pic}^0(X)$ into the corresponding abelian variety. We choose the spin structure on the abelian varieties so that the $\Theta$-divisors correspond. By pullback we have a line bundle $\mathcal{O}(\Theta) \to Y$ with its $\vartheta$-function. By definition, the Dirac operator $D_P$ vanishes precisely on the $\Theta$-divisor. The line bundles $\mathcal{L}$ and $\mathcal{O}(\Theta)^{-1}$ are isomorphic, and an isomorphism is unique up to a constant.[13] However, this isomorphism is not an isometry. Faltings denotes the ratio of the metrics on the zero section of $q$ by $\exp(\delta(X)/8)$. Over the zero section we compare $\det D$ with the determinant $\det\bar\partial$ of the $\bar\partial$ operator, which is a section of a line bundle $\mathcal{K}$. The curvature formula (1.30) shows that $\mathcal{L}^{\otimes 2}$ and $\mathcal{K}^{-1}$ have the same curvature. hence their metrics agree up to a constant. Equating the norms of the determinants, Bost and Nelson [**BN**] prove

$$|\det D| = C_g \left(\frac{\det'\Delta}{\det(\mathrm{Im}\, Z)\cdot \mathrm{Area}(X)}\right)^{-1/4} |\vartheta(0,Z)|, \tag{4.23}$$

[13]This uses a compactness principle on moduli space—any holomorphic function is constant—which is valid for genus $g > 2$.

where $Z$ is the period matrix of $X$ (relative to some basis of $H_1(X)$), and $\Delta$ is the Laplacian on $X$. Notice that Equation (4.23) was derived here for the Arakelov metrics, though extensions to other metrics are possible.

Equation (4.23) corresponds to "bosonization" in the physics literature. Alvarez-Gaumé, Bost, Moore, Nelson, and Vafa [**ABMNV**] use these ideas to obtain other bosonization formulæ by coupling Dirac to a power of the canonical bundle (c.f. [**Bo**]). Their bosonization program goes beyond the determinant. They build a Lagrangian for the bosonic theory, but we do not pursue those ideas here. Also, Bost [**Bo**] uses (4.23) to determine Falting's constant $\delta(X)$.

Finally, consider bundles $L \in \mathrm{Pic}^0(X)$ of finite order $N$. Suppose $\varphi$ is a diffeomorphism of $X$ preserving both the spin structure and the topological class $[L] \in H^1(X;\mathbb{Z}/N\mathbb{Z})$. Form the 3-manifold $P$ by gluing the ends of $X \times [0,1]$ via $\varphi$; then $P$ fibers over $S^1$. Furthermore, a lift $\tilde{\varphi}$ acts on spinors, so determines a spin structure along the fibers. Also, $L$ extends to a flat bundle $L \to P$. Let $\phi(\tilde{\varphi})$ denote the ratio of the holonomies of the determinant line bundles $\mathcal{L}_L$ and $\mathcal{L}$ for the Dirac operator coupled to $L$ and the ordinary Dirac operator, respectively. This is the phase in a certain transformation of $\vartheta$-functions.

PROPOSITION 4.24 [**FV**],[**LMW**]. *Fix a canonical basis $a_j, b_j$ of $H_1(X)$ and let $A \in Sp(2g;\mathbb{Z})$ be the matrix of the induced action by $\varphi$ on homology. Let $u_j, v_j \in \mathbb{Q}/\mathbb{Z}$ be the value of $[L]$ on $a_j, b_j$, and choose lifts $u_j, v_j \in \mathbb{Q}$. Suppose*

$$(1-A^*)\begin{pmatrix} u_j \\ v_j \end{pmatrix} = \begin{pmatrix} s_j \\ t_j \end{pmatrix}$$

*for integers $s_j, t_j$. Then $\phi_L(\tilde{\varphi})$ depends only on $A$ and is given by*

$$\phi_L(A) = \exp \pi i \left\{ q\left(\begin{pmatrix} \bar{s} \\ \bar{t} \end{pmatrix}\right) + \sum_j (s_j v_j - t_j u_j) \right\}, \tag{4.25}$$

*where $q$ is the quadratic form on $H_1(X;\mathbb{Z}/2\mathbb{Z}) = H^1(X;\mathbb{Z}/2\mathbb{Z})$ determined by the spin structure, and $\bar{s}, \bar{t}$ are the reductions of $s, t$ modulo 2.*

As before, the phase $\phi_L(A)$ is a $(2N)^{\text{th}}$ root of unity. Proposition 4.24 was proved in [**FV**,Corollary 7.45] for the case $[L] \equiv 0 \pmod 2$. The argument there also proves (4.25) up to a sign in the general case. The approach of Lee, Miller, and Weintraub [**LMW**], based on [**BM**], relates this sign to the quadratic form $q$, and so proves the proposition. A direct argument in $K$-theory would be nice here; the quadratic form should appear as in (4.2). (The bundle $L$, together with its trivialization, determines an element of $K^{-1}(P;\mathbb{Q}/\mathbb{Z})$. Then $\phi_L(A)$ is the image of this element under the direct image $K^{-1}(P;\mathbb{Q}/\mathbb{Z}) \to \mathbb{Q}/\mathbb{Z}$ determined by the spin structure. One needs to compute this direct image in terms of the matrix $A$.) Also, an explicit comparison of (4.25) with the transformation law for the $\vartheta$-function may prove interesting. Proposition 4.24 reduces to Proposition 4.12 for elliptic curves with the odd spin structure.

For $N = 2$ the bundle $L$ is real, and $\phi_L(A)$ expresses the ratio of holonomies for different spin structures. That this is of order 4 was proved in [**AMV**], [**F1**,Appendix], [**LMW**], and [**W2**] by four distinct methods! (Notice that in this case $L$ is a real 2-plane bundle, and the Pfaffian of Dirac is defined. Then the ratio of the holonomies of the Pfaffians for different spin structures is of order 8.) Lee, Miller, and Weintraub [**LMW**] identify $\phi_L(A)$ as a difference of *Rohlin invariants*.

## §5 Anomalies and Orientation

A vector space $V$ is oriented by choosing a basis (nonzero element) of its determinant $\det V$. A family of vector spaces, say a *real* vector bundle $V \to M$, is oriented in ordinary cohomology by a trivialization of its determinant bundle $\det V$. There is a single topological obstruction to orientability—the first Stiefel-Whitney class $w_1(V) \in H^1(M;\mathbb{Z}/2\mathbb{Z})$. In extraordinary cohomology there are extraordinary orientations [**ABS**], [**Ad**]. Suppose $w_1(V) = 0$. Then $V$ is orientable in $KO$-theory if and only if $w_2(V) \in H^2(M;\mathbb{Z}/2\mathbb{Z})$ vanishes, while for $K$-theory it is the integral Stiefel-Whitney class $W_3(V) \in H^3(M;\mathbb{Z})$ which obstructs orientability. Recently, new topological constraints have entered string theory through work of Ed Witten [**W2**], [**W3**], [**F1**], [**FV**]. Roughly speaking, $V$ is "oriented" if its first Pontrjagin class $p_1(V) \in H^4(M;\mathbb{Z})$ is zero.[14] (This ignores a factor of two, as we explain below.) In string theory this vanishing is a sufficient condition for modular invariance, i.e., for the absence of anomalies in the theory. Topologists are familiar with this condition in a different theory—they call it $MO\langle 8\rangle$. (See [**Da**], [**Pe**], [**St**], for

[14] The setup of heterotic string theory is somewhat more complicated than the situation we consider here.

example. The 7-connected cover $BO\langle 8\rangle$ of $BO$ classifies certain bundles, and topologists study the bordism theory represented by the associated Thom spectrum $MO\langle 8\rangle$.) The link [**W3**] between string theory and the burgeoning "elliptic cohomology" [**La**] suggests that the new orientability condition will enter the definitive version of this most extraordinary cohomology theory. One should compare the situation here with Atiyah's interpretation [**A3**] of Witten's assertion [**W4**], [**W2**] that the free loop space $\mathfrak{L}M$ is orientable if and only if $M$ is spin. Ultimately, that too involves real $K$-theory. Killingback [**K**] takes a different (and perhaps more fundamental) approach to this new orientation condition, which is related to the Hamiltonian viewpoint of anomalies [**Seg**].

The anomaly in physics has a topological interpretation due to Atiyah and Singer [**AS2**]—it is the obstruction to trivializing the determinant line bundle $\mathcal{L}$ of a family of Dirac operators. (This topological obstruction is the Chern class $c_1(\mathcal{L})$.) Let $X$ be the two dimensional torus $\mathbb{R}^2/\mathbb{Z}^2$, and $\mathfrak{L}^2M = \mathrm{Map}(X, M)$ the double free loop space of $M$. Fix a spin structure on $X$. There is a natural evaluation map

$$e\colon \mathfrak{L}^2M \times X \longrightarrow M, \tag{5.1}$$

by which we pull back the real vector bundle $V \to M$. Let $\pi^X\colon \mathfrak{L}^2M \times X \to \mathfrak{L}^2M$ be projection onto the second factor, and choose differential geometric data as in (1.1). This determines a family of Dirac operators (coupled to $e^*V$) along the fibers of $\pi^X$, and by Theorem 3.1 its Pfaffian is a section of a complex line bundle $\mathcal{L}^{1/2} \to \mathfrak{L}^2M$. Proposition 3.15 computes the Chern class of $\mathcal{L}^{1/2}$:

$$c_1(\mathcal{L}^{1/2}) = \mathrm{Pfaff}\left(\pi^X_! e^![V]\right). \tag{5.2}$$

Here $[V] \in KO(M)$ is the $KO$ class, $e^!$ is the pullback in $KO$-theory, $\pi_!\colon KO(\mathfrak{L}^2M\times X) \to KO^{-2}(\mathfrak{L}^2M)$ is the direct image determined by the spin structure on $X$, and Pfaff is the map (3.13). Of course, we will denote the determinant line bundle, which is the square of $\mathcal{L}^{1/2}$, by $\mathcal{L}$.

An important refinement occurs by noting that Diff $X$ acts on $\mathfrak{L}^2M\times X$ preserving the evaluation map (5.1). We restrict our attention to the group of components $\pi_0\,\mathrm{Diff}\,X = SL(2;\mathbb{Z})$. Introduce the upper half-plane $H$ with its $SL(2;\mathbb{Z})$ action. (This essentially amounts to working with equivariant $KO$-theory.) Let $\Gamma \subseteq SL(2;\mathbb{Z})$ be the subgroup preserving the given spin structure on $X$. Then a double cover $\tilde{\Gamma}$ lifts the $\Gamma$ action to spinors, and we obtain a diagram

$$(5.3)\qquad \begin{array}{ccc} e^*V & & V \\ \downarrow & & \downarrow \\ Z = H \times \mathfrak{L}^2 M \times X \,/\, \Gamma & \xrightarrow{\;e\;} & M \\ \downarrow{\scriptstyle \pi^X} & & \\ Y = H \times \mathfrak{L}^2 M \,/\, \Gamma & & \end{array}$$

with a $KO$ orientation along the fibers of $\pi$ (the symbol of the Dirac family). Equation (5.2) applies to this setup, and so expresses $c_1(\mathcal{L}^{1/2})$ in $KO$-theoretic terms. Over the space $H/\Gamma$ we also have the Pfaffian line bundle $\mathcal{L}_0^{1/2}$ of the ordinary Dirac operator (not coupled to any extrinsic data). This is the bundle discussed in §4. Recall from Proposition 4.5 that the Pfaffian of the Dirac operator is a theta function divided by the Dedekind $\eta$-function. The line bundle $\mathcal{L}_0^{1/2}$ lifts to a bundle over $Y$.

Our work in this section concerns the passage from (5.2) to characteristic classes in cohomology. Over the rationals this is standard, but over the integers there are not usually good general formulæ. In low dimensions, however, the situation is quite tractable. Let $\lambda \in H^4(B\,\mathrm{Spin};\mathbb{Z})$ denote the generator such that $2\lambda$ lifts $p_1 \in H^4(BSO;\mathbb{Z})$. For a real vector bundle $V$ with $w_1(V) = w_2(V) = 0$ the characteristic class $\lambda(V)$ is well-defined. Our proof of the following proposition is in the spirit of index theory, though surely other (homotopy-theoretic) arguments can be given instead.

PROPOSITION 5.4. *In the situation described above,*

$$(5.5)\qquad c_1(\mathcal{L}) - c_1(\mathcal{L}_0^{(\dim V)}) = \pi_*^X e^* p_1(V),$$

*where $\pi_*$ is integration along the fiber. If $w_2(V) = 0$, then*

$$(5.6)\qquad c_1(\mathcal{L}^{1/2}) - c_1(\mathcal{L}_0^{(\dim V)/2}) = \pi_*^X e^* \lambda(V).$$

PROOF: A cohomology class is determined by its periods on integral homology classes and $\mathbb{Q}/\mathbb{Z}$ homology classes. (It can also be specified by these periods [**MS**,§2].) In two dimensions we can represent these classes by manifolds. Suppose $\Sigma \to Y$ is a closed Riemann surface mapping to $Y$, and let $\pi^X: Q \to \Sigma$ be the pullback of $\pi^X$ in (5.3). Here $Q$ is a closed 4-manifold. It inherits a spin structure along the fibers and a bundle $V$ by pullback. Let $\pi_!^\Sigma: KO^{-2}(\Sigma) \to KO^{-4}(\text{point}) \cong \mathbb{Z}$ denote the direct image for some fixed spin structure on $\Sigma$. We claim that the diagram

(5.7)

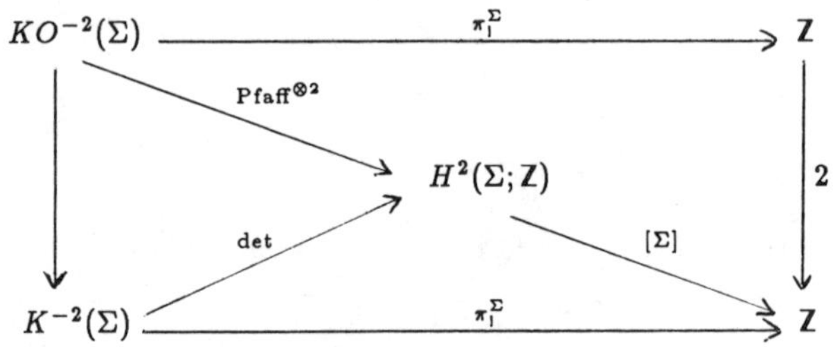


commutes. The left triangle is (3.14), and the bottom triangle commutes by the standard formula for the direct image in $K$-theory. The outer rectangle

$$\begin{array}{ccc} KO^{-2}(\Sigma) & \xrightarrow{\pi_!^\Sigma} & KO^{-4}(\text{point}) \\ \downarrow & & \downarrow \\ K^{-2}(\Sigma) & \xrightarrow{\pi_!^\Sigma} & K^{-4}(\text{point}) \end{array} \tag{5.8}$$

commutes since the spin structure defines both the $KO$ orientation of the top arrow and the $K$ orientation of the bottom arrow. The right arrow is multiplication by 2 on the generators (c.f. the discussion of Rohlin's theorem in §3). Hence the upper right triangle in (5.7) also commutes, from which

$$\text{Pfaff}(x)[\Sigma] = \pi_!^\Sigma(x) \tag{5.9}$$

for any $x \in KO^{-2}(\Sigma)$. Applied to $x = \pi_!^X e^!([V] - \dim V)$ using (5.2) we obtain

$$\{c_1(\mathcal{L}^{1/2}) - c_1(\mathcal{L}_0^{(\dim V)/2})\}[\Sigma] = \pi_!^\Sigma \pi_!^X e^!([V] - \dim V). \tag{5.10}$$

But the transitivity of direct images [**AS3**] implies

$$\{c_1(\mathcal{L}^{1/2}) - c_1(\mathcal{L}_0^{(\dim V)/2})\}[\Sigma] = \pi_!^Q e^!([V] - \dim V), \tag{5.11}$$

where $\pi_!^Q \colon KO(Q) \to KO^{-4}(\text{point}) \cong \mathbb{Z}$ is the direct image on $Q$. Now $\pi_!^Q(y) = \frac{1}{2}\hat{A}(Q)\,\mathrm{ch}(y)[Q]$ for $y \in KO(Q)$ (see [**AS4**]). Hence

$$\{c_1(\mathcal{L}^{1/2}) - c_1(\mathcal{L}_0^{(\dim V)/2})\}[\Sigma] = e^*\lambda(V)[Q] = \pi_*^X e^*\lambda(V)[\Sigma], \tag{5.12}$$

which proves (5.5) and (5.6) over Q.

Following Sullivan [**Su**], [**MS**] we use $\mathbb{Z}/k$-manifolds to represent $\mathbb{Z}/k\mathbb{Z}$ cycles.[15] Let $\overline{\Sigma} \to Y$ be a $\mathbb{Z}/k$-Riemann surface in $Y$. So now $\Sigma$ is a smooth surface whose boundary consists of $k$ disjoint circles, which we glue together to obtain $\overline{\Sigma}$. From this description it follows that a $\mathbb{Z}/k$-manifold carries a fundamental class in $\mathbb{Z}/k\mathbb{Z}$ homology. Now the pullback of $\pi^X$ in (5.3) is $\pi^X \colon \overline{Q} \to \overline{\Sigma}$ for some four dimensional $\mathbb{Z}/k$-manifold $\overline{Q}$. Here $\overline{Q}$ is obtained from a manifold $Q$ with boundary by gluing together the $k$ diffeomorphic boundary components. (See Figure 2.) Pull back the spin structure along the fibers and the $KO$ class $[V]$. Let $P$ be the *Bockstein* of $\overline{Q}$; it is diffeomorphic to each boundary component of $Q$ and fibers over the circle (which is diffeomorphic to each boundary component of $\Sigma$).

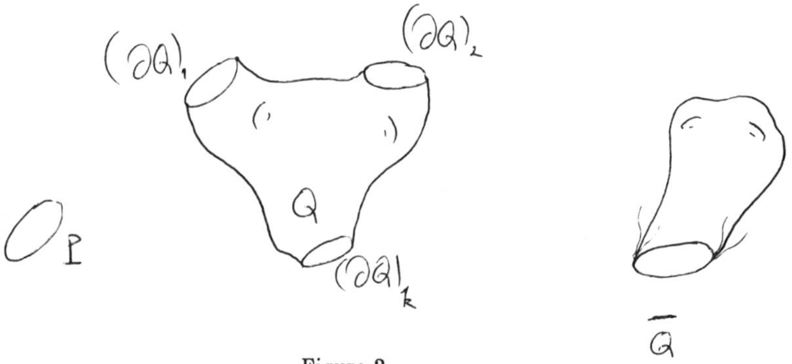


Figure 2

A spin structure on a $\mathbb{Z}/k$-manifold determines a fundamental class in $KO$ theory. The resulting direct image has no direct cohomological formula in general, but in two dimensions we claim (compare (5.9))

$$\mathrm{Pfaff}(x)[\overline{\Sigma}] = \pi_!^{\overline{\Sigma}}(x) \tag{5.13}$$

[15] Our arguments here follow [**F2**]. See also [**F1**,§3], [**FM**].

for $x \in KO^{-2}(\overline{\Sigma})$. To prove (5.13) write the Bockstein of $\overline{\Sigma}$ (which is a circle) as the boundary of some spin manifold $R$. Then $\Sigma - kR$ is closed, and $x$ extends to a class $\tilde{x} \in KO^{-2}(\Sigma - kR)$. By (5.9) we obtain the desired result:

$$(5.14) \qquad \pi_!^{\overline{\Sigma}}(x) \equiv \pi_!^{\Sigma - kR}(\tilde{x}) \equiv \mathrm{Pfaff}(\tilde{x})[\Sigma - kR] \equiv \mathrm{Pfaff}(x)[\overline{\Sigma}] \pmod{k}.$$

The direct images obey a multiplicative law for fiberings of closed manifolds over $\mathbb{Z}/k$-manifolds. Applied to $\pi^X : \overline{Q} \to \overline{\Sigma}$, imitating (5.10)–(5.11),

$$(5.15) \qquad \{c_1(\mathcal{L}^{1/2}) - c_1(\mathcal{L}_0^{(\dim V)/2})\}[\overline{\Sigma}] = \pi_!^{\overline{Q}} e^!([V] - \dim V).$$

To evaluate the right hand side of (5.15) we repeat the argument of (5.14), now using (5.12). But for this we must write the Bockstein $P$ of $\overline{Q}$ as a spin boundary $P = \partial S$ and extend the bundle $V$ over $S$. In case $w_2(V) = 0$ we can lift $V$ to a spin bundle. Then the relevant bordism group $\Omega_3^{\mathrm{spin}}(B\,\mathrm{Spin})$ vanishes, by the Atiyah-Hirzebruch spectral sequence, whence the bounding manifold $S$ exists. Hence by (5.12),

(5.16)

$$\pi_!^{\overline{Q}} e^!([V] - \dim V) \equiv \pi_!^{Q-kS} e^!([V] - \dim V) \equiv \pi_*^X e^* \lambda(V)[\Sigma - kR] \equiv \pi_*^X e^* \lambda(V)[\overline{\Sigma}] \pmod{k},$$

which completes the proof of (5.6).

The assertion in (5.5) is obtained by applying (5.6) to the bundle $V \oplus V$.

## REFERENCES


[**Ad**] F. Adams, "Chicago Notes," 1974.

[**Al1**] O. Alvarez, *Theory of strings with boundaries: fluctuations, topology and quantum geometry*, Nuc. Phys. **B216** (1983), 125–184.

[**Al2**] O. Alvarez, *Conformal anomalies and the index theorem*, to appear in Nucl. Phys.

[**AW**] O. Alvarez, P. Windy, *The energy momentum tensor as geometrical datum*, preprint LBL-22558, UCB-PTH-86/36.

[ASZ] O. Alvarez, I. M. Singer, B. Zumino, *Gravitational anomalies and the families index theorem*, Commun. Math. Phys. **96** (1984), 409–417.

[ABMNV] L. Alvarez-Gaumé, J.-B. Bost, G. Moore, P. Nelson, C. Vafa, *Bosonization in arbitrary genus*, Phys. Lett. **B178** (1986), 41–47.

[AMV] L. Alvarez-Gaumé, G. Moore, C. Vafa, *Theta functions, modular invariance, and strings*, Commun. Math. Phys. **106** (1986), 1–40.

[AgW] L. Alvarez-Gaumé, E. Witten, *Gravitational anomalies*, Nucl. Phys. **B234** (1983), 269–330.

[Ar] S. Arakelov, *An intersection theory for divisors on an arithmetic surface*, Izv. Akad. Nauk. **38** (1974), 1179–1192.

[ACGH] E. Arbarello, M. Cornalba, P. A. Griffiths, J. Harris, "Geometry of Algebraic Curves, Volume 1," Springer-Verlag, New York, 1985.

[A1] M. F. Atiyah, *Riemann surfaces and spin structures*, Ann. Scient. Éc. Norm. Sup. 4ᵉ série, t. 4, fasc. 1 (1971), 47–62.

[A2] M. F. Atiyah, *The logarithm of the Dedekind $\eta$-function*, preprint.

[A3] M. F. Atiyah, *Circular symmetry and stationary-phase approximation*, in "Colloque en L'Honneur de Laurent Schwartz," Volume 1, Astérisque **131** (1985), 43–59.

[ABS] M. F. Atiyah, R. Bott, A. A. Shapiro, *Clifford modules*, Topology **3** (1964), 3–38.

[APS] M. F. Atiyah, V. K. Patodi, I. M. Singer, *Spectral asymmetry and Riemannian geometry. I*, Math. Proc. Cambridge Philos. Soc. **77** (1975), 43–69; *II*, Math. Proc. Cambridge Philos. Soc. **78** (1975), 405–432; *III*, Math. Proc. Cambridge Philos. Soc. **79** (1976), 71–99.

[AS1] M. F. Atiyah, I. M. Singer, *Index theory for skew-adjoint Fredholm operators*, Publ. Math. Inst. Hautes Etudes Sci. (Paris) **37** (1969), 5–26.

[AS2] M. F. Atiyah, I. M. Singer, *Dirac operators coupled to vector potentials*, Proc. Nat. Acad. Sci. **81** (1984), p. 259.

[AS3] M. F. Atiyah, I. M. Singer, *The index of elliptic operators. I*, Ann. Math. **87** (1968), 484–530.

[AS4] M. F. Atiyah, I. M. Singer, *The index of elliptic operators. IV*, Ann. Math. **93** (1971), 119–138.

[BK] A. A. Belavin, V. G. Knizhnik, *Algebraic geometry and the geometry of quantum strings*, Phys. Lett. **B168** (1986), 201–206.

[B] J. M. Bismut, *The Atiyah-Singer Index Theorem for families of Dirac operators: two heat equation proofs*, Invent. Math. **83** (1986), 91–151.

[BF1] J. M. Bismut, D. S. Freed, *The analysis of elliptic families I: Metrics and connections on determinant bundles*, Commun. Math. Phys. **106** (1986), 159–176.

[BF2] J. M. Bismut, D. S. Freed, *The analysis of elliptic families II: Dirac operators, eta invariants, and the holonomy theorem of Witten*, Commun. Math. Phys. **107** (1986), 103–163.

[BGS] J.-M. Bismut, H. Gillet, C. Soulé, *Analytic torsion and holomorphic determinant bundles*, in preparation.

[Bo] J.-B. Bost, *Conformal and holomorphic anomalies on Riemann surfaces and determinant line bundles*, preprint.

[BJ] J.-B. Bost, Th. Jolicœur, *A holomorphy property and the critical dimension in string theory from an index theorem*, Phys. Lett. **B174** (1986), 273–276.

[BN] J.-B. Bost, P. Nelson, *Spin-1/2 bosonization on compact surfaces*, Phys. Rev. Lett. **57** (1986), 795.

[BM] G. Brumfield, J. Morgan, *Quadratic functions, the index mod 8, and a $\mathbb{Z}/4\mathbb{Z}$-Hirzeb Formula*, Topology **12** (1973), 105–122.

[CCMR] R. Catenacci, M. Cornalba, M. Martellini, C. Reina, *Algebraic geometry and path integrals for closed strings*, preprint.

[C] J. Cheeger, *$\eta$-invariants, the adiabatic approximation, and conical singularities*, J. Diff. Geom. **26** (1987).

[Da] D. M. Davis, Trans. AMS **276** (1983), 671–683.

[De] P. Deligne, *Le déterminant de la cohomologie*, preprint.

[D] S. K. Donaldson, *Infinite determinants, stable bundles, and curvature*, preprint.

[Fa] G. Faltings, *Calculus on arithmetic surfaces*, Ann. of Math. **119** (1984), 397–424.

[F1] D. S. Freed, *Determinants, torsion, and strings*, Commun. Math. Phys **107** (1986), 483–513.

[F2] D. S. Freed, *$\mathbb{Z}/k$-manifolds and families of elliptic operators*, preprint.

[FM] D. S. Freed, R. Melrose, in preparation.

[FV] D. S. Freed, C. Vafa, *Global anomalies on orbifolds*, preprint.

[Fr] D. H. Friedan, *Introduction to Polyakov's string model*, in "Les Houches Summer School, 1982," edited by J. B. Zuber and R. Stora, North Holland, Amsterdam, 1984.

[GH] P. A. Griffiths, J. Harris, "Principles of Algebraic Geometry," Wiley-Interscience, New York, 1978.

[H] N. Hitchin, *Harmonic spinors*, Adv. Math. **14** (1974), 1–55.

[JM] D. Johnson, J. J. Millson, *Modular lagrangians and the theta multiplier*, preprint.

[K] T. P. Killingback, *World-sheet anomalies and loop geometry*, Princeton preprint PUPT-1035.

[KM] F. F. Knudsen, D. Mumford, *The projectivity of the moduli space of stable curves I: Preliminaries on "det" and "Div"*, Math. Scand. **39** (1976), 19–55.

[La] P. S. Landweber, *Elliptic cohomology and modular forms*, to appear in Proceedings of a Conference on Elliptic Curves and Modular Forms in Algebraic Topology.

[L] S. Lang, "Introduction to Modular Forms," Springer-Verlag, New York, 1977.

[LMW] R. Lee, E. Y. Miller, S. H. Weintraub, *Spin structures: their relationship to quadratic pairings, Arf invariants, Rohlin invariants, and the holonomy of determinant line bundles*, in preparation.

[MS] J. W. Morgan, D. P. Sullivan, *The transversality characteristic class and linking cycles in surgery theory*, Ann. Math. **99** (1974), 461–544.

[OPS] B. Osgood, R. Phillips, P. Sarnak, *Extremals of determinants of Laplacians*, preprint.

[Pe] D. J. Pengelley, *$H^*(MO\langle 8\rangle; \mathbb{Z}/2)$ is an extended $A_2^*$-coalgebra*, Proc. AMS **87** (1983), 355–356.

[P] A. M. Polyakov, *Quantum geometry of bosonic strings*, Phys. Lett. **B103** (1981), 207–210; *Quantum geometry of fermionic strings*, Phys. Lett. **B103** (1981), 211–213.

[PS] A. Pressley, G. Segal, "Loop Groups," Oxford University Press, Oxford, 1986.

[Q] D. Quillen, *Determinants of Cauchy-Riemann operators over a Riemann surface*, Funk. Anal. i ego Prilozhenya **19** (1985), 37–41.

[RS1] D. B. Ray, I. M. Singer, *R-torsion and the Laplacian on Riemannian manifolds*, Adv. Math. **7** (1971), 145–210.

[RS2] D. B. Ray, I. M. Singer, *Analytic torsion for complex manifolds*, Ann. of Math. **98** (1973), 154–177.

[Se] R. T. Seeley, *Complex powers of an elliptic operator*, Proc. Symp. Pure Math. **10** (1967), 288–307.

[Seg] G. Segal, *Fadeev's anomaly in Gauss's law*, preprint.

[S] I. M. Singer, *The $\eta$-invariant and the index*, This Volume.

[Si] C. L. Siegel, "Lectures on advanced analytic number," Tata Institute, Bombay, 1961.

[St] R. E. Stong, Trans. AMS **107** (1963), 526–544.

[Su] D. Sullivan, *Geometric topology*, Princeton University.

[V] C. Vafa, *Modular invariance and discrete torsion on orbifolds*, Nucl. Phys. **B273** (1986), 592–60.

[W1] E. Witten, *Global gravitational anomalies*, Commun. Math. Phys. **100** (1985), 197–229.

[W2] E. Witten, *Global anomalies in string theory*, in "Anomalies, Geometry, and Topology," edited by A. White, World Scientific, 1985, pp. 61–99.

[W3] E. Witten, *Elliptic genera and quantum field theory*, Commun. Math. Phys. (to appear).

[W4] E. Witten, *Supersymmetry and Morse theory*, J. Diff. Geom. **17** (1982), 661–692.

# THE $\eta$-INVARIANT AND THE INDEX

by

I. M. SINGER

1. Introduction. Elliptic boundary value problems occur at two critical places in index theory. First, in the original proof of the index theorem for elliptic operators [3a, 7], one had to prove that the index was a cobordism invariant. For the index formula states that index $\partial\!\!\!/_B = \int_B \hat{A}(B)$, where $\partial\!\!\!/_B$ is the Dirac operator from positive $S_+$ to negative $S_-$ spinor sections over B, and $\hat{A}$ is the $\hat{A}$ polynomial in the Pontrjagin classes. If B is a boundary, then the Pontrjagin number $\int_B \hat{A}(B)$ vanishes, and we had to prove analytically that index $\partial\!\!\!/_B = 0$ directly.

To do so, we considered a boundary value problem. Let N be an odd dimensional compact oriented spin Riemannian manifold with boundary B. Let $\partial\!\!\!/_N : C^\infty(S) \to C^\infty(S)$ be the Dirac operator of N. If n(b) denotes the outward unit normal at $b \in B$, then Clifford multiplication by n(b), which we denote by $\gamma_o(b)$, splits spinors of N at b into $S_b^+ + S_b^-$ the $\pm i$ eigenspaces of $\gamma_o(b)$. At the boundary, $\partial\!\!\!/_N = \gamma_o(D_o + \partial\!\!\!/_B)$ where $D_o$ is the Riemannian covariant derivative in the outward normal direction and $\partial\!\!\!/_B = \begin{bmatrix} 0 & \partial\!\!\!/_B^- \\ \partial\!\!\!/_B^+ & \end{bmatrix}$ is the Dirac operator on B acting on $C^\infty(B, S|_B) = C^\infty(B, S_+) \oplus C^\infty(B, S_-)$.

Let $P_\pm$ denote the projection of $C^\infty(B, S|_B) \to C^\infty(B, S_\pm)$. The pairs $\{\partial\!\!\!/_N, P_\pm\}$ are well posed elliptic boundary value problems.

We showed that

Theorem 1. $\text{index}\{\partial\!\!\!/_N, P_+\} - \text{index}\{\partial\!\!\!/_N, P_-\} = \text{index}\ \partial\!\!\!/_B^+$.

{We also showed that $\text{index}\{\partial\!\!\!/_N, P_\pm\} = 0$. [See 7, Chap. 7.]}

The "signature defect theorem" [2] also uses an elliptic boundary value problem in an essential way. Now let M be an <u>even</u> dimensional compact oriented spin Riemannian manifold with boundary N. Assume the metric is a product near the boundary. Again $\partial\!\!\!/_M^+ = \gamma_o(D_o + \partial\!\!\!/_N)$. Let $P_\pm$ denote projection of $L_2(N, S|_N)$ on the space spanned by the eigenvectors with $\{\begin{matrix}\text{positive}\\ \text{negative}\end{matrix}\}$ eigenvalues of the self adjoint elliptic operator $\partial\!\!\!/_N$. Though $P_+$ is not local, $\{\partial\!\!\!/_M^+, P_+\}$ is a well posed elliptic boundary value problem and we have

Theorem 2. $\text{index}\{\partial\!\!\!/_M^+, P_+\} = \int_M A(M) - \frac{h+\eta(0)}{2}$

where h is the dimension of the null space of $\partial\!\!\!/_N$ and $\eta(0)$ is the $\eta$ invariant of $\partial\!\!\!/_N$. In particular, when $\partial\!\!\!/_N$ is nonsingular, $\int_M \hat{A}(M) \equiv \frac{\eta(0)}{2} \bmod Z$.

The $\eta$ invariant of $\partial\!\!\!/_N$ is defined as the meromorphic extension of the function $\eta(s) = \sum_{\lambda_j \neq 0} (\text{sign}\ \lambda_j)\ |\lambda_j|^{-s} =$

$\frac{1}{\Gamma(\frac{s+1}{2})} \int_0^\infty t^{\frac{s-1}{2}}\ \text{tr}(\partial\!\!\!/_N\ e^{-t\partial\!\!\!/_N^2})dt$. Here $\lambda_j$ are eigenvalues of $\partial\!\!\!/_N$ and $\eta(s)$ is holomorphic in a half plane rl $s > \frac{\dim N}{2}$. It has a meromorphic extension to the entire plane with no pole at $s = 0$. For simplicity, we have not included the extensions of Theorems 1 and 2 to other elliptic first order operators.

The $\eta$-invariant for self-adjoint elliptic operators can be viewed as the analogue in odd dimensions to the index invariant for elliptic operators in even dimensions. The index can be expressed in terms of characteristic classes, the $\eta$-invariant (mod integers) in terms of secondary characteristic classes. The index of a family of elliptic operators parameterized by a space P gives elements in $K^0(P)$ with Chern character in $H^*(P,\mathbb{R})$; in particular the $H^0$ component is the index, and the $H^2$ component the Chern class of the determinant line bundle. A family of self-adjoint operators gives an element of $K^1(P)$; the $H^1$ component of its Chern character in the $\eta$-invariant mod Z.

It is therefore natural to investigate the boundary value problems above, but with the $\eta$-invariant replacing index in theorems 1 and 2. We find a new boundary contribution in Theorem 1' and the appropriate $\hat{A}$ in 2' below.

Although we have just explained why it is reasonable to replace index by $\eta$, our original motivation for Theorem 1' was different. It was suggested by a theorem of Witten [9].

Witten interprets the covariant anomaly (which involves the variation of log determinants; see the appendix) as a natural connection on the determinant line bundle for a family of Dirac operators parameterized by metrics on B. The global anomaly is holonomy of this connection along a curve $\gamma$ from $\rho$ to $\phi\cdot\rho$ where $\phi \in$ Diff(B). To compute it, Witten constructs $\tilde{\rho}$ on the mapping cylinder N, a fibre bundle over $S^1$ with fibre B and metric determined by $\gamma$. He found that the holonomy is the exponential of $\lim_{R\to\infty} \frac{2\pi}{i}\, \eta(N)$ where the circle $S^1$ is parameterized by $R\theta$ $0 \le \theta \le 2\pi$, (the so-called adiabatic limit).

The curve of metrics $\gamma$ gives a metric on $B\times I$. To understand how log determinants enter, we concentrated the adiabatic limit at the boundary, leading to the cylinder computation below. This method does not require a product space and it was natural to consider an arbitrary N with boundary B.

2. Statement of results. Consider then the set up of theorem 1, with the two well posed elliptic boundary value problems $\{\not\partial_N, P_\pm\}$. Let $\tilde{\not\partial}_\pm$ be the corresponding operators with domains $[f \in C^\infty(N,S),\ P_\mp f|_B = 0]$. $\tilde{\not\partial}_\pm$ have closed extensions $\mathcal{D}_\pm = (\tilde{\not\partial}_\mp)^*$ with domains $[f \in \text{Sobolev } H_1;\ P_\mp f|_B = 0]$.

According to Gilkey and Smith [6], $\mathcal{D}_\pm$ have well defined $\eta$ series $\eta_\pm(s) = \frac{1}{\Gamma(\frac{s+1}{2})} \int_0^\infty t^{\frac{s-1}{2}} \operatorname{tr}(e^{-t(\mathcal{D}_\pm)^2})dt$; the operators $\mathcal{D}_\pm$ have all but a finite number of generalized eigenvalues lying in a cone about the real axis centered at 0. In contrast with the case of no boundary, $\eta_\pm(s)$ can have poles at $s = 0$. Let $\tilde{\eta}_\pm = \frac{1}{2}\{\eta_\pm(s) - \frac{1}{s}\text{residue}-N_\pm\}|_{s=0}$ where $N_\pm$ is the dimension of the span of eigenvectors of $\mathcal{D}_\pm$ with purely imaginary generalized eigenvalues.

Note that $\mathscr{D}_{\mp} = \mathscr{D}_{\pm}^{*}$, so that $\eta_- = \overline{\eta}_+$ and $\tilde{\eta}_+ - \tilde{\eta} =$ Imaginary part of the finite part of $\eta_+(o)$.

Assume that N is a product near the boundary so that $N = [0,R)\times B$ in a neighborhood of B. In this neighborhood, $\partial_N = \gamma_o(\frac{\partial}{\partial u} + \partial_B)$ with $\gamma_o = \pm i$ on $S_{\pm}$ and u is the coordinate on $[0,\mathbb{R})$. See Fig. 1.

Theorem 1'. Assume that $\partial_B$ is nonsingular. Then

$$\lim_{R\to\infty} (\tilde{\eta}_+ - \tilde{\eta}_-) = \frac{1}{4\pi i} \log\det \partial_B^2.$$

We use here a modified zeta function definition of determinant [8]. That is, usually $\log\det \partial_B^2 =$
$-\frac{d}{ds}\Big|_{s=0} \{\frac{1}{\Gamma(s)} \int_o^\infty t^{s-1} \operatorname{tr}(e^{-t\partial_B^2})dt\}$. Let $k(s) = \int_o^\infty t^{s-1} \operatorname{tr}(e^{-t\partial_B^2})dt$ and note that $\Gamma(s) = \frac{1}{s} + h(s)$, $h(s)$ holomorphic with $h(o)$ equal to Euler's constant. Then $\log\det \partial_B^2 =$
$-\frac{d}{ds}\Big|_{s=0} \frac{k(s)}{\Gamma(s)} = \frac{\rho(o)}{s} - k(s)\Big|_{s=0} - h(o)\rho(o)$ where $\rho(o) = \frac{k(s)}{\Gamma(s)}\Big|_{s=0}$.
We drop the term $h(o)\rho(o)$. It is the term that represents the scaling change of the metric. Both the $\eta$-invariant and the term $\frac{\rho(o)}{s} - k(s)\Big|_{s=0}$ are invariant under a scale change.

To extend Theorem 2, we note that $N\times[-1,0]$ is a manifold with boundary $B\times[-1.0] \cup (N,-1) \cup (N,0)$. The boundary can be identified with the double of the manifold $\tilde{N} = N \cup B\times[-\frac{1}{2},0]$ with N and $B\times[-\frac{1}{2},0]$ identified along $B \subset N$ and $(B,0)$ in $B\times[-\frac{1}{2},0]$. See Fig. 2.

On $N\times[-1,0]$ we have the reflection $(n,x) \to (n,-1-x)$ with fix points $[(n,-\frac{1}{2});\ n \in N]$. The manifold $M = N\times[-\frac{1}{2},0]$ has boundary $B\times[-\frac{1}{2},0] \cup (N,0) \cup (N,-\frac{1}{2}) = \tilde{N} \cup N$ joined along $(B,-\frac{1}{2})$. See Fig. 3.

We extend the metric on $N$ to $\tilde{N}$ in the obvious way and we use reflection to get a metric on the double of $\tilde{N}$, the boundary of $N\times[-1,0]$. We extend the metric from the boundary to the interior of $N\times[-1,0]$ so that it is a product near the boundary. Note that $N\times -\frac{1}{2}$ intersects $\tilde{N}$ orthogonally at $B$ (Fig. 3). Finally, let $\hat{A}(M)$ denote the $\hat{A}$ polynomial in the Pontrjagin classes $p_j$ expressed in terms of the metric on $N\times[-1,0]$, but restricted to $M = N\times[-\frac{1}{2},0]$.

Although near the boundary $N\times[-1,0]$ is $B\times[0,\epsilon)\times[-1,0]$, the product metric is not being used. In figure 3, $[0,\epsilon]\times 0 \cup 0\times[-\frac{1}{2},0]$ is a straight line and not the two sides of a square.

Theorem 2'. Assume $\partial_B$ is nonsingular. $\lim_{R\to\infty}\{\tilde{\eta}_+ + \tilde{\eta}_- - \int_M \hat{A}(M)\}$ is a constant $c(N)$ independent of the metric on $N$.

3. <u>Sketch of proofs</u> We follow the argument in [2] and construct the heat kernels for $\mathscr{D}^2_\pm$.

One can compute the heat kernels on $B\times[0,\infty)$ satisfying the boundary conditions above. They are $e_1^\pm(t,u,v) = e^{-tA^2}\{f(t,u,v) + A^{-1}P_\pm g(t,u,v)\}$ where $f(t,u,v) = \frac{1}{\sqrt{4\pi t}}\{e^{-(u-v)^2/4t} - e^{-(u+v)^2/4t}\}$

and $g(t,u,v) = \frac{1}{\sqrt{4\pi t}} \frac{1}{t} (u+v) e^{-(u+v)^2/4t}$. Let $e_2(t)$ denote the heat kernel for $\not{\partial}^2$ on the double of N, restricted to N.

We construct a parametrix for $e^{-t\mathcal{D}_\pm^2}$ as $q^\pm(t,u,v) = \phi_1(u)e_1^\pm \psi_1(v) + \phi_2(u)e_2 \psi_2(v)$ with $\phi_2 = \rho(\frac{1}{4}R, \frac{1}{2}R)$, $\phi_1 = 1 - \rho(\frac{3}{4}R, 1)$, $\psi_2 = \rho(\frac{1}{2}R, \frac{3}{4}R)$ and $\psi_1 = 1 - \psi_2$, where $\rho(a,b)$ is an increasing $C^\infty$ function of u vanishing for $u \leq a$ and equal to 1 for $u \geq b$. Note that $q^\pm(o,u,v) = \delta(u,v)$. Hence

$$e^{-t\mathcal{D}_\pm^2} - q^\pm(t) = \int_o^t \frac{d}{ds} e^{-t\mathcal{D}_\pm^2} q(t-s)ds = \int_o^t e^{-t\mathcal{D}_\pm^2} \{-\mathcal{D}_\pm^2 + \frac{d}{ds} q^\pm(t-s)\}.$$

Let $c_\pm(s) = (-\mathcal{D}_\pm^2 + \frac{d}{ds})q^\pm(t-s)$. Then

$$e^{-t\mathcal{D}_\pm^2} = q^\pm(t) + \int_o^t e^{-t\mathcal{D}_\pm^2} c_\pm(t-s)ds \text{ and } e^{-t\mathcal{D}_\pm^2} =$$

$$q^\pm(t) + \sum_{k=1}^{\infty} (-1)^k c^k \# q^\pm(t) \text{ where } (\alpha * q)(t) = \int_o^t \alpha(s)q(t-s)ds.$$

Note that $c_\pm(s) = -\frac{\partial^2\phi_1}{\partial u^2} e_1^\pm \psi_1(v) - 2\frac{\partial\phi_1}{\partial u}\frac{\partial e_1^\pm}{\partial u}\psi_1(v)$ $- \frac{\partial^2\phi_2}{\partial u} \rho_2\psi_2(v) - 2\frac{\partial\phi_2}{\partial u}\rho_2\psi_2(v)$ and vanishes except when $u \in [\frac{1}{4}R, R]$. Thus, when computing $tr(\mathcal{D}_\pm c_\pm)$ we can take $\mathcal{D}_\pm = \gamma_o(\frac{\partial}{\partial u} + \not{\partial}_B)$, and $\frac{1}{2} tr(\mathcal{D}_+ c_+ - \mathcal{D}_- c_-)$ $= tr(\gamma_o(\frac{\partial}{\partial u} + \not{\partial}_B) \frac{c_+(s)-c_-(s)}{2})$.

To compute $\tilde{\eta}_+ - \tilde{\eta}_-$, we first compute $tr(\mathcal{D}_+ q_+) - tr(\mathcal{D}_- q_-)$ and find the boundary contribution. Since $\phi_2$ vanishes on $[0, \frac{1}{4} R]$, we can again take $\mathcal{D}_+ = \mathcal{D}_- = \gamma_o(\frac{\partial}{\partial u} + \not{\partial}_B)$. The $e_2$ term cancels and a direct computation gives $tr(\gamma_o(\frac{\partial}{\partial u} + \not{\partial}_B)(\frac{q_+ - q_-}{2}) =$ $i\, tr(e^{-t\not{\partial}_B^2}) \int_o^R \phi_1(u)g(t,u,u)\psi_1(u)$. But $\phi_1\psi_1 = \psi_1$ so that

$\int_o^R \phi_1(u)g(t,u,u)\psi_1(u) = -\frac{1}{\sqrt{4\pi}}\int_o^R \frac{1}{\sqrt{t}}(e^{-u^2/t})'\,\psi_1(u)du = \frac{1}{\sqrt{4\pi t}} - \frac{1}{\sqrt{4\pi t}}$
$\int_{R/2}^{3R/4} e^{-u^2/t}\,\psi_1'(u)du \le \frac{1}{\sqrt{4\pi t}}(1 + e^{-R^2/4t}\cdot\frac{R}{4})$. Hence the contribution to $\eta_+ - \eta_-$ due to $q_\pm$ is

$\frac{i}{4\sqrt{\pi}}\frac{1}{\Gamma(\frac{1}{2})}\int_o^\infty t^{\frac{s}{2}-1}\,\mathrm{tr}(e^{-t\partial_B^2})dt\Big|_{s=0}$ plus an error term bounded by

$\frac{1}{4\sqrt{\pi}}\frac{R/4}{\Gamma(\frac{1}{2})}\int_o^\infty t^{|\frac{s}{2}|-1}\,\mathrm{tr}(e^{-t\partial_B^2})e^{-R^2/4t}\,dt$. By writing $\int_o^\infty$ as $\int_o^1 + \int_1^L + \int_L^\infty$ and estimating each term separately one can show that the error $\to 0$ as $R \to \infty$.

For the remainder term, the critical estimate is for the term $\gamma_o(\frac{\partial}{\partial u} + \partial_B)\frac{c_+(s)-c_-(s)}{2}$, an integral operator with kernel $k(s,u,v)$. A direct computation shows that the kernel vanishes in a uniform neighborhood of the diagonal in $[0,R]\times[0,R]$ and also in $[0,\frac{1}{4}R]\times[0,R]$. Again, the factor $e^{-R^2/4t}$ forces the remainder to go to zero as $R \to \infty$.

Qualitatively, one might say that as the length of the cylinder goes to $\infty$, the heat kernels separate into the interior piece plus the cylinder piece. Taking differences cancels the interior piece leaving only a boundary contribution, directly computable.

The qualitative argument for Theorem 2' can be understood from Fig. 4. Let N' be the manifold N with another metric in the interior agreeing with the given metric on $B\times[-\frac{1}{2},R]$

Then $\hat{N} = N' \underset{B}{\cup} N$ is of dim 2k+1, B is of dimension 2k and $\hat{M} = M' \cup M$ is of dimension 2k+2. Theorem 2 gives

$$\int_M \hat{A}(M) - \int_{M'} \hat{A}(M') = \frac{\eta(\hat{N})+h(\hat{N})}{2} \bmod Z.$$

($N\times -\frac{1}{2}$) However, as $R \to \infty$, the heat kernel of $\hat{N}$ separates into three pieces. One is $\ell_2$ of N', one is $\ell_2'$ of N', and one due to the cylinder. Substituting into the formula for $\eta$, one finds, as $R \to \infty$, that the cylinder contribution is zero and $\eta(\hat{N}) - \frac{1}{2}\{\eta_+(N) + \eta_-(N) - \eta_+(N') - \eta_-(N')\} \to 0$. Moreover, the assumption that $\partial_B$ is nonsingular implies that $\partial_{\hat{N}}$ is nonsingular if R is sufficiently large. Hence as $R \to \infty$

$$\int_M \hat{A}(M) - \left(\tilde{\eta}_+(N)+\tilde{\eta}_-(N)\right) - \left[\int_{M'} \hat{A}(M') - \left(\tilde{\eta}_+(N')+\eta_-(N')\right)\right] \to 0.$$

This proves a weak form of theorem 2' with $c(B,\rho_B)$ depending on a metric $\rho_B$ on B.

To eliminate the dependence on $\rho_B$, consider a one parameter family $\rho_t$ of metrics on N which restricted to a fixed neighborhood of B is a one parameter family of product metrics on $B\times[0,R]$. The variation $\dot{\eta}(\hat{N})$ of $\eta(\hat{N})$ with respect to the parameter t at t = 0 can be expressed as an integral over $\hat{N}$ of terms local in the geometric data given by $\rho_o$ and $\dot{\rho}_t\big|_{t=0}$. In fact, because $\hat{A}(\hat{M}) = d(\mathcal{S})$, one has $\eta(\hat{N}) = \int_{\hat{N}} \mathcal{S}$. As before, the cylinder contribution is zero as $R \to \infty$. Now $\dot{\eta}_\pm(N)$ are integrals over N of the same local expressions in $\rho_o$ and $\dot\rho_t\Big|_{t=0}$ as for $\eta(\hat N)$, since as $R \to \infty$ we can replace the heat kernel of N by the heat kernel of $\hat N$ restricted to N. Hence we conclude that $\int_M \dot{\hat A}(M) = \left[\dfrac{\eta_+(N)+\eta_-(N)}{2}\right]^{\bullet}$.

We now give a more complicated argument for Theorem 2' because it leads to a conjecture concerning the constant c(N). Consider Fig. 5.

As $R \to \infty$, our integrals and $\eta_+ + \eta_-$ split into three parts. One must show that there is no contribution for the interior part. So consider the manifold $N_2$ in Figure 6 which is the interior part extended to be infinite cylinders in both directions.

The metric on $N_2$ is obtained by choosing a path of metrics on B from $\rho_o$ to $\rho_1$. Let $M_2$ be N×R so that the boundary of $M_2$ is $N_2$. One must show that $\int_{M_2} \hat A = \frac{1}{2}\,\eta_{N_2}$.

The equality is a mild extension of [2, Theorem 3.10]. We note that $\hat A$ has compact support because near infinity, $M_2$ is metrically a product. Also near infinity, the contribution to $\operatorname{tr}(\partial\!\!\!/_{N_2} e^{-t\partial\!\!\!/^2_{N_2}})$ decays rapidly because the heat kernel is practically the one for the cylinder $B\times[0,\pm\infty)$. The index of $\partial\!\!\!/_{M_2}$ is zero because where $M_2$ is a product $B\times(a,b)\times(c,d)$, $\partial\!\!\!/_{M_2} = \partial\!\!\!/_B \otimes I + I \otimes \partial\!\!\!/_{R^2}$. Our assumption that $\partial\!\!\!/_B$ is invertible implies that elements of the kernel (and cokernel) vanish on $B\times(a,b)\times(c,d)$

and hence everywhere (in the real analytic case ). One repeats the proof in [2]; there are no contributions because of $\pm\infty$.

The argument we have just sketched makes it plausible to conjecture that the universal constant $\mathcal{L}(N)$ is an integer, the index of an elliptic boundary value problem. The interior manifold is M with operator $\partial_M$. The boundary $\tilde{N} \cup (N \times -\frac{1}{2})$ has a corner B×C with C the boundary near a corner of a square. One can still define a self-adjoint operator A on the boundary, with A the tangential part of $\partial_M$ and matching values at the corner. Then $(\partial_M, P_-)$ will be a well posed elliptic operator with $P_-$ the projection on the negative spectrum of A. Alternatively, one could consider $\tilde{\tilde{N}} = B\times[0, \infty) \cup N$, the boundary of $\tilde{\tilde{M}} = N \times [0,\infty)$ and the $L_2$-index of $\partial_{\tilde{\tilde{M}}}$. We conjecture that $\mathcal{L}(N)$ is the index of either of these operators.

4. Concluding remarks. Our motivation has been the consideration of the $\eta$-invariant for self-adjoint operators as the analogue in odd dimensions of the index invariant in even dimensions. We conclude with one further example.

Consider a family of elliptic operators parameterized by a compact manifold Y [3b]. Let Z be the total space of a fibre bundle over Y with fibres compact manifolds $X_y$. Let $\{B_y\}_{y\in Y}$ be a family of elliptic operators on $\{X_y\}_{y\in Y}$ and let A be an elliptic operator on Y.

Theorem 3. $\text{Index}(A\#B) = \text{index } A \otimes I_{\text{Ind } B}$ where

$$A\#B \text{ is the elliptic operator } \begin{bmatrix} A \otimes I & I \otimes B^* \\ I \otimes B & -A^* \otimes I \end{bmatrix}$$

on Z and Ind B is the index of the family $\{B_y\}_{y \in Y}$ in K(Y).

When $Z = X \times Y$ and $B_y$ is constant, index A#B = index A·index B.

We note that when A is self adjoint, so is A#B. What happens when index is replaced by $\eta$?

For index purposes, only the symbol of A#B is important. For the $\eta$-invariant, we need to be more precise. We assume given a metric on Z which projects to a metric on Y, of dim 2k+1. Let R be a scaling factor for the metric on Y. The bundle of frames on Z can be reduced to the subbundle of frames tangent to the fibre and orthogonal to the fibre. The Riemannian connection on Z induces a connection $\mu$ on the subbundle.

We assume A and $B_y$ are first order elliptic differential operators. Their symbols and the Riemannian connection makes A#B into one as well. We can also lift A, using $\mu$, to a first-order, self-adjoint differential operator on Z which we denote by $\tilde{A} = \sigma_A \circ D_\mu$. If $A : C^\infty(E) \to C^\infty(E)$, $\tilde{A}$ differentiates the lift of E to Z in the horizontal directions. We obtain a modified first-order self-adjoint operator

$$\widetilde{A\#B} = \begin{bmatrix} \tilde{A} \otimes I & I \otimes B^{*} \\ I \otimes B & -\tilde{A} \otimes I \end{bmatrix}$$

For simplicity, we shall also assume that index $B_y \geq 0$ and that $B_y$ is surjective, $y \in Y$. Let $B_y^{-1}$ be the map onto the orthogonal complement of the null space of $B_y$. (Modifications can be made for the general situation, which are simple when the null spaces of $B_y^{*}$ are of constant dimension.) The metric on Z induces a metric and connection on the vector bundle $\{\ker B_y\}_{y\in Y}$, which under our assumptions is Ind B. Finally, the local trace of $e^{-tA^2}$ has an asymptotic expansion as $t \to 0$ equal to $\frac{1}{(4\pi t^{2k+1})^{1/2}} \sum_{r=0}^{\infty} c_r(y,y)t^r$.

Theorem 3'. $\lim_{R\to\infty} \eta(A\#B) - \eta(A \otimes I_{\text{Ind } B}) =$

$$\text{Imaj. part } \frac{1}{2\pi} \int_Y dy\ c_k(y,y)\ \mathrm{tr}(B_y^{*} B_y)^{-s} B_y^{-1}[A, B_y]) \Big|_{s=0} .$$

The product case $Z = X \times Y$ is the well known $\eta(A\#B) =$ index $B \cdot \eta(A)$. When $k = 0$, $Y = S^1$, and $A = \frac{1}{i}\frac{d}{d\theta}$ then Theorem 3' reduces to Witten's Theorem; that is $\eta(A\#B) = \int_{S^1}$ (covariant anomaly).

The method of proof uses two essential ingredients in ADS [1], where the main theorem can be considered as a corollary of Witten's Theorem. First one replaces the Riemannian connection on Z by $\mu$ and notes that as $R \to \infty$, $\eta(A\#B) - \eta(\widetilde{A\#B}) \to 0$. Secondly one uses the parametrix $e^{-t\tilde{A}^2} e^{-tB_y^2}$ for the heat kernel of $(A\#B)^2$. Duhamel's principle, i.e., the Levi sum expansion, implies that the error term goes to zero as $R \to \infty$. The formula in Theorem 3' is obtained by computing $\lim_{R\to\infty} \operatorname{tr}(\widetilde{A\#B}\, e^{-t\tilde{A}^2} e^{-tB_y^2})$.

Witten's argument using the adiabatic limit has been formalized by several mathematicians: Bismut and Freed [4] and Cheeger [5]. Aside from Theorem 3', one could obtain Witten's formula by extending Theorem 1' to the case where the metric is not a product at the boundary. The common feature in all these proofs is the choice of appropriate parametrix for the heat kernel and the use of Duhamel's principle to show that the error $\to 0$ as $R \to \infty$.

Appendix

Chiral anomalies take different forms and as a result are a source of confusion to mathematicians. We hope this appendix will be useful. They all represent log det $\not\partial_\rho$, with $\not\partial_\rho$ the Dirac operator in a metric $\rho$ from spinor fields of positive to negative chirality. For this exposition, we shall assume $\not\partial_\rho$ is invertible.

1. We might define log det $\partial_\rho$ as $\frac{1}{2}$ log det $\not\partial_\rho^* \not\partial_\rho$, defined via zeta function regularization as $-\frac{1}{2}\frac{d}{ds}\Big|_{s=0} \operatorname{tr}((\not\partial_\rho^* \not\partial_\rho)^{-s})$. This definition gives a real valued answer and is not applicable to anomalies which reflect phase effects.

2. We can fix a $\rho_0$ and define det $\not\partial_\rho$ as det $\not\partial_{\rho_0}^* \not\partial_\rho$, a function from $\tilde{\mathcal{M}}$ = [metrics $\rho$; $\not\partial_\rho$ is invertible] to $\mathbb{C}$ - (0). In particular, we get a map from Diff(M) $\to \mathbb{C}$ - (0) when $\rho$ is restricted to the orbit [$\phi\cdot\rho_0$; $\phi \in$ Diff(M)]. As for the definition of det $\not\partial_{\rho_0}^* \not\partial_\rho$, all but a finite number of eigenvalues $\lambda_1,\ldots,\lambda_k$ of $\not\partial_{\rho_0} \not\partial_\rho$ lie in a positive cone about the positive real axis. Let $Z$ = det $\not\partial_\rho^* \not\partial_\rho = \lambda_1,\ldots,\lambda_k\, e^{\zeta'}$ where $\zeta' = -\frac{d}{ds}\Big|_{s=0} \sum_{j>k} e^{-s\log\lambda_j}$. The <u>anomaly which satisfies the consistency</u> conditions is the closed 1-form on $\tilde{\mathcal{M}}$ or Diff(M) which is $\frac{1}{2\pi i}\frac{dZ}{Z}$ pulled back from $\mathbb{C}$ - 0. Formally, it is the variation of log det $\not\partial_{\rho_0}^* \not\partial_\rho$ and can be written

as $\operatorname{tr}((\mathscr{D}^*_{\rho_o}\mathscr{D}_\rho)^{-s}\mathscr{D}_\rho^{-1}\delta\mathscr{D}_\rho)\big|_{s=0}$, where $\delta\mathscr{D}_\rho = \frac{d}{du}\big|_{u=0}\mathscr{D}_{\rho+u\delta\rho}$ and $\delta\rho$ is a tangent vector to $\tilde{\mathcal{M}}$ at $\rho$.

3. The anomaly in 2 is not invariant under Diff(M) because of the arbitrary choice $\rho_o$. An invariant 1-form can be obtained by regulating the determinant as a function of $\rho$. Specifically, one has the 1-form $\omega(\delta\rho)$ at $\rho$ which equals $\frac{1}{2}\frac{d}{du}\big|_{u=0}\log\det\mathscr{D}^*_{\rho-u\delta\rho}\mathscr{D}_{\rho+u\delta\rho}$. It is easy to see that this equals the imaginary part of $\operatorname{tr}((\mathscr{D}_\rho^2)^{-s}\mathscr{D}_\rho^{-1}\delta\mathscr{D}_\rho)\big|_{s=0}$, and is invariant under Diff(M). This 1-form is the <u>covariant form</u> of the anomaly.

# REFERENCES

1. Atiyah, M F., Donnelly, H., Singer, I.M., Eta invariants, signature defects of cusps, and values of L-functions, Ann. of Math., 118 (1983), 131-177.

2. Atiyah, M. F, Patodi, V. K., and Singer, I. M., Spectral asymmetry and Riemannian geometry, Math. Proc. Cambridge Philos. Soc. 77(1975), 43-69.

3(a). Atiyah, M. F. and Singer, I. M., The index of elliptic operators on compact manifolds, Bull. Amer. Math. Soc. 69(1963), 423-433.

3(b). Atiyah, M. F. and Singer, I. M., The index of elliptic operators (IV), Ann. Math. 93(1971), 119-138.

4. Bismut, J.M. and Freed, D.S., The analysis of elliptic families: Dirac operators, eta invariants, and the holonomy theorem of Witten, to appear in Comm. Math. Phys.

5. Cheeger, Jeff, On the formulas of Atiyah-Patodi-Singer and Witten.

6. Gilkey, Peter B. and Smith, Lance, The Eta invariant for a class of elliptic boundary value problems, Comm. Pure and Appl. Math. 36 (1983), 85-131.

7. Palais, Richard S., Seminaron the Atiyah-Singer Index Theorem, Annals of Mathematical Studies 57(1965), Princeton University Press.

8. Ray, D. B. and Singer, I. M., R-torsion and the Laplacian on Riemannian manifolds, Adv. in Math. 7(1971), 145-210.

9. Witten, E., Global gravitational anomalies, Comm. Math. Phys. 100(1985), 197-229.

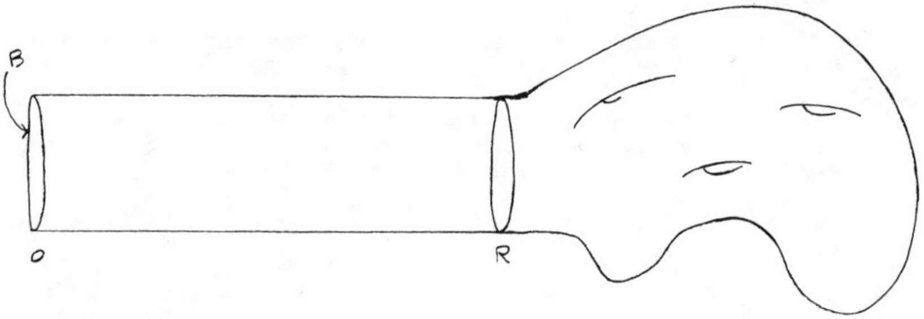


Fig. 1

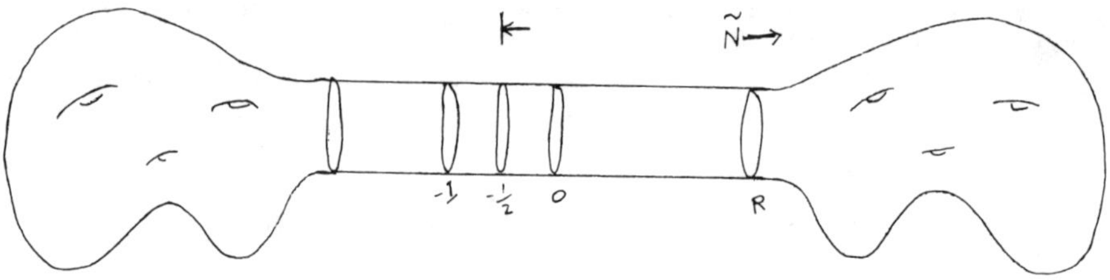


Fig. 2

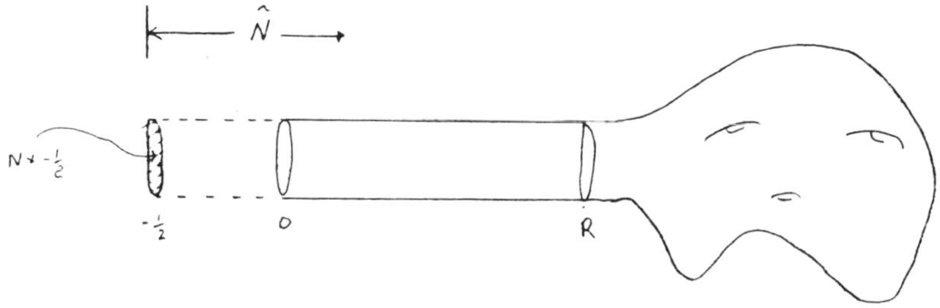


Fig. 3

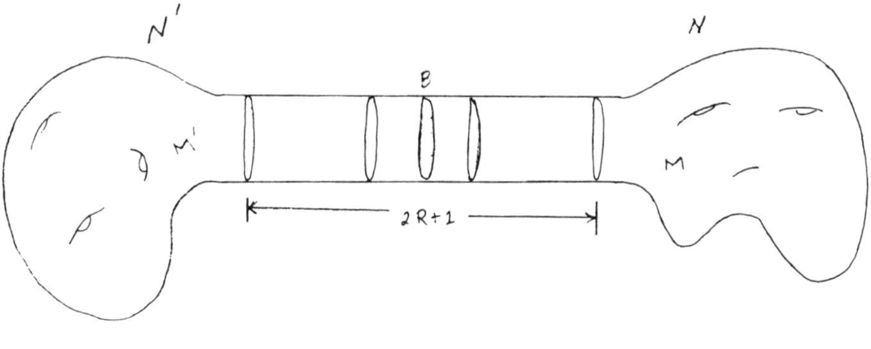


Fig. 4

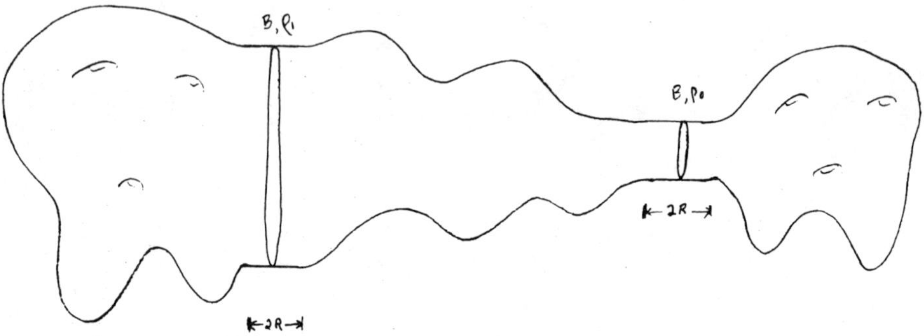


Fig. 5

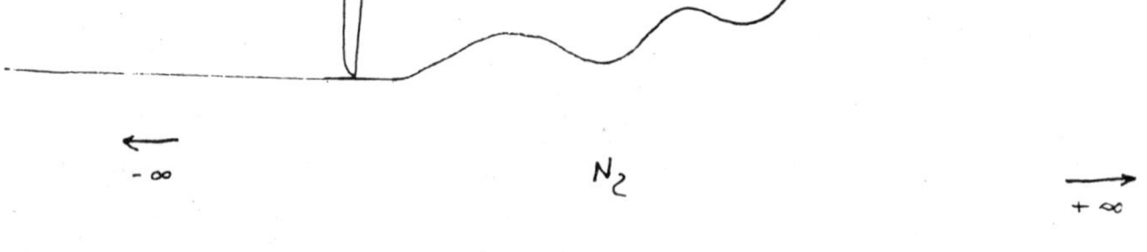


Fig. 6

# ACTION PRINCIPLES AND GLOBAL GEOMETRY

Gregg J. Zuckerman*

Yale University
Department of Mathematics
New Haven, CT 06520

ABSTRACT

A new universal conserved current, depending on a field and two variations of the field, is defined and shown to provide a general approach to the covariant description of the phase space of a Lagrangian field theory. The global symplectic -- and in certain cases Kählerian -- geometry of phase space is examined for particular models. The moduli space of Riemann surfaces makes an unexpected appearance in three dimensional field theory; in particular, gravity. A brief discussion of quantization concludes a mainly classical treatment of field theory and action principles.

*Research supported by the National Science Foundation under Grant #DMS-8401377.

## INTRODUCTION

Modern physics has achieved a high degree of mathematical coherence through its virtually universal application of certain abstract principles. Hamilton's principle of least action is basic throughout mechanics, field theory, particle physics, and, lately, string theory. Noether's principle -- a continuous symmetry of the action leads to a conservation law -- is equally basic. Time translation invariance, for example, yields the law of conservation of energy.

We have discovered a new universal conservation law that holds for any system governed by a least action principle. We have investigated and exploited this new law in a wide variety of physical models and related mathematical problems. Our conservation law endows the phase space of the physical system with a mathematical structure generalizing the canonical structure of Hamiltonian mechanics[1]. Moreover, this generalized canonical structure becomes, for various field theories, the starting point for a quantum mechanical investigation. Finally, our approach to phase space preserves all the symmetries of the original variational problem. Hence, whenever the action is relativistically invariant, both the classical and quantum structures are explicitly compatible with Einstein's special and/or general principles of relativity.

E. Noether's famous 1918 paper, "Invariant Variational Problems"[2], crystallized essential mathematical relationships among symmetries, conservation laws, and identities for the variational or "action" principles of physics. Modern accounts of Noether's paper appear in 3], 4], 5]. Noether herself was motivated by the then new theory of general relativity[6]. In the last decade, physicists have been attempting to unify general relativity with quantum particle physics. Supergravity theory[7] and superstring theory[8] are related approaches to unified physics (see 9] for an introductory account for mathematicians.) String theory has spawned string field theory, which, unlike previous models in physics, involves infinite component fields on space-time (see 10]). Virtually all physical models begin with the formulation of an action principle invariant with respect to a Lie group and/or Lie algebra, usually infinite dimensional. Thus, Noether's abstract analysis in 2] continues to be relevant to contemporary physics (as well as to applied mathematics (see 5])).

A basic notion in Noether's theory is that of a conserved current: if the space-time $X$ is n-dimensional, then such a current is an (n-1)-form $J = J(\psi)$ which is "local" in the field $\psi$ (see below), and which is closed whenever $\psi$ is an extremal for the action principle. Extremal fields are the solutions to the Euler-Lagrange field equations derived from the action. Linearization of the field equations about a given extremal leads to the Jacobi equations and their solutions, the Jacobi fields. One can generalize the notion of a conserved current as follows: consider an (n-1)-form $J = J(\psi, \delta_1\psi, \ldots, \delta_m\psi)$ which is local in $\psi$ and a finite number $\delta_1\psi, \ldots, \delta_m\psi$ of infinitesimal variations of $\psi$; assume further that $J$ is closed whenever $\psi$ is extremal and the variations $\delta_1\psi, \ldots, \delta_m\psi$ are Jacobi fields for $\psi$. Our universal conservation law, alluded to above, amounts to a completely general construction of a universal conserved current $U = U(\psi, \delta_1\psi, \delta_2\psi)$ which is alternating bilinear in $\delta_1\psi$ and $\delta_2\psi$. Particular examples of the universal current $U$ appear in 11], 12], 13], 14]. In the next section, we formulate a theory of conserved currents (alternating multilinear in the variations) in terms of differential forms on the product $X \times \mathcal{S}$ of space-time, $X$, with the infinite dimensional manifold $\mathcal{S}$ of field configurations over $X$.

## 1. LOCAL DIFFERENTIAL FORMS

Let $\pi: Y \to X$ be a smooth fibration of a manifold $Y$ over the space-time manifold $X$ (incidentally, we are using the term space-time in a rather vague way; we are not assuming $X$ has a semi-Riemannian metric, as in general relativity). Let $\mathcal{S}$ be the manifold of smooth sections of $Y$ over $X$. The de Rham complex $\Omega(X \times \mathcal{S})$ of smooth differential forms on $X \times \mathcal{S}$ has a number of nontrivial special features:

1) $\Omega(X \times \mathcal{S})$ is bigraded according to the product structure of $X \times \mathcal{S}$. We will write

$$\Omega(X \times \mathcal{S}) = \coprod_{p,q} \Omega^{p,q}(X \times \mathcal{S}).$$

Corresponding to this bigradation, the exterior derivative $d$ on $X \times \mathcal{S}$ breaks into two operators: $D$, of type $(1,0)$, and $\partial$, of type $(0,1)$. We have $d = D + \partial$, $d^2 = D^2 = \partial^2 = D\partial + \partial D = 0$.

2) If $J \in \Omega^{p,0}(X \times \mathcal{S})$ and $\psi \in \mathcal{S}$, define a p-form $J(\psi)$ on $X$ by $J(\psi)(x) = J(x,\psi)$. $DJ$ will be in $\Omega^{p+1,0}(X \times \mathcal{S})$, and we will have

$$(DJ)(\psi) = d(J(\psi)),$$

where the $d$ on the right is the exterior derivative on $X$.

3) More generally, if $J \in \Omega^{p,q}(X \times \mathcal{S})$, $\psi \in \mathcal{S}$,

and $\delta_1\psi,\ldots,\delta_q\psi$ are vectors in $T_\psi\mathcal{S}$ -- the tangent space of $\mathcal{S}$ at $\psi$ --, then we can define a p-form $J(\psi,\delta_1\psi,\ldots,\delta_q\psi)$ on X by

$$J(\psi,\delta_1\psi,\ldots,\delta_q\psi)(x) = \{i(\delta_1\psi)\cdots i(\delta_q\psi)J\}(x,\psi)$$

where $i(\delta\psi)$ is the operation of interior multiplication (contraction) of a tangent vector against a form. We will have

$$(DJ)(\psi,\delta_1\psi,\ldots,\delta_q\psi) = d(J(\psi,\delta_1\psi,\ldots,\delta_q\psi))$$

where again d is the exterior derivative on X.

4) $\Omega(X\times\mathcal{S})$ has a canonical sub-bicomplex $\Omega_{loc}(X\times\mathcal{S})$ defined as follows: let $J^\infty Y$ be the manifold of infinite jets of sections of Y at points of X (see [5]). Let $\varepsilon_\infty$ be the evaluation map from $X\times\mathcal{S}$ to $J^\infty Y$: $\varepsilon_\infty(x,\psi)$ = the $\infty$-jet of $\psi$ at x. We have the induced map $\varepsilon_\infty^*$ from $\Omega(J^\infty Y)$ to $\Omega(X\times\mathcal{S})$. The image, $\varepsilon_\infty^*\Omega(J^\infty Y)$, is stable under both D and $\partial$, and hence is a sub-bicomplex, which we call $\Omega_{loc}(X\times\mathcal{S})$. We will write

$$\Omega_{loc}(X\times\mathcal{S}) = \coprod_{p,q} \Omega_{loc}^{p;q}(X\times\mathcal{S}).$$

Takens[5] defines a canonical bicomplex structure on $\Omega(J^\infty Y)$. It turns out that the map $\varepsilon_\infty^*$ yields an isomorphism of bicomplexes between $\Omega(J^\infty Y)$ and $\Omega_{loc}(X\times\mathcal{S})$.

We call a form K on $X\times\mathcal{S}$ <u>local</u> if K lies in $\Omega_{loc}(X\times\mathcal{S})$. Thus, if $K\in\Omega_{loc}^{p;q}(X\times\mathcal{S})$, $\psi\in\mathcal{S}$, and $\delta_1\psi,\ldots,\delta_q\psi\in T_\psi\mathcal{S}$, the space-time p-form $K(\psi,\delta_1\psi,\ldots,\delta_q\psi)$ (see 3) above) depends on $\psi,\delta_1\psi,\ldots,\delta_q\psi$ in a local fashion, i.e. $K(\psi,\delta_1\psi,\ldots,\delta_q\psi)(x)$ depends only on (finite) jets of $\psi,\delta_1\psi,\ldots,\delta_q\psi$ at x. (Note: we can regard $T_\psi\mathcal{S}$ as the sections of an appropriate vector bundle over $X$; hence, we can speak of a "jet at $x$" of an element $\delta\psi$ of $T_\psi\mathcal{S}$).

5) The space $\Omega^{n,1}_{loc}$ of local, type $(n,1)$, forms on $X \times \mathcal{S}$ has a distinguished subspace[15]: we call $K \in \Omega^{n,1}_{loc}$ a <u>source form</u> if for $\psi \in \mathcal{S}$ and $\delta\psi \in T_\psi\mathcal{S}$, the n-form $K(\psi,\delta\psi)(x)$ depends only on a finite jet of $\psi$ and the <u>zero-jet</u> of $\delta\psi$ at $x$.

We write $\Omega^{n,1}_{source}$ for the space of source forms. $\Omega^{n,1}_{loc}$ also contains the subspace $D\Omega^{n-1,1}_{loc}$.

LEMMA 1[15]:

$$\Omega^{n,1}_{loc} = \Omega^{n,1}_{source} \oplus D\Omega^{n-1,1}_{loc},$$

i.e. if $K$ is a local $(n,1)$ form, there exists a <u>unique</u> source form $G$ and a local $(n-1,1)$ form $H$ such that

$$K = G + DH.$$

6) We have a useful cohomological result for the $D$ operator on $\Omega_{loc}$:

THEOREM 2[15]: If $K \in \Omega^{p,q}_{loc}$ with $p < n$ and $q \geq 1$, then $DK = 0$ implies $K = DF$ for a local form in $\Omega^{p-1,q}_{loc}$. However, D-closed $(p,0)$-forms need <u>not</u> be D-exact; moreover, $(n,q)$-forms, which are always D-closed, need not be D-exact. For example, a nonzero source form (type $(n,1)$) is never D-exact.

7) The entire theory of local forms on $X \times \mathcal{S}$ can be written in suitable local coordinates: let $Z$ denote the fiber of $Y$ over $X$ -- thus, $X$ can be covered by open coordinate charts $\mathcal{U}$ such that $\pi^{-1}(\mathcal{U})$ is diffeomorphic to $\mathcal{U} \times Z$; moreover, $Y$ can be covered by coordinate charts of the form $\mathcal{U} \times \mathcal{V}$, with

$\mathcal{U}$ as above and with $\mathcal{V}$ an open coordinate chart in $Z$. We label by $x_1,\ldots,x_n$ the coordinates on $\mathcal{U}$ and by $u_1,\ldots,u_m$ the coordinates on $\mathcal{V}$. Let $\mathfrak{W}(\mathcal{U} \times \mathcal{V})$ be the set of pairs $(x,\psi)$ such that $\psi(x)$ is in $\mathcal{U} \times \mathcal{V}$ -- then $\mathfrak{W}(\mathcal{U} \times \mathcal{V})$ is an open subset of $X \times \mathfrak{S}$, and $X \times \mathfrak{S}$ is covered by these special open subsets. For any such $\mathfrak{W} = \mathfrak{W}(\mathcal{U} \times \mathcal{V})$ we can explicitly describe $\Omega_{loc}(\mathfrak{W})$ as follows: define functions $x_i$ on $\mathfrak{W}$ by $x_i(x,\psi) = x_i(x)$; then define functions $u_{j,I}$ on $\mathfrak{W}$ by

$$u_{j,I}(x,\psi) = D_I[u_j(\psi(x))],$$

where $D_I$ is the partial derivative with respect to the $x_i$'s and corresponding to the multi-index $I = (i_1,\ldots,i_n)$, $i_1,\ldots,i_n \geq 0$. The functions $x_i$ and $u_{j,I}$ are type $(0,0)$ local forms; $dx_i$ and $\partial u_{j,I}$ are local forms of type $(1,0)$ and $(0,1)$ respectively. Any local $(0,0)$ form $F$ on $\mathfrak{W}$ is smooth function of finitely many of the variables $x_i$, $u_{jI}$. Any local $(p,q)$ form is expressible as a finite sum

$$\sum F_{i_1,\ldots,i_p,(j_1,I_1),\ldots,(j_q,I_q)} \wedge$$

$$dx_{i_1} \wedge \cdots \wedge dx_{i_p} \wedge \partial u_{j_1,I_1} \wedge \cdots \wedge \partial u_{j_q,I_q}$$

where each $F_{(\ldots)}$ is a local $(0,0)$ form.

For the differentials $D$ and $\partial$ we have the formulae:

$$Dx_i = dx_i, \ \partial x_i = 0;$$

$$Du_{j,I} = \sum u_{j,I \cup \{i\}} dx_i;$$

$$\partial u_{j,I} = \partial u_{j,I};$$

$$DF = \sum \frac{\partial F}{\partial x_i} dx_i + \sum \frac{\partial F}{\partial u_{j,I}} Du_{j,I};$$

$$\partial F = \sum \frac{\partial F}{\partial u_{j,I}} \partial u_{j,I};$$

where $F$ is a local $(0,0)$ form.

A variation $\delta\psi$ of $\psi$, with $(x,\psi)$ in $\mathfrak{N}$, determines a function $\overline{\delta\psi}$ from a neighborhood of $x$ into $\mathbb{R}^m$. We find that

$$\partial u_{j,I}(x,\psi,\delta\psi) = D_I[u_j(\overline{\delta\psi}(x))].$$

Thus, for any $K \in \Omega^{p,q}_{loc}(\mathfrak{N})$, $(x,\psi) \in \mathfrak{N}$, and $\delta_1\psi,\ldots,\delta_q\psi \in T_\psi\mathcal{S}$, we can in principle write an explicit expression for the evaluation at $x$ of the p-form $K(\psi,\delta_1\psi,\ldots,\delta_q\psi)$.

## 2. THE UNIVERSAL CONSERVED CURRENT

Suppose we now fix $L$ in $\Omega^{n,0}_{loc}(X \times \mathcal{S})$. Then $\partial L$ is in $\Omega^{n,1}_{loc}(X \times \mathcal{S})$, and by 1,5) we can write $\partial L = E + DM$, where $E$ is a source form in $\Omega^{n,1}_{loc}(X \times \mathcal{S})$ and $M$ is a form in $\Omega^{n-1,1}_{loc}(X \times \mathcal{S})$. We can then define $U$ in $\Omega^{n-1,2}_{loc}(X \times \mathcal{S})$ by setting $U = \partial M$. We will have $\partial U = 0$ and $DU = D\partial M = -\partial DM = -\partial(\partial L - E) = + \partial E$. We summarize our construction below:

THEOREM 3 (The Fundamental Formulae):*

1) To any $L \in \Omega^{n,0}_{loc}(X \times \mathcal{S})$ there is a system of forms $\{E,M,U\}$ satisfying

---

*During the preparation of this paper we learned of a July, 1986 letter from P. Deligne to D. Kazhdan in which Deligne sketches a result essentially equivalent to our Theorem 3, Part 1).

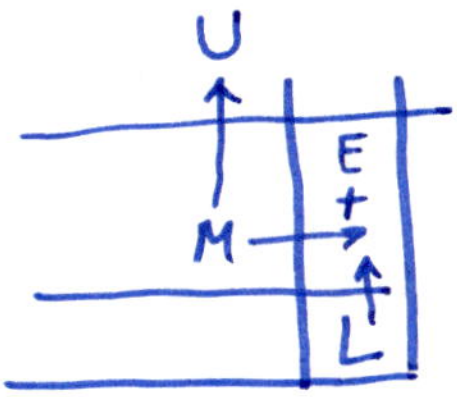


a) $E \in \Omega^{n,1}_{source}(X \times \mathcal{S})$

b) $M \in \Omega^{n-1,1}_{loc}(X \times \mathcal{S})$

c) $U \in \Omega^{n-1,2}_{loc}(X \times \mathcal{S})$

d) $\partial L = E + DM$

e) $U = \partial M$

f) $\partial U = 0$

g) $DU = \partial E$

2) a) $E$ is determined uniquely by $L$; moreover, if we replace $L$ by $L + DK$, with $K \in \Omega^{n-1,0}_{loc}(X \times \mathcal{S})$, $E$ does not change.

b) $M$ is not uniiquely determined by $L$, however, the class of $M$ modulo addition of forms $DN$, $N \in \Omega^{n-2,1}_{loc}(X \times \mathcal{S})$, <u>is</u> uniquely determined by $L$.

c) $U$ is uniquely determined by $L$, modulo the addition to $U$ of a term $DV$, where $V$ is in $\Omega^{n-2,2}_{loc}(X \times \mathcal{S})$; moreover, if we replace $L$ by $L + DK$, $K \in \Omega^{n-1,0}_{loc}(X \times \mathcal{S})$, then $U$ modulo D-exact forms does not change. In other words, the association of $U$ to $L$ defines a linear map $\tilde{U}$:

$$\tilde{U} : \Omega^{n,0}_{loc}/D\Omega^{n-1,0}_{loc} \to \Omega^{n-1,2}_{loc}/D\Omega^{n-2,2}_{loc}$$

PROOF: We have explained part 1); 2 a) follows from Lemma 1; 2 b) and c) follow from (Taken's) Theorem 2.

Theorem 3 above is quite abstract and needs to be interpreted in the more concrete terms of the calculus of variations and Lagrangian field theory:

1) The form $L \in \Omega^{n,0}_{loc}$ leads to an n-form $L(\psi)$

on X. $L(\psi)$ is called the Lagrangian density (strictly speaking, $L(\psi)$/(fixed volume form on X) is often referred to as the Lagrangian density; but we have no specific choice of volume form in general, so we work here with $L(\psi)$ instead). $L(\psi)$ is local in $\psi$ -- almost all Lagrangians in physics are local. The action of $\psi$ in a space-time domain $\mathfrak{U} \subseteq X$ is defined as the integral

$$\int_{\mathfrak{U}} L(\psi).$$

2) Suppose $\psi \in \mathcal{S}$, $\mathfrak{U}$ is a domain in X with smooth boundary $\partial\mathfrak{U}$, and $\delta\psi \in T_\psi\mathcal{S}$ is a variation that vanishes along $\partial\mathfrak{U}$. We consider the variation of the action:

$$\delta\int_{\mathfrak{U}} L(\psi) = \int_{\mathfrak{U}} \partial L(\psi,\delta\psi)$$

Applying our abstract formula $\partial L = E + DM$, we obtain

$$\partial L(\psi,\delta\psi) = E(\psi,\delta\psi) + d(M(\psi,\delta\psi)),$$

and hence, by Stokes Theorem,

$$\delta\int_{\mathfrak{U}} L(\psi) = \int_{\mathfrak{U}} E(\psi,\delta\psi)$$

DEFINITION 4: $\psi \in \mathcal{S}$ is an extremal for the Lagrangian field theory determined by L if $\delta\int_{\mathfrak{U}} L(\psi) = 0$ for all pairs $\mathfrak{U}$ and $\delta\psi$ such that $\mathfrak{U}$ is a (relatively compact) domain in X and $\delta\psi$ is a variation of $\psi$ that vanishes on the boundary of $\mathfrak{U}$.

PROPOSITION 5: $\psi$ is an extremal for L if and only if $\psi$ satisfies the equations $E(\psi,\delta\psi) = 0$, where $\delta\psi$ runs over all variations of $\psi$. In local coordinates, the system $E(\psi,\delta\psi) = 0$ is equivalent to the standard Euler-Lagrange equations (see 5]).

DEFINITION 6: The solution variety $\mathcal{S}_L$ associated to $L$ is the set of extremals for $L$.

$\mathcal{S}_L$ is not in general a submanifold of $\mathcal{S}$. However, we can still define, for each $\psi \in \mathcal{S}_L$, a tangent space to $\psi$ in $\mathcal{S}_L$, which we denote by $T_\psi \mathcal{S}_L$:

DEFINITION 7: The subspace $T_\psi \mathcal{S}_L$ of $T_\psi \mathcal{S}$, for $\psi \in \mathcal{S}_L$, consists of those $\delta\psi$ that satisfy the Jacobi equations -- the linearization about $\psi$ of the Euler-Lagrange equations.

It is possible to write the Jacobi equations in our abstract global formalism. We skip the details here, but we do mention an important property of Jacobi fields:

LEMMA 8: If $\psi \in \mathcal{S}_L$ and $\delta_1\psi$ and $\delta_2\psi$ are in $T_\psi \mathcal{S}_L$, then

$$\partial E(\psi, \delta_1\psi, \delta_2\psi) = 0.$$

3) Apparently the form $M$ of Theorem 3 has disappeared from our discussion of Lagrangian field theory. It was E. Noether[2] who observed (in local coordinates) that $M$ is essential to the construction of conservation laws associated to $L$.

DEFINITION 9: Fix $L$: a form $J \in \Omega^{n-1,0}_{loc}$ is called a conserved current for $L$ if whenever $\psi \in \mathcal{S}_L$, $J(\psi)$ is a closed $(n-1)$-form on $X$; more generally, a form $K \in \Omega^{n-1,q}_{loc}$ is a conserved current for $L$ if whenever $\psi \in \mathcal{S}_L$ and $\delta_1\psi, \ldots, \delta_q\psi \in T_\psi \mathcal{S}_L$, then $K(\psi, \delta_1\psi, \ldots, \delta_q\psi)$ is a closed $(n-1)$-form on $X$.

Following the ideas of Sophus Lie, Noether introduced what are now called (see 5]) generalized symmetries of a Lagrangian L. Moreover, she proved in 2] her celebrated theorem that to every generalized symmetry of L there is an associated conserved current J in $\Omega^{n-1,0}_{loc}$. The form M is an essential ingredient in Noether's construction of J. We give some details of Noether's theory in part 5) of this section.

4) Of course, since U is defined as $\partial M$, the form M is essential to our own work. Our main result is the following:

THEOREM 10: a) For any Lagrangian L, the associated form $U \in \Omega^{n-1,2}_{loc}(X \times \mathcal{S})$ is a conserved current.

b) The restriction of U to the manifold portion (smooth locus) of $X \times \mathcal{S}_L$ defines a closed (n+1)-form.

PROOF; Combine the equations $\partial U = 0$, $DU = \partial E$, and $dU = \partial U + DU$ with Lemma 8.

DEFINITION 11: We call U the universal conserved current associated to L.

Suppose now that C is a compact oriented (n-1)-dimensional submanifold of X. Define a 1-form $\theta_C$ and a 2-form $\omega^0_C$ on $\mathcal{S}$ via

$$\theta_C = \int_C M, \quad \omega^0_C = \int_C U;$$

more explicitly, if $\psi \in \mathcal{S}$ and $\delta_1\psi, \delta_2\psi \in T_\psi\mathcal{S}$, then

$$\theta_C(\psi, \delta_1\psi) = \int_C M(\psi, \delta_1\psi), \quad \text{and}$$

$$\omega_C^0(\psi, \delta_1\psi, \delta_2\psi) = \int_C U(\psi, \delta_1\psi, \delta_2\psi).$$

Then $\theta_C$ and $\omega_C^0$ are smooth on $\mathbb{S}$, and $\omega_C^0 = d\theta_C$. In particular, $d\omega_C^0 = 0$.

COROLLARY OF THEOREM 10: The restriction of $\omega_C^0$ to the smooth part of $\mathbb{S}_L$ is a closed 2-form, denoted by $\omega_{[C]}$, which depends only on the homology class of C in X.

QUESTION: Given X, $Y \xrightarrow{\pi} X$, L, and C as above: when is $\omega_{[C]}$ symplectic -- nondegenerate at every point of the smooth part of $\mathbb{S}_L$?

Of course, the manifold X may not contain any homologically nontrivial submanifolds C of codimension one. We can try in that case to integrate $U(\psi, \delta_1\psi, \delta_2\psi)$ over a noncompact submanifold -- but then we need appropriate boundary conditions to guarantee convergence of the integral as well as ensure the invariance of the integral under a deformation of the noncompact manifold -- contour -- inside X. In reference 12] we discuss linear field equations on an anti-de Sitter space-time X, which is diffeomorphic to $D^3 \times \mathbb{R}$. The noncompactness of the 3-ball $D^3$ leads to nontrivial difficulties with the two form $\omega_{D^3} = \int_{D^3} U$.

If we choose not to integrate at all, then Theorem 10 b) tells us that the restriction of U to $X \times \mathbb{S}_L^{\prime}$, where $\mathbb{S}_L^{\prime}$ is the smooth locus of $\mathbb{S}_L$, defines a cohomology class in $H^{n+1}(X \times \mathbb{S}_L^{\prime}, \mathbb{R})$. We can ask under what conditions this "universal" cohomology class is nonzero.

5) Suppose next that X <u>does</u> contain a homologically nontrivial (n-1)-cycle, C -- a compact

oriented (n−1)-submanifold without boundary. Even for such a space $X$, the nondegeneracy of $\omega_{[C]}$ depends nontrivially on symmetry properties of $L$.

DEFINITION 12: a) A smooth vector field $\xi$ on $\mathcal{S}$ is <u>local</u> if when we regard $T_\psi\mathcal{S}$ as sections of a vector bundle over $X$ depending on $\psi \in \mathcal{S}$, then $\xi(\psi)(x)$ is a function of some jet of $\psi$ at $x$.

b) A local vector field $\xi$ is a generalized symmetry of a Lagrangian $L$ if for some $R$ in $\Omega^{n-1,0}_{loc}$, we have

$$(\partial L)(\psi,\xi(\psi)) = (DR)(\psi)$$

for all $\psi$ in $\mathcal{S}$. In other words, the variation of $L$ by $\xi$ does not change the class of $L$ modulo D-exact local forms.

c) A family of local vector fields $\xi(\varepsilon)$ which depend locally and linearly on an arbitrary smooth section $\varepsilon$ of some fixed vector bundle over $X$ -- $\xi(\varepsilon)(\psi)(x)$ depends only on some jets of $\psi$ and $\varepsilon$ at $x$ -- is a generalized gauge symmetry of a Lagrangian $L$ if for some family $R(\varepsilon)$ in $\Omega^{n-1,0}_{loc}$ that depends locally on $\varepsilon$, we have

$$\partial L(\psi,\xi(\varepsilon)(\psi)) = DR(\varepsilon)(\psi)$$

for all $\psi$ in $\mathcal{S}$ and all sections $\varepsilon$.

THEOREM 13: a) Suppose a Lagrangian $L$ admits a generalized symmetry $\xi$: then for any $\psi$ in $\mathcal{S}_L$, $\xi(\psi)$ is in $T_\psi\mathcal{S}_L$ -- $\xi(\psi)$ is a Jacobi field for $\psi$; moreover, the Lie derivative of $\omega_{[C]}$ along $\xi$ vanishes on $\mathcal{S}_L$.

b) Suppose a Lagrangian $L$ admits a generalized gauge symmetry $\xi(\varepsilon)$: then for any extremal $\psi$ in $\mathcal{S}_L$, the Jacobi field $\xi(\varepsilon)(\psi)$ is in the radical of the two-form $\omega_{[C]}(\psi)$ -- for any Jacobi field $\delta\psi$ in $T_\psi \mathcal{S}_L$, we have

$$\omega_{[C]}(\psi, \xi(\varepsilon)(\psi), \delta\psi) = 0$$

for all sections $\varepsilon$.

PROOF: The details will appear in a forthcoming preprint[16]. A sketch of the proof follows:

We remark here that our Definition 12 is a globalization of concepts that play a key role in Noether's paper[2]. In fact, it is easy at this point to state Noether's main results.

DEFINITION 14: Suppose $\xi$ is a generalized symmetry of a Lagrangian L. The Noether current $J_\xi$ associated to L and $\xi$ is given by

$$J_\xi(\psi) = R(\psi) - M(\psi, \xi(\psi))$$

for $\psi$ in $\mathcal{S}$. (Recall that the form R occurs in the condition that $\xi$ be a generalized symmetry.)

THEOREM 15[2]: a) Suppose $\xi$ is a generalized symmetry of L: then the Noether current $J_\xi$ is conserved -- $dJ_\xi(\psi) = 0$ for any extremal $\psi$.

b) Suppose $\xi(\varepsilon)$ is a generalized gauge symmetry of L: then the Noether current $J_{\xi(\varepsilon)}$ has the following triviality property: for any extremal $\psi$, $J_{\xi(\varepsilon)}(\psi)$ is an exact $(n-1)$-form on X.

The relationship between Noether's theorem and our Theorem 13 is clear from the following:

PROPOSITION 16: Suppose $\xi$ is a generalized symmetry of $L$: then for any $\psi$ in $\mathcal{S}_L$ and $\delta\psi$ in $T_\psi \mathcal{S}_L$ we have

$$(dQ_\xi)(\psi, \delta\psi) = \omega_{[C]}(\psi, \xi(\psi), \delta\psi),$$

where

$$Q_\xi(\psi) = \int_C K_\xi(\psi) \text{ --}$$

the Noether charge associated to $\psi$, $\xi$, $L$, and $C$.

Thus, the argument for Theorem 13 b) is simple: If $\xi(\varepsilon)$ is a generalized gauge symmetry of $L$, and $\psi$ is an extremal, then $J_\xi(\psi)$ is exact on $X$, by (Noether's) Theorem 15 (part b)). Therefore, the Noether charge $Q_{\xi(\varepsilon)}(\psi)$ vanishes by Stoke's theorem, since $C$ is compact without boundary. Finally, by Proposition 16 we conclude that $\omega_{[C]}(\psi, \xi(\varepsilon)(\psi), \delta\psi)$ is zero for any Jacobi field $\delta\psi$ in $T_\psi \mathcal{S}_L$.

Part a) of Theorem 13 is a consequence of the theory of Jacobi fields, which will be discussed in 16].

6) Suppose $G$ is a connected Lie group that operates smoothly on the fiber bundle $Y \xrightarrow{\pi} X$: thus $G$ operates on both $Y$ and $X$, and for any $y \in Y$, $g \in G$ we have $\pi(g \cdot y) = g \cdot \pi(y)$. We have induced actions of $G$ on $\mathcal{S}$, $X \times \mathcal{S}$, and the bicomplex of local forms $\Omega_{loc}(X \times \mathcal{S})$.

DEFINITION 17: $G$ is a symmetry group of a Lagrangian $L$ if for any $g \in G$, then $g \cdot L - L = DK$ for some $K \in \Omega^{n-1,0}_{loc}$.

Suppose $C$ is a compact $(n-1)$-cycle in $X$. Because $G$ is connected, the homology class $[C]$ of $C$ will be $G$-invariant. Because $G$ is a symmetry group of $L$ mod D-exact local forms, the Euler-Lagrange equations of $L$ will be left invariant by the action of $G$. Thus, $G$ operates on $\mathcal{S}_L$ and on the differential forms on $\mathcal{S}_L$.

THEOREM 18: Given $L$, $G$, and $C$ as above:

a) the closed 2-form $\omega_{[C]}$ on $\mathcal{S}_L$ is $G$-invariant;

b) Suppose $\psi \in \mathcal{S}_L$ and we denote by $G_\psi$ the stability group of $\psi$ for the action of $G$ on $\mathcal{S}_L$. Then the tangent space $T_\psi \mathcal{S}_L$ carries a linear representation of $G_\psi$; moreover, the alternating bilinear form $\omega_{[C]}(\psi, \delta_1\psi, \delta_2\psi)$ is invariant for this $G_\psi$ representation; finally, $T_\psi \mathcal{S}_L/\text{radical } \omega_{[C]}(\psi)$ carries a symplectic representation of $G_\psi$.

REMARK: In the above Theorem 18, $G$ and $G_\psi$ may both be noncompact, and the various representations of $G_\psi$ may be infinite dimensional (see for example [2]).

## 3. EXAMPLES

We give now some examples of variational principles:

1) Suppose $X = \mathbb{R}$ with coordinate $t$, $Y = \mathbb{R} \times Z$, and $\pi$ is projection onto the first factor. A section $\psi \in \mathcal{S}$ is just a path in $Z$. Suppose at first that $L(\psi)$ is first order in $\psi$. Working in local coordinates $u_i$ in a patch on $Z$, we have

$$\begin{aligned} \partial L &= \left[\frac{\partial \tilde{L}}{\partial u_{i,t}} \, \partial u_{i,t} + \frac{\partial \tilde{L}}{\partial u_i} \partial u_i\right] \wedge dt \\ &= \left[\frac{\partial \tilde{L}}{\partial u_i} - \frac{d}{dt}\left[\frac{\partial \tilde{L}}{\partial u_{i,t}}\right]\right] \partial u_i \wedge dt \\ &\quad + D\left[-\frac{\partial \tilde{L}}{\partial u_{i,t}} \, \partial u_i\right] , \end{aligned}$$

where $\tilde{L} = L/dt$. If we define $p_i$ as $\partial\tilde{L}/\partial u_{i,t}$, we see that $M = -p_i \partial u_i$, and $U = -\partial p_i \wedge \partial u_i$. An (n-1)-cycle $C$ in $X$ is just a point, $t_0$, so that we have $\omega_{[C]} = (-\partial p_i \wedge \partial u_i)(t_0)$; this form is nondegenerate when we can solve for $u_{i,t}$ in terms of the "canonical variables" $\{u_k, p_k\}$. Relabeling $u_i$ as $q_i$ we see that our general theory reproduces the symplectic structure of classical mechanics.

If $Z$ is a Riemannian or more generally, semi-Riemannian manifold, the arclength action provides a nontrivial case of a first order action with gauge invariance -- in this case reparameterization invariance. Let $\mathcal{E}$ be the variety of complete parameterized geodesics: $\mathcal{E}$ is a subvariety of $\mathcal{S}_L$. The set $\tilde{\mathcal{E}}$ of unparameterized geodesics in $Z$ can be constructed as the quotient of the unit tangent bundle of $Z$ by the action of the geodesic flow. By Theorem 13, our 2-form $\omega_{[t_0]}$ drops down to a 2-form on $\tilde{\mathcal{E}}$. However, $\tilde{\mathcal{E}}$ may not be a manifold at <u>any</u> point: consider the case when $Z$ is a compact surface with negative curvature: the geodesic flow is ergodic, and $\tilde{\mathcal{E}}$ is a nasty object (perhaps approachable by Connes' noncommutative differential geometry[17]). If $Z$ is a sphere, $\tilde{\mathcal{E}}$ is quite nice: it is a homogeneous compact Kähler variety, for which the Kähler form is our form $\omega_{[t_0]}$, pushed forward from $\mathcal{E}$ to $\tilde{\mathcal{E}}$.

Specific higher-order "mechanical" actions arise in elasticity[18] and in the theory of the KdV equation[11]. Our universal current leads again to the known symplectic structures.

2) Suppose $X$ is a three-manifold $M_3$ and $G$ is a Lie group, regarded as a real matrix group, with matrix Lie algebra $\mathfrak{G}$. Let $\mathbb{Q}$ be the space of

connections: $\mathfrak{G}$-valued 1-forms on X. Define the Chern-Simons[19] action of $A \in \mathcal{Q}$ by

$$\int_{\mathcal{U}} \mathrm{Tr}(A \wedge dA + \frac{2}{3} A \wedge A \wedge A),$$

$\mathcal{U}$ a domain in $M_3$. The Euler-Lagrange equation is $F = 0$, where F is the curvature of A. The gauge group is enormous: the semi-direct product of $\mathrm{Diff}(M_3)$ with $C^{\infty}(M_3, G)$. Consequently, $\mathfrak{S}_L$ modulo gauge transformations is small: it is the space $\mathfrak{F}$ of geometrical classes of flat G-connections over $M_3$. If $\pi_1(M_3)$ is finitely presented, $\mathfrak{F}$ is the quotient of a finite dimensional real algebraic variety by the discrete group $\Gamma$ of components of $\mathrm{Diff}(M_3)$. When $G = SL(2,\mathbb{C})$ we are studying Kleinian groups, i.e. homomorphisms of $\pi_1(M_3)$ to $SL(2,\mathbb{C})$.

If $G = SL(2,\mathbb{R})$ and we assume X is diffeomorphic to $S \times \mathbb{R}$, with S a compact oriented surface, then an open subset of $\mathfrak{F}$ is identifiable with the moduli space of Riemann surface structures on S.

The 2-form $\omega_{[S]}$, pushed forward to $\mathfrak{F}$ via Theorem 13, yields the Weil-Petersson symplectic form on moduli space. This symplectic form is in turn a multiple of the Kähler form for the known Kählerlian structure on moduli space[20].

3) Suppose X is a two-dimensional Riemannian or Lorentzian manifold and M is a given background space-time manifold. The energy of a map from X to M defines an "action" familiar in mathematics. If we regard the energy as a functional in both the map <u>and</u> the metric on X we obtain instead the Polyakov action for bosonic string theory[8]. The symplectic structure for this physical model is studied in the physics literature by means of conformal gauge fixing[8].

4) If we consider the Yang-Mills action on an arbitrary space-time, our universal current U specializes to the simple form

$$U(A, \delta_1 A, \delta_2 A) == Tr(\delta_1 A \wedge * \delta_2 F - \delta_2 A \wedge * \delta_1 F);$$

here $A$ is a Yang-Mills potential with variations $\delta_1 A$ and $\delta_2 A$, and corresponding curvature variations $\delta_1 F$ and $\delta_2 F$; "$*$" is the Hodge $*$ operator induced by the background space-time metric. Our formula agrees with the formula in Crnkovic-Witten[13].

For instantons on a compact closed four-manifold $X$, all periods of $U(A, \delta_1 A, \delta_2 A)$ vanish, for in this case the universal form is exact. It is unclear what happens when $X$ is a compact manifold with boundary, as in 21].

5) Suppose $X$ is an arbitrary manifold and we consider the Einstein action with cosmological constant for (semi-)Riemannian metrics on $X$. In this case our universal current U specializes to the Crnkovic-Witten current[13] and the current in Ashtekar et. al[14]. If the space-time $X$ is three-dimensional, the extremal gravitational metrics have constant curvature. In the case of negative curvature we are led back to flat $SL(2,\mathbb{C})$ connections over three manifolds, as in 2) above.

In 16] we will compare our "covariant" symplectic geometry for Yang-Mills and Einstein theory with the more standard non-covariant "3 + 1" formalism of Fischer and Marsden[22].

6) If $X$ is $\mathbb{R}^{26}$ $(\mathbb{R}^{10})$, Witten's open string field (open superstring field) action[10](21]) offers the most sophisticated and mathematically involved variational principle ever considered. The string (superstring) field has infinitely many components.

The Witten action is a kind of noncommutative differential geometry[17] analog of the Chern-Simons action discussed in 2) above. Witten proposes in 21] a covariant symplectic structure for the solution variety modulo string gauge transformations. We plan to investigate the relationship between our universal current U and Witten's symplectic 2-form.

## 4. CONCLUDING REMARKS

We have limited our discussion to classical bosonic field theory. Fermionic fields may be introduced by extending our formalism to fibrations $Y \xrightarrow{\pi} X$ of a supermanifold $Y$ over a superspace-time $X$. It should then be possible to incorporate superparticles[24], superstrings[8], and supergravity fields[7],[25] into our collection of physical examples.

Quantum fields -- in a covariant operator formalism -- would be the ultimate goal of our covariant approach to Lagrangian field theory. Under reasonable conditions, the form $\omega_{[C]}$ should push-forward to a symplectic form on the quotient of $\mathbb{S}_L$ by gauge transformations. (When the gauge algebra has field-dependent structure coefficients, as in supergravity theory[25], we have a new difficulty with taking such a quotient.) We would then regard this quotient as our covariant phase space, $\mathbb{P}$. Under ideal conditions (see Example 2) in Section 3) our phase space $\mathbb{P}$ would be Kählerian with the Kähler form equal to the push-forward of our $\omega_{[C]}$. Suppose this Kähler form represented an integral cohomology class of $\mathbb{P}$: then, following the ideas of geometric quantization[26],[27], there would be a holomorphic Hermitian line bundle $\mathcal{L}$ with connection such that the curvature 2-form would equal the Kähler form. The Hilbert space $\mathcal{H}$ of square-integrable global holomorphic sections of $\mathcal{L}$ would serve as the quantum state space for our quantum field theory. There would be two main difficulties: defining "square-integrable" when $\mathbb{P}$ is infinite dimensional; and identifying the classical observables -- functions on $\mathbb{P}$ --, the quantum observables -- operators on $\mathcal{H}$ --, as well as the connection between these two sets of observables.

An ideal testing ground for geometric quantization is gravitational theory on topologically nontrivial three-manifolds, X. If X is diffeomorphic to $S \times \mathbb{R}$ with S a compact oriented surface, and we choose the cosmological constant to have the proper sign, then our phase space $\mathbb{P}$ will be the space of hyperbolic structures on $X = S \times \mathbb{R}$. If we let $\mathfrak{M}$ denote the moduli space of Riemann surface structures on S, then it is known[20] that $\mathbb{P}' = \mathfrak{M} \times \overline{\mathfrak{M}}$, where $\mathbb{P}'$ is the subset of complete hyperbolic metrics on X, and $\overline{\mathfrak{M}}$ is the space of Riemann surface structures on $S^{opp}$, i.e. S with the opposite orientation.

Our form $\omega_{[S]}$ becomes a Kähler form for the complex structure[20] on $\mathfrak{M} \times \overline{\mathfrak{M}}$. The theory of Teichmüller modular forms should imply that our quantum state space $\mathcal{H}$ for three-dimensional gravity is finite dimensional!

One last remark: the appearance of Teichmüller theory above has nothing obvious to do with the current applications of that theory to strings and superstrings[28]. In the latter theories, the moduli space $\mathfrak{M}$ shows up in the quantum theory -- the Feynman path integral reduces to an integral over $\mathfrak{M}$. Clearly, the correspondence between deep mathematical objects and deep physical models is not one-to-one!

## REFERENCES

1] Abraham, R. and Marsden, J.E.,"Foundations of Mechanics", Benjamin, New York (1967)

2] Noether, E., "Invariante Variationsprobleme", Nachr. König. Gesell. Göttingen, Math.-Phys. Kl., 235-257 (1918); English translation in Transport Theory and Stat. Phys. 1, 186-207 (1971).

3] Takens, F., "Symmetries, Conservation Laws, and Variational Principles", Lecture Notes in Math. 597, Springer-Verlag, Berlin, 581-604 (1977).

4] Uhlenbeck, K., "Conservation Laws and Their Application to Global Differential Geometry", in Emmy Noether in Bryn Mawr, ed. by B. Srinivasan, Springer-Verlag, 103-116 (1983).

5] Olver, P., "Applications of Lie Groups to Differential Equations", Graduate Texts in Math 107, Springer-Verlag, Berlin (1986).

6] Einstein, A., "Die Grundlage der Allgemeinen Relativitats Theorie", Annalen der Physik, 49 (1916).

7] Julia, B., "Kac-Moody Symmetry of Gravitation and Supergravity Theories", in Lectures in Applied Math. Vol. 21, ed. by P. Sally, A.M.S., Providence, 355-374 (1985).

8] Schwarz, J.H., "Mathematical Issues in Superstring Theory", in Lectures in Applied Math. Vol. 21, ed.

by P. Sally, A.M.S., Providence, 117-138 (1985).

9] Witten, E., "Physics and Geometry", address to the International Congress of Mathematicians, Berkeley, (Aug. 1986) -- Princeton Physics Dept. preprint (Oct. 1986).

10] Witten, E., "Noncommutative Geometry and String Field Theory", Nucl.Phys. B268, 253- (1986).

11] Sternberg, S., "Some Preliminary Remarks on the Formal Variational Calculus of Gelfand and Dikii", in Lecture Notes in Math. 673, ed. by K. Bleuler, Springer-Verlag, Berlin, 399-408 (1978).

12] Zuckerman, G.J., "Quantum Physics and Semisimple Symmetric Spaces", in Lecture Notes in Math 1077, ed. by R. Lipsman, Springer-Verlag, Berlin, 437-454 (1984).

13] Crnkovic, C. and Witten, E., "Covariant Description of Canonical Formalism in Geometrical Theories", Princeton Physics Dept. preprint (Sept. 1986).

14] Ashtekar, A., Bombelli, L., and Kour, R., "Phase Space Formulation of General Relativity Without a 3+1 Splitting", Syracuse Physics Dept. preprint (1986).

15] Takens, F., "A Global Version of the Inverse Problem in the Calculus of Variations", J. Diff. Geom. 14, 543-562 (1979).

16] Zuckerman, G.J., "Invariant Variational Problems", Yale Math. Dept. preprint (to appear).

17] Connes, A., "Noncommutative Differential Geometry", Publ. Math. I.H.E.S. 62, 257-360 (1985).

18] Massey, W., "Elasticae in $\mathbb{R}^3$", Yale Univ. Math. Dept. preprint (1985).

19] Chern, S.-S., and Simons, J., "Characteristic Forms and Geometric Invariants", Annals of Math. 99, 48-69 (1974).

20] Bers, L., "Finite Dimensional Teichmüller Spaces and Generalizations", Bull. A.M.S. Vol. 5, No. 2, 131-172 (1981).

21] Fintushel, R., and Stern, R., "Rational Cobordisms of Space Forms", Tulane Univ. Math. Dept. preprint (1986).

22] Fischer, A.E., and Marsden, J.E., "The Initial Value problem and the Dynamical Formulation of General Relativity", in General Relativity, ed. by S.W. Hawking, Cambridge U. Press, 138-211 (1979).

23] Witten, E., Interacting Theory of Open Superstrings", Nucl. Phys. B276, 291- (1986).

24] Brink, L., and Schwarz, J., Phys. Lett. 100B, 310- (1981).

25] Freedman, D., and van Nieuwenhuisen, P., "Properties of Supergravity Theory", Phys. Rev. D, Vol. 14, No. 4, 912- (1976).

26] Kostant, B., Quantization and Unitary Representations", in Lecture Notes in Math. 170, Springer-Verlag, Berlin, 87-208 (1970).

27] Segal, I.E., "Functional Integration and Interacting Quantum Fields", in Functional

Integration and its Applications, ed. by A.M. Arthurs, Clarendon Press, Oxford, 157-168 (1975).

28] Manin, Y.U., "Algebraic Curves and Quantum Strings", address to the International Congress of Mathematicians at Berkeley (Aug. 1986).

## An Introduction to the Moduli Space of Curves

Joe Harris*
Department of Mathematics
Brown University
Providence RI 02912



* Supported by NSF grant DMS-84-02209

THE SMITHSONIAN ASSOCIATES
COMPLIMENTARY GUEST MEMBERSHIP
NON-TRANSFERABLE
REGISTERED NUMBER: SI-89148
MR JAMES STASHOFF
EXPIRES: DECEMBER 31, 1989
NOTE: Holder entitled to same regional and national privileges as Smithsonian National Associates until date printed above.
Secretary, Smithsonian Institution
Washington, D.C. 20560

The basic object we will be considering here is the moduli space of curves of genus g, that is,

$$\mathcal{M}_g = \left\{ \begin{array}{l} \text{compact Riemann surfaces of} \\ \text{genus } g, \text{ up to isomorphism} \end{array} \right\}$$

This is an object with a long history, whose study we can only begin to describe here (for more detail and references see [M2], [ACGH] Vol. 2 and [H]). We can probably best get a sense of what sort of object it is and what we would like to know about it by considering the first nontrivial case; we will start with this.

## §1: An example.

In the case of surfaces of genus 0, the moduli space is trivial: in either the analytic theory or the algebraic one a basic theorem says that any Riemann surface homeomorphic to a sphere is in fact isomorphic to the Riemann sphere $\mathbb{P}^1$, so that $\mathcal{M}_0$ consists just of one point. The next case, $g = 1$, is far more interesting, and indeed represents one of the most beautiful confluence of ideas in geometry. Basically, there are two approaches to describing the set of isomorphism classes of curves of genus one, which we will now sketch.

i). the analytic approach. In the analytic theory, we would describe any Riemann surface E homeomorphic to a torus as the quotient $E_\Lambda$ of its universal covering space, the complex plane $\mathbb{C}$, by a lattice $\Lambda \subset \mathbb{C}$. Two such tori $E_\Lambda$ and $E_{\Lambda'}$ are isomorphic if and only if there is an automorphism $z \mapsto az + b$ of $\mathbb{C}$ carrying $\Lambda$ to $\Lambda'$; if the lattice $\Lambda$ is generated by $\omega_1, \omega_2 \in \mathbb{C}$ and $\Lambda'$ by $\omega_1', \omega_2' \in \mathbb{C}$ this will be the case if and only if the ratios $\tau = \omega_1/\omega_2$ and $\tau' = \omega_1'/\omega_2'$ are related by

$$\tau' = \frac{a\tau + b}{c\tau + d}, \qquad \begin{bmatrix} a & b \\ c & d \end{bmatrix} \in SL_2\mathbb{Z}$$

Since we can take the ratios $\tau$ to lie in the upper half plane, this means that the set of isomorphism classes of elliptic curves is naturally identified with the quotient of the upper half plane by the action of the group $SL_2\mathbb{Z}$; the famous fundamental domain for this action is pictured below.

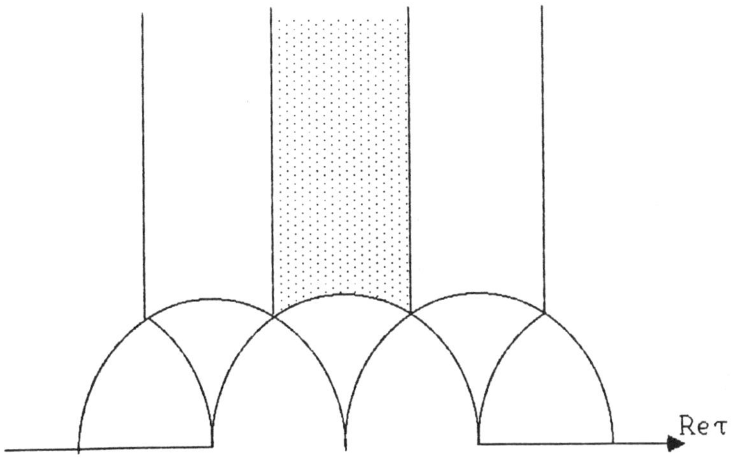


The quotient, it turns out, is isomorphic as Riemann surface to the complex plane $\mathbb{C}$ via the automorphic function $j = j(\tau)$, so that we have $\mathcal{M}_1 \cong \mathbb{C}$.

ii). <u>the algebraic approach</u>. Within the realm of algebraic varieties and maps the universal cover $\varphi:\mathbb{C} \to E$ does not exist, so we need another approach. To describe isomorphism classes of tori, we represent each elliptic curve $E$ as a plane cubic curve; it's not hard to see that in fact we can choose the defining equation of $E$ to be of the special form $y^2 = x(x-1)(x-\lambda)$ for some $\lambda \neq 0, 1$ (equivalently, we are expressing the curve $E$ as a double cover of the Riemann sphere $\mathbb{P}^1$, branched at the four points $0, 1, \infty$ and $\lambda$). Calling the curve given by this equation $E_\lambda$ we see that two such curves $E_\lambda$, $E_{\lambda'}$ are isomorphic if and only if there exists an automorphism of $\mathbb{P}^1$ (that is, a linear fractional transformation) carrying the set $\{0,1,\infty,\lambda\}$ to the set $\{0,1,\infty,\lambda'\}$, which is to say if and only if $\lambda' \in \{\lambda, 1-\lambda, \lambda^{-1}, 1-\lambda^{-1}, (1-\lambda)^{-1}, -\lambda(1-\lambda)^{-1}\}$. The set of isomorphism classes of curves of genus one is in this way seen to be the quotient of the twice punctured plane $\mathbb{C} - \{0,1\}$ by the group $S_3$ obtained in this way; and it's easy to see that the rational function

$$j(\lambda) = 256\,\frac{(\lambda^2-\lambda+1)^3}{\lambda^2(\lambda-1)^2}$$

maps this quotient isomorphically to the complex plane $\mathbb{C} = \mathbb{A}^1$ (the factor of $256$ is there for reasons having to do with the appearance of the j-function in arithmetic). Of course, the j-invariant introduced in each of these two examples is the same, so that the identifications $\mathfrak{M}_1 = \mathbb{C} = \mathbb{A}^1$ agree with one another.

This example was a crucial one in the development of the classical theory of abelian integrals. Even if we start out knowing only algebraic functions, we quickly find that to evaluate indefinite integrals such as

$$(1) \qquad \varphi(x) = \int_{x_0}^{x} \frac{dt}{\sqrt{(1-t)(1+t)}}$$

we have to introduce transcendental functions -- the function $\varphi(x)$ above is of course just the inverse of the sine function. Similarly, to evaluate further integrals, such as

$$(2) \qquad \Psi(x) = \int_{x_0}^{x} \frac{dt}{\sqrt{t(t-1)(t-\lambda)}}$$

we have to introduce even more functions -- $\Psi(x)$ here is just the inverse of the Weierstrass $\wp$ function associated to the torus $E_\lambda$ with equation $y^2 = x(x-1)(x-\lambda)$, which we may in turn express in terms of theta functions -- that are not in the field of "elementary" functions expressible in terms of algebraic and trigonometric ones.

A key distinction between integrals of the form (2) and the earlier ones (1) is the further fact that having introduced the elliptic functions $\theta(x)$ that solve a particular indefinite integral such as (2), we still cannot express the indefinite integral (2) corresponding to other values of $\lambda$ unless the two elliptic curves $E_\lambda$ and $E_{\lambda'}$ in question have the same moduli (if we are willing to allow algebraic functions of the theta function the two curves need only be isogenous, and their moduli correspondingly related, but the point remains). To solve the general integral of the form (2), then, we really have to introduce a function $\theta(x,\tau)$ of two variables, where the second variable is in effect the modulus of the elliptic curve $E_\lambda$; it is then natural to try and analyze the dependency of this function on the second variable, and hence to ask questions about the global geometry of the moduli space $\mathcal{M}_1$. These ideas appeared fairly early in the development of the theory of abelian integrals, albeit in somewhat disguised form, so that the study of the moduli space of curves of genus one may be said to have been around for a while, and to be relatively well understood.

## §2. Constructions of the moduli space.

We would like now to describe the moduli space $\mathcal{M}_g$ in general. Of course, to begin with, $\mathcal{M}_g$ is just a set; but it will be noticed that in the above example $\mathcal{M}_1$ seems to come to us equipped with natural and compatible structures of analytic and algebraic variety, and we would like among other things to see in what sense it has this additional structure in general. It is worth observing that this additional structure is characterized by a very natural property. Specifically, we define a family of compact Riemann surfaces of genus $g$ to be a map

$$\pi : \mathcal{X} \to B$$

of analytic varieties such that the fibers $X_b = \pi^{-1}(b)$ of $\pi$ are all compact Riemann surfaces of genus $g$ (of course we have a similar definintion in the algebraic category). Given any such family, we have an associated map of sets

$$\varphi : B \to \mathcal{M}_g$$

defined by sending each point $b \in B$ to the isomorphism class $[X_b]$ $\in \mathcal{M}_g$ of the fiber $X_b$. If we are to give the set $\mathcal{M}_g$ the structure of an analytic variety, then, we would like to require that

(3) <u>for any family</u> $\pi$ <u>of Riemann surfaces of genus</u> $g$, <u>the associated map</u> $\varphi$ <u>is holomorphic</u>.

Thus, in the case of genus one, any time we write down an elliptic curve whose coefficients we may consider as variable the $j$ invariants of these curves should be algebraic functions of these coefficients; for example, for the family of curves $E_m$ given as the loci $X^3 + Y^3 + Z^3 - 3mXYZ = 0$ in $\mathbb{P}^2$, we have

$$j(E_m) = 27\,\frac{m^3(m^3+8)}{(m^3-1)^3}$$

It's not hard to see that the structure of analytic variety on $\mathcal{M}_g$ satisfying this condition is unique if it exists at all; so that what we really have to do first is to show that it exists.

It would be nice if we could simply use either of the two approaches above to the construction of $\mathcal{M}_1$ to give an explicit description of $\mathcal{M}_g$. However, the situation changes dramatically when we pass from tori to Riemann surfaces of higher genus. Here the simple fact of the matter is that, for a variety of reasons, no such explicit description of $\mathcal{M}_g$ as that given above of $\mathcal{M}_1$ seems possible; difficulties seem to confront any attempt to explicitly generalize the constructions above. We will review briefly what is known; in each case, the construction described assures us that the space $\mathcal{M}_g$ does in fact exist, either as an analytic or as an algebraic variety, and gives us some information about the geometry of $\mathcal{M}_g$, but falls far short of the explicit description we have of $\mathcal{M}_1$.

i). <u>the analytic approach</u>. In the case of genus one, we realized elliptic curves $E$ as quotients of their universal covering spaces $\mathbb{C} \to E$, and explicitly descibed the moduli of the groups of covering transformations of $\mathbb{C}$ over $E$. For curves $X$ of higher genus, of course, the universal cover is the disc or half plane rather than the plane, and it is much harder to describe the moduli of subgroups of $\mathrm{Aut}(\Delta) \cong \mathrm{PSL}_2\mathbb{R}$ isomorphic to $\pi_1(X)$ up to conjugacy. The upshot is that not much information about moduli of curves of general genus has been obtained using this point of view that has not been obtained more readily by other means.

Probably a better way of generalizing the analytic approach to $\mathfrak{M}_1$ to higher genera is to view the data of the universal cover $\varphi: \mathbb{C} \to E$ of an elliptic curve as the data of $E$ together with a suitably normalized set of generators for the fundamental group $\pi_1(E)$, or, equivalently, for the first homology group $H^1(E, \mathbb{Z})$. For higher genus surfaces, of course, these two notions no longer coincide, so this leads to two further approaches.

To begin with, we can consider the data of a curve $X$ of genus $g$ together with a standard set of generators $\delta_1, \ldots, \delta_{2g}$ for $\pi_1(X)$ (i.e., represented by loops whose complement is a 4g-sided polygon); this is the same as giving $X$ together with an isotopy class of homeomorphism with a fixed compact orientable surface $X_0$ of genus $g$. It was found that the set of isomorphism classes of such pairs $(X; \delta_1, \ldots, \delta_{2g})$ can be parametrized by an open subset $\mathbb{T}_g$ of $\mathbb{C}^{3g-3}$, called <u>Teichmüller space</u>. So the moduli space $\mathfrak{M}_g$ of smooth curves of genus $g$ is Teichmüller space modulo the mapping class group of the topological surface $X_0$ (that is, the group of isotopy classes of homeomorphisms of $X_0$ with itself).

Teichmüller's construction has been especially important of late because of the handle it gives us on the toplogy of the moduli space: because Teichmüller space $\mathbb{T}_g$ is itself contractible, and the fixed points of the action of the mapping class group on it (corresponding to Riemann surfaces with automorphisms) occur in codimension $g-2$ and greater, this approach allows us for example to identify the low-dimensional cohomology of the moduli space with that of the mapping class group. This is the basis of the theorem of John Harer we will discuss later; c.f. [H] for a discussion of other theorems and references.

The other generalization of the analytic approach to constructing $\mathfrak{M}_g$ is to look at pairs $(X; \delta_1, \ldots, \delta_{2g})$ consisting of a curve $X$ and a basis $\delta_1, \ldots, \delta_{2g}$ for the first homology $H_1(X, \mathbb{Z})$, again suitably normalized (in the case of homology, this amounts to saying that the intersection pairing on the cycles $\delta_1, \ldots, \delta_{2g}$ is normalized). To the pair $(X; \delta_1, \ldots, \delta_{2g})$ we can then associate the period matrix $\Omega$ consisting of the integrals, over the cycles $\delta_{g+1}, \ldots, \delta_{2g}$, of the $g$ holomorphic differential forms $\omega_\alpha$ on $C$ determined by the condition that for $1 \le \alpha, \beta \le g$ the integral of $\omega_\alpha$ over the cycle $\delta_\beta$ is the Kronecker delta $\delta_{\alpha,\beta}$. The period matrix then varies in an analytic subvariety of the Siegel upper half space $\mathcal{H}_g$ of $g \times g$ symmetric matrices with positive definite imaginary part; and since by a basic theorem of Torelli a curve is characterized by its period matrix, we can in this way identify the space of pairs $(X; \delta_1, \ldots, \delta_{2g})$ with a subvariety $\mathfrak{J}_g$ of $\mathcal{H}_g$. The moduli space is then realized as the quotient of $\mathfrak{J}_g$ by the action of the group $Sp(2g, \mathbb{Z})$, corresponding to the choice of basis $\delta_1, \ldots, \delta_{2g}$.

This construction had, until recently, the defect that the space $\mathfrak{J}_g$ was extremely mysterious; for example, no one knew how to tell whether a given period matrix $\Omega \in \mathcal{H}_g$ came from a Riemann surface (this was known as the Schottky problem). It still is pretty mysterious, but less so, thanks to the efforts of Arbarello-DeConcini

[A-D] and Shiota [Sh] who have given explicit characterizations of period matrices arising from curves. Thanks to work of Satake, Baily-Borel and others, however, it is possible to see from this construction that $\mathfrak{M}_g$ is indeed an algebraic variety; in fact, by compactifying the quotient of $\mathcal{H}_g$ by Sp(2g, $\mathbb{Z}$), Satake and Baily-Borel give a compactification $\mathfrak{M}$ of $\mathfrak{M}_g$ in which the boundary $\mathfrak{M}_g - \mathfrak{M}_g$ has codimension 2 in $\mathfrak{M}_g$; showing in particular that <u>there are no nonconstant holomorphic functions on $\mathfrak{M}_g$</u>.

ii). <u>the algebraic approach</u>. The two analytic constructions described above conform to one general pattern: they both consist of first constructing a moduli space for curves with some additional structure, and then realizing the moduli space $\mathfrak{M}_g$ of curves alone as a quotient of this space. The algebraic approach follows the same lines: in the case of genus one, we considered first the moduli of a curve together with a particular choice of defining equation; and in general we similarly consider first the moduli of a curve $X$ of genus $g$ together with a choice of projective embedding.

The projective realization necessarily comes about in different fashion for curves of higher genus than for genus one. Specifically, for any $m \geq 3$ we consider curves $X$ embedded in projective space $\mathbb{P}^{(2m-1)(g-1)-1}$ by a basis $\omega_1, \ldots, \omega_{(2m-1)(g-1)}$ for the vector space of m-fold differentials on $X$. Just as plane curves may be parametrized in a fairly elementary way by the points of a projective space, such images -- the set of curves $X \subset \mathbb{P}^{(2m-1)(g-1)-1}$ of degree $m(2g-2)$ and genus $g$ whose hyperplane sections are divisors of zeroes of m-fold differentials -- are naturally parametrized by a subvariety $\mathcal{H}$ of a projective variety, called the <u>Hilbert scheme</u>; the moduli space $\mathfrak{M}_g$ is thus the quotient of this subvariety by the action of the group $PGL_{(2m-1)(g-1)}\mathbb{C}$.

This is the construction that most directly shows us the algebraic (as opposed to analytic) structure of $\mathcal{M}_g$, and in fact can be used to give $\mathcal{M}_g$ the structure of a variety over $\mathbb{Q}$. Of course, we have absolutely no idea here what the variety $\mathcal{H}$ looks like. There is, however, one feature in particular that makes this construction essential: because we are taking the quotient of a quasiprojective variety by a reductive group we may apply Geometric Invariant Theory to arrive at a compactification of $\mathcal{M}_g$, discussed below.

## §3. Basic properties of the moduli space

i). dimension. Of course, the moduli space was studied long before any of the constructions above were given; it was simply assumed that $\mathcal{M}_g$ possessed the structure of a variety. Thus, Riemann first computed the dimension of $\mathcal{M}_g$ to be $3g-3$ by showing that a general Riemann surfaces of genus $g$ could be represented as a $g$-sheeted branched cover of the Riemann sphere with $3g$ branch points in finitely many ways. Since such a branched cover is determined up to a finite number by its branch points (which may be any collection of $3g$ points in $\mathbb{P}^1$), it followed that the dimension of $\mathcal{M}_g$ was the dimension $3g-\dim(PGL_2) = 3g-3$ of the space of $3g$-tuples of points on $\mathbb{P}^1$ up to isomorphism. More generally, a general Riemann surface of genus $g$ can be represented as a branched cover of $\mathbb{P}^1$ with degree $d$, and hence $b = 2d+2g-2$ branch points, in a $(2d-g+2)$-dimensional family of ways, so that the dimension of the moduli space is the dimension $b-3 = 2d+2g-5$ of the space of b-tuples of points on the Riemann sphere $\mathbb{P}^1$ up to isomorphism, minus $(2d-g+2)$.

The dimension of $\mathcal{M}_g$ can also be determined by looking at representations of a general Riemann surface as a plane curve: it's not hard to see that the space of plane curves of degree d and genus g has dimension $3d+g-1$, while a general curve X can be represented as such a plane curve in a $(3d-2g+6)$-dimensional family of ways, so we have $\dim(\mathcal{M}_g) = (3d+g-1) - (3d-2g+6) - \dim(PGL_3) = 3g-3$.

ii). <u>irreducibility</u>. Likewise, it was found back in the $19^{th}$ century that the moduli space was irreducible. Again, this was done by examining branched covers of $\mathbb{P}^1$ with given degree and given number of branch points and showing that the space of such covers was irreducible (since the space of branch divisors is obviously irreducible, this just amounts to the essentially combinatorial fact that by moving the branch points around we can go from any one cover with given branch divisor to any other); and again it can also be done by examining the space of plane curves of given degree and genus.

iii). <u>singularities of</u> $\mathcal{M}_g$. It's not hard to describe the local character of the moduli space. Specifically, for $g \geq 4$ it is the case that a point $X \in \mathcal{M}_g$ is a singular point if and only if the corresponding Riemann surface has automorphisms. More generally, the tangent space T to $\mathcal{M}_g$ at any smooth point $X \in \mathcal{M}_g$ can be naturally identified with the dual of the vector space $H^0(X, \omega^2)$ of quadratic differentials on X; and for an arbitary point $X \in \mathcal{M}_g$ we can see that an analytic neighborhood of X is isomorphic to the quotient of a neighborhood of the origin in T by the action of the automorphism group of $X$ on $T$.

## §4. Compactifying the moduli space

Probably the most significant development for the study of the moduli space of curves was the description by Mumford and Deligne of the stable compactification $\overline{\mathcal{M}}_g$ of $\mathcal{M}_g$ (see [M3] for a construction of this space via geometric invariant theory). There are many reasons why having an explicit compactification of $\mathcal{M}_g$ is important. For one thing, most of the techniques developed over the last century for studying varieties have been directly relevant only to projective, or complete, varieties; the most natural thing to do these days with an open variety is to try and express it as an open subset of a projective variety, study that variety, and then go back and say something about the open subset. Perhaps more importantly in the case of moduli spaces of curves, a good compactification is desirable simply because families of curves do degenerate -- in the case of curves of genus one, we can see that any family of elliptic curves with nonconstant modulus must degenerate; and families of curves of higher genus that do not degenerate, while they do exist, are not common. When a family of curves $\pi: X \to B$ does have singular fibers $X_{b_0}$ we would like to be able to describe the family in terms of a map $\varphi$ from $B$ to some compactified moduli space, rather than work with the restricted family $\pi': X' = X - X_{b_0} \to B - \{b_0\}$ and in effect pretend that the singular fiber didn't exist.

What constitutes a "good" compactification of the moduli space of curves? The first of the reasons above for compactifying $\mathfrak{M}_g$ requires simply that we should be able to describe the added points, and that the singularities of the compactification, as in the interior, should be under control (the Satake compactification mentioned above satisfies the first, but not the second, of these criteria). The second reason requires that the points we add should correspond to isomorphism classes of geometric objects naturally arising as limits of smooth curves, and that the compactified space should have the fundamental property (3) above with respect to families of curves including such degeneracies. This in turn suggests that we simply add to $\mathfrak{M}_g$ a point corresponding to each isomorphism class of singular curve (of course, this means extending the notion of genus to singular curves, which we do by introducing the arithmetic genus, about which more below). This, however, forces us into an untenable position: if we do this, the moduli space will no longer be Hausdorff. For example, consider once more the moduli of elliptic curves. We would clearly want to add to $\mathfrak{M}_1$ a point $C_0$ corresponding to the isomorphism class of the nodal cubic $y^2 = x^2(x+1)$: in the real (x,y) plane

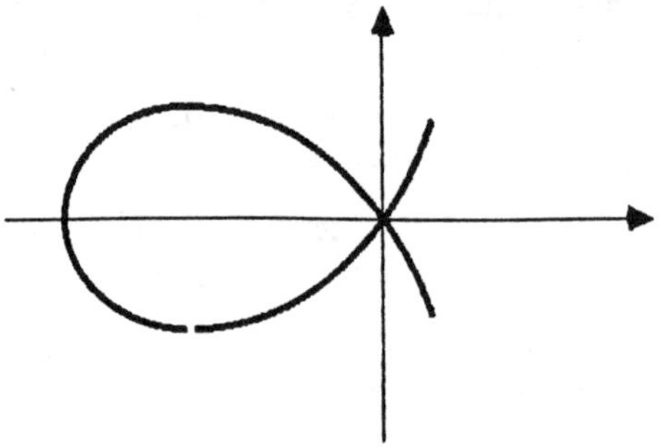

or, as complex analytic variety

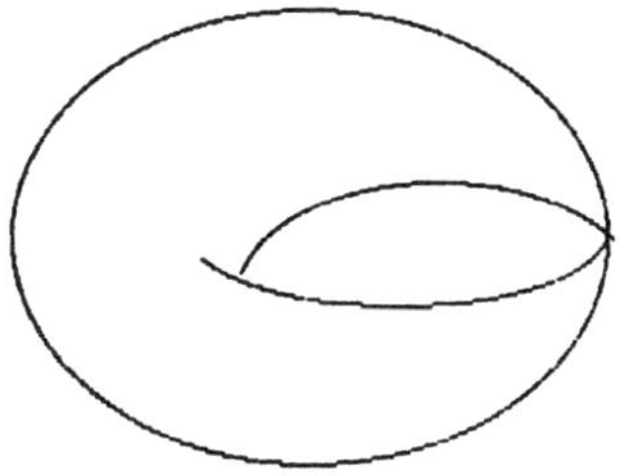

which will correspond to the limit of a family of curves with j-invariant tending to $\infty$. But in the family $y^2 = x^2(x-t)$ parametrized by $t \in \mathbb{C}$, all the fibers except the one corresponding to $t=0$ are isomorphic to $C_0$, while the fiber corresponding to $t=0$ is the cuspidal curve $C_1$. In order for the induced map $\varphi:\mathbb{C} \to \overline{\mathfrak{M}}_1$ to be continuous, then, the point $C_1$ would have to lie in the closure of the point $C_0$. To make matters worse, in the family $y^2 = x^3 - t$ all the fibers except for $t=0$ are isomorphic to the elliptic curve $E$ with j-invariant $0$, while the fiber $E_t$ is isomorphic to $C_1$, so that the point $C_1$ must also lie in the closure of the point $E$ as well; in the same fashion one can see that $C_1$ must lie in the closure of every point in $\mathfrak{M}_1$ as well.

The solution is to throw into the moduli space the isomorphism classes of some, but not all, singular curves; specifically, we include the isomorphism classes of what are called <u>stable</u> curves. By a stable curve we mean a connected curve $C$ with only nodes as singularities -- that is, it is obtained from a disjoint union of compact Riemann surfaces $C_i$ by taking $2\delta$ distinct points $p_1, \ldots, p_\delta, q_1, \ldots, q_\delta$ and identifying $p_i$ with $q_i$; we make the further requirement that the automorphism group of the resulting curve be finite, which is equivalent to requiring that each of the components $C_i$ isomorphic to the Riemann sphere contains at least three of the points $\{p_1, \ldots, p_\delta, q_1, \ldots, q_\delta\}$. The arithmetic genus of such a curve $C$ is given by $p_a(C) = \sum(g(C_i)-1) + \delta - 1$; in effect this means that such a curve can arise as a limit of smooth curves of genus $g$ with smooth total space only if $g = p_a(C)$, so that the compactification $\overline{\mathfrak{M}}_g$ of $\mathfrak{M}_g$ will include a point for each isomorphism class of stable curve of arithmetic genus $g$. The remarkable fact is that these isomorphism classes do naturally form the points of a projective variety $\overline{\mathfrak{M}}_g$ satisfying the fundamental property (3).

(Having said this much, we may appear to have a slight paradox on our hands: if $\overline{\mathfrak{M}}_g$ is compact, that would suggest that every family of curves of genus $g \geq 2$ will have a stable limit; but this is clearly not the case. The resolution of this paradox lies in base change and birational modifications of a family: the correct statement is that for any family $\pi: X \to \Delta$ of curves parametrized by the disc $\Delta$ whose fibers $X_t$ are stable for $t \neq 0$, we can, by replacing the variable $t$ in $\Delta$ by a power $t^m$ and blowing up or down the total space of this family, arrive at a family $\pi': X' \to \Delta$ all of whose fibers are stable; and moreover the fiber $X'_0$ of our new family over $t=0$ will be uniquely determined by the restricted family $X^* = X - X_0 \to \Delta^*$. This process is called stable reduction; it can usually be carried out very explicitly in cases.)

The stable compactification of $\mathfrak{M}_g$ satisfies the criteria for a good compactification to a degree that is almost unique among nontrivial moduli problems. To begin with, the description of the points we have added on is clearly explicit; for example, we can count parameters to see that the locus $\Gamma_\delta$ in $\overline{\mathfrak{M}}_g$ corresponding to stable curves with $\delta$ nodes has pure codimension $\delta$ in $\overline{\mathfrak{M}}_g$. Since it is also clear that $\Gamma_\delta$ lies in the closure of $\Gamma_{\delta-1}$, this in turn tells us that the boundary $\Delta = \overline{\mathfrak{M}}_g - \mathfrak{M}_g$ consists of the union of $[g/2] + 1$ divisors (codimension one subvarieties) $\Delta_0, \Delta_1, \ldots, \Delta_{[g/2]}$, where $\Delta_0$ is the closure of the locus of irreducible curves with one node

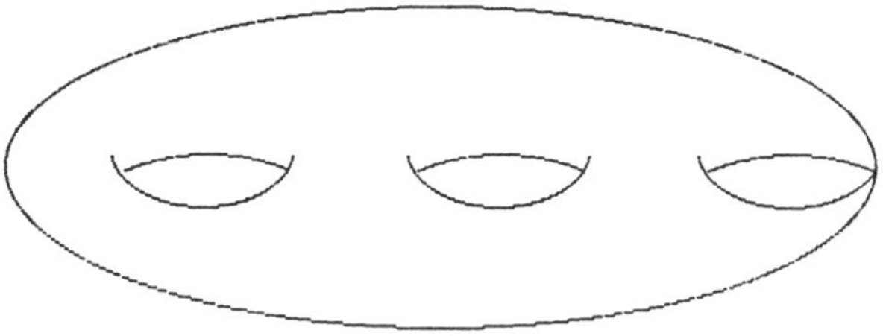

or, digrammatically,

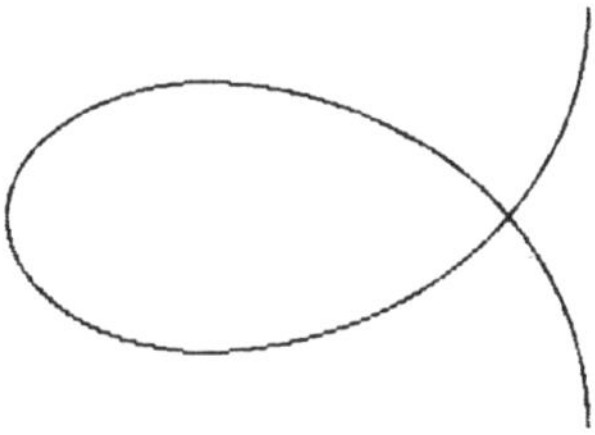

and for $i > 0$, $\Delta_i$ is the closure of the locus of curves consisting of a curve of genus $i$ and a curve of genus $g-i$ joined at one point:

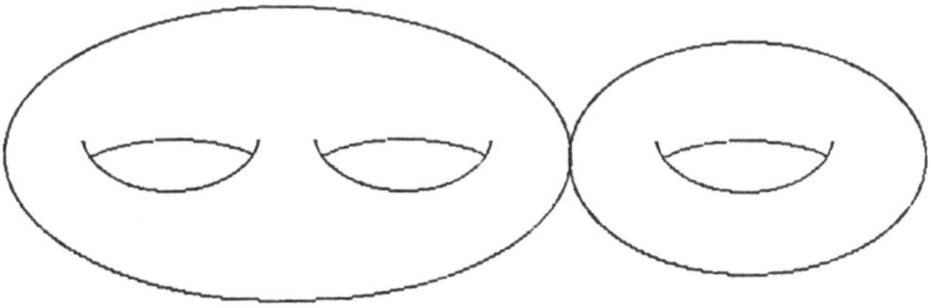

or, diagrammatically,

Likewise, the tangent spaces to and singularities of the variety $\overline{\mathcal{M}}_g$ are very similar to those of $\mathcal{M}_g$. Just as for smooth curves, $\overline{\mathcal{M}}_g$ is smooth at any point $C$ corresponding to a stable curve without automorphisms; and it may be seen that the natural identification of the cotangent space to $\mathcal{M}_g$ at such a point with the space $H^0(C, K^{\otimes 2})$ of quadratic differentials extends naturally to an identification of the cotangent space to $\overline{\mathcal{M}}_g$ at a point $C$ with the space $H^0(C, K \otimes \omega)$ of sections of the tensor product of the sheaf $K$ of holomorphic differentials on $C$ with the dualizing sheaf $\omega$. Likewise, the singularities of $\overline{\mathcal{M}}_g$ are all finite quotient singularities; specifically, the germ of $\overline{\mathcal{M}}_g$ at any point $C$ is isomorphic to the quotient of the dual of the vector space $H^0(C, K \otimes \omega)$ by the action of $\mathrm{Aut}(C)$.

## §5. Divisor theory of the moduli space

Having been given a compactification of the moduli space of curves, we are in a position to reap the rewards and apply many of the techniques developed in this century for dealing with projective varieties, most of which center around the theory of divisors and line bundles. The basic idea is a simple one: on a projective variety any meromorphic function is characterized up to scalars by its zeroes and poles, which are of course linearly equivalent divisors; and we can hope in turn to use geometry to say something about these divisors. Thus, to understand the function theory of a projective variety, we ask two questions. First, we want to describe the Picard group of the variety -- the group of divisors modulo linear equivalence, or equivalently the group of line bundles, on the variety -- and to be able to identify in these terms a given line bundle or the class of a given divisor. Second, we would like to know which of these divisor classes are represented by effective divisors, and by how many. The answers to these two questions will enable us to say when there exists a function (or section of a given line bundle) with given singularities.

In the case of the moduli space of curves, we have in response to the first of these questions the celebrated theorem of John Harer [Hr], who calculated (using the description of $\mathcal{M}_g$ as the quotient of the Teichmüller space by the action of the mapping class group) that

$$H^2(\mathcal{M}_g, \mathbb{Z}) = \mathbb{Z}$$

for $g \geq 4$; it follows that the group $\mathrm{Pic}(\mathcal{M}_g)$ is likewise $\mathbb{Z}$, being identified with $H^2(\mathcal{M}_g, \mathbb{Z})$ via the first Chern class. There is a natural way to define an element of this group (which Arbarello and Cornalba [A-C] showed was in fact a generator), roughly as follows: we define the <u>Hodge vector bundle</u> on $\mathcal{M}_g$ to be the vector bundle of rank $g$ whose fiber over any point $C \in \mathcal{M}_g$ is the vector space of holomorphic differentials on $C$, and the <u>Hodge line bundle</u> to be the determinant of this vector bundle; the <u>Hodge class</u>, usually denoted $\lambda$, is then either the Hodge line bundle, viewed as an element of the divisor class group $\mathrm{Pic}(\mathcal{M}_g)$, or its first Chern class in $H^2(\mathcal{M}_g, \mathbb{Z})$.

This carries over readily to a description of the divisor class group $\mathrm{Pic}(\overline{\mathcal{M}}_g)$ of the compactified moduli space. To begin with, let $C$ be the stable curve obtained from the possibly disconnected but smooth curve $\tilde{C}$ by identifying points $p_i$ and $q_i$ for $i = 1, \ldots, \delta$.

We define the dualizing sheaf $\omega_C$ of $C$ to be the sheaf whose sections on any open set in $C$ is the space of differentials on the corresponding open subset of $\tilde{C}$, possibly with simple poles at the points $p_i$ and $q_i$, but always having opposite residues at $p_i$ and $q_i$. It's not hard to see that the dualizing sheaf is the sheaf of sections of a line bundle on $C$ and that if $C = C_0$ belongs to a family $\pi : \mathcal{C} \to B$ of curves with smooth total space the dualizing sheaves/canonical bundles of the fibers fit together to form a single line bundle on $\mathcal{C}$. Indeed, $\omega_C$ may be characterized by this property: the relative cotangent bundle of $\pi$ is a line bundle outside the singularities of $C_0$, and hence has a unique extension to a line bundle on all of $\mathcal{C}$; $\omega_C$ is the restriction of this line bundle to $C_0$. Alternately, $\omega_C$ may be readily seen to be the unique sheaf for which the statement of Kodaira-Serre duality $H^1(C,L) = H^0(C,\omega_C \otimes L^*)^*$, and hence the Riemann-Roch formula $h^0(C,L) = \deg(L) - g + 1 + h^1(C,\omega_C \otimes L^*)$, hold for any line bundle $L$. It is true in particular that the dualizing sheaf on a stable curve $C$ of arithmetic genus $g$ has exactly a g-dimensional vector space of global holomorphic sections; and since the dualizing sheaf behaves so well in families we can extend the Hodge vector bundle over the compactified moduli space by taking its fiber over any $C$ to be this vector space. The Hodge line bundle and the class $\lambda$ may thus be extended to all of $\overline{\mathcal{M}}_g$; we continue to use the symbol $\lambda$ to denote this bundle/class.

Given that $\mathrm{Pic}(\mathcal{M}_g)$ is generated by the class $\lambda$, it's not hard to see that the Picard group $\mathrm{Pic}(\overline{\mathcal{M}}_g)$ of the compactification is freely generated by $\lambda$ and the classes $\delta_\alpha$ of the boundary components $\Delta_\alpha$. This then answers the first of our two questions. Before going on to the second, we consider here three ways in which divisors and/or line bundles on $\mathcal{M}_g$ arise.

To begin with, the divisor theory of $\overline{\mathcal{M}}_g$ naturally arises whenever we want to discuss modular forms on $\overline{\mathcal{M}}_g$. Thus, one way of describing a function on the moduli space might be to find a function on the Siegel upper half space $\mathcal{H}_g$ (or just on the Jacobian locus $\mathcal{J}_g \subset \mathcal{H}_g$) invariant under the action of the symplectic group $Sp(2g,\mathbb{Z})$. Of course, no such function exists on $\mathcal{J}_g$, since as we have observed for $g \geq 3$ there are no nonconstant holomorphic functions of $\mathcal{M}_g$. What we do have, on the other hand, are <u>modular forms</u>; these are functions $f$ on $\mathcal{H}_g$ that satisfy a functional equation

$$f\left(\frac{A\Omega+B}{C\Omega+D}\right) = \det(C\Omega+D)^m \cdot f(\Omega), \qquad \begin{bmatrix} A & B \\ C & D \end{bmatrix} \in Sp(2g,\mathbb{Z});$$

the number $m$ is called the <u>weight</u> of the modular form. Needless to say, a modular form is not a well-defined function on $\mathcal{M}_g$; but in view of its functional equation it <u>is</u> a well-defined section of a suitable line bundle: the $m^{th}$ power of the Hodge line bundle $\lambda$. Thus questions about the existence of modular forms with specified singularities at the boundary of the moduli space are in effect questions about divisors on $\overline{\mathcal{M}}_g$.

Another way in which sections of line bundles on $\overline{\mathcal{M}}_g$ arise is by considering natural holomorphic tensors on $\overline{\mathcal{M}}_g$, such as m-fold pluricanonical differential forms (these are the tensors written in terms of local coordinates $z_1, \ldots, z_{3g-3}$ as

$$\varphi(z) = \varphi(z)\cdot(dz_1 \wedge \ldots \wedge dz_{3g-3})^m).$$

These are by definition sections of a line bundle $(K_{\mathcal{M}_g})^{\otimes m}$, which we may hope to describe in terms of the set of generators we have already given for $\mathrm{Pic}(\overline{\mathcal{M}}_g)$; in fact, given the description of the cotangent space to $\mathcal{M}_g$ at a point $C$ as the vector space of quadratic differentials on $C$ (and its extension to $\overline{\mathfrak{M}}_g$ above), it is a straightforward consequence of Grothendieck's version of the Atiyah Singer Hirzebruch Riemann Roch theorem that we have

(4) $$K_{\mathfrak{M}_g} \sim 13\cdot\lambda - 2\delta.$$

One further ingredient in this stew is the presence of geometrically defined divisors in the moduli space. At first, the moduli space may seem like a very abstract sort of object; after all, none of the constructions really tells us much more about it than the fact that it exists. For example, we can't hope to write down equations for it in any projective embedding and so we certainly can't expect to be able to describe, for example, subvarieties of $\overline{\mathfrak{M}}_g$ by giving their equations. But there is another way. Since the points of $\mathfrak{M}_g$ correspond to curves, any time we write down a geometric property shared by some but not all curves of genus $g$ we arrive at a subset of $\mathfrak{M}_g$; in most cases this is a (locally) closed subvariety. Thus, for example, the locus of hyperelliptic curves is a subvariety of, and hence describes a cycle in, $\mathfrak{M}_g$; as are the loci of trigonal (or more generally k-gonal) curves, smooth plane curves of degree $d$ (or more generally curves realizable as curves of degree $d$ in $\mathbb{P}^r$), curves with a given automorphism group, curves with a given type of Weierstrass point, and so on.

The study of these subvarieties of $\overline{\mathfrak{M}}_g$ is a fascinating end in itself. Since, moreover, many of these subvarieties happen to have codimension one, it carries the additional reward of giving us information about the divisor theory of $\overline{\mathfrak{M}}_g$. To find divisors, we just have to look for those geometric conditions that define subvarieties of codimension one; for example, the locus of curves with a non-normal Weierstrass point is a union of two divisors whose classes have been determined by Diaz [D] and Cukierman [C];

and the locus of curves with a vanishing theta-null has a component of codimension one whose class has been determined by Mumford and Teixidor [T].

We turn now to the second of the questions posed above in relation to the divisor theory of a variety in general: determining the effective divisor classes. In our present circumstances we would like to know when the divisor class $a\cdot\lambda - b_0\cdot\delta_0 - \ldots - b_{[g/2]}\cdot\delta_{[g/2]}$ is effective. For simplicity we often restrict ourselves to consideration just of the divisor classes $a\cdot\lambda - b\cdot\delta$, where $\delta = \Sigma\delta_\alpha$ is the class of the entire boundary; these are in some sense the most important classes. As another simplification, we usually ask only what classes $D$ are rationally effective; that is, positive rational linear combinations of effective divisors, or equivalently such that a positive power of the corresponding line bundle has a holomorphic section. It is known that powers of the line bundle $\lambda$ have lots of sections, and $\delta$ is effective by definition; on the other hand, we see from the Satake compactification of $\mathfrak{M}_g$ that $\mathfrak{M}_g$ is swept out by complete curves on which $\lambda$ has positive degree, so that no linear combination of $\lambda$ and $\delta$ with negative coefficient of $\lambda$ can possibly be effective. The cone of rationally effective divisor classes on $\overline{\mathfrak{M}}_g$ thus looks like

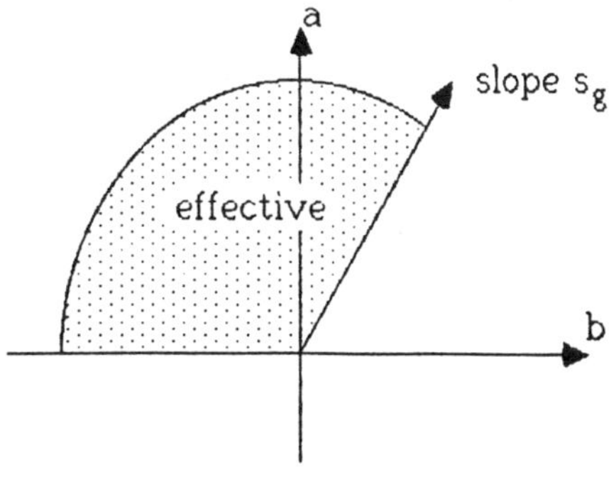

and our question amounts to asking what is the slope $s_g$ of the right hand edge of this cone.

As observed earlier, the question of effective divisors on $\overline{\mathfrak{M}}_g$ may be rephrased as a question about the existence of modular forms of given weight a vanishing to given order b at the boundary of the moduli space. This recasting of the problem does not, however, give any new information: while it has been possible to write down modular forms on the Siegel upper half space, it is not clear when such forms will vanish identically on the Jacobian locus. (Perhaps the recent solutions of the Schottky problem will make this more feasible.) For the time being, most of the flow of information has been in the other direction: information about the effectivity of certain divisor classes has been used to assert the existence or nonexistence of corresponding modular forms on $\mathfrak{M}_g$.

On the other hand, looking at divisors in $\mathfrak{M}_g$ characterized by geometric properties of the curves represented does give us a start towards describing the cone of effective divisor classes on $\overline{\mathfrak{M}}_g$ for the simple reason that it gives us a source of divisor classes that are known to be effective. A class of examples of geometrically defined divisors in $\overline{\mathfrak{M}}_g$ that is important in this respect is provided by the Brill-Noether theorem. This theorem says that a general curve C of genus g will possess a nondegenerate map of degree d to $\mathbb{P}^r$ if and only if the Brill-Noether number $\rho = g - (r+1)(g-d+r)$ is nonnegative; thus, whenever $\rho < 0$, the locus $\mathfrak{M}_{g,r,d}$ of curves realizable as curves of degree d in $\mathbb{P}^r$ (or as d-sheeted branched covers of the Riemann sphere in case $r=1$) is a proper subvariety of $\mathfrak{M}_g$. More generally, it seems sometimes to be the case, especially when $\rho$ is small negative, that $\mathfrak{M}_{g,r,d}$ will have codimension $-\rho$; in particular, it turns out that whenever $\rho = -1$, $\mathfrak{M}_{g,r,d}$ will have a (unique, as it turns out) component of codimension one. The class of this divisor -- or rather its closure in $\overline{\mathfrak{M}}_g$ -- was first determined in case $r = 1$ by Harris and Mumford [H-M], and for arbitrary $\rho$ by Eisenbud and Harris [E-H]; the class was found to be

$$\overline{\mathfrak{M}}_{g,r,d} \sim c_{g,r,d} \cdot \left[ \left(6 + \frac{12}{g+1}\right) \cdot \lambda - \delta_0 - b_1 \cdot \delta_1 - \dots \right]$$

where $c_{g,r,d}$ is a positive constant and all the $b_i$ are greater than one.

We deduce from the above that the divisor class $a \cdot \lambda - b \cdot \delta$ is rationally effective whenever the ratio $a/b$ exceeds $6 + 12/(g+1)$, i.e., in terms of the diagram above we have

$$s_g \leq 6 + \frac{12}{g+1}$$

This is the best known estimate, and it's conceivable that it is sharp.

One consequence of this is that the moduli space $\overline{\mathfrak{M}}_g$ possesses global holomorphic pluricanonical forms whenever the ratio $6 + 12/(g+1)$ is less than the ratio $13/2$ of the coefficients of $\lambda$ and $\delta$ in the expression (4) above for the class of $K_{\overline{\mathfrak{M}}_g}$; since holomorphic tensors pull back to holomorphic tensors under rational maps and rational varieties possess no such global tensors we deduce in particular that for $g \geq 23$ there does not exist a surjective rational map from projective space to $\overline{\mathfrak{M}}_g$.

This result is often expressed by saying that for $g \geq 23$ <u>the moduli space $\mathfrak{M}_g$ cannot be unirationally parametrized</u>. This is in contrast to the case of $g$ small, where such parametrizations do exist (for example, in case $g = 2$ we have the family of curves

$$y^2 = x^5 + a_4x^4 + \ldots + a_0$$

in which the parameters $a_0, \ldots, a_4$ are free to vary in an open subset of $\mathbb{C}^5$, and which includes an open subset of curves of genus 2); so far, unirational parametrizations of $\mathfrak{M}_g$ have been found for all genera $g \leq 13$ (c.f. [C-R1] and [C-R2] for the most recent results and bibliography)

Finally, we may ask what is known in the other direction, about the nonexistence of effective divisors in certain divisor classes; unfortunatly, all we have at present are the very partial results of [H-Mo] and [C-R3]. The former deal primarily with divisors linearly equivalent to $a \cdot \lambda - \sum b_\alpha \cdot \delta_\alpha$ with $b_\alpha \geq b$ (alternately, divisors linearly equivalent to $a \cdot \lambda - b \cdot \delta$ possibly containing some of the boundary components, or equivalently modular forms vanishing to order at least $b$ along the boundary), while the latter deal in part with the existence of divisors of the form $a.\lambda - b \cdot \delta$ not containing any of the boundary components (equivalently, modular forms of weight $a$ vanishing to order exactly $b$ along every boundary component). The results themselves are too complicated to write down here and in any event are not believed to be sharp. A typical consequence of what is known is the statement that there does not exist a modular form of weight $8$ vanishing to order two or more along the boundary components of $\overline{\mathfrak{M}}_g$ for any $g \leq 19$, and that there does not exist a modular form of weight $8$ vanishing to order exactly $2$ on each boundary component of $\overline{\mathfrak{M}}_g$ for any $g$; physical consequences of these assertions are given in [MHNS] and [C-R3] respectively.

References

[A-C] E. Arbarello and M. Cornalba, The Picard groups of the moduli spaces of curves, to appear, Topology

[ACGH] E. Arbarello, M. Cornalba, P. Griffiths and J. Harris, Geometry of Algebraic Curves, Springer-Verlag 198?

[A-D] E. Arbarello and C. DeConcini, On a set of equations characterizing Riemann matrices, Annals of Math. 120 (1984); see also: Another proof of a conjecture of S. P. Novikov on periods of integrals on Riemann surfaces, preprint

[C-R1] M. Chang and Z. Ran, Unirationality of the moduli spaces of curves of genus 11, 13 (and 12), Invent. Math. 76 (1984)

[C-R2] M. Chang and Z. Ran, The Kodaira dimension of the moduli space of curves of genus 15, J. Diff. Geom. 24 (1986)

[C-R3] M. Chang and Z. Ran, Divisors on $\overline{\mathcal{M}}_g$ and the cosmological constant, preprint

[C] F. Cukierman, Brown University Ph.D. thesis, 1987

[D] S. Diaz, Exceptional Weierstrass points and the divisor on moduli space that they define, Mem. Amer. Math. Soc. 56 (1985)

[E-H] D. Eisenbud and J. Harris, The Kodaira dimension of the moduli space of curves of genus ≥23, to appear, Invent. Math.

[H] J. Harris, Curves and their moduli, to appear, Proceedings of the A.M.S. Summer Institute in Alegbraic Geometry, Bowdoin, 1986

[Hr] J. Harer, The second homology group of the mapping class group of an orientable surface, Invent. Math. 72 (1983)

[H-Mo] J. Harris and I. Morrison, Slopes of effective divisors on the moduli space of curves, preprint

[H-M] J. Harris and D. Mumford, On the Kodaira dimension of the moduli space of curves, Invent. Math. 67 (1982)

[M1] D. Mumford, Geometric Invariant Theory, Springer-Verlag

[M2] D. Mumford, Curves and their Jacobians, U. of Mich. Press, Ann Arbor 1976

[M3] I. Morrison and D. Mumford, Stability of Projective Varieties, L'Enseignement Math. 23 (1977)

[MHNS] G. Moore, J. Harris, P. Nelson and I. Singer, Modular forms and the cosmological constant, preprint

[S] T. Shiota, Characterization of jacobian varieties in terms of soliton equations, Invent. Math. 83 (1986)

[T] M. Teixidor, preprint

# The Moduli Space of Punctured Surfaces

R.C. Penner

Math Department
University of Southern California
Los Angeles, CA 90089

# The Moduli Space of Punctured Surfaces

In this paper, we begin with a survey of results from [Pe1], [Pe2], and [EP] on the Teichmüller spaces of punctured surfaces. In these papers, coordinates on a certain foliated fiber bundle over Teichmüller space are derived, the action of the modular group on coordinates is computed, and a modular-invariant cell decomposition of the foliated bundle is developed. These results are then used to translate the computation of a certain cohomological invariant of moduli space into a perturbative series, which is then evaluated in closed form. Analogous cell decompositions are described in [Ha] and [BE], and the cohomology computation agrees with [HZ].

In the final section §5, we describe three new results: a construction of p-forms on moduli space; a (computationally effective) description of a fundamental domain for the action of the modular group; an explicit formula for the Weil-Petersson Kähler form on moduli space. The first two results follow from elaborations on material from §1 - §4, and an explicit scheme of integration over moduli space is proposed. The expression for the Kähler form is then considered as a test case for this integration scheme, and we describe a computation of the Weil-Petersson volume of moduli space; in particular, for the once-punctured torus, this computation is performed, and we reproduce a result of Wolpert.

The first four sections are basically expository and cover the material in somewhat more detail than did my lecture at U.C.S.D. . Certain further details are described as needed for application in §5.

It is a pleasure to thank Bill Harvey and Scott Wolpert for fruitful discussions about the new results of §5 at U.C.S.D. and Francis Bonahon, Bob Brooks, and Paul Yang for similar discussions at U.S.C. .

## §1 Coordinates on the Decorated Teichmüller Space

Let $\mathcal{T}_g^s$ denote the usual Teichmüller space of marked conformal types of hyperbolic metrics on the surface $F_g^s$ of genus g with s points removed, $s > 1$, $g > 0$, $2g - 2 + s > 0$. For each conformal type $\Gamma \in \mathcal{T}_g^s$, we consider the space $\mathcal{B}_\Gamma$ of horocycles on $F_g^s$, one horocycle about each cusp, and define the *decorated Teichmuller space*

$$\tilde{\mathcal{T}}_g^s = \{\Gamma \times \mathcal{B}_\Gamma : \Gamma \in \mathcal{T}_g^s\} \approx \mathcal{T}_g^s \times \mathbb{R}_+^s$$

to be the Cartesian product of $\mathcal{T}_g^s$ and the space of horocycles. The usual mapping class group $MC_g^s$ of diffeotopy classes of orientation-preserving diffeomorphisms of $F_g^s$ acts on $\mathcal{T}_g^s$ in the usual way, and we extend this action to a diagonal action on $\tilde{\mathcal{T}}_g^s$ by the action on horocycles of a representative map f which respects Poincaré metrics. If $\varphi \in MC_g^s$, let $\varphi_* : \tilde{\mathcal{T}}_g^s \circlearrowleft$ denote the corresponding diffeomorphism.

We seek a natural parameterization for $\tilde{\mathcal{T}}_g^s$ and work in Minkowski three-space $\mathbb{M}$ to accomplish this. Recall that $\mathbb{M}$ is a real vector space of dimension three which supports a non-degenerate pairing

$$\begin{aligned} \langle\cdot,\cdot\rangle : \ & \mathbb{M} \times \mathbb{M} \to \mathbb{R} \\ & :(x,y,z) \times (x',y',z') \to xx' + yy' - zz' \ . \end{aligned}$$

The sheet

$$\mathbb{H} = \{v \in \mathbb{M} : \langle v,v\rangle = -1\} \cap \{v = (x,y,z) : z > 0\}$$

of the hyperboloid (inherits a Riemmanian metric from the pairing and) gives a model for the hyperbolic plane; the group $SO^+(1,2)$ of linear automorphism of $\mathbb{M}$ which respect the pairing and preserve the sheet $\mathbb{H}$ with its orientation is identified with the group of Möbius tranformations. The (positive) light-cone

$$L^+ = \{v \in \mathbb{M} : \langle v,v\rangle = 0\} \cap \{v = (x,y,z) : z > 0\}$$

is identified with the horocycle bundle over the circle at infinity in the hyperbolic place by the correspondence

$$v \in L^+ \longleftrightarrow \{w \in \mathbb{H} : \langle v,w\rangle = -1\}.$$

If $v,v' \in L^+$ and and $h,h' \subset \mathbb{H}$ denote the corresponding horocycles, we let $\delta$ denote the signed hyperbolic distance between $h,h'$, taken with positive sign if $h \cap h' = \phi$ and with negative sign if $h \cap h' \neq \phi$. We have the equality [Lemma 2.1 of Pe 1]

$$-\langle v,v'\rangle = 2e^{\delta}.$$

A point of $\mathfrak{T}_g^s$ gives rise to an isomorphism $\pi_1 F_g^s \to \Gamma < SO^+(1,2)$, where $\Gamma$ is a discrete group defined up to conjugacy in $SO^+(1,2)$, and there is a corresponding covering map $\mathbb{H} \to \mathbb{H}/\Gamma = F_g^s$. Using our identification of $L^+$ with the horocycle bundle, a choice of horocycle about a cusp of $F_g^s$ give rise to a $\Gamma$-orbit of points in $L^+$. A point of $\tilde{\mathfrak{T}}_g^s$ therefore gives rise to (a conjugacy class of) $\Gamma < SO^+(1,2)$ together with s $\Gamma$-orbits $\mathcal{B}_1,\ldots,\mathcal{B}_s \subset L^+$. We let $\mathcal{B}$ denote $\bigcup_{i=1}^{s} \mathcal{B}_i$.

Fix $\Gamma \times B \in \tilde{\mathcal{T}}_g^s$ and let c be a homotopy class of (not necessarily simple paths on $F_g^s$ connecting (not necessarily distinct) punctures. Such a homotopy class is called on <u>ideal arc</u>. Orient c and choose a lift $\tilde{c}$ of c to $\mathbb{H}$. $\tilde{c}$ is asymptotic to a pair of rays $r, r' \subset L^+$, and there are unique points

$$v = r \cap B \in L^+$$

$$v' = r' \cap B \in L^+.$$

We define the $\lambda$-<u>length</u> of c with respect to $\tilde{\Gamma} = \Gamma \times B$ to be

$$\lambda(c;\tilde{\Gamma}, B_1,\ldots,B_s) = \sqrt{-\langle v,v'\rangle} .$$

One checks easily that this quantity is independent of the orientation on c and choice of lift of c; from remarks above, the $\lambda$-length is essentially the exponential of hyperbolic length between horocycles.

For each ideal arc c on $F_g^s$, the $\lambda$-length gives a positive real-valued function $\lambda(c)$ defined on $\tilde{\mathcal{T}}_g^s$, and by fixing an appropriate collection $c_1,\ldots,c_q$ of ideal arcs, we get a map $\prod_{i=1}^{q} \lambda(c_i) : \tilde{\mathcal{T}}_g^s \to \mathbb{R}_+^q$ which is a surjective homeomorphism. Such a collection $\Delta$ is given by taking a maximal family of disjointly embedded simple arcs connecting punctures so that no component of $F_g^s - \Delta$ is a mono-gon or bi-gon. The homotopy class of such a collection is called an <u>ideal triangulation</u> (i.t.) of $F_g^s$.

<u>Parametrization Theorem</u> PT[Theorem 3.1 & 3.2 of Pe 1]: <u>If</u> $\Delta = (c_1,\ldots,c_q)$ <u>is an ideal triangulation of</u> $F_g^s$, <u>then</u>

$$\prod_{i=1}^{q} \lambda(c_i) : \tilde{\mathfrak{T}}_g^s \to \mathbb{R}_+^q$$

is a homeomorphism. Furthermore, this coordinitization is natural in the sense that i $\varphi \in MC_g^s$, $\tilde{\Gamma} \in \tilde{\mathfrak{T}}_g^s$ and c is an ideal arc on $F_g^s$, then

$$\lambda(c;\tilde{\Gamma}) = \lambda(\varphi c;\varphi_*\tilde{\Gamma})$$

## §2 The Cell Decomposition of the Decorated Teichmüller Space, Cell Parametrization, and Modular Group Action

Fix $\tilde{\Gamma} = \Gamma \times \mathcal{B} \in \tilde{\mathfrak{T}}_g^s$. A key fact [Lemma 4.1 of Pe 1] is that the subset $\mathcal{B} \subset L^+$ is a discrete set in $L^+ \cup \{0\} \subset \mathbb{M}$. Let K denote the euclidean closed convex hull of $\mathcal{B}$ in $\mathbb{M}$; since $\Gamma < SO^+(1,2)$ acts linearly, K is a $\Gamma$-invariant convex body. Using discreteness of $\mathcal{B}$, one shows [Proposition 4.4 of Pe 1] that the boundary of K "inside" the light-cone consists of a locally finite countable set $\Phi_1, \Phi_2, \ldots$ of codimension-one "faces", each of which is the convex hull of a finite set of points in $\mathcal{B}$ ; furthermore, the plane of a face of $\partial K$ always intersects $L^+$ in a ellipse. Each edge of a face determines a pair of points in $\mathcal{B}$, hence a geodesic in $\mathbb{H}$, and hence an arc in $F_g^s$ connecting punctures. The collection of arcs in $F_g^s$ arising in this way from all the edges of faces of $\partial K$ can then easily be shown to give a disjointly embedded collection $\Delta$ of arcs in $F_g^s$ connecting punctures so that each component of $F_g^s - \Delta$ is simply connected [Theorem 4.5 of Pe 1]. The homotopy class of such a family of arcs is called an ideal cell decomposition (i.c.d.). The convex hull construction just described thus canonically associates an i.c.d. of $F_g^s$ to a point $\tilde{\Gamma} \in \tilde{\mathfrak{T}}_g^s$. Let the i.c.d. $\Delta(\tilde{\Gamma})$ arise from $\tilde{\Gamma}$.

We define a decomposition $\mathcal{C}$ of $\tilde{\mathfrak{J}}_g^s$ itself by considering the collection of all

$$C(\Delta) = \{\tilde{\Gamma} \in \tilde{\mathfrak{J}}_g^s : \Delta(\tilde{\Gamma}) \subset \Delta\}$$

as $\Delta$ ranges over i.c.d.'s of $F_g^s$ .

Decomposition Theorem DT [Corollary 6.3, 6.4 and Theorem 6.8 of Pe 1]: Each $C(\Delta)$ is non-empty and contractible. $\mathcal{C}$ is a $MC_g^s$-invariant cell decomposition of $\tilde{\mathfrak{J}}_g^s$ , and the isotropy subgroup of $C(\Delta)$ in $MC_g^s$ is the subgroup of symmetries of $(F_g^s, \Delta)$.

We will comment briefly on the proof of the decomposition theorem since we require some further details below. The assertion of DT which is difficult is the contractibility of each $C(\Delta)$. To prove this, one develops a system of coupled non-linear inequalities on $\lambda$-length coordinates and proves contractibility of the corresponding subset of $\mathbb{R}_+^q$ . To avoid later complications and because this case will suffice for application below, we restrict to the case that $\Delta$ is an i.t. .

Cell Paramterization Theorem CPT [Lemma 6.1 of Pe 1]: Suppose that $\Delta$ is an i.t. of $F_g^s$ , $\Lambda$ an assisgnment of positive numbers to the ideal arcs of $\Delta$, and $(\Delta,\Lambda)$ corresponds to $\tilde{\Gamma} \in \tilde{\mathfrak{J}}_g^s$ . If $a,\ldots,e \in \Delta$, let $\alpha,\ldots,\varepsilon$ denote their values under $\Lambda$.

(i) If $\{a,b,c\} \subset \Delta$ are the edges of a triangle in $F_g^s$ , then

$$\alpha < \beta + \gamma,\ \beta < \alpha + \gamma,\ \gamma < \alpha + \beta$$

(ii) _If_ $\{a,\ldots,e\} \subset \Delta$ _and_ $(a,b,e)$, $(c,d,e)$ _are each the set of consecutive edges of a triangle in_ $F_g^s$ , _then_

$$0 < \operatorname{sgn}(\alpha^2 + \beta^2 - \varepsilon^2)\, K(\alpha,\beta,\varepsilon) + \operatorname{sgn}(\gamma^2 - \delta^2 - \varepsilon^2)\, K(\gamma,\delta,\varepsilon),$$

_where_

$$\alpha\beta\gamma\ K(\alpha,\beta,\gamma) = \sqrt{(\alpha+\beta-\gamma)(\alpha+\gamma-\beta)(\beta+\gamma-\alpha)(\alpha+\beta+\gamma)}$$

Remarks: 1) Condition (i) guarantees positivity of the radical in (ii). 2) We comment on the formula in (ii). The faces of the convex hulls used to defined $\Delta(\tilde{\Gamma})$ lie in elliptic planes, and the pairing on $\mathbb{M}$ induces a euclidean structure on this plane so that the intersection with $L^+$ is a round circle. Given a triple $(\alpha,\beta,\gamma)$ of positive numbers satisfying the triangle inequalities in (i), we construct a unique (up to congruence) euclidean triangle $\tau$ of these edge lengths. $K(\alpha,\beta,\gamma)$ is the reciprocal of the radius of the circle circumscribing $\tau$ and is called the "scaling" function.

The proof of contractibility of the subset of $\mathbb{R}_+^q$ determined by (i) and (ii) would take us far afield, as would the extension to arbitrary i.c.d.'s. Furthermore, it is the parametrization in this "top dimensional" case which we will require below.

We close this section with a brief discussion of the action of $MC_g^s$ on $\lambda$-length coordinates with respect to a fixed i.t. $\Delta$ . If $\varphi \in MC_g^s$ and $\tilde{\Gamma} \in \tilde{\mathcal{J}}_q^s$, then there is a smooth path p in $\tilde{\mathcal{J}}_g^s \approx \mathcal{J}_g^s \times \mathbb{R}_+^s$ from $\tilde{\Gamma}$ to $\varphi_*\tilde{\Gamma}$. General position with respect to $C$ allows us to conclude that $\Delta$ and $\varphi^{-1}\Delta$ are related by a sequence of the following

Elementary Move: If $\{a,\ldots,e\} \subset \Delta$ satisfy the hypotheses of condition (ii) in CPT, then add to $\Delta - e$ the other diagonal of the quadrilateral to obtain an i.t. $\Delta'$.

It follows from naturality of $\lambda$-lengths that the coordinates of $\varphi_*\tilde{\Gamma}$ with respect to $\lambda$-lengths on a fixed $\Delta$ are related by iterates of the following algebraic transformation.

Ideal Ptolemy Theorem IPT [Proposition 2.6a of Pe 1]:

Suppose $\{u_i\}_1^4 \subset L^+$ are so that any three are linearly independent, $u_1$ and $u_4$ lie on different sides of the plane spanned by $u_2, u_3$, and let

$$-\lambda_{ij}^2 = \langle u_i, u_j \rangle \text{ for } i < j .$$

Then

$$\lambda_{14}\lambda_{23} = \lambda_{12}\lambda_{34} + \lambda_{13}\lambda_{24} .$$

Remark: 1) If one pictures the edges of the convex hull of $\{u_i\}_1^4$ (from below) as a convex planar Euclidean quadrilateral, then the assertion above is Ptolemy's theorem on the lengths of a quadrilateral which inscribes in a circle. Noting that the formula in IPT is invariant under scaling the $u_i \in L^+$ separately and projecting to a horizontal plane in $\mathbb{M}$, one easily proves IPT from the euclidean version.

2) The fact that the algebraic transformation in IPT "generates" the action of $MC_g^s$ on $\lambda$-lengths will have later consequence. In §7,8 of [Pe 10], we give an algebraic representation of $MC_g^s$ using these ideas.

## §3 Fat Graphs and Perturbative Series

We wish to reformulate the cell decomposition $C$ of $\tilde{\mathcal{T}}_g^s$ in order to capture some of the combinatorics in a perturbative series. To accomplish this, we introduce some machinery from [Pe 2].

By a "graph" we mean simply a (not necessarily connected) CW complex consisting of (open) edges and vertices. We emphasize that a pair of edges may have the same vertices, and the endpoints of an edge need not be distinct. A *half - edge* of a graph G is an edge of the first barycentric subdivision of G. A vertex of G is said to be k-*valent* if there are k half-edges incident on it; we will always assume that each vertex of G is at least tri-valent. A *fattening* of G is a specification of cyclic order on the half-edges about each vertex, and a graph together with a fattening is called a *fat graph*.

A fat graph can be described by giving a generic projection (with under/over crossings) of a graph in space. The half-edges incident on each vertex inherit a clockwise order from the orientation on the plane. For instance, Figure 1a depicts three distinct fat graphs, two of which have isomorphic underlying graphs.

A fat graph G determines an oriented surface F(G) with spine G as follows: represent G in a planar projection as above, and take a regular neighborhood of the vertex set in the plane of projection; attach orientation-preserving bands to the neighborhood along the edges of G. See Figure 1b for examples.

A *marking* on a fat graph G is a marking in the sense of Riemann surfaces on F(G). If e is an edge of G with endpoints $v \neq w$, we define a new fat graph G' by collapsing e and combining the cyclic orders about v and w in the obvious manner. We say G' arises from G by a *Whitehead collapse*. F(G) is homeomorphic to F(G'), and a marking on G induces a marking on G'.

An *isomorphism* of fat graphs is an isomorphism of the underlying graphs which respects the fattening. An isomorphism $f : G_1 \to G_2$ gives rise to a homeomorphism of pairs $(F(G_1), G_1) \to (F(G_2), G_2)$. A *marked isomorphism* is an isomorphism which respects marking.

Returning to the setting §2 of the decomposition theorem DT, if $\Delta$ is an i.c.d. of $F_g^s$, then there is a dual fat graph $G(\Delta) \subset F_g^s$, which has one vertex for each component of $F_g^s - \Delta$ and one edge for each arc in $\Delta$, and the dual edge of the arc $c \in \Delta$ connects the vertices of $G(\Delta)$ corresponding to the components of $F_g^s - \Delta$ on either side of c. The cyclic order on half-edges of $G(\Delta)$ is induced from the orientation of $F_g^s$.

It follows that cells in $\mathcal{C}$ are indexed by isomorphism classes of marked fat graphs G with F(G) homeomorphic to $F_g^s$. In addition to taking duals of i.c.d.'s to get fat graphs, we take the dual $\tilde{\Sigma}_g^s$ (in the usual sense) of the decomposition $\mathcal{C}$ of $\tilde{\mathcal{J}}_g^s$ itself. A pleasant abstract description of the combinatorics of $\tilde{\Sigma}_g^s$ is as follows: a cell in $\tilde{\Sigma}_g^s$ is indexed by an isomorphism class of marked fat graph G so that $F(G) \approx F_g^s$. The cell corresponding to G has dimension

$$\sum_{k>0} (k-3)v_k(G),$$

where

$$v_k[G] = \#\{k\text{-valent vertices of } G\}.$$

In this same notation, the Euler characteristic of G is

$$\chi(G) = \chi(F(G)) = \frac{1}{2} \sum_{k>0} (2-k)v_k[G].$$

The cell of $\tilde{\Sigma}_g^s$ corresponding to G' is a face of the cell corresponding to G if G arises from G' by a Whitehead collapse (since the dual of inclusion of i.c.d.'s is Whitehead collapse).

The complex $\tilde{\Sigma}_g^s$ is contractible (since $\tilde{\mathcal{J}}_g^s$ is) and $MC_g^s$ acts cellularly (by change of marking) on $\tilde{\Sigma}_g^s$. Moreover, the isotropy group of the cell in $\tilde{\Sigma}_g^s$ corresponding to the fat graph G is the group

$$\Gamma[G] = \{\text{fat graph automorphisms of } G\},$$

and $\Gamma[G]$ is a finite group; indeed, we have the estimate

$$\#\Gamma[G] < \prod_{k>0} (v_k[G]!\ k^{v_k[G]}).$$

Furthermore, if we let $[G]$ denote unmarked fat graph isomorphism class of G. Then

$$\#\{[G] : F(G) = F_g^s\} < \sum \Big( \sum_{k>0} kv_k - 1\Big)!!\ ,$$

where the outside sum is over all positive tuples $(v_k)_{k>0}$ so that $v_0 = v_1 = v_2 = 0$ and $\Sigma(k-2)v_k = 4g - 4 + 2s$.

The situation we have is that of a group $A = MC_g^s$ acting cellularly on a contractible complex $X = \tilde{\Sigma}_g^s$ of dimension m with finite isotropy and a finite list

$$\{\sigma_i^p : i = 1,\ldots, \mu_p,\ p = 1,\ldots,m\}$$

of cells of X so that $\sigma_i^p$ is of dimension p and any p-cell is A-equivalent to exactly one element of the list. In this case, the virtual Euler characteristic (see [Br]) is

$$\chi_v A = \sum_{i,p} \frac{(-1)^p}{\#\Gamma(\sigma_i^p)}\ ,$$

where $\Gamma(\sigma_i^p) < A$ is the isotropy group of $\sigma_i^p$.

We define

$$s[G] = \#\{\text{boundary components of } F(G)\}$$

and form

$$\Phi_c(I,N) = \sum_{\substack{[G] \text{ so that } G \\ \text{is connected and} \\ I = -2\chi[G]}} \frac{(-1)^{\Sigma v_k[G]} N^{s[G]}}{\#\Gamma[G]} .$$

For each I, $\Phi_c(I,N)$ is a rational polynomial in N, where the index I determines $\chi[G]$ and the exponent of N determines $s[G]$. It follows from the remarks above that

$$\Phi_c(I,N) = \sum_{\substack{g \geq 0,\ s \geq 1 \text{ with} \\ 4g - 4 + 2s = I}} \chi_v MC_g^s N^s ,$$

and we take this identity as motivation for studying the function $\Phi(I,N)$ defined analogously but summing over all (not necessarily connected) fat graph isomorphism classes.

To compute $\Phi(I,N)$ in closed form, we make use of the following identity.

<u>Perturbative Series Theorem PST</u> [Theorem 2.1 of Pe 2]:

<u>Suppose</u> $\{v_k\}_3^K \subset \mathbb{Z}_+$ <u>and define</u>

$$P_{v_3,\ldots,v_K}(N) = \sum_{\substack{[G] \text{ so that} \\ v_k[G] = v_k \text{ for all } k}} \frac{(-1)^{\Sigma v_k[G]} N^{s[G]}}{\#\Gamma[G]} .$$

<u>Then</u>

$$p_{v_3,\ldots,v_K}(N) = \frac{1}{\mu_N \Pi_k v_k!} \int_{\mathcal{H}^N} \Pi_k \left[\frac{-\mathrm{tr}\, H^k}{k}\right]^{v_k} e^{-\frac{\mathrm{tr} H^2}{2}} \, dM(H),$$

<u>where</u> tr <u>denotes</u> <u>trace</u>, dM(H) <u>is</u> <u>the</u> <u>unitary-invariant</u> <u>product</u>

$$dM(H) = \prod_{i<j} (d\mathrm{Re}\, H_{ij})\, d(\mathrm{Im}\, H_{ij}) \prod_{i=1}^{n} dH_{ii}$$

<u>of</u> <u>Lebesgue</u> <u>measures</u> <u>on</u> <u>the</u> N <u>by</u> N <u>hermitians</u> $\mathcal{H}^N$, <u>and</u>

$$\mu_N = 2^{N/2}\,\pi^{N^{2/2}} .$$

<u>Remark</u>: The situation is analogous to [BIZ]. It is remarkable that this perturbative series technique so effectively captures the combinatorics of the action of $MC_g^s$ on the dual $\tilde{\Sigma}_g^s$ of $C$ .

## §4 <u>The Computation of Virtual Euler Characteristics</u>

It follows from PST that

$$\Phi(I,N) = \sum_{(*)} (-1)^{\Sigma v_k}\, p_{v_3,\ldots,v_K}(N)$$

$$= \sum_{(*)} \frac{1}{\mu_N \Pi_k v_k!} \int_{\mathcal{H}^N} \Pi_k \left[\frac{-\mathrm{tr} H^k}{k}\right]^{v_k} e^{-\frac{1}{2}\mathrm{tr} H^2} \, dM(H),$$

where both sums are over all tuples as before with $\Sigma(k-2)v_k = I$. Unitary invariance of the integral and a change of variables allows us to compute

$$\Phi(I,N) = \nu_N \sum_{(*)} \frac{1}{\Pi_k v_k!} \int_{\mathbb{R}^N} \Pi_k \left[\frac{-p_k}{k}\right]^{v_k} dL ,$$

where

$$p_k = \sum_{i=1}^{N} x_i^k ,$$

$$dL = \prod_{i \neq j} (x_i - x_j)\, e^{-\frac{1}{2} p_2} \prod_{i=1}^{N} dx_i ,$$

and

$$\nu_N = (2\pi)^{-N/2} \prod_{p=1}^{N} \frac{1}{p!} .$$

We seek a generating function

$$Z(t,N) = \sum_{I>0} \Phi(I,N)\, (i\sqrt{t})^N$$

in the $\Phi(I,N)$. $\Phi(I,N)$ does not converge at zero, but a formal computation gives

$$Z(t,N) = \nu_N \int_{\mathbb{R}^N} \prod_{i \neq j} (x_i - x_j) \prod_{i=1}^{N} d\mu_t(x_i),$$

where

$$d\mu_t(x) = (1 - ix\sqrt{t})^{-t^{-1}}\, e^{-ix\sqrt{t}}\, dx .$$

The next result then allows us to determine the integral in terms of the measure $d\mu_t$.

Lemma: [Lemma 3.3 of Pe 2] Suppose $\mu$ is a measure on $\mathbb{R}$ whose moments $\bar{\mu}_n = \int_{\mathbb{R}} x^n d\mu(x)$ exist and are finite for $n > 0$. In this case,

$$\int_{\mathbb{R}^N} \prod_{i \neq j} (x_i - x_j) \prod_{i=1}^{N} d\mu_t(x_i) = N! [\bar{\mu}_0]^N \prod_{r=1}^{N} \rho_n^{N-n} ,$$

where the $\rho_n$ are the coefficients in the three - term recursion

$$k_{n+1}(x) = (x + \alpha_n) \, i_n(x) - \rho_n \, k_{n-1}(x)$$

of the orthogonal polynomials of $\mu$ .

The similarity of

$$d\mu_t(x) = -i\sqrt{t}\,(et)^{-t^{-1}} (-z)^{-t^{-1}} e^{-z} dz, \quad z = \frac{1}{t}(ix\sqrt{t} - 1),$$

to the classical Laguerre measure allows us to compute the relevant data about $d\mu_t$, and we find [Theorem 3.8 of Pe 2]

$$Z(t,N) = \left[ \frac{\sqrt{2\pi t}\,(et)^{-t^{-1}}}{\Gamma(t^{-1})} \right]^N \prod_{p=1}^{N-1} (1 - pt)^{N-p} .$$

Unfortunately, the formal power series $Z(t,N)$ does not converge at 0, but if one spatially truncates the $\Phi(I,N)$, then the computation above is valid. Thus, $Z(t,N)$ gives an asymptotic series at 0 for the (analytic on Re $t > 0$) function above.

Finally, a standard generating function trick allows us to conclude that $\Phi_c(I,N)$ gives an asymptotic series for $\log Z(t,N)$. This is easily computed using Stirling's and Bernoulli's formula. The result of this computation is

Theorem [Theorem 3.9 of Pe 2]: If $g > 0$, $s > 1$, and $2g - 2 + s > 0$, then

$$\chi_v MC_g^s = (-1)^s \frac{(s+2g-3)!\,(2g-1)}{s!\,(2g)!} B_{2g} ,$$

*where* $B_{2g}$ *denotes the* $2g^{th}$ *Bernoulli number.*

*Remark*: This is in agreement with the result derived in [HZ].

## §5 Integration over Moduli Space and the Kähler Form

Part of the purpose of this section is to propose a method of explicit integration over the moduli space $\mathcal{M}_g^s = \mathfrak{J}_g^s / MC_g^s$. We begin with a construction of a p-form on $\mathcal{M}_g^s$ from a p-form on $\tilde{\mathfrak{J}}_g^s \approx \mathbb{R}_+^q$ and require a reformulation of results from §§1,2.

There is a $MC_g^s$-equivariant codimension - s foliation of $\tilde{\mathfrak{J}}_g^s$ determined by taking level sets of the function

$$F(\tilde{\Gamma}) = \prod_{i=1}^{s} \rho_i(\tilde{\Gamma}) : \tilde{\mathfrak{J}}_g^s \to \mathbb{R}_+^s$$

where $\rho_i(\tilde{\Gamma})$ is the hyperbolic length of the horocycle about the $i^{th}$ cusp. In particular, the leaf $F \equiv \prod_{i=1}^{s} 1$ is invariant under the action of $MC_g^s$, giving rise to a $MC_g^s$-equivariant inclusion

$$i : \mathfrak{J}_g^s \subset \tilde{\mathfrak{J}}_g^s$$

which is a right inverse to the canonical "forgetful" fibration

$$\varphi : \tilde{\mathfrak{J}}_g^s \to \mathfrak{J}_g^s .$$

In fact, if $\Delta$ is a fixed i.t. of $F_g^s$ , then a little hyperbolic geometry allows one to compute

$$\rho_i(\tilde{\Gamma}) = \sqrt{2} \sum \frac{a^2+b+c^2}{abc} ,$$

where the sum is over ends of triangles T of $F_g^s - \Delta$ which limit on the $i^{th}$ cusp, and a,b,c denote the $\lambda$-lengths of edges of T. We remark that the foliation on $\tilde{\mathcal{T}}_g^s$ descends to a non-trivial foliation of $\tilde{M}_g^s = \tilde{\mathcal{T}}_g^s / MC_g^s$ if $s > 1$.

Trivially, a form $\tilde{\omega}$ on $\tilde{\mathcal{T}}_g^s$ arises as $\varphi^* \omega$ for some form $\omega$ on $\mathcal{T}_g^s$ if $\tilde{\omega} u = \tilde{\omega} v$ whenever u and v are tangent p-vectors on $\tilde{\mathcal{T}}_g^s$ so that $\varphi_* u = \varphi_* v$. The question of modular invariance of forms on $\mathcal{T}_g^s$ and $\tilde{\mathcal{T}}_g^s$ is somewhat more involved. A "Ptolemy transformation" $\tilde{\mathcal{T}}_g^s \circlearrowleft$ is defined as follows. Choose an i.t. of $\Delta$ and an edge $e \in \Delta$. Suppose $\{a,b,c,d\} \subset \Delta$ and (a,b,e), (c,d,e) are each the set of edges of distinct triangles in $F_g^s - \Delta$. If $\tilde{\Gamma} \in \tilde{\mathcal{T}}_g^s$ , let $\alpha = \lambda(a\ ; \tilde{\Gamma}), \ldots, \varepsilon = \lambda(e; \tilde{\Gamma})$, and replace the $\lambda$-length of e by $\frac{\alpha\gamma + \beta\delta}{\varepsilon}$ to define a new $\tilde{\Gamma}' \in \tilde{\mathcal{T}}_g^s$ . The transformation $\tilde{\Gamma} \to \tilde{\Gamma}'$ is the $(\Delta,e)$ <u>Ptolemy transformation on</u> $\tilde{\mathcal{T}}_g^s$ . Ptolemy transformations preserve each level set of F. We argued in §3 that these transformations "generate" $MC_g^s$ and have

<u>Proposition 1</u>: <u>Suppose</u> $\tilde{\omega}$ <u>is a p-form on</u> $\tilde{\mathcal{T}}_g^s$ <u>so that the following conditions hold.</u>

(i) $\tilde{\omega} u = \tilde{\omega} v$ whenever u and v are tangent p-vectors to $\tilde{\mathcal{T}}_g^s$ so that $\varphi_* u = \varphi_* v$.

(ii) $\tilde{\omega}$ <u>is invariant under all Ptolemy transformations.</u>

<u>Then</u> $\tilde{\omega}$ <u>descends to a form</u> $\omega$ <u>on</u> $M_g^s$ .

Remarks: 1) $MC_g^s$ is a sub-group (of finite-index) in the semi-group of all Ptolemy transformations on $\tilde{\mathcal{J}}_g^s$ .

2) To shed some light on condition (i), we may choose a pleasant i.t. $\Delta$ of $F_g^s$ . An i.t. $\Delta$ "based" at the first puncture x is an i.t. so that each arc in $\Delta$ has at least one endpoint at x. Let $\Delta' \subset \Delta$ consist of all the ideal arcs in $\Delta$ so that both endpoints are at x ; $\Delta = \Delta'$ consists of (s-1) arcs $d_2,\ldots,d_s$ (indexed by puncture). It follows from the construction in the PT of §1 that scaling $\lambda(d_i ; \tilde{\Gamma})$ alone leaves invariant $\varphi\tilde{\Gamma}$ and $\rho_j\tilde{\Gamma}_k$, $j \neq i$, and scales $\rho_i\tilde{\Gamma}$. Furthermore, homothety on $\lambda$-lengths leaves $\varphi\tilde{\Gamma}$ invariant and projectively scales all the $\rho_i\tilde{\Gamma}$. Thus, any of the following conditions implies condition (i).

(i') $\Delta$ is an i.t. based at x with $\Delta'$ as above, and $\omega$ is invariant under homothety (on $\lambda$-lengths with respect to $\Delta$). Furthermore, as an expression in $\lambda$-lengths, $\omega$ does not involve any $\lambda(c, \tilde{\Gamma})$ or its differential for $c \in \Delta - \Delta'$.

(i") $\Delta$ is an i.t. of $F_g^1$ and $\omega$ is invariant under homothety.

(i‴) $\Delta$ is an i.t. of $F_g^s$ and $\omega$ is a form which is invariant under scaling each $\lambda(c;\tilde{\lambda})$, $c \in \Delta$, separately. (This may seem very special, but the Weil-Petersson Kähler form is of this type.)

To integrate a form over moduli space, we require some description of a fundamental domain for the action of $MC_g^s$ on $\mathcal{J}_g^s$ . A crude but computationally effective method is to employ the cell decomposition $\mathcal{J}_g^s \cap C$ of $\mathcal{J}_g^s$ described in §2. Every fat graph G with $F(G) = F_g^s$ is represented by a cell in $\mathcal{J}_g^s \cap C$ .

We define

$$\varepsilon[G] = \begin{cases} 1, & \text{if there is } 1 \neq f \in \Gamma[G] \\ & \text{which fixes each edge of } G \\ 0, & \text{otherwise.} \end{cases}$$

If $\varepsilon[G] = 1$, then we say G is a "hyperelliptic" fat graph; a hyperelliptic fat graph has exactly two vertices, and each edge has distinct endpoints.

If $\sigma$ is a p-cell in $\mathcal{T}_g^s \cap C$ with projection $\sigma' \subset \mathcal{M}_g^s$ and $\omega'$ is a p-form on $M_g^s$, then we claim

$$\int_{\sigma'} \omega' = \frac{2^{\varepsilon[G]}}{\#\Gamma[G]} \int_{\sigma} \omega ,$$

where $\omega$ is a lift of $\omega'$ to $\mathcal{T}_g^s$. To see this, suppose first that G is not hyperelliptic. We must show that the branch locus of $\sigma$ over $\sigma'$ has measure zero in $\lambda$-length coordinates on $\sigma$. In case G is further trivalent (so $\sigma$ is top dimensional), this follows easily from CPT and naturality of $\lambda$-lengths. The general case (where G is not trivalent) is similarly handled by the analogue [Theorem 6.2 of Pe 1] of CPT. Finally, if G is hyperelliptic, then there is an involutive change of marking on G which respects $\lambda$-lengths, accounting for the additional factor 2. We have

<u>Theorem 2</u>: <u>If $\omega'$ is a top-dimensional form on $M_g^s$, then</u>

$$\int_{M_g^s} \omega' = \sum_{\substack{[G] \text{ of unmarked} \\ \text{trivalent } G \text{ with} \\ F(G) = F_g^s}} \frac{2^{\varepsilon[G]}}{\#\Gamma[G]} \int_{D_G} \varphi^* \omega ,$$

where $\omega$ is the lift of $\omega'$ to $\mathcal{T}_g^s$ and $D_G$ denotes the set of all positive functions $\Lambda$ on the edges of G so that the the following conditions hold. If a,...,e are edges of G, let $\alpha = \Lambda(a),\ldots,\varepsilon = \Lambda(e)$.

(a) If (a,b,c) are (half-) edges of G incident on a common vertex, then

$$\alpha < \beta + \gamma,\ \beta < \alpha + \gamma,\ \gamma < \alpha + \beta$$

(b) If (a,...,e) are edges of G as pictured in Figure 2, then

$$0 < \operatorname{sgn}(\alpha^2 + \beta^2 - \varepsilon^2)\, K(\alpha,\beta,\varepsilon) + \operatorname{sgn}(\gamma^2 + \delta^2 - \varepsilon^2)\, K(\gamma,\delta,\varepsilon),$$

where K is the scaling function described in CPT.

(c) $\rho_i(G, \Lambda) = 1$, for $i = 1,\ldots,s$.

Remarks: 1) The analogue of the theorem to the case of arbitrary dimension is identical except that condition (b) becomes more complicated.

2) If $\coprod \omega_g^s$ is an appropriate family of forms on $\coprod M_g^s$, then the perturbative series technique of §4 should be a viable means of computing all the $\int_{M_g^s} \omega_g^s$ simultaneously; the PST gives an analytic expression for a sum over fat graph classes [G] weighted by $\frac{1}{\#\Gamma[G]}$.

3) The numerical integration of a form over moduli space would be routine, at least for $|g| + |s|$ reasonably small.

Of course, the facility of our proposed integration scheme depends on being able to express the form $\varphi^* \omega$ reasonably in terms of $\lambda$-lengths on $\tilde{\mathcal{T}}_g^s$.

*Theorem 3: Fix a trivalent fat graph G with $F(G) = F_g^s$, and let $\omega$ denote the Weil-Petersson Kähler form. Parameterizing $\tilde{\mathcal{T}}_g^s$ by (all positive) real functions $\Lambda$ on the edges of G, we have*

$$\varphi^* \omega = -2 \sum d\ln a \wedge d\ln b + d\ln b \wedge d\ln c + d\ln c \wedge d\ln a,$$

*where the sum is over vertices of G whose consecutive incident (half) edges take the values (a,b,c) under $\Lambda$.*

*Proof* (sketch): It is straight-forward to check the conditions of Proposition 1, so w descends to *some* closed two-form on $M_g^s$, and we may choose any convenient triangulation to prove the theorem.

Our strategy is as follows. Let H denote $F_g^s$ with horoball neighborhoods of the cusps removed. The double $\ddot{F}$ of H inherits a pants decomposition from $(H, H \cap \Delta)$. Since the Kähler form extends to pinched surfaces, if we pinch the curves $\partial H \subset \ddot{F}$, then the Kähler form on $\ddot{F}$ limits to $\omega$ on each side of $\partial H$. To recognize the Kähler form on $\ddot{F}$, we recall

*Wolpert's Theorem* [Theorem 3.3 and Lemma 4.5 of W1]: *If the geodesic length functions $\{\ell_i\}_1^n$ of geodesic curves $\{c_i\}_1^n$ give local coordinates on $\mathcal{T}_g^s$, then*

$$\omega_{ij} = \omega(d\ell_i, d\ell_j) = \sum_{p \in c_i \cap c_j} \cos \alpha_p,$$

*where $\alpha_p$ is the angle between $c_i$ and $c_j$ at p. Furthermore, if $(W_{ij}) = (w_{ij})^{-1}$, then*

$$\omega = - \sum_{i<j} W_{ij} \, d\ell_i \wedge d\ell_j .$$

If $c \in \Delta$, we let c also denote the corresponding curve on F. To apply the theorem, we define the "detecting" curve $c^*$ to c as follows. An elementary move (see §2) on the ideal arc c produces an ideal arc $c^*$ whose double $c^*$ is a curve in F. If $c \in \Delta$ is oriented, $c^*$ inherits an orientation from the orientation on $F_g^s$ (see Figure 3).

If c,d are oriented geodesic curves or arcs, let

$$\alpha(c,d) = \sum_{p \in c \cap d} \cos \alpha_p .$$

The matrix $(w_{ij})$ of Kähler pairings has the form

$$\begin{pmatrix} 0 & D \\ -D & A \end{pmatrix} \text{ with inverse } \begin{pmatrix} D^{-1}AD^{-1} & -D^{-1} \\ D^{-1} & 0 \end{pmatrix},$$

where $D_{ii} = \alpha(c_i, c_i^*)$ and $A_{ij} = \alpha(c_i^*, c_j^*)$. Applying Wolpert's Theorem, we find

$$(*) \qquad -\omega = \sum_{i<j} \frac{\alpha(c_i^*, c_j^*)}{\alpha(c_i, c_i^*)\,\alpha(c_j, c_j^*)}\, dc_i \wedge dc_j$$

$$- \sum_{i=1}^{g} \frac{1}{D_{ii}}\, dc_i \wedge dc_i^* .$$

The sum over intersection points of $c_i^*$ and $c_j^*$ in the first term breaks naturally into a sum over triangles as indicated in Figure 4.

The geodesic representative for $\partial H$ in F separates each $c \in \{c_i, c_i^*\}$, say of hyperbolic length $\ell$, into right and left halves, say of hyperbolic lengths $\ell_R$ and $\ell_L$. We define "λ-lengths"

$$\lambda_R(c) = \sqrt{2e^{\ell_R}}$$

$$\lambda_L(c) = \sqrt{2e^{\ell_L}} \quad ,$$

and easily express $\omega$ in terms of $\lambda$-lengths, since

$$\begin{aligned} d\ell &= d\ell_R + d\ell_L \\ &= 2[d\ell n\ \lambda_R(c) + d\ell n\ \lambda_L(c)]. \end{aligned}$$

Furthermore, adopting the notation of Figure 3 for ideal arcs (and identifying an arc with its $\lambda$-length), IPT gives

$$d\ell n\ e^* + d\ell n\ e$$

$$= \frac{ac}{ac+bd}\,[d\ell n\, a + d\ell n\, c] + \frac{bd}{ac+bd}\,[d\ell n\, b + d\ell n\ d] \ .$$

Thus, the second term in the expression (*) also decomposes into a sum over triangles in the natural way.

Adhering to the notation of Figure 3, we find that $-\frac{1}{4}\omega$ has the following projection onto the $d\ell n a_R \wedge d\ell n e_R$ subspace

$$\frac{\alpha(a^*,e^*)}{\alpha(a,a^*)\,\alpha(e,e^*)} - \frac{1}{\alpha(a,a^*)}\,\frac{e_R f_R}{e_R f_R + d_R g_R} + \frac{1}{\alpha(e,e^*)}\,\frac{e_R f_R}{a_R c_R + b_R d_R} \ .$$

Some hyperbolic geometry gives

$$\alpha(e_R,e_R^*) = \frac{a_R c_R - b_R d_R}{a_R c_R + b_R d_R} \ ,$$

and several applications of IPT finishes the computation.

Since the pair a,e was arbitrary (with orientations induced from the surface as in Figure 3), the result follows.

Example: We reproduce Wolpert's result [W2] that the Weil-Petersson mass of $M_1^1$ is $\pi^2/6$. The unique relevant fat graph is the first one in Figure 1a, and conditions (a) and (b) of Theorem 2 simplify to the non-strict triangle inequalities amongst the squares $\{a^2, b^2, c^2\}$ of the $\lambda$-lengths (a,b,c) which parametrize $\tilde{\mathcal{J}}_1^1 \approx \mathbb{R}_+^3$. Furthermore, condition (c) of Theorem 2 gives

$$a^2 + b^2 + c^2 = abc$$

(for the level set $\rho_1 = 2^{-\frac{1}{2}}$), and we change variables

$$\alpha = \frac{a}{bc}, \quad \beta = \frac{b}{ac}, \quad \gamma = \frac{c}{ab}.$$

The domain of integration is therefore

$$D_G = \{ (\alpha,\beta,\gamma) : \alpha,\beta < \frac{1}{2}, \alpha + \beta > \frac{1}{2}\}.$$

Application of Theorem 3 and a short computation with forms gives

$$\omega = \frac{-d\alpha \wedge d\beta}{\alpha\beta(1-\alpha-\beta)},$$

and we find the dilogarithm

$$\int_{D_G} \omega = \int_0^{\frac{1}{2}} \int_{\frac{1}{2}-y}^{\frac{1}{2}} \frac{-dxdy}{xy(1-x-y)} = \frac{\pi^2}{2}.$$

Finally, the fat graph G of Figure 1a is hyperelliptic with $\#\Gamma[G] = 6$, and we find the Weil-Petersson mass $\pi^2/6$.

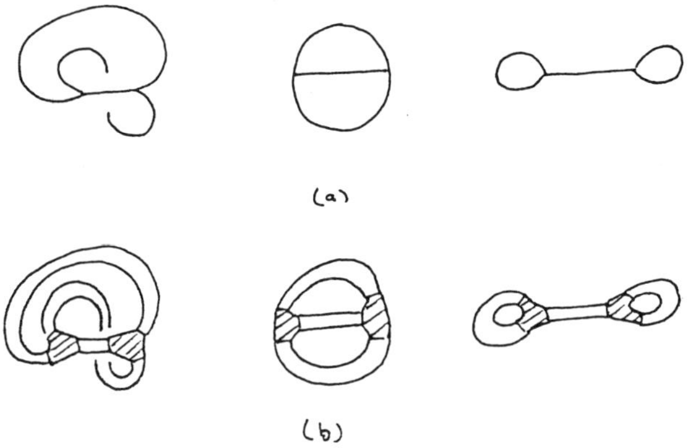

Figure 1

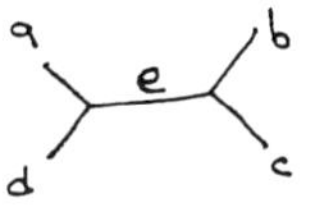


Figure 2

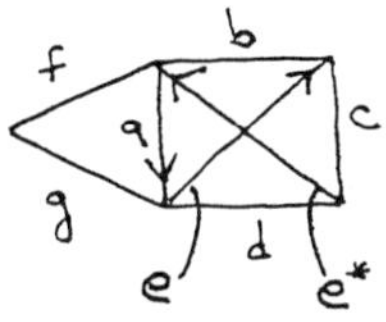


Figure 3

Figure 4

## Bibliography


[BE] B. Bowditch and D.B.A. Epstein, "Triangulations associated to punctured surfaces," Warwick reprint (1986).

[BIZ] D. Bessis, C. Itzykson, and J.B. Zuber, "Quantum field theory techniques in graphical enumeration," Adv. Appl. Math 1 (1980), 109-157.

[Br] K.S. Brown "Euler characteristics of discrete groups and G-spaces," Invent. Math. 27 (1974), 229-264.

[EP] D.B.A. Epstein and R.C. Penner, "Euclidean decompositions of non-compact hyperbolic manifolds," to appear in Jour. Diff. Geom. (1986).

[Ha] J. Harer, "The virtual cohomological dimension of the mapping class group of an oriented surface," to appear in Invent. Math. (1986).

[HZ] J. Harer and D. Zagier, "The Euler characteristic of the moduli space of stable curves," preprint (1985).

[Pe1] R.C. Penner, "The Teichmüller space of a punctured surface," preprint (1986).

[Pe2] R.C. Penner, "Perturbative series and the moduli space of Riemann surfaces," preprint (1986).

[W1] S. Wolpert, "On the symplectic geometry of deformations of a hyperbolic surface," Ann. of Math.

[W2] S. Wolpert, "On the Kähler form of the moduli space of once punctured tori", Comentarii Math Helv. 58 (1983), 246-256.

# Geometric Complex Coordinates for Teichmüller Space

by

Albert Marden

## 1. Introduction.

The purpose of this note is to report on a system of complex coordinates for Teichmüller space that Clifford Earle and I found several years ago. The first complex coordinate system was established by Ahlfors [1], with earlier work by Rauch. Ahlfors' coordinates are local ones, and they are given in terms of the variation of a holomorphic differential, its zeros, and its periods. The most widely used coordinates are those of Bers [2] which are given indirectly in terms of what is now called Bers slices of the quasifuchsian space of kleinian groups. Other coordinates that are given in terms of kleinian groups are those of Maskit [11]; see also Kra [8].

Here we will describe coordinates that are at once global, geometric, and, like Ahlfors', directly involve the Riemann surfaces. In fact they arise from a long used device for blowing up punctures. They are of special value in computing asymptotic expansions for the effect of the degeneration of a surface on the analytic expressions on the surface.

The coordinates also determine the analytic structure on moduli space and its compactification. There is a different approach by Bers [3], and an approach through algebraic geometry, see [13].

At the same time the coordinates also appear as parameters in the generators of a class of kleinian groups. This relates them to the geometry of the associated class of hyperbolic 3-manifolds which in turn plays an essential role in establishing their properties.

Kleinian groups have entered deeply into the study of Teichmüller space. This has occurred because the associated hyperbolic 3-manifolds carry the Riemann surfaces as boundary components. Thus kleinian group parameters may also serve as surface parameters. Certain classes of groups are particularly useful for this purpose. Even more importantly, the surfaces may degenerate in horrible ways while the interiors of the 3-manifolds retain geometric structure. It is the interiors that have well determined limits which are again hyperbolic 3-manifolds. Thus the degenerated surface may be realized in some sense as an "ideal" boundary component and this permits finer analysis of the process of degeneration itself.

For general background see for example [5], [9], [10].

## 2. Review of Teichmüller theory.

Fix a compact Riemann surface $R$ of genus $g>1$. The Teichmüller space $T_g$ is the set of all pairs $(S,f)$ where $S$ is another Riemann surface and $f: R \to S$ is an orientation preserving homeomorphism, subject to the identification $(S,f) \equiv (S_1,f_1)$ if and only if $f_1 f^{-1}: S \to S_1$ is homotopic to a conformal map.

An orientation preserving homeomorphism $h: R \to R$ acts on $T_g$ according to the rule $(S,f) \mapsto (S,fh)$. This bijection depends only on the homotopy class of $h$. In view of this action, the homotopy classes of homeomorphisms $[h]$ of $R$ determine a group of bijections of $T_g$ which is called the Teichmüller modular group or the mapping class group and we will denote it by $\Gamma_g$.

The modular group is generated by Dehn twists. We recall that the Dehn twist about a simple loop $\alpha$ (not contractible to a point) is determined by removing an annular neighborhood of $\alpha$, holding one contour fixed while twisting the other by $360°$, and then sewing it back into the surface by the identity map on its boundary. The homotopy type of the twist (modulo orientation) is uniquely determined by the loop.

Neighborhoods of points of $T_g$ can be simply described as follows: $(S_1,f_1)$ is close to $(S,f)$ if $f_1 f^{-1}$ is homotopic to a homeomorphism close to a conformal map (technically, this is measured by its dilatation as a quasiconformal map). In this topology the action of the modular group is properly discontinuous. The quotient, $T_g/\Gamma_g = M_g$ is called the moduli space.

There are two principal compactifications, and consequently boundaries, of $T_g$ , the Thurston boundary and the Bers boundary. The former has the advantage that the elements of the modular group extend to be homeomorphisms of the compactified space. From the holomorphic point of view however, the Bers boundary is the more important, although the extension property does not hold [7]. There are certain points on the Bers boundary which will be called r-points because they play the analogous role in the action of the Teichmüller modular group as the rational points on the real axis do in the action of the classical modular group $PSL(2,\mathbb{Z})$ in the upper half plane. An r-point is a Riemann surface with nodes, that is, the result of pinching at least one but possibly as many as $3g-3$ simple, mutually disjoint, non-parallel loops. We will refer to these as the pinching loops for the r-point.

Let $T_g^+$ denote the Teichmüller space which has been augmented by the addition of the r-points. To describe the neighborhood structure of this space it suffices to describe the neighborhoods of the r-points. The exact analogy is the upper half plane augmented by the rational points with the neighborhoods of the rational points taken to be the horodisks at those points. A "horodisk" at an r-point is the following set: An r-point $P$ is the union of a finite number of finitely punctured Riemann surfaces. Choose a (small) neighborhood about each puncture and let $N$ denote their union. Corresponding to the choice of $N$ is the set of all Riemann surfaces of genus $g$ (which may have nodes) into which $P\setminus N$ can be conformally embedded. The points of this set comprise the N-neighborhood of the r-point $P$ . Automatically, this neighborhood is invariant under the abelian group generated by the Dehn twists about the pinching curves, and the Dehn twists themselves fix the r-point. In fact, the modular group extends to a group of homeomorphisms of the augmented space.

## 3. The coordinates.

The starting point is the choice of a trivalent graph representing the combinatorial type of a compact surface of genus $g > 1$ . That is, we start with a surface of genus $g$ that has been degenerated into triply punctured spheres by pinching $3g-3$ simple, non-parallel loops. The graph has

$2g-2$ vertices representing triply punctured spheres and $3g-3$ edges representing a pairing of the punctures. (It is possible to start with partially degenerated surfaces, or to work in the Teichmüller space of finitely punctured surfaces, but in this exposition we will stick to the simplest case.) When $g = 2$ there are two possibilities:

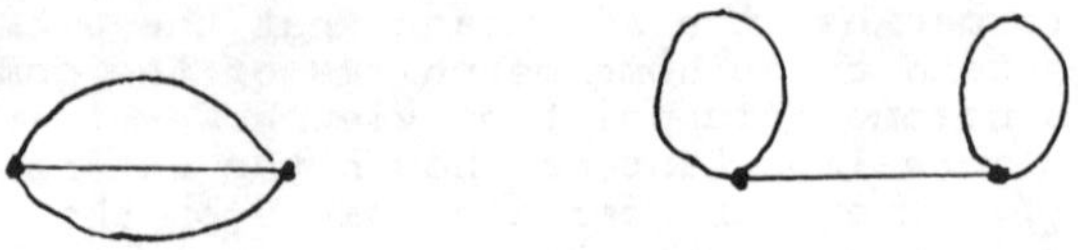

About each puncture of the configuration of triply punctured spheres we choose an explicit local coordinate which we call the (normalized) horocyclic local coordinate. This is canonically determined as follows.

We may take the triply punctured sphere $P$ to be the z-plane punctured at $z = 0$ and $z = 1$; we will describe the horocyclic local coordinate about $z = 0$. Choose the universal covering surface of $P$ to be the upper half w-plane $\mathbb{H}_+$ and normalize so that $\infty$ corresponds to $z = 0$ and the subgroup of the group of cover transformations that fixes $\infty$ is generated by $w \mapsto w+1$. A renormalization of the famous elliptic modular function,

$$z = \lambda(w)$$

projects $\mathbb{H}_+$ back onto $P$. The horodisks $\{w \mid \operatorname{Im} w > c > 1/4\}$ project to neighborhoods of $z = 0$. These neighborhoods are round disks in the local coordinate $z_1 = \exp(2\pi i w + \pi/2)$ about $z = 0$. The $z_1$ coordinate is related to the z-coordinate by the formula,

$$z_1 = - z \exp(\pi/2)/16 \quad + \quad \ldots .$$

Next consider a pair of punctures, the pairing having been dictated by the trivalent graph. Let now $z$ and $w$ denote the local horocyclic parameters about the two punctures and consider the equation

$$z\, w = t, \quad t \in \mathbb{C}$$

The normalization of the horocyclic coordinates insures that the equation is meaningful for $0 < |t| < 1$.

We regard the formula as a prescription for identifying the two circles $|w| = \sqrt{|t|}$ and $|z| = \sqrt{|t|}$ .

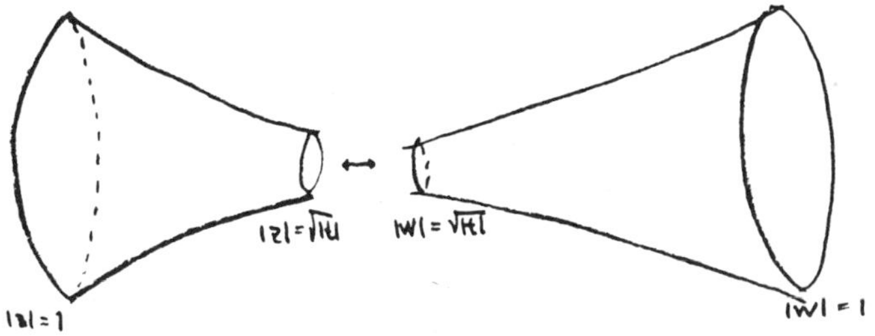


This construction can be carried out independently for all 3g-3 pairs of punctures since the unit disks in the horocyclic local parameters are mutually disjoint. This gives a map $\vec{t} \to S_t$

$$\{t \in \mathbb{C} \mid 0 < |t| < 1\}^{3g-3} \to T_g / F,$$

where $F$ denotes the abelian group generated by the Dehn twists about the $3g-3$ pinching curves; the marking itself of $S_t$ is determined by the trivalent graph up to the action of $F$.

surface with nodes

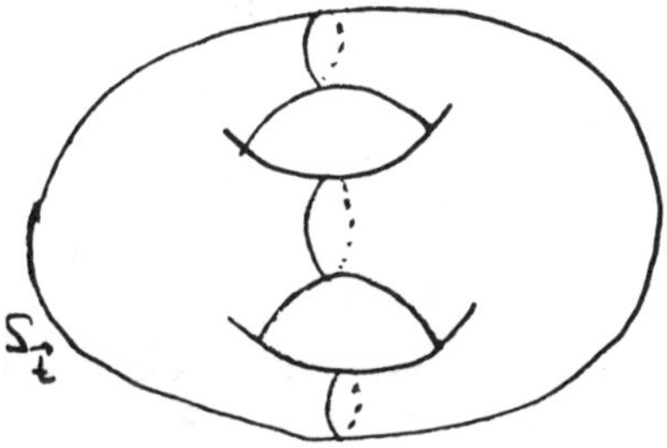


blow-up of punctures

## 4. The main result.

THEOREM. The map $\vec{t} \to S_t$ extends to a holomorphic bijection of a subdomain of the poly-punctured disk,

$$\{t \in \mathbb{C} \mid 0 < |t| < e^{\pi}\}^{3g-3}$$

onto

$$T_g \ /F .$$

Near $\vec{t} = \vec{0}$, $\vec{t}$ gives local holomorphic coordinates about the "ideal" point of $T_g / \Gamma_g = M_g$ corresponding to the original trivalent graph.

Instead of choosing horocyclic local coordinates about the pairs of punctures, choose any pair $z,w$ of local coordinates and carry through the same construction.

COROLLARY. This more general procedure gives a local injection into $T_g$ /F .

Such a technique has long been used by complex analysts to form degenerating families of Riemann surfaces.

Some variational formulas are known (see Fay [6]) for period matrices and canonical holomorphic differentials on the Riemann surfaces as $\vec{t} \to 0$, but more needs to be done: the general case has not yet been treated. The coordinates have been used to analyze the Weil-Petersson metric near the "ideal" points of $M_g$ (Masur [12], Wolpert [15]).

The blowing up process determines a (3g-3)-parameter holomorphic family of surfaces. To prove the Theorem, it is necessary to show that the mapping to $T_g$ / F is locally and then globally injective, in particular that the construction has an analytic continuation beyond the immediate geometric picture. All this is done by tying the construction to the theory of holomorphic deformations of geometrically finite kleinian groups.

It is very instructive to work out the corresponding study for tori vis-à-vis the twice punctured sphere.

## 5. Kleinian groups in a nutshell.

A Möbius transformation acts in the complex $z = x + iy$ plane according to

$$z \mapsto (az + b)/(cz + d), \quad ad - bc = 1.$$

It extends uniquely to upper half space $\{(x,y,t) \mid t > 0\}$ so as to be an orientation preserving isometry in the hyperbolic metric $|d\vec{x}|/t$. The group of all Möbius transformations in this way extends to become the group of orientation preserving isometries of hyperbolic 3-space $\mathbb{H}^3$.

A parabolic transformation is one conjugate to $z \mapsto z+1$, an elliptic transformation is one conjugate to $z \mapsto \exp(i\theta)z$, $\theta \neq 0$, and a loxodromic transformation is one conjugate to $z \mapsto k\exp(i\theta)z$, $k>0$, $k \neq 1$. In particular, the inversion $z \mapsto 1/z$ is elliptic of order two. Each of these has an obvious extension to $\mathbb{H}^3$ and the general transformation is a composition of these simple models.

Let $G$ be a group of Möbius transformations acting in $\mathbb{H}^3$. The limit set $\Lambda(G)$ is the set of accumulation points of the orbit $\{ T(\vec{x}) \mid T \in G \}$ for $\vec{x} \in \mathbb{H}^3$; it is independent of the choice of $\vec{x}$. The group G is discrete (no sequence in G converges to the identity) if and only if $\Lambda(G)$ lies entirely in $\mathbb{C} \cup \{\infty\}$. The group $G$ is called kleinian if it is discrete and non-elementary ($\Lambda(G)$ has more than three points; it is then a perfect set which is either nowhere dense or is the entire plane).

The complement $\Omega(G)$ of the limit set in the extended plane is called the ordinary set; possibly $\Omega(G) = \emptyset$.

If $G$ is a kleinian group the quotient is well defined and it is an oriented 3-manifold with boundary,

$$\mathcal{M}(G) = \mathbb{H}^3 \cup \Omega(G)/G, \quad \partial \mathcal{M}(G) = \Omega(G)/G$$

Note that there are two structures that play a role: the hyperbolic structure on the interior $\mathbb{H}^3/G$, and the complex structure on the boundary $\Omega(G)/G = \partial\mathcal{M}(G)$ which is then a union of Riemann surfaces, if $\Omega(G) \neq \emptyset$. In addition, if there are elliptic transformations in $G$, $\mathbb{H}^3$ will be branched over $\mathcal{M}(G)$ which is then called an orbifold, and $\Omega(G)$ may be branched over $\partial\mathcal{M}(G)$. However for the remainder of this report, we may assume that the kleinian groups have no elliptic elements.

The kleinian group $G$ is called geometrically finite if the 3-manifold $\mathcal{M}(G)$ is compact except for a finite number of "cusps" (see below). The fundamental uniqueness property [9] of geometrically finite groups $G$ is as follows.

Suppose $G_1$ is another kleinian group, not necessarily geometrically finite, and $f$ is a conformal map of $\Omega(G)$ into $\Omega(G_1)$ such that for an isomorphism $\theta: G \to G_1$,

$$f\,T = \theta(T)\,f$$

Then, G is conjugate to G: $\theta(T) = ATA^{-1}$ , for some A.

This result says that the hyperbolic 3-manifold (and corresponding kleinian group) is determined by the conformal type of its bounding surfaces. Therefore in appropriate situations, the parameters in a set of generators of the group, or related geometric quantities, can be used to parameterize Riemann surfaces.

6. Carriers of Riemann surfaces: quasifuchsian groups.

Start with a fuchsian group $\Gamma$ acting on the upper $\mathbb{H}_+$ and lower $\mathbb{H}_-$ half planes, and such that the quotient $\mathbb{H}_-/\Gamma = R$ is a compact surface of genus $g \geq 2$. The associated 3-manifold has the topological structure

$$\mathcal{M}(\Gamma) \simeq R \times [0,1].$$

We will refer to the boundary component $\mathbb{H}_+/\Gamma$ as the top one, and R as the bottom.

The generators $A_1, B_1, \ldots, A_g, B_g$ of have real entries, and in the usual presentation satisfy the relation $A_1B_1A_1^{-1}B_1^{-1}\cdots A_gB_gA_g^{-1}B_g^{-1}$ = identity. Deform $\Gamma$ by deforming through complex entries in such a way that the relation remains satisfied. This process generates the space of quasifuchsian groups { G }. That is, corresponding to each group G is a quasiconformal homeomorphism f of the extended plane such that

$$f\,T = \theta(T)\,f$$

where $\theta : \Gamma \to G$ is an isomorphism. In particular the limit set $\Lambda(G)$ of G is a simple closed curve, the image of the extended real line under a quasiconformal mapping. The quasiconformal map f also induces a quasiisometry between the quotient 3-manifolds. In other words the quotients $\mathcal{M}(G)$ are also homeomorphic to $R \times [0,1]$. The quasifuchsian groups fill out a connected open set and a complex submanifold of dimension 6g - 3 of the representation variety $\mathrm{Hom}(\Gamma, PSL(2,\mathbb{C}))$.

Quasifuchsian space can be characterized geometrically by use of Bers' notion of simultaneous uniformization. Let G be a quasifuchsian group. Consider the associated 3-manifold $\mathcal{M}(G)$. Its boundary has two components, each a compact surface of genus g. By means of the quasiconformal isomorphism to the base fuchsian group $\Gamma$ and its quotient 3-manifold $\mathcal{M}(\Gamma)$ it is natural to refer to the two components of $\partial\mathcal{M}(G)$ as the top component and the bottom. There is an orientation reversing involution of $\mathcal{M}(G)$ that interchanges the top and bottom component and fixes each element of the fundamental group.

The principle of simultaneous uniformization is this. Given two compact Riemann surfaces R, S of genus g, and an orientation reversing map J between them, there is a uniquely determined (up to normalization) quasifuchsian group G for which the bottom surface is (conformally equivalent to ) R, the top surface is S, and the associated involution $R \to S$ is homotopic to J.

Consider all quasifuchsian groups whose bottom surface R is fixed. This set is called the Bers slice based on R . It is a complex submanifold of quasifuchsian space of dimension 3g, and it provides a (unnormalized) model of Teichmüller space: the points of Teichmüller space are represented by the top surfaces and their relation via the involution to the fixed bottom one. Elements of this set of quasifuchsian groups are characterized by the Riemann map $h: \mathbb{H}_- \to \Omega$ where $\Omega$ is the component of $\Omega(G)$ for which $\Omega/G = R$ . Bers' coordinates for Teichmüller space are given in terms of the schwarzian derivative of h.

A Bers slice is relatively compact in the representation variety $\mathrm{Hom}(\Gamma, PSL(2,\mathbb{C}))$. Its boundary is precisely the Bers boundary. As we range through a slice, we range through hyperbolic 3-manifolds whose bottom boundary component is conformally fixed, but whose top component varies. As we head toward the edge of the slice, the top component must degenerate in some way. There are two ways it can degenerate: one or more curves can become pinched or the relation (given by the involution) to the bottom surface can become worse and worse topologically. In particular, if we head "non-tangentially" toward an r-point, the top surface becomes pinched along the pinching curves associated with the r-point. This and all boundary points are represented by kleinian groups. It is now known, through work of Thurston [14] and Bonahon [4], that the interiors of the quotient manifolds for all boundary groups are homeomorphic to $R \times (0,1)$; earlier this had been known only for the geometrically finite boundary groups. But the geometrically finite ones are precisely the ones that correspond to the r-points.

## 7. The role of parabolic transformations.

Our coordinates are naturally associated with a certain class of kleinian groups. To describe this class, it is necessary to understand the role of parabolic transformations in a geometrically finite group $G$ (without elliptics).

Corresponding to each parabolic fixed point $\zeta$ is its stabilizer

$$\mathrm{Stab}(\zeta) = \{\, T \in G \mid T(\zeta) = \zeta \,\}.$$

By conjugating the group we may assume that $\zeta = \infty$. There are two cases according to whether $\mathrm{Stab}(\infty)$ has two generators or one.

Case 1. $\mathrm{Stab}(\infty)$ has rank two so that there exists $\omega_1, \omega_2$ such that $\mathrm{Stab}(\infty)$ is generated by

$$z \mapsto z + \omega_1, \quad z \mapsto z + \omega_2, \quad \mathrm{Im}(\omega_2/\omega_1) > 0.$$

For all sufficiently large $c > 0$ the horoball,

$$\mathcal{H} = \{\, (x,y,t) \mid t > c \,\},$$

has the property that $T(\mathcal{H}) \cap \mathcal{H} = \phi$ unless $T \in \mathrm{Stab}(\infty)$. This means that in the quotient manifold,

$$\mathcal{H}/\mathrm{Stab}(\infty) \simeq \{\, z \mid 0<|z|<1 \,\} \times S^1,$$

is embedded in $\mathcal{M}(G)$. It is called a solid cusp torus. In fact there is a nested family in $c$ as $c \to \infty$.

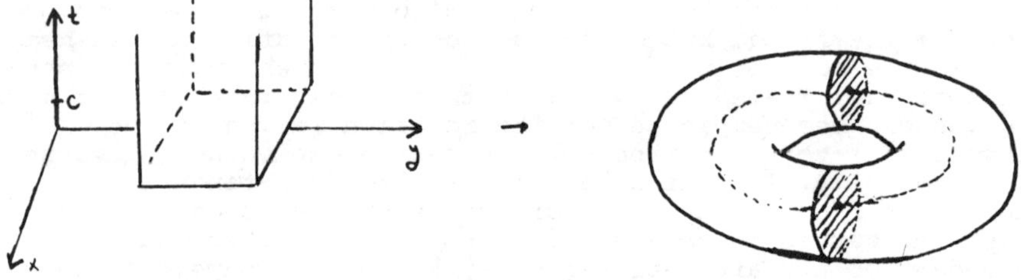


Its boundary is a torus, called a cusp torus, which is identical to $\mathbb{C}\,/\,\mathrm{Stab}(\infty)$. That is, the induced conformal structure on $\{(x,y,t) \mid t = c\}/\,\mathrm{Stab}(\infty)$ is independent of $c$ and is completely determined by $\mathrm{Stab}(\infty)$ alone. Conversely, the conformal structure of these embedded tori determine a 2-generator parabolic subgroup up to conjugation.

Case 2. Stab($\infty$) is infinite cyclic. We may assume that Stab($\infty$) is generated by $Tz = z+1$. In this case there is a pair of horodisks in $\Omega(G)$ at $\infty$. That is, for sufficiently large $s > 0$,

$$\{ z \mid \operatorname{Im} z < -s \} \cup \{ z \mid \operatorname{Im} z > s \},$$

is mapped disjoint to itself by everything not in Stab($\infty$). Interpreted in $\mathfrak{M}(G)$ this means that on $\partial\mathfrak{M}(G)$, there corresponds to $\infty$ a pair of punctures, a pair of disks about those punctures, and in $\mathfrak{M}(G)$ itself a cylinder connecting the boundaries of the disks and which bounds a "pairing tube" $\mathfrak{J} \cong \{z \mid 0<|z|<1\}\ [0,1]$.

For example, the kleinian group manifold that represents, on the boundary of a Bers slice, the r-point corresponding to a degeneration of a surface of genus $g = 2$ into two triply-punctured spheres has the structure as indicated in the following diagram:

trivalent graph

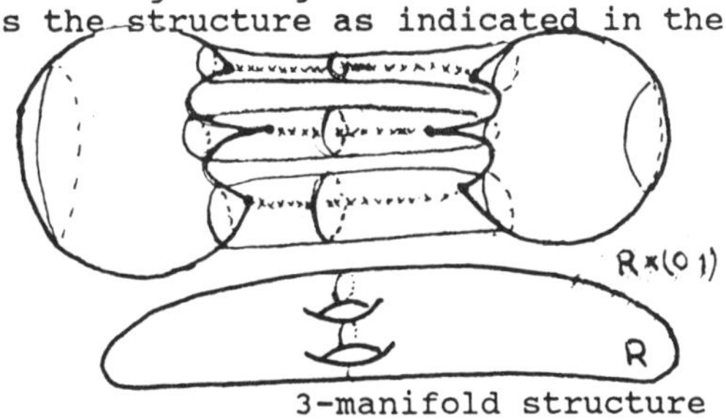


3-manifold structure

Indeed, it follows from Thurston's work that one generator parabolic subgroups of geometrically finite groups can always be viewed as resulting from the pinching of a boundary component of a hyperbolic 3-manifold. Likewise, two generator parabolic subgroups can always be viewed as the limiting case of a sequence of Dehn surgeries. We will not pursue these matters here.

## 8. Carriers of Riemann surfaces: another class of groups.

The set of hyperbolic 3-manifolds which are modelled on the product manifold $R\times[0,1]$, where $R$ is a compact surface of genus $g > 1$, consists of the manifolds associated with the groups of quasifuchsian space.

Consider from a topological point of view the product

manifold $R\times[0,1]$. To every simple loop on one boundary component corresponds a parallel simple loop on the other (actually the correspondence is between free homotopy classes). The two loops together bound a cylinder in the 3-manifold. Similarly, a set of mutually disjoint, non-parallel, simple loops on one boundary component corresponds to such a set on the other. Pairwise, the loops bound mutually disjoint cylinders in the 3-manifold. The homotopy class of an orientation reversing homeomorphism which sends one boundary surface to the other, each loop to the corresponding one, is uniquely determined up to Dehn twists about the loops.

Starting with the product manifold we will form a new manifold as follows. Given a set of $3g-3$ mutually disjoint, non-parallel, simple loops (called distinguished loops) on one, and hence on the other, boundary component, choose a set of $3g-3$ simple loops in the interior of the manifold each of which is parallel to one of the distinguished loops on the boundary. Now remove from the product manifold these $3g-3$ interior loops. This results in a new 3-manifold $M'$ which has the same two boundary components but which is no longer compact. Note that there is an orientation reversing involution of $M'$ which interchanges the boundary components, extends to fix the removed loops, and otherwise preserves the product structure. The homotopy class of its restriction to the boundary is uniquely determined up to Dehn twists about the distinguished boundary loops.

There is a family of necessarily geometrically finite kleinian groups whose associated 3-manifolds are modelled on $M'$. They have the following additional structure. Corresponding to each of the $3g-3$ removed loops is a nested family of solid cusp tori. On each cusp torus is a uniquely determined (up to homotopy) simple loop called the longitude that is parallel to a distinguished loop on each of the two bounding surfaces. Each of the bounding surfaces is a compact Riemann surface of genus g. Each of the cusp tori has a conformal type as well.

Conversely, given two compact Riemann surfaces $R$, $S$ of genus $g$, a set of $3g-3$ mutually disjoint, non-parallel simple loops on $R$, and an orientation reversing homeomorphism $J$ from $R$ to $S$, there exists a kleinian group $G$ whose corresponding quotient manifold is topologically equivalent to $M'$, whose bounding surfaces are conformally equivalent to $R$, $S$ respectively, and whose natural involution restricted to $R$ is homotopic to $J$, modulo Dehn twists about the distinguished boundary curves. Such a group is geometrically finite and it follows from the uniqueness theorem that $G$ is uniquely determined up to conjugation by the given data. This is the analogue of simultaneous uniformization for this new class.

Now consider the cusp tori. Their conformal types are determined by the conformal structure of the surfaces bounding the 3-manifold. Furthermore, each one corresponds to the conjugacy class of a two generator parabolic subgroup where the longitude corresponds to one of the generators.

Call one of the two bounding surfaces the bottom one and the other the top. Fix the conformal type of the bottom surface. Range through all possible conformal types of the top surface. In this set-up, the relation to the bottom surface is determined only up to Dehn twists about the distinguished curves. Thus this model most naturally is associated with the quotient $T_g/F$. The holomorphic deformation theory for kleinian groups shows that the corresponding groups sweep out a "slice" of their representation variety which is an analytic manifold.

The uniqueness theorem for geometrically finite groups shows that there is a one-to-one correspondence between the conformal type of the top surface and the ordered (3g-3)-tuple of conformal types of the cusp tori, modulo twists about their longitudes (the bottom surface is fixed). This is the key fact. The exponentiated torus moduli sweep out a domain in the poly-punctured plane $(C\backslash\{0\})^{3g-3}$. When the proper normalizations are made, these exponentiated moduli are precisely the parameters $t$ used in the construction of section 3.

This slice is (essentially) relatively compact in its representation variety. The resulting boundary is probably equivalent to the Bers boundary. In any case geometrically finite groups correspond to r-points. The new space is interesting in its own right.

References.

[1] L.V. Ahlfors, The complex analytic structure of the space of closed Riemann surfaces, in Analytic Functions, Ahlfors et al, eds., Princeton Univ. Press, 1960.

[2] L. Bers, Correction to "Spaces of Riemann surfaces as bounded domains," BAMS 67 (1961), 465-466.

[3] L. Bers, to appear.

[4] F. Bonahon, Bouts des variétés hyperboliques de dimension 3, Ann. of Math. 124 (1986), 71-158.

[5] C. Earle, Teichmuller theory, in Discrete Groups and Automorphic Functions, W. Harvey, ed., Academic Press, London, 1974.

[6] J. Fay, Theta Functions on Riemann Surfaces, Lecture Notes in Math. 352, Springer, 1973.

[7] S. Kerckhoff and W.P. Thurston, Discontinuities at Bers' boundary, preprint.

[8] I. Kra, Non-variational global coordinates for Teichmüller spaces, preprint.

[9] A. Marden, Geometriclly finite kleinian groups and their deformation spaces, in Discrete Groups and Automorphic Functions, W. Harvey, ed., Academic Press, London, 1974.

[10] A. Marden, Geometric relations between homeomorphic Riemann surfaces, BAMS 3 (1980), 1001-1017.

[11] B. Maskit, Moduli of marked Riemann surfaces, BAMS 80 (1974), 773-777.

[12] H. Masur, The extension of the Weil-Petersson metric to the boundary of Teichmuller space, Duke Math. J. 43 (1976), 623-635.

[13] D. Mumford, The structure of the moduli space of curves and Abelian varieties, in Proc. Internat. Congress Math., Nice, 1970, Gauthier-Villars, Paris, 1971.

[14] W.P. Thurston, The Geometry and Topology of 3-Manifolds, to appear.

[15] S. Wolpert, On the homology of the moduli space of stable curves, Ann. of Math. 118 (1983), 491-523.

# ASYMPTOTICS OF THE SELBERG ZETA FUNCTION AND THE POLYAKOV BOSONIC INTEGRAND

by

Scott A. Wolpert*
Department of Mathematics
University of Maryland
College Park, Maryland 20742
U.S.A.

* Partially supported by the National Science Foundation

# ASYMPTOTICS OF THE SELBERG ZETA FUNCTION AND THE POLYAKOV BOSONIC INTEGRAND

Scott A. Wolpert

At this time there are three methods for analyzing the divergence of the Polyakov string integrand. The first is sketched in the paper of Gava-Iengo-Jayaraman-Ramachandran [9] and is based on two observations: 1) the degeneration of a hyperbolic metric localizes into the collars about short geodesics (see Section 2) and 2) the vanishing of the conformal anomaly. In brief the argument is: given 1), the surface may effectively be replaced by its collars and given 2), the metrics on the collars may be replaced by Euclidean metrics. Then, the calculation is made for the Euclidean case, an elementary exercise. The second approach is based on an extension of Quillen's theorem [3] and the structure of the Picard group for the moduli space of stable curves. This was initiated with the work of Belavin and Knizhnik and is truly impressive, [1]. A third approach is motivated by the general question of the degeneration of the spectrum of the Laplacian, not just the string integrand. In this lecture we describe our results on this last problem, [16]. Our method is to estimate the basic quantities.

The discussion consists of three parts. Section 2 contains a review of the degeneration of a hyperbolic metric and Section 3, the estimates for the spectral degeneration. In Section 4 the results are put together to obtain an estimate for the Polyakov bosonic, $d = 26$, integrand. In complete agreement with [1] we find the *double-pole* divergence.

The underlying issue is the limiting behavior of the Selberg Zeta function $Z(s)$ on the moduli space. By the work of [5,8,10] the string determinants for a constant curvature -1 metric are evaluated either in terms of the length spectrum or the eigenvalue spectrum. We believe that the eigenvalue spectrum is better suited for the current problem. This view is based on three observations. First, the length formula for $Z(s)$ is valid only for $\operatorname{Re}s > 1$, while $Z'(1)$ and even $Z^{(n)}(\frac{1}{2})$ are of interest. Already the analytic continuation of $Z(s)$ involves the eigenvalue spectrum. Second the behavior under deformation of the length of the *average geodesic* is quite different than that of the maximally stretched geodesic. A sharp estimate would have to allow for the different growth rates. And finally by comparing one eigenvalue problem to another it is possible to estimate the entire eigenvalue spectrum in one step. A propoer analogue of this method does not exist for the length spectrum.

## §1. The degeneration of hyperbolic metrics.

There is a single phenomena for the degeneration of a hyperbolic metric: the formation of collars about short geodesics. Recall that a collar $C(\ell)$ about a geodesic of length $\ell$ is the set of points within $\log 2/\ell$ of the core geodesic and that the geometry of $C(\ell)$ is completely determined by $\ell$. Of course the basic result is the following.

Collar Lemma [12,14]. $R$ has a complete hyperbolic metric. There exists a positive $c_*$, the collar constant, such that for $\gamma$ a closed geodesic of length $\ell \le c_*$ then $C(\ell)$ embeds in $R$.

We would like to explain the fundamental principle: the degeneration of a hyperbolic metric localizes into collar neighborhoods about the short $(\ell \le c_*)$ geodesics. A good place to start is with an example.

Consider the family of surfaces $\mathcal{F} = \{zw = t \mid |z|, |w|, |t|, < 1\} \subset \mathbb{C}^3$ as a fibration over $D = \{|t| < 1\}$. The t-fibre $(t \neq 0)$ is $\{|t| < |z| < 1\}$ with hyperbolic metric $ds_t^2 = \left(\frac{\pi}{\log |t|} \csc\left(\pi \frac{\log |z|}{\log |t|}\right) \left|\frac{dz}{z}\right|\right)^2$. The 0-fibre after removing $(0,0,0)$ is $\{0 < |z| < 1\} \cup \{0 < |w| < 1\}$ with hyperbolic metric $ds_0^2 = \left(\frac{|d\zeta|}{|\zeta|\log|\zeta|}\right)^2$, $\zeta = z,w$. For $z,w$ fix $ds_t^2 \to ds_0^2$ as $t \to 0$ in fact the family is real analytic in $\frac{1}{\log 1/|t|}$. On the other hand the convegence is not uniform since $ds_t^2$ and $ds_0^2$ have distinct asymptotics for $|z| = |t|^{\frac{1}{2}}$, $t \to 0$. The non-uniform convergence is actually the source of all difficulties in treating the spectrum. For example the Green's function $(\Delta_t - c)^{-1}$, $c > 0$ is not locally but rather globally determined on the surface and thus the effect of the non-uniform convergence will propagate outside the collar.

This basic example is in analogy with the general case. Specifically $\mathcal{F}$ is analagous to the universal curve (the natural family of Riemann surfaces over, say, Teichmüller space) and $D$ is analogous to Teichmüller space itself. The *Teichmüller theory* can be worked out for this example [16]. For instance the tangent vector $\frac{\partial}{\partial t}$ at $t \in D$ is represented by the Beltrami differential $\frac{1}{2t \log |t|} \frac{z}{\bar{z}}$ on $\{|t| < |z| < 1\}$. Similary the cotangent vector $\frac{d}{dt}$ to $D$ is represented by the quadratic differential $-\frac{t}{\pi}\left(\frac{dz}{z}\right)^2$. Also the W-P metric can be found, $ds_{WP}^2 = \frac{2\pi^3|dt|^2}{|t|^2(\log 1/|t|)^3}$.

The family $\mathcal{F}/D$ is more than an interesting example. In fact the general case is constructed from this family. A diagram suggests the procedure

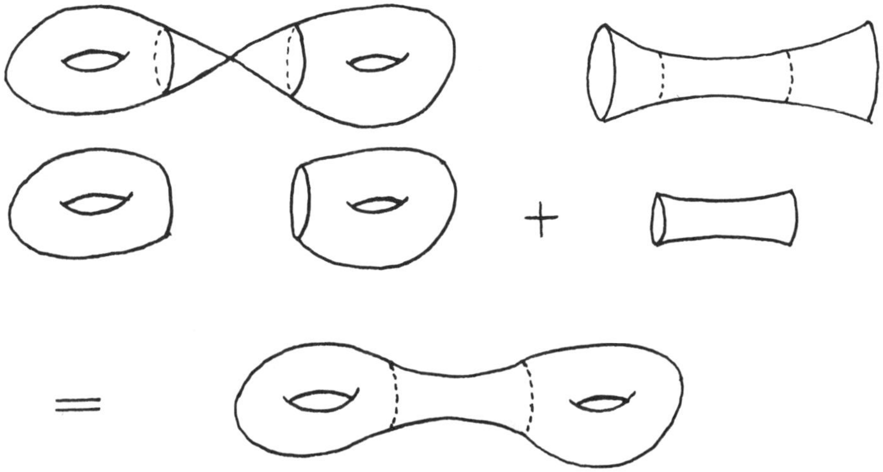

An explanation is in order. Let $R$ be a surface with a simple node and charts $z_1$, $w_1$ on each side of the node.

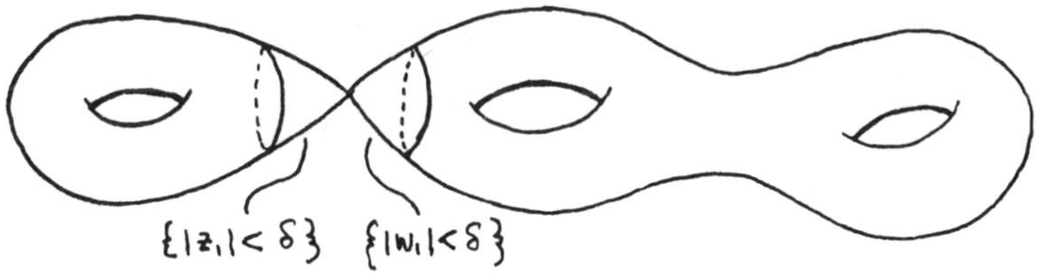


The plan is to identify $z_1$ and $w_1$ with the coordinates $z$ and $w$ for $\mathcal{F}$ and use the coordinate $t$ of $\mathcal{F}$ to parametrize opening the node. Specifically choose $0 < \delta < 1$ and consider the subfamily $\mathcal{S}\mathcal{F} = \{zw = t \mid |z|, |w| \le \delta, |t| \le \delta^2\}$ with the restriction fibre metrics $ds_*^2$. Now for each $t$ the fibre of $\mathcal{PF}$ can be glued to $R - \{|z_1| < \delta\} \cup \{|w_1| < \delta\}$ by identifying $\{|z_1| = \delta\}$ with $\{|z| = \delta\}$ and $\{|w_1| = \delta\}$ with $\{|w| = \delta\}$ to obtain a surface $R_t$. We point out that $R_t$ is a smooth Riemann surface and $\{R_t\}$ is a holomorphic family in $t$, [2,7,11].

There are two hyperbolic metrics on $\mathcal{PF} \subset R_t$, first the restriction metrics $ds_t^2$ and then $d\sigma^2(R_t)$, the restriction of the complete $R_t$ hyperbolic metric. An improved version of the standard comparison argument gives the following.

Convergence Lemma, [16]. $ds_t^2 / d\sigma^2(R_t) \to 1$ uniformly as $t \to 0$.

I expect that the convergence is real analytic in $\dfrac{1}{\log 1/|t|}$ but such a result is not known.

There is a second part of the fundamental principle: the behavior of the metrics in the collar complements. The desired result follows from a simple construction. Given a surface $S$ with short geodesics $\gamma_1, \ldots, \gamma_n$, remove the collars $C(\ell(\gamma_1))$, $\ldots, C(\ell(\gamma_n))$ and glue the boundaries in pairs to obtain the *associated surface* $\hat{S}$.

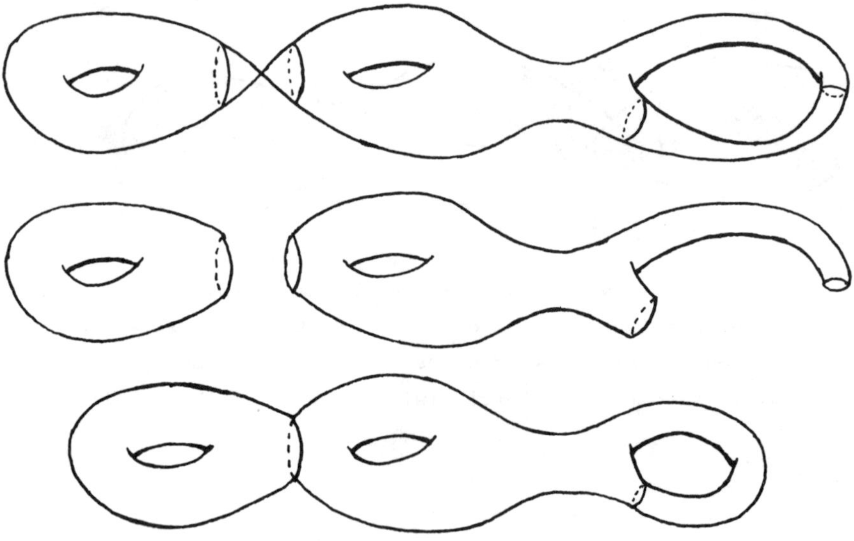

The natural conformal structure on $\hat{S}$ is smooth and the glued metric is piecewise $C^1$. We are interested in the space of all $\hat{S}$ structures, since the moduli space of $\hat{S}$ structures is the moduli space for the geometry of the collar complements.

Collar Complement Lemma.

1) $\{\hat{S}\text{-metrics}\} / \text{Diff}$ is a compact moduli space.

2) $\{\hat{S}\text{-conformal structures}\} / \text{Diff}$ is a compact moduli space.

We warn the reader that the above is a slight simplification of the actual result [16] and is not quite correct as stated. Although we have only sketched the degeneration theory for a hyperbolic metric, the general idea should be clear. The behavior of a degenerating family of hyperbolic metrics consists of two phenomenon: that of the collar, which closely approximates the basic example and that of the collar complement, where the geometry is totally bounded.

## §2. Degeneration of the spectrum.

One might reason as follows: the geometry in the collar complement is bounded, thus its contribution to the spectrum is bounded and hence Spec(R) = Spec(collars) + uniformly bounded term. For the eigenvalues (positive by convention) above $\frac{1}{4}$ this is exactly the case. By contrast already the conclusion is false for eigenvalues below $\frac{1}{4}$. Recall that the representation theory of $SL(2;\mathbb{R})$ already suggests that there can be a qualitative change in the spectrum at $\frac{1}{4}$.

The behavior of the small eigenvalues is already known. For instance Buser [4] showed that $\lambda_{4g-2} \geq \frac{1}{4}$ and thus the spectrum cannot accumulate in the interval $[0,\frac{1}{4}]$. To describe the actual behavior of the small eigenvalues consider the following.

Definition. A j-disconnect is a collection $\mathcal{E}$ of disjoint simple closed geodesics such that $R - \mathcal{E}$ has j-components. Let $L_j(R)$ be the minimal length of a j-disconnect for $R$.

Theorem 1, [6,15]. There exists $\alpha_1$ and $\alpha_2$ depending only on the genus such that

$$\alpha_1 L_j(R) \le \lambda_{j-1} \le \alpha_2 L_j(R) , \qquad 1 \le j \le 2g-2$$

and

$$\alpha_1 \le \lambda_{2g-2} \le \alpha_2 .$$

In particular if a simple geodesic $\gamma$ is pinched (length $\to$ 0) then $\lambda_1 \to 0$ if and only if $\gamma$ separates the surface.

Now we shall turn our attention to the behavior of the eigenvalues above $\frac{1}{4}$. It is easiest to describe the result in terms of the trace $\theta(t)$ of the heat kernel.

Theorem 2, [16]. There exists a constant $\rho$, $0 < \rho < \frac{1}{4}$ depending only on the genus such that for $\{\ell_j\}_{j=1}^{n}$ being pinched

for $t \ge 1$,
$$\theta(t) = \sum_{\lambda_k < \frac{1}{4}} e^{-\lambda_k t} + \sum_{j=1}^{n} \frac{e^{-t/4}}{\sqrt{\pi t}} \log 1/\ell_j + O(e^{-\rho t})$$

and for $0 \le t \le 1$,
$$\theta(t) = \frac{\text{Area}(R)}{4\pi t} + O(t^{-1/2})$$

where the error terms are uniformly bounded in $R$.

The uniform boundedness of the error terms means that we have identified the divergence of the spectrum modulo terms of this lower growth rate. There is a direct relation between the exponent $\rho$ and the domain of estimation for $Z(s)$: by a Laplace transform an $O(e^{-\rho t})$ estimate is equivalent to an estimate for $Z(s)$, $\text{Re } s > \sigma$, $\sigma = \frac{1}{2} + (\frac{1}{4} - \rho)^{\frac{1}{2}}$. For instance the easier $O(1)$ bound for $\theta(t)$, $t$ large, only produces an estimate for $Z(s)$ with $\text{Re } s > 1$.

The theorem is the consequence of three basic estimates for the heat trace. Further notation will simplify the statement of the results.

Notation: $\theta_*(t, P) = \sum_{\lambda_k \geq \frac{1}{4}} e^{-\lambda_k t}$, $\lambda_k$ are for an eigenvalue problem $P$.

Examples: $DC(\ell)$, the Dirichlet problem for a collar, $NC(\ell)$, the Neumann problem for a collar and $\hat{R}$ the eigenvalue problem for the associated surface $\hat{R}$.

The standard Dirichlet-Neumann comparison argument gives the following.

Estimate 1: let $R$ have short ($\ell \leq c_*$) geodesic lengths $\{\ell_j\}_{j=1}^{n}$, then

$$\theta_*(t,DC(\ell)) + \theta_*(t,\hat{R}) \leq \theta_*(t,R) \leq \theta_*(t,NC(\ell)) + \theta_*(t,\hat{R}).$$

We note that $\theta_*(t,\hat{R})$ is uniformly bounded by the Collar Complement Lemma and thus $\theta_*(t,R)$ is comparable to the heat trace of a collar. The necessity of excluding the small eigenvalues is underscored since $\lambda_1(DC(\ell)) > \frac{1}{4}$ while $\lambda_1(NC(\ell)) = 0$ and $\lambda_2(NC(\ell)) \sim \ell$. Next we find that the Dirichlet and Neumann spectrum of a collar are actually of the same order.

Estimate 2: $\theta(t,DC(\ell)) \leq \theta_*(t,NC(\ell)) \leq \theta(t,DC(\ell)) + 4 \sum_{k \neq 0} e^{-8\pi^2 k^2 t}$.

And finally we find the expansion.

Estimate 3: $\theta(t,DC(\ell)) = e^{-t/4} \left(\frac{\log 1/\ell}{\sqrt{\pi t}} + O(1)\right)$.

An immediate application is to consider the spectral counting function $N(x,R) = \#\{\text{eigenvalues of } R \leq x\}$. For $x$ fixed a heuristic arugment shows

$$N(x,R) \Big/ \sum_j 2 \log 1/\ell \longrightarrow \frac{1}{\pi}\sqrt{x - \tfrac{1}{4}}$$

in particular the spectrum accumulates everywhere in the interval $(\frac{1}{4}, \infty)$.

§3. Estimating Z(s) and the Polyakov bosonic integrand.

McKean observed that

$$\frac{Z'(s)}{Z(s)} = (2s-1)\int_0^\infty e^{-\sigma t}(\theta(t) - k(t)\,\mathrm{Area}(R))\,dt$$

for $\sigma = s(s-1)$, $\mathrm{Re}\,\sigma > 0$ and $k(t)$ the elementary solution of the heat equation, evaluated at the source, [13]. Naively the integral converges in the largest half plane where the (analytic continuation of the) transform is pole free. In fact the poles correspond to the eigenvalues. By subtracting the contribution of the small eigenvalues we can obtain an integral convergent in a larger half plane (for instance $\mathrm{Re}\, s > \frac{1}{4} + \varepsilon$). Of course the whole problem is to estimate the new integrand $(\theta_*(t) - k(t)$ Area) to order $O(e^{-\rho t})$, $0 < \rho < \frac{1}{4}$, $t$ large. By contrast the comparison $0 \le \theta_*(t) = \theta(t) - \sum_{\lambda_k<\frac{1}{4}} e^{-\lambda_k t}$, $\theta_*(t) \le \theta(t)$ produces an estimate of $O(1)$ for large $t$.

A suitable estimate is given by Theorem 2 and unraveling the bounds gives the following.

Theorem 3, [16]. With the above notation. Given a compact subset $K$ of $\{\mathrm{Re}(s^2-s) > -\rho\}$ there exists a constant $A$ such that

$$A^{-1} \le \left| Z(s) \Big/ \prod_{\ell_j \le c_*} \exp(-\pi^2/3\ell_j)\ell_j^{-1} \prod_{0\le\lambda_n<\frac{1}{4}} (s^2-s+\lambda_n) \right| \le A$$

for $s \in K$.

By [5,8,10] $Z'(1)$ is essentially the functional determinant $\det'(\Delta)$ and we have the bound.

Corollary **4**. There exists a positive constant B depending only on the genus such that

$$B^{-1} \le Z'(1) \Big/ \prod_{\ell_j \le c_*} \exp(-\pi^2/3\ell_j)\ell_j^{-1} \prod_{0<\lambda_n<\frac{1}{4}} \lambda_n \le B.$$

D'Hoker and Phong have already observed that the Polyakov bosonic, $d = 26$, integrand is $e^{-c\chi} Z'(1)^{-13} Z(2)$ dW-P for $\chi$ the Euler characteristic, c an absolute constant and dW-P the Weil-Petersson volume form, [5]. Substituting from Theorem 3 we have

$$Z'(1)^{-13} Z(2) \approx \prod_j \exp(4\pi^2/\ell_j)\ell_j^{10} \Big(\prod_{0<\lambda_n<\frac{1}{4}} \lambda_n\Big)^{-13}.$$

By the example of Section 1 and the Convergence Lemma $|t_j| \approx \exp(2\pi^2/\ell_j)$, where $t_j$ is a holomorphic parameter for the pinching. Substituting in the above we find $Z'(1)^{-13} Z(2) \approx \prod_j |t_j|^{-2} \times$ (lower order terms) and the example suggests Masur's result $dW\text{-}P \approx \prod_j \dfrac{|dt_j|^2}{|t_j|^2(\log 1/|t_j|)^3}$, [11]. In conclusion the Polyakov integrand behaves as $\prod_j \dfrac{|dt_j|^2}{|t_j|^4} \times$ (lower order terms) in agreement with the Belavin-Knizhnik *double-pole* result, [1]. Unfortunately the Convergence Lemma is not sufficient to determine the lower order terms.

## REFERENCES


[1] A. A. Belavin and V. G. Knizhnik, Algebraic geometry and the Geometry of Quantum Strings, Landau Institute, preprint.

[2] L. Bers, Spaces of degenerating Riemann surfaces, Ann. of Math. Studies, 79 (1974), 43-55.

[3] J. M. Bismut and D. S. Freed, Geometry of Elliptic Families: Anomalies and Determinants, preprint.

[4] P. Buser, Riemannsche flächen mit eigenwwerten in $(0,\frac{1}{4})$, Comm. Math. Helv., 52 (1977), 25-34.

[5] E. D'Hoker and D. H. Phong, Multiloop amplitudes for the bosonic Polyakov string, Nucl. Phys., B269 (1986), 205-234.

[6] J. Dodzuik, T. Pignataro, B. Randol and D. Sullivan, Estimating small eigenvalues on Riemann surfaces, preprint.

[7] C. Earle and A. Marden, manuscript.

[8] D. Fried, Analytic torsion and closed geodesics on hyperbolic manifolds, preprint.

[9] E. Gava, R. Iengo, T. Jayarman and R. Ramachandran, Multiloop divergences in the closed bosonic string theory, Phys. Lett., 168B (1986), 207-211.

[10] M. Kierlanczyk, Determinants of Laplacians, M.I.T. Thesis, 1986.

[11] H. Masur, The extension of the Weil-Petersson metric to the boundary of Teichmüller space, Duke Math. J., 43 (1976), 623-635.

[12] J. P. Matelski, A compactness theorem for Fuchsian groups of the second kind, Duke Math. J., 43 (1976), 829-840.

[13] H. P. McKean, Selberg's trace formula as applied to a compact Riemann surface, Comm. Pure App. Math., 25 (1972), 225-246.

[14] B. Randol, Cylinders in Riemann surfaces, Comm. Math. Helv., 54 (1979), 1-5.

[15] R. Schoen, S. Wolpert and S. T. Yau, Geometric bounds on the low eigenvalues of a compact surface, Proc. Symp. Pure Math., 36 (1980), 279-285.

[16] S. Wolpert, Asymptotics of the spectrum and the Selberg zeta function on the space of Riemann surfaces, preprint.

# Super Riemann Surfaces*

*Jeffrey M. Rabin*

Enrico Fermi Institute and Department of Mathematics
The University of Chicago
Chicago, IL 60637

*ABSTRACT*

A mathematically rigorous notion of supermanifold is presented and used to develop the theory of super Riemann surfaces. Super Riemann surfaces describe superconformal equivalence classes of supergravity geometries in two dimensions and therefore provide a powerful formalism for studying fermionic string theory. They also provide a convenient description of spin structures and half-forms on ordinary Riemann surfaces. I prove the uniformization theorem for super Riemann surfaces, determine the dimension and global structure of super Teichmüller space, and show that the super modular group is isomorphic to the ordinary modular (mapping class) group. Quasisuperconformal mappings and the super Beltrami equations are briefly discussed.

*Research supported by the NSF (PHY 83-01221) and DOE (DE-AC02-82-ER-40073).

# Super Riemann Surfaces*

*Jeffrey M. Rabin*

Enrico Fermi Institute and Department of Mathematics
The University of Chicago
Chicago, IL 60637

## 1. Introduction

The theory of Riemann surfaces [1-3] has recently become an important mathematical tool for string theorists. Bosonic string theory generates questions concerning line bundles on moduli space and the behavior of their sections near the boundary, determination of fundamental regions for the modular (mapping class) group acting on Teichmüller space, cell decompositions of moduli space, and many others. But physicists are more interested in fermionic strings and superstrings, since these theories may be finite and realistic models of particle physics [4]. Although superstrings can be studied using ordinary Riemann surfaces [5,6], a more elegant formulation involves the notion of a "super Riemann surface" (SRS) [7-10]. In this paper I will develop the theory of super Riemann surfaces in a mathematically rigorous way, explaining the relation between SRS's and ordinary Riemann surfaces with spin structure, and determining the dimension and global structure of super Teichmüller space. For physicists this provides a rigorous foundation for applications of SRS's to superstrings, while mathematicians may find new tools for the study of ordinary Riemann surfaces. Most of the work described here was done with Louis Crane [11].

Recall the connection between closed bosonic strings and Riemann surfaces [12,13]. The string is described by an immersion $X^\mu(\sigma^a)$ of its (compact) two-dimensional world sheet with coordinates $\sigma^a$ in Euclidean spacetime $R^{26}$ with coordinates $X^\mu$, and by a metric $h^{ab}$ on the world sheet itself. The action functional is,

$$S = \frac{1}{2} \int d^2\sigma \sqrt{h}\ h^{ab}\ \partial_a X^\mu\ \partial_b X_\mu . \tag{1.1}$$

In classical mechanics the equations of motion for the string are the Euler-Lagrange equations for extrema of $S$. Quantum mechanics is more interesting: here we need to compute

*Research supported by the NSF (PHY 83-01221) and DOE (DE-AC02-82-ER-40073).

functional integrals such as

$$Z = \sum_g \int DX\ Dh\ \exp -S[X,h], \tag{1.2}$$

over the space of all immersions and all metrics, including a sum over the genus $g$ of the world sheet. The integral over $X^\mu$ is Gaussian and gives

$$Z = \sum_g \int Dh\ (2\pi\ det'\,\Delta_h)^{-13}\ (\int d^2\sigma\ \sqrt{h}\,)^{13}, \tag{1.3}$$

where $\Delta_h$ is the Laplacian for the metric $h$, and $det'$ is the determinant excluding the zero eigenvalue. To define the integral over metrics, observe that the action (1.1) has a very large symmetry group consisting of diffeomorphisms of the world sheet as well as conformal transformations

$$h^{ab} \to \Omega^2\, h^{ab}. \tag{1.4}$$

Metrics which are related by these symmetries are physically equivalent, and therefore we should integrate over the space of equivalence classes only. But the equivalence classes of metrics under diffeomorphisms and conformal transformations are precisely the Riemann surfaces, so that the integration is over the moduli space of Riemann surfaces of genus $g$. For present purposes we need not continue with the definition of the integration measure [13] or the definition of the determinant in (1.3) as a section of the determinant line bundle [14].

The fermionic string has a similar formulation in terms of an immersion of a world sheet equipped with a certain geometry, but the geometry in this case is that of two-dimensional supergravity. A discussion of 2D supergravity is beyond the scope of this paper, but it is a non-Riemannian geometry on a supermanifold having two anticommuting dimensions in addition to two commuting ones [15,16]. The action analogous to (1.1) is invariant under superdiffeomorphisms and superconformal transformations, and the equivalence classes of geometries under these symmetries are super Riemann surfaces. The functional integral will be over a super moduli space whose structure will be determined in this paper.

The theory of SRS's, like that of Riemann surfaces, makes use of several topological constructions including universal covering spaces, quotient spaces, and homotopy groups. Therefore it requires a theory of supermanifolds which is sufficiently rigorous to give meaning to these constructions and to ensure that they have their usual mathematical properties. My discussion of SRS's will employ a theory of supermanifolds due to A. Rogers [17,18], which is closer to physicists' intuitive notion of superspace than the other theories in the literature

[19,20]. Rogers' theory is much more general than physical applications require, however, so at several points in the discussion it will be necessary to restrict attention to a subset of SRS's which are physically relevant.

In Section 2 I define super Riemann surfaces and explain their relation to Riemann surfaces with spin structure. Section 3 proves the uniformization theorem which lists all simply connected SRS's, and shows how to represent the others by Fuchsian supergroups. In Section 4 I determine the dimension and global structure of super Teichmüller space and show that the super modular group is isomorphic to the ordinary modular group. The case of super tori (genus 1) is worked out explicitly. Section 5 concludes with a brief discussion of the super Beltrami equations, and open problems.

## 2. Definitions

A SRS is to be a complex manifold with one commuting local coordinate $z$ and one anticommuting local coordinate $\theta$. To make sense of this idea we introduce a fixed Grassmann (exterior) algebra $B_L$ having $L$ generators $v_1, v_2, \cdots, v_L$ which satisfy $v_i v_j = -v_j v_i$. General even and odd elements of this algebra have the form,

$$z = z_0 + z_{ij} v_i v_j + \cdots ,$$

$$\theta = \theta_i v_i + \theta_{ijk} v_i v_j v_k + \cdots , \tag{2.1}$$

where the coefficients $z_{ij \cdots l}$ and $\theta_{ij \cdots l}$ are ordinary complex numbers and all subscripts are in increasing order, $i<j< \cdots <l$. The number $z_0$ is called the body of $z$ while $z - z_0$ is the soul; $\theta$ has no body and is pure soul [21]. The fact that there are $2^L$ independent coefficients in the expansions (2.1) suggests the following idea. Let $M$ be an ordinary complex manifold of dimension $2^L$. In each chart it has $2^L$ complex coordinates, which we call the body coordinate $z_0$ and the soul coordinates $\theta_i$, $z_{ij}$, $\cdots$. Then Eqs. (2.1) define an even Grassmann coordinate $z$ and an odd Grassmann coordinate $\theta$ in each chart. Obviously this can be done on any complex manifold $M$. However, we will now demand that the transition functions have special analytic properties in terms of the Grassmann coordinates. Not every $M$ will admit charts with these properties; the ones that do will be SRS's.

Motivated by the properties of superfields in physics, we demand that the transition functions be superanalytic, meaning that they have the form,

$$\tilde{z} = f(z) + \theta\zeta(z), \quad \tilde{\theta} = \psi(z) + \theta g(z), \tag{2.2}$$

where the $B_L$-valued component functions must have Taylor expansions in powers of the soul of $z$, for example,

$$f(z) = f(z_0) + (z - z_0)f'(z_0) + \cdots, \tag{2.3}$$

with $f(z_0)$ analytic. If we were discussing real rather than complex supermanifolds, $f(z_0)$ would be required to be smooth rather than analytic, and the transition functions would be called supersmooth or $G^\infty$. The functions $f$ and $g$ take even values in $B_L$ while $\zeta$ and $\psi$ take odd values.

Eq. (2.3) is not as restrictive as one might suppose. Any ''reasonable'' function, such as $z^2$ or $\sin z$, satisfies it because (2.3) is the only sensible way to *define* such functions of a Grassmann variable. The functions which are ruled out by (2.3) are those with explicit dependence on the components of $z$, such as $(z_{12})^2$. Eq. (2.3) should be thought of as a definition of analytic continuation of a function of $z_0$ to a function of $z$. Because the series terminates, no convergence problems arise. In the physics literature this analytic continuation is implicitly assumed, and no distinction is made between $z$ and $z_0$.

Differentiation of a superanalytic function $F(z,\theta) = f(z) + \theta\zeta(z)$ is defined by

$$\partial_\theta F = \zeta(z),$$

$$\partial_z F = f'(z) + \theta\zeta'(z), \tag{2.4}$$

where $f'(z)$ is defined as the Grassmann analytic continuation of $f'(z_0)$. There is an ambiguity in the definition of $\partial_\theta$ because $F$ only determines $\zeta$ up to addition of a multiple of $v_1 v_2 \cdots v_L$. This ambiguity can be removed by requiring the components $f(z_0)$ and $\zeta(z_0)$ of all superanalytic functions to take values in the subalgebra $B_{L-1}$ obtained by deleting the generator $v_L$ [22]. Differential forms on a supermanifold can be defined as formal expressions in $dz$ and $d\theta$ which transform in the obvious way under changes of coordinates.

Superanalyticity of the transition functions only makes $M$ into a complex supermanifold. To get a SRS one more condition is necessary. In each chart there is a derivative operator,

$$D = \partial_\theta + \theta\partial_z, \quad D^2 = \partial_z, \tag{2.5}$$

which transforms by

$$D = (D\tilde{\theta})\tilde{D} + (D\tilde{z} - \tilde{\theta}D\tilde{\theta})\tilde{D}^2. \tag{2.6}$$

By analogy with the properties of $\partial_z$ on a Riemann surface, we ask that $D$ transform homogeneously, so that the second term in (2.6) vanishes. This requires $D\tilde{z} = \tilde{\theta} D\tilde{\theta}$, or

$$\zeta = g\psi, \quad g^2 = f' + \psi\psi'. \tag{2.7}$$

Transition functions with these properties are called superconformal and have the form,

$$\begin{aligned} \tilde{z} &= f + \theta\psi\sqrt{f'}, \\ \tilde{\theta} &= \psi + \theta\sqrt{f' + \psi\psi'}. \end{aligned} \tag{2.8}$$

A SRS is a complex manifold $M$ with local coordinates $z$ and $\theta$ whose transition functions are superconformal.

For physical applications it is necessary to place restrictions on the global topology of $M$. We say that $M$ has the DeWitt topology if each coordinate chart is the Cartesian product of an open set in the $z_0$ plane with the entire complex planes of the soul coordinates [21]. This effectively trivializes the topology in all but the $z_0$ directions and evidently makes $M$ a fiber bundle of some kind. Unless otherwise stated, all SRS's will be assumed to have the DeWitt topology.

In physics it is also necessary to work with an infinite-dimensional Grassmann algebra. For example, Green's functions of more than $L$ fermionic fields vanish identically in a finite-dimensional algebra, because the product of more than $L$ odd elements is zero. The cleanest way to handle the $L \to \infty$ limit is as a direct limit of categories, by showing that SRS's over $B_L$ embed as a subcategory of SRS's over $B_{L+1}$ [22]. This issue will not be pursued here.

Associated to each (DeWitt) SRS $M$ there is an ordinary Riemann surface $M_0$ called its body. The charts for $M_0$ are the projections on the $z_0$ plane of the charts of $M$, and its transition functions are the bodies $\tilde{z}_0 = f_0(z_0)$ of the transition functions of $M$. A SRS determines more than just a Riemann surface structure on its body, however. The derivatives $f'_0(z_0)$ are the transition functions for the holomorphic tangent bundle of $M_0$, and a consistent choice of signs for the square roots $\sqrt{f'_0(z_0)}$ defines a square root of this bundle, a spin structure on $M_0$ (physically, a consistent transformation law for spinors on $M_0$). Such a choice of signs is implicit in Eqs. (2.8), because $\sqrt{f'_0(z_0)}$ is the body of $\sqrt{f'(z)}$. Therefore a SRS determines one of the $2^{2g}$ spin structures on its body of genus $g$. Conversely, given a Riemann surface $M_0$ with spin structure, a SRS $M$ can be constructed in a canonical way. The charts on $M$ are the Cartesian products of the charts of $M_0$ with the entire complex planes in the soul coordinates. Its transition functions have $\psi(z) = 0$ and $f(z)$ the Grassmann analytic continuation of the transition functions of $M_0$, with the square roots in Eqs. (2.8) defined according to the given spin structure. An important question which will be answered in the negative by the Teichmüller theory to be developed here is whether every SRS is equivalent to one of these canonical ones. In fact, the souls of the super moduli will measure the difference between a given SRS and a canonical one.

Evidently a SRS is a fiber bundle over its body with a complex vector space as fiber. It is not a vector bundle however, since Eqs. (2.2) and (2.3) show that its transition functions are generally polynomial rather than linear in the fiber coordinates.

## 3. Uniformization

Let $M$ be a SRS with body $M_0$. By the general theory of covering spaces, the universal covering space $\hat{M}$ of $M$ can be given a SRS structure such that the covering group $G = \pi_1(M)$ acts by superconformal transformations, and $M = \hat{M}/G$. The body of $\hat{M}$ will be the universal covering space $\hat{M}_0$ of $M_0$. Furthermore, since $M$ is a bundle with contractible fibers we have $G = \pi_1(M_0)$. The classification of SRS's now splits into two parts: classifying all simply connected SRS's, and determining the groups $G$ which can act on them.

By the uniformization theorem for Riemann surfaces, $\hat{M}_0$ must be the plane $C$, the Riemann sphere $C^*$, or the upper half plane $U$. Since each of these surfaces has a unique spin structure, they determine via the canonical construction described in Section 2 three simply connected SRS's, to be denoted $SC$, $SC^*$, and $SU$. I will now prove that any simply connected SRS $\hat{M}$ is necessarily one of these three. If $\{V^\alpha\}$ is a set of charts for $\hat{M}_0$, then the cylinders over the $V^\alpha$ are charts on $\hat{M}$. The structure of $\hat{M}$ is completely specified by the functions $f^{\alpha\beta}(z_0^\beta)$ and $\psi^{\alpha\beta}(z_0^\beta)$ which give the transition functions (2.8) relating $z^\alpha$ to $z^\beta$ in the overlap region above $V^\alpha \cap V^\beta$. If the souls of these functions all vanish, then $\hat{M}$ is canonical and must be $SC$, $SC^*$, or $SU$. If they do not vanish, it may be possible to make them vanish by superconformal coordinate redefinitions, which do not change the SRS structure. The standard tool for investigating such questions is sheaf cohomology or deformation theory [23].

The analysis begins from the consistency condition satisfied by the transition functions (2.8) on triple overlaps $V^\alpha \cap V^\beta \cap V^\gamma$. Changing coordinates from $z^\gamma$ to $z^\beta$ and then to $z^\alpha$ must give the same result as changing from $z^\gamma$ to $z^\alpha$ directly. If we expand $f$ and $\psi$ in the generators $v_i$ of $B_L$, there will be consistency conditions at each order of the expansion. At first order, the coefficients $\psi_i$ of $v_i$ in the expansion of $\psi$ must satisfy

$$\psi_i^{\alpha\gamma}(z_0^\gamma) = \psi_i^{\alpha\beta}(z_0^\beta) + \psi_i^{\beta\gamma}(z_0^\gamma)\sqrt{f_0^{\alpha\beta\,\prime}(z_0^\beta)}, \tag{3.1}$$

(no sum over the Greek indices). To see the significance of this equation, multiply both sides by $e^\alpha$, a local section over $V^\alpha$ for the spin bundle whose square is the tangent bundle of $\hat{M}_0$. The transformation law of $e^\alpha$ absorbs the square root in (3.1), yielding

$$e^\alpha\psi_i^{\alpha\gamma}(z_0^\gamma) = e^\alpha\psi_i^{\alpha\beta}(z_0^\beta) + e^\beta\psi_i^{\beta\gamma}(z_0^\gamma). \tag{3.2}$$

This says that $e^\alpha\psi_i^{\alpha\beta}$ is a cocycle in the Cech cohomology of $\hat{M}_0$ with coefficients in the sheaf of sections of the spin bundle. This cohomology is known to be trivial for simply connected Riemann surfaces [24], so there are functions $\eta_i^\alpha$ in each chart with

$$e^\alpha\psi_i^{\alpha\beta} = e^\beta\eta_i^\beta - e^\alpha\eta_i^\alpha, \tag{3.3}$$

or

$$\psi_i^{\alpha\beta} = \sqrt{f_0^{\alpha\beta\,\prime}}\,\eta_i^\beta - \eta_i^\alpha. \tag{3.4}$$

This implies that all the $\psi_i^{\alpha\beta}$ can be set to zero by the superconformal coordinate redefinitions,

$$\begin{aligned} \tilde{z}^\alpha &= z^\alpha + \theta^\alpha\eta_i^\alpha(z^\alpha)v_i, \\ \tilde{\theta}^\alpha &= \theta^\alpha\sqrt{1 + \eta_i^\alpha\eta_j^{\alpha\prime}v_iv_j} + \eta_i^\alpha(z^\alpha)v_i. \end{aligned} \tag{3.5}$$

At second order in the $v_i$ we consider the functions $f_{ij}^{\alpha\beta}$. One finds that they define a Cech cocycle with coefficients in the holomorphic vector fields on $\hat{M}_0$ and can also be set to zero by coordinate changes. Continuing inductively, no new cohomology groups appear and we find that $\hat{M}$ can be brought to canonical form by coordinate redefinitions as claimed. In particular, the coordinate redefinition at each order does not affect the lower-order terms which were already set to zero.

I now turn to the determination of the groups $G$ of superconformal automorphisms of the simply connected SRS's. Begin with $SC^*$. A superconformal automorphism (2.8) of $SC^*$ is specified by two functions $f(z_0)$ and $\psi(z_0)$. Certainly the body of $f$ must be a Möbius transformation,

$$\tilde{z}_0 = f_0(z_0) = \frac{a_0 z_0 + b_0}{c_0 z_0 + d_0}. \tag{3.6}$$

At first it appears that there are no restrictions on $\psi$ or the soul of $f$. This is because a superanalytic map is always invertible if its body is: the inverse can be found as a terminating expansion around the inverse of the body in powers of the soul [21]. However, the situation is more subtle because of the presence of the pole in (3.6). An expansion in powers of the soul can only be valid if the soul is finite where the body is, and if it is not much more singular than the body at poles. So the soul can only have a pole at $z_0 = -d_0/c_0$. To study the behavior of the map at the pole, we use the transition functions of $SC^*$ to replace $(\tilde{z}, \tilde{\theta})$ by $(-1/\tilde{z}, \tilde{\theta}/\tilde{z})$. This replacement must make both body and soul finite, which constrains $f$ and $\psi$ to have the form,

$$f(z_0) = \frac{az_0 + b}{cz_0 + d}, \quad \psi(z_0) = \frac{\gamma z_0 + \delta}{cz_0 + d}. \tag{3.7}$$

The most general superconformal automorphism of $SC^*$ is then,

$$\begin{aligned} \tilde{z} &= \frac{az + b}{cz + d} + \theta \frac{\gamma z + \delta}{(cz + d)^2}, \\ \tilde{\theta} &= \frac{\gamma z + \delta}{cz + d} + \frac{\theta}{cz + d}(1 + \frac{1}{2}\delta\gamma), \end{aligned} \tag{3.8}$$

where we can normalize so that $ad - bc = 1$. The transformations (3.8) form a supergroup (a Lie group which is also a supermanifold) with three even and two odd parameters. This supergroup generalizes $PSL(2,C)$ and will be denoted $SPL(2,C)$. Its Lie superalgebra is $osp(1,2)$, the subalgebra of the super Virasoro or Neveu-Schwarz algebra whose generators are conventionally denoted $L_0$, $L_{\pm 1}$, and $G_{\pm 1/2}$ [25].

The SRS's $SC$ and $SU$ do not contain the point at infinity, so their automorphisms may not have poles at finite points but may be very singular as $z \to \infty$. Their automorphism groups are much larger than $SPL(2,C)$. For example, if $\varepsilon$ is odd and $n > 1$ then

$$\tilde{z} = z + \theta \varepsilon z^n, \quad \tilde{\theta} = \theta + \varepsilon z^n \tag{3.9}$$

is a superconformal automorphism of $SC$ which does not belong to $SPL(2,C)$. However, only SRS's of the form $\hat{M}/G$ with $G$ a subgroup of $SPL(2,C)$ are physically relevant. It is a classical result that the three simply connected Riemann surfaces admit metrics invariant under all their automorphisms, so that any other Riemann surface inherits a metric from its covering space. This result is false for SRS's. On $SC$, for example, there is a metric

$$ds = |dz + \theta d\theta|, \tag{3.10}$$

which is invariant under a certain subgroup of $SPL(2,C)$ to be determined in Section 4, but not under other superconformal automorphisms. The analogous metric on $SU$,

$$ds = (\mathrm{Im}\, z + \frac{1}{2}\theta\bar{\theta})^{-1}|dz + \theta d\theta|, \tag{3.11}$$

is only invariant under the subgroup $SPL(2,R)$ of $SPL(2,C)$ for which the even parameters are real and $\bar{\gamma} = i\gamma$, $\bar{\delta} = i\delta$ (note: complex conjugation reverses the order of anticommuting factors in a product, so that $\overline{v_1 v_2} = v_2 v_1$). Since physics makes use of a geometry on a SRS, only the "metrizable" SRS's admitting these metrics are relevant.

[Because supergravity is not a Riemannian geometry, the expressions (3.10) and (3.11) do not actually play the role of metrics in this geometry. Instead, they are the norms of one of the components $E^z$ of a frame field or local basis for 1-forms. The other component of the frame field is $E^\theta = d\theta$ for $SC$ and

$$E^\theta = (\mathrm{Im}\, z + \frac{1}{2}\theta\bar{\theta})^{-1/2}d\theta + \frac{1}{2}(i\theta - \bar{\theta})(\mathrm{Im}\, z + \frac{1}{2}\theta\bar{\theta})^{-3/2}(dz + \theta d\theta) \tag{3.12}$$

for $SU$. Invariance of the complete frame up to $U(1)$ phase rotations is the same condition as metrizability, and guarantees that a supergravity geometry exists on any quotient SRS because $U(1)$ is the structure group of the frame bundle in two dimensions.]

At this point we know that any metrizable SRS $M$ is the quotient of $SC$, $SC^*$, or $SU$ by a subgroup $G$ of $SPL(2,C)$. In order for the quotient space to be a manifold, $G$ must act properly discontinuously: each point must have a neighborhood which does not intersect any of its images under group elements other than the identity. However, this does not guarantee that $M$ will have the DeWitt topology. It is necessary that the neighborhoods be cylinders over neighborhoods in the body. This in turn is true if the group $G_0$ of Möbius transformations which are the bodies (3.6) of elements of $G$ acts properly discontinuously. A simple example of what goes wrong if this condition is not met is provided by supersymmetry. The group generated by a supersymmetry transformation,

$$\tilde{z} = z + \theta\delta, \quad \tilde{\theta} = \theta + \delta, \tag{3.13}$$

for fixed odd $\delta$, acts properly discontinuously on $SC$, but $G_0$ consists of the identity alone.

Taking the quotient of $SC$ by this group will identify points having different souls but the same body, curling up the fibers and violating the DeWitt topology.

## 4. Teichmüller Theory

The results of the last section show that any metrizable DeWitt SRS $M$ is $SC^*$ or the quotient of $SC$ or $SU$ by a subgroup $G$ of $SPL(2,C)$. The subgroup $G$ is isomorphic to $\pi_1(M)$ and is unique up to a conjugation $G \to q^{-1}Gq$. Furthermore, the space of all physically distinct supergravity geometries in two dimensions is the super moduli space of metrizable SRS's. As in the theory of Riemann surfaces, it is convenient to introduce a related but topologically simpler space called super Teichmüller space, denoted $ST_g$ for genus $g$. This is the space of metrizable SRS's $M$ with a specific choice of generators for $\pi_1(M)$. In this section the structure of this space will be determined. Super moduli space will be the quotient of $ST_g$ by a super modular group which acts by changing the choice of generators. Since $\pi_1(M) = \pi_1(M_0)$, it is obvious that the super modular group is isomorphic to the ordinary modular group for genus $g$. This would not be true without the DeWitt topology.

I will begin with a concrete example, the classification of super tori, SRS's of genus 1. A super torus must be the quotient of $SC$ by an Abelian group $G$ having two generators of the form (3.8). If the super torus is to be metrizable, then invariance of the metric (3.10) requires $c = 0$, $a^2 = 1$, and $\gamma = 0$. If $G_0$ acts properly discontinuously then also $b_0 \neq 0$. The generators of $G$ then take the form,

$$\tilde{z} = z + ab + \theta\delta, \quad \tilde{\theta} = a(\theta + \delta), \tag{4.1}$$

with $a = \pm 1$. I represent the two generators by the ordered triples $A = (a, b, \delta)$ and $A' = (a', b', \delta')$. Their commutator is

$$A'^{-1}A^{-1}A'A = [1, (aa' + a + a' - 1)\delta\delta', (1 - a')\delta - (1 - a)\delta']. \tag{4.2}$$

Note that this commutator is a pure soul transformation. This means that if the requirement that $G$ be Abelian were relaxed, the quotient SRS would not have the DeWitt topology.

The four choices of signs for $a$ and $a'$ describe the four spin structures on a torus. For a nontrivial spin structure, $a$ and $a'$ are not both 1; assume for definiteness that $a' = -1$. Then by a conjugation it is possible to bring the generators into a standard form with $\delta' = 0$ and $b' = 1$, whereupon the vanishing of the commutator (4.2) forces $\delta = 0$ as well. Therefore the SRS's with nontrivial spin structures are characterized by a single even modulus parameter $b$. Because a specific choice of generators for $G$ was made, $b$ is a global coordinate on $ST_1$ rather than super moduli space. $ST_1$ has four disconnected sheets corresponding to the spin structures, three of which are complex supermanifolds of one even and no odd dimensions.

For the trivial spin structure, $a = a' = 1$. Again a conjugation can set $\delta'$ to zero and $b'$ to unity, but the commutator (4.2) then vanishes identically. It would appear that the fourth sheet of $ST_1$ is a complex supermanifold with one even and one odd coordinate $(b,\delta)$, but this is not quite true. Conjugation by the special $SPL(2,C)$ element

$$I: \tilde{z} = z, \quad \tilde{\theta} = -\theta, \tag{4.3}$$

changes the sign of $\delta$ without affecting $b$. Therefore both signs of $\delta$ describe the same super torus and must be identified. The trivial sheet of $ST_1$ is a $Z_2$ super orbifold, since the points with $\delta = 0$ are singular. The body of $ST_1$ consists of four copies of the ordinary Teichmüller space of a torus, with coordinate $b_0$, and points of the body represent SRS's of the canonical type.

The action of the (super) modular group on $ST_1$ can be determined by choosing a new pair of generators for the group $G$. The modular group is generated by the two transformations which replace the generators $A$ and $A'$ by $AA'$ and $A'$, or by $A'^{-1}$ and $A$. By conjugating the new generators into standard form, the effect of these transformations on $b$ and $\delta$ can be read off. For example, the modular group fixes the trivial sheet of $ST_1$, and its action on this sheet is generated by

$$b \to b + 1, \quad \delta \to \delta, \tag{4.4}$$

and

$$b \to -1/b, \quad \delta \to \delta b^{-3/2}. \tag{4.5}$$

Because $\delta$ is identified with $-\delta$, (4.5) makes sense despite the sign ambiguity in the square root of $b$.

The results for higher genus are similar. A SRS of genus $g>1$ is the quotient of $SU$ by a subgroup $G$ of $SPL(2,R)$ having $2g$ generators $q_1, q_2, \cdots, q_{2g}$. The choice of overall sign for the even parameters $a$, $b$, $c$, $d$ of each generator determines a spin structure. By a conjugation in $SPL(2,R)$ two odd parameters can be set to zero and three even parameters can be fixed. Then the relation

$$q_1 q_2 q_1^{-1} q_2^{-1} \cdots q_{2g-1} q_{2g} q_{2g-1}^{-1} q_{2g}^{-1} = 1 \tag{4.6}$$

satisfied by the generators of the fundamental group of any surface imposes three even and two odd conditions [26]. Conjugation by $I$ shows that the same SRS is described by changing the signs of all the odd parameters together. $ST_g$ therefore has $2^{2g}$ disconnected pieces corresponding to the spin structures, each of which is a $Z_2$ super orbifold with $6g-6$ even and $4g-4$ odd dimensions. The only constraint on the group parameters is that $G_0$ act properly discontinuously on $U$, which affects only the bodies of the parameters. Since the soul coordinates are unrestricted, $ST_g$ has the DeWitt topology. Its body consists of $2^{2g}$ copies of the ordinary Teichmüller space $T_g$. Because the parameters of $SPL(2,R)$ satisfy reality conditions, this analysis only produces real coordinates on $ST_g$. Its complex structure can be exhibited using more advanced techniques to be described in Section 5.

In principle the structure of super moduli space is known given that of $ST_g$ and of the modular group. Because any two even (odd) spin structures can be related by a modular transformation [27,28], super moduli space has two connected components rather than $2^{2g}$. Its body is the "spin-moduli space", a $2^{2g}$-sheeted cover of the Riemann moduli space for genus $g$. A more explicit description requires more knowledge about the action of the modular group on spin structures and on the odd coordinates of $ST_g$. This is an important open problem.

## 5. Further Results and Open Problems

To go beyond the results established so far, some heavy machinery is needed: the super-generalization of the Beltrami equation and the theory of quasiconformal mappings. With this machinery a Bers embedding theorem can be proven, establishing the existence of a complex structure on $ST_g$ for $g>1$. I will briefly sketch the results; the proofs are generalizations of the classical ones [2,3] and can be found in [11].

Let $(z,\theta)$ and $(w,\phi)$ be two sets of coordinates on $SC$ (or $SU$), related by a $G^\infty$ but not necessarily superconformal or superanalytic transformation $w = w(z,\bar{z},\theta,\bar{\theta})$, $\phi = \phi(z,\bar{z},\theta,\bar{\theta})$. Then the invariant metric $|dw + \phi d\phi|$ appears in the other coordinates as

$$\lambda(z,\bar{z},\theta,\bar{\theta})\,|dz + \mu(z,\bar{z},\theta,\bar{\theta})d\bar{z} + \nu(z,\bar{z},\theta,\bar{\theta})d\theta + \sigma(z,\bar{z},\theta,\bar{\theta})d\bar{\theta}|, \tag{5.1}$$

and by using

$$d = dz\, \partial_z + d\bar{z}\, \partial_{\bar{z}} + d\theta\, \partial_\theta + d\bar{\theta}\, \partial_{\bar{\theta}}, \tag{5.2}$$

we obtain the super Beltrami equations,

$$\begin{aligned} w_{\bar{z}} + \phi\phi_{\bar{z}} &= \mu(w_z + \phi\phi_z), \\ -w_\theta + \phi\phi_\theta &= \nu(w_z + \phi\phi_z), \\ -w_{\bar{\theta}} + \phi\phi_{\bar{\theta}} &= \sigma(w_z + \phi\phi_z), \end{aligned} \tag{5.3}$$

where subscripts indicate differentiation.

The first question is whether Eqs. (5.3) have a unique solution for the map $(w,\phi)$ when the Beltrami coefficients $(\mu, \nu, \sigma)$ are prescribed. Because there are three equations for two unknowns, the Beltrami coefficients must satisfy a certain integrability condition (which I do not know explicitly) if solutions are to exist. If this condition is satisfied, then there are exactly two solutions satisfying Eqs. (5.3) and appropriate boundary conditions. If $(w,\phi)$ is one solution, the other is $(w,-\phi)$. Furthermore, the solutions of the trivial super Beltrami equations with $\mu = \sigma = 0$, $\nu = \theta$ are precisely the superconformal maps.

Now fix a sheet of $ST_g$, $g>1$ corresponding to a particular spin structure, and a base point in this sheet representing a SRS $M = SU/G$. Let $M$ be canonical, so that $G = G_0$. For any other $M' = SU/G'$ in this sheet, a fundamental theorem due to Batchelor [29] says that $M$ and $M'$ are equivalent as $G^\infty$ supermanifolds, if not as SRS's. Then there is a $G^\infty$ diffeomorphism $f: M \to M'$ which lifts to $SU$ as a "quasisuperconformal" map $(z,\theta) \to (w,\phi)$. If the metric of $M'$ lifts to the standard form $(\mathrm{Im}\, w + \frac{1}{2}\phi\bar{\phi})^{-1} |dw + \phi d\phi|$, then the metric of $M$ lifts to a $G$-invariant Beltrami differential (5.1) on $SU$, and the super Beltrami equations are satisfied on $SU$. Now extend the Beltrami coefficients into the region $SL = SC - \overline{SU}$ above the lower half plane trivially. A solution to the super Beltrami equations on all of $SC$ will be superconformal in $SL$, so it makes sense to compute its super Schwarzian derivative, defined for any superconformal map as

$$S(z,\theta;\, w,\phi) = \frac{D^4\phi}{D\phi} - 2\frac{(D^3\phi)(D^2\phi)}{(D\phi)^2}. \tag{5.4}$$

The super Schwarzian derivative is valued in the odd elements of $B_L$, vanishes iff the map belongs to $SPL(2,C)$, and has the following behavior under composition of maps:

$$S(z,\theta;\, \tilde{w},\tilde{\phi}) = S(z,\theta;\, w,\phi) + (D\phi)^3 S(w,\phi;\, \tilde{w},\tilde{\phi}). \tag{5.5}$$

These properties and the $G$-invariance of the Beltrami differential imply that

$$S(z,\theta;\, w,\phi)(dz + \theta d\theta)^{3/2} \tag{5.6}$$

is a $G$-invariant differential on $SL$. If we write $S = S^1 + \theta S^0$, then $S^0(z_0)$ is an even $G_0$-invariant quadratic differential, and $S^1(z_0)$ an odd $G_0$-invariant 3/2-differential on the lower half plane. One proves that the association of $M'$ with $(S^0, \pm S^1)$ is in fact an embedding of the chosen sheet of $ST_g$ in the space of pairs of even quadratic differentials and odd 3/2-differentials modulo sign. This space is a complex super orbifold of dimension $(3g-3, 2g-2)$, and the embedding defines such a structure on $ST_g$ as well. The sign ambiguity in $S^1$ comes from the sign ambiguity in solutions of the super Beltrami equations.

The super Beltrami equations have other applications as well. They can be used to represent SRS's by Schottky, rather than Fuchsian, supergroups. Infinitesimal deformations of SRS's are described by the linearized super Beltrami equations, which can be used to study the tangent space to $ST_g$ [30]. No doubt other aspects of the geometry of moduli space can be generalized to super moduli space with their help.

Many problems about SRS's remain open. The following list includes mathematically interesting questions as well as physically important ones bearing on the finiteness of the fermionic string functional integral expressed as an integral over super moduli space. This is controlled by the behavior of the integrand near the boundary of super moduli space, and by possible cancellations between the contributions of different sheets representing different spin structures.

(1) Classify SRS's without the simplifying assumptions of metrizability and DeWitt topology. This probably has no relevance to physics *except* that the embedding of super moduli space in the larger space of all SRS's may aid in characterizing its boundary.

(2) Describe as explicitly as possible the action of the super modular group on the spin structures and on the odd coordinates of $ST_g$. What constraints on the string integrand (viewed as a form on super Teichmüller space) follow from invariance under the super modular group?

(3) Ordinary moduli space can be compactified by adding points at infinity representing Riemann surfaces with nodes. Wolpert has shown that Chern classes of line bundles over moduli space can be expressed in terms of cohomology classes associated to the compactification divisor [31]. How do spin structures degenerate at nodes? Does the spin-

moduli space (the body of super moduli space) have a similar compactification, and do the results about Chern classes generalize [32]?

(4) What is the Riemann-Roch theorem for SRS's? How many holomorphic forms of given degree are there on a SRS of genus $g$? Note that on the super torus $dz + \theta d\theta$ is an odd holomorphic form which is not closed, because $(d\theta)^2 \neq 0$. If the spin structure is trivial then there is also the even form $d\theta$. Can SRS's be characterized by the entries of a period matrix? What are super Jacobian varieties and theta functions (the name will have to be changed!)?

(5) Is every SRS an algebraic variety in a super projective space? What is the algebraic condition that a set of polynomials represent a SRS rather than merely a complex supermanifold?

**References**


[1] H.M. Farkas and I. Kra, *Riemann Surfaces,* Springer-Verlag, New York 1980.

[2] L. Bers, ''Uniformization, moduli, and Kleinian groups'', Bull. London Math. Soc. **4,** 257-300 (1972).

[3] C.J. Earle, ''Teichmüller theory'', in *Discrete Groups and Automorphic Functions,* ed. W.J. Harvey, Academic Press, London, 1977.

[4] J.H. Schwarz, *Superstrings: The First 15 Years of Superstring Theory,* World Scientific, Singapore, 1985.

[5] E. D'Hoker and D.H. Phong, ''Loop amplitudes for the fermionic string'', to appear in Nucl. Phys.

[6] G. Moore, P. Nelson, and J. Polchinski, ''Strings and supermoduli'', Phys. Lett. **169B,** 47-53 (1986).

[7] D. Friedan, ''Notes on string theory and two-dimensional conformal field theory'', in the Proceedings of the Workshop on Unified String Theories, ed. D. Gross and M. Green, World Press, Singapore, 1986.

[8] E. Martinec, ''Conformal field theory on a (super)Riemann surface'', Princeton preprint (physics) 1986.

[9] M.A. Baranov and A.S. Shvarts, ''Multiloop contribution to string theory'', JETP Lett. **42,** 419-421 (1986).

[10] P. Nelson and G. Moore, ''Heterotic geometry'', Nucl. Phys. **B274,** 509-519 (1986).

[11] L. Crane and J.M. Rabin, ''Super Riemann surfaces: uniformization and Teichmüller theory'', University of Chicago preprint EFI 86-25, submitted to Commun. Math. Phys.

[12] O. Alvarez, ''Theory of strings with boundaries'', Nucl. Phys. **B216,** 125-184 (1983).

[13] E. D'Hoker and D.H. Phong, ''Multiloop amplitudes for the bosonic Polyakov string'', Nucl. Phys. **B269,** 205-234 (1986).

[14] D.S. Freed, ''Determinants, torsion, and strings'', MIT preprint (mathematics), April 1986.

[15] L. Smolin, ''Superspace unmasked: the physical interpretation of supergravity theory'', Ann. Phys. **131,** 398-425 (1981).

[16] P.S. Howe, ''Super Weyl transformations in two dimensions'', J. Phys. A **12,** 393-402 (1979).

[17] A. Rogers, ''A global theory of supermanifolds'', J. Math. Phys. **21,** 1352-1365 (1980).

[18] J.M. Rabin and L. Crane, ''Global properties of supermanifolds'', Commun. Math. Phys. **100,** 141-160 (1985).

[19] D. Leites, ''Introduction to the theory of supermanifolds'', Russ. Math. Surv. **35,** 1-64 (1980).

[20] M.J. Rothstein, ''The axioms of supermanifolds and a new structure arising from them'', Trans. AMS **297,** 159-180 (1986).

[21] B.S. DeWitt, *Supermanifolds,* Cambridge University Press, Cambridge, 1984.

[22] A. Rogers, ''Graded manifolds, supermanifolds, and infinite dimensional Grassmann algebras'', Commun. Math. Phys. **105,** 375-384 (1986).

[23] K. Kodaira, *Complex Manifolds and Deformation of Complex Structures,* Springer-Verlag, New York, 1986.

[24] O. Forster, *Lectures on Riemann Surfaces,* Springer-Verlag, New York, 1986.

[25] Z. Qiu, ''Supersymmetry, two-dimensional critical phenomena, and the tricritical Ising model'', Nucl. Phys. **B270,** 205-234 (1986).

[26] L.V. Ahlfors, ''Some remarks on Teichmüller's space of Riemann surfaces'', Ann. Math. **74,** 171-191 (1961).

[27] L. Alvarez-Gaumé, G. Moore, and C. Vafa, ''Theta functions, modular invariance, and strings'', Commun. Math. Phys. **106,** 1-40 (1986).

[28] C.J. Earle and I. Kra, ''Half-canonical divisors on variable Riemann surfaces'', J. Math. Kyoto Univ. **26,** 39-64 (1986).

[29] M. Batchelor, ''The structure of supermanifolds'', Trans. AMS **253,** 329-338 (1979); ''Two approaches to supermanifolds'', ibid. **258,** 257-270 (1980).

[30] J.M. Rabin, ''Teichmüller deformations of super Riemann surfaces'', in preparation.

[31] S. Wolpert, ''On the homology of the moduli space of stable curves'', Ann. Math. **118,** 491-523 (1983).

[32] R. Catenacci, M. Cornalba, M. Martellini, and C. Reina, ''Algebraic geometry and path integrals for closed strings'', Phys. Lett. **172B,** 328-332 (1986).

# DIVISORS ON $\overline{\mathcal{M}}_g$ AND THE COSMOLOGICAL CONSTANT

M. C. Chang*
Department of Mathematics
University of South Carolina
Columbia, SC 29208

Z. Ran*
Department of Mathematics & Comp Sci
University of California
Riverside, CA 92521

## ABSTRACT

*We prove the nonexistence of certain effective divisors on the moduli space* $\overline{\mathcal{M}}_g$ *of stable curves of genus g and deduce the vanishing of the cosmological constant of the heterotic string theory, as well as the existence of a large fixed part in the pluricanonical systems on* $\overline{\mathcal{M}}_g$.

* Partially supported by NSF.

The purpose of this paper is to prove a result in algebraic geometry which, thanks to the work of Moore, Harris, Nelson and Singer [6], has as its consequence the vanishing, to all orders of perturbation theory, of the vacuum amplitude for the heterotic string in a flat background.

In order to state our result, we need some notation. Denote by $\overline{\mathcal{M}}_g$ the moduli space of stable curves of genus $g$, with boundary $\Delta = \bigcup_{i=0}^{[g/2]} \Delta_i$. Here $\Delta_i$ is the locus of curves of the form $C_1 \bigcup_p C_2$, where $C_1$ and $C_2$ have genus $i$ and $g-i$ and meet at $p$; except that we reserve the right to denote $\Delta_i$ by $\Delta_{g-i}$ when convenient; and $\Delta_0$ is the locus of irreducible singular curves. The divisor class of $\Delta$ is denoted by $\delta$ (more precisely, the class of $\Delta_1$ appears in $\delta$ with coefficient $\frac{1}{2}$). We denote by $\lambda$ the Hodge line bundle (or the corresponding divisor class); that is, the fibre of $\lambda$ at $[C]$ is $\Lambda^g H^0(C, \omega_C)$.

**THEOREM.** *Fix integers $i = 1, 2, 3, 4, 5$ or $15$ and $g \geq i+1$ and rational numbers $a, b > 0$, and let $A$ be an effective divisor in the class $a\lambda - b\delta$ on $\overline{\mathcal{M}}_g$. Define numbers $c_i = c_i(a,b)$ and $\sigma_i$ by the following table.*

| $i$ | $c_i$ | $\sigma_i$ |
|---|---|---|
| 1 | $11b - a$ | 11 |
| 2 | $19b - 2a$ | $9\frac{1}{2}$ |
| 3 | $26b - 3a$ | $8\frac{2}{3}$ |
| 4 | $33b - 4a$ | $8\frac{1}{4}$ |
| 5 | $38b - 5a$ | $7\frac{3}{5}$ |
| 15 | $\frac{1473}{11}b - \frac{1113}{55}a$ | $6.6172\ldots$ |

Table 1

*Then if $c_i > 0$, i.e., if $a/b < \sigma_i$, then $A$ must contain $\Delta_i$ with multiplicity $\geq c_i$.*

It follows, of course, that if $a/b < 11$ then there is no section of $a\lambda$ having zeroes of order exactly $b$ at (the generic point of) every component of $\Delta$; this, for $a = 8, b = 2$, is the result needed in physics. Let us now give a precise statement of the physical consequence of our theorem (for physical background, see [6] and the references therein, or, for more elementary account, [3]).

**Corollary 1.** *Under the physical factorization hypothesis of [6], the vacuum-to-vacuum amplitude of the heterotic string with flat background vanishes to all orders of string perturbation theory.*

The details of why Corollary 1 follows from the Theorem, with the possible exception of one point which is perhaps not fully stressed, are in [6], in which the same result is proven for all orders $g \leq 15$ by using the noneffectiveness of $8\lambda - 2\delta$ proved, in these genera, by Harris and Morrison [4]. Let us merely give a very brief sketch. The above vacuum amplitude is given by a perturbation theory series $\sum Z_g$, where $Z_g$ is the integral of a certain density on $\mathcal{M}_g$. One of the factors of this density yields a holomorphic section $\xi\varsigma$ of the bundle $8\lambda$, and it was shown in [6] that the hypothesis of physical factorization implies that $\xi\varsigma$ extends to $\overline{\mathcal{M}}_g$ with $2^{\text{nd}}$ order zeros on $\Delta$. Moreover, it is clear from [6]—though this is not stated explicitly—that $\xi\varsigma$, if not identically zero, must in fact vanish to order exactly 2 along any boundary component. Hence by the theorem we have $Z_g = 0$ for $g \geq 2$. By [4] and [6], we have $Z_g = 0$ for $g \leq 15$, so the whole series vanishes. Actually, it has been known for some time (see references in [6]) that $Z_0 = Z_1 = 0$, so our Theorem is really logically independent of [6].

Besides the string theory case, another interesting case of the theorem is that of pluricanonical divisors, and it yields the following.

**Corollary 2.** *Let $K$ denote the canonical divisor of $\overline{\mathcal{M}}_g$. Then for $i, g$ as in the Theorem and any $n \geq 1$, any divisor in the $n$-canonical system $|nK|$ must contain $\Delta_i$ with multiplicity $\geq nc_i(13, 2)$.*

*Proof:* By Mumford's formula [7], $K \sim 13\lambda - 2\delta$, so it suffices to plug $a = 13n, b = 2n$ into the theorem.

This corollary is particularly interesting when $g$ is large because then, by the work of Harris-Mumford, etc., $\mathcal{M}_g$ is of general type, so that for large $n$, $|nK|$ is big enough to yield a birational map; by Corollary 2, $|nK|$ nonetheless has a large fixed part. Harris and Morrison [4] have shown that any pluricanonical system $|nK|$ must have a base locus containing the hyperelliptic and trigonal loci, but Corollary 2 shows that $|nK|$ actually has fixed components (albeit in the boundary).

We turn now to the proof of the theorem, beginning with the most complicated case $i = 15$. This case will be a consequence of the methods and results of our earlier paper [1] which was concerned with curves of genus 15 (see also [2]). The idea is, assuming a divisor as in the theorem exists, to study its restriction to $\Delta_{15}$.

We will write out the proof for $g \geq 17$, and leave to the reader the modification needed to treat the case $g = 16$ (this is mainly a matter of working with the moduli *functor* instead of the moduli *space*). We now proceed to recall the necessary information from [1]. There is a surface $\mathcal{Y} \subset \mathbb{P}^1 \times \mathbb{P}^3$, a locally complete intersection with finite singular locus, which gives rise to a family $\{Y_t : t \in \mathbb{P}^1\}$ of stable, irreducible, automorphism-free curves of genus 15 and degree 14 in $\mathbb{P}^3$. Let $f : \mathbb{P}^1 \to \overline{\mathcal{M}}_{15}$ be the associated map, with image cycle $F$. Then $F \cdot \lambda = 318$, $F \cdot \delta = F \cdot \delta_0 = 2120$. Moreover by varying the choices determining $\mathcal{Y}$, $F$ fills up a divisor $D \subset \overline{\mathcal{M}}_{15}$ with $F \cdot D > 0$. Finally $\mathcal{Y}$ projects birationally to a surface of degree 16 in $\mathbb{P}^3$.

Now let $H \subset \mathbb{P}^3$ be a general plane and $B = p_2^{-1}(H) \subset \mathcal{Y}$, a smooth irreducible curve, and let $\tilde{\mathcal{Y}} \subset B \times \mathbb{P}^3$ be induced from $\mathcal{Y}$ via the base-change $B \to \mathbb{P}^1$, with the corresponding family $\{\tilde{\mathcal{Y}}_b\}$, map $\tilde{f} : B \to \overline{\mathcal{M}}_{15}$ and cycle $\tilde{F} = deg(Y_t) \cdot F = 14F$. Note that $\tilde{\mathcal{Y}} \to B$ has a section, say $s$, and put $\tilde{B} = s(B)$.

Now we will fix $g \geq 17$ and consider pencils of the form $\{X_b\}$ where $X_b = \tilde{Y}_b \cup C$, with $C$ a fixed general curve of genus $g - 15$ attached to $\tilde{Y}_b$ by identifying a fixed general point $c \in C$ with $s(b) \in \tilde{Y}_b$. Note that all the $X_b$'s are automorphism-free. Let $F_g \subset \Delta_{15} \subset \overline{\mathcal{M}}_g$ denote the corresponding pencil. Now it is clear first of all that by varying the various choices involved, $F_g$ fills up a divisor $D_g$ in $\Delta_{15}$, which is just the pullback of $D$ via the natural map $\Delta_{15} \to \overline{\mathcal{M}}_{15}$. Also $F_g \cdot D_g = \tilde{F} \cdot D > 0$.

Now note that $F_g \cdot \lambda = \tilde{F} \cdot \lambda$, $F_g \cdot \Delta_0 = \tilde{F} \cdot \Delta_0$, $F_g \cdot \Delta_i = 0$ for $i \neq 0, 15$, and we next compute $F_g \cdot \Delta_{15}$. As is well known, [7], this equals $-\tilde{B} \cdot \omega$, where $\tilde{\omega} = \omega_{\tilde{\mathcal{Y}}/B}$. Now $B \cdot \tilde{\omega} = B \cdot \omega$ where $\omega = \omega_{\mathcal{Y}/\mathbb{P}^1}$, and we have $B \cdot \omega = B \cdot (K + 2Y_t) = B \cdot K + 28$ where $K$ is the canonical bundle of $\mathcal{Y}$. By the adjunction formula,

$$B \cdot K = 2g(B) - 2 - B^2 = 2g(B) - 18$$

(as $B^2 = deg(im(\mathcal{Y} \to \mathbb{P}^3)) = 16$). Now $g(B)$ may be computed using the formulas of [5], but for our purposes it suffices to estimate it by noting that $B$ maps birationally to a plane curve of degree 16, hence $g(B) \leq \frac{15 \cdot 14}{2} = 105$. Putting everything together, we get $B \cdot \omega \leq 220$.

The theorem for the case $i = 15$ now follows quite formally, as in [1] or [2]: namely, let $A \sim a\lambda - b\delta$ be effective and let $m$ be the multiplicity of $\Delta_{15}$ in $A$. Then we can write $(A - m\Delta_{15}) \cdot \Delta_{15} = dD_g + \sum b_i B_i$ with $B_i$ prime divisors different from $D_g$ and $d$, $b_i \geq 0$. As $F_g \not\subset B_i$ in general, clearly $F_g \cdot B_i \geq 0$ and we have seen $F_g \cdot D_g > 0$. Hence $F_g \cdot (A - m\Delta_{15}) = F_g \cdot ((A - m\Delta_{15}) \cdot \Delta_{15}) \geq 0$, i.e.,

$$F_g \cdot A \geq mF_g \cdot \Delta_{15} \geq -220m.$$

On the other hand,

$$\begin{aligned} F_g \cdot A = aF_g \cdot \lambda - bF_g \cdot \delta &= aF_g \cdot \lambda - bF_g \cdot \delta_0 - bF_g \cdot \delta_{15} \\ &\leq a \cdot 14 \cdot 318 - b \cdot 14 \cdot 2120 + b \cdot 220 \\ &= 4452a - 29460b. \end{aligned}$$

Combining the two inequalities proves the theorem for $i = 15$.

Now the cases $1 \leq i \leq 5$ are similar in principle but much more elementary and are indicated in the table below. In each case we start with a pencil $F_i$ of curves of given $i$, which is a general pencil in a suitable linear system on a suitable rational surface, and construct a pencil $F_{i,g}$ in $\Delta_i$ on $\overline{\mathcal{M}}_g$ by attaching a fixed curve of given $g-i$ along the section corresponding to one of the base points of the pencil $F_i$.

| $i$ | type of pencil $F_i$ | # of base points | $\lambda \cdot F_i$ | $\delta \cdot F_i$ | $\delta \cdot F_{i,g}$ |
|---|---|---|---|---|---|
| 1 | plane cubics | 9 | 1 | 12 | 11 |
| 2 | type (2,3) on quadric | 12 | 2 | 20 | 19 |
| 3 | plane quartics | 16 | 3 | 27 | 26 |
| 4 | type (3,3) on quadric | 18 | 4 | 34 | 33 |
| 5 | quadric sections on del Pezzo quartic | 16 | 5 | 39 | 38 |

Table 2

The values of $\lambda \cdot F_i$, $\delta \cdot F_i$ and $\delta \cdot F_{i,g}$ are easily computed (see the formulas in [1], §3) and moreover we have $\lambda \cdot F_{i,g} = \lambda \cdot F_i$. Using the fact that $F_{i,g}$ fills up $\Delta_i$ on $\overline{\mathcal{M}}_g$, whence $A \cdot F_{i,g} \geq 0$ for any divisor $A$ not containing $\Delta_i$, we can easily complete the proof as above.

Incidentally, the pencils $F_i$ constructed above already yield some bounds on the slopes of effective divisors on $\overline{\mathcal{M}}_i$, namely:

**Corollary 3.** *Let $A \sim a\lambda - b\delta$ be effective on $\overline{\mathcal{M}}_i$, $2 \leq i \leq 5, a, b > 0$. Then we have $a/b \geq 10$ ($i = 2$), $9$ ($i = 3$), $8\frac{1}{2}$ ($i = 4$) or $7\frac{4}{5}$ ($i = 5$).*

Related bounds were obtained by Harris and Morrison [4]: namely, they get the values 10,9,8.42... and 8, respectively. Their bounds are sharp for $i = 2, 4, 5$.

With a little more work, these methods may be pushed a little further to give a bound for genus 6 as well.

**Corollary 4.** *If $A \sim a\lambda - b\delta$ is effective on $\overline{\mathcal{M}}_6$ then $a/b \geq 7\frac{2}{5}$.*

*Proof:* Suppose $a/b < 7\frac{2}{5}$. By looking at the pencils $F_{i,6}$ for $i = 1, 2, 3$ we see that $A$ contains $\Delta_1$, $\Delta_2$ and $\Delta_3$, each with multiplicity $\geq 1$. In addition, we may make a pencil $F_{0,6}$ in $\Delta_0$ by taking the pencil $F_5 \subset \overline{\mathcal{M}}_5$ and gluing together two sections corresponding to base points. It is easy to check that $F_{0,6} \cdot \lambda = F_5 \cdot \lambda = 5, F_{0,6} \cdot \delta = F_5 \cdot \delta - 2 = 37$. Since $a/b < \frac{37}{5}$, we have $F_{0,6} \cdot A < 0$, hence $A$ must contain $\Delta_0$, too. Thus we can write $A = A' + \Delta$, $A'$ effective, and the same argument is applicable to $A'$ in place of $A$. Hence $A$ must contain arbitrarily high multiples of $\Delta$, which is obviously a contradiction.

Corollary 4 again slightly improves on a result of Harris and Morrison [4], who obtain the bound 7.328. . . .

*Remark:* We have now obtained additional lower bounds on $a/b$ in the genera $i = 6, 7, 8, 9$ and 16; these are, respectively: $\frac{47}{6} \sim 7.83, \frac{72}{7} \sim 7.43, 7, 7, 6.56$. Each of these bounds is an improvement on the corresponding Harris-Morrison bound. It follows in particular that $\mathcal{M}_{16}$ has Kodaira dimension $-\infty$. The proof will be presented elsewhere. Also, we hope to be able to prove the bound $a/b \geq 7$ in genus 11, which is sharp. Each of these bounds yields an analogue of the Theorem for the corresponding boundary component.

**Acknowledgements.** Our first—and, so far, only—exposure to string theory came at the conference on "Mathematical Aspects of String Theory," organized in July 1986 by S. T. Yau. We are grateful to Professor Yau for this experience. In addition, we wish to thank R. Treger for showing us a copy of [6] and G. Horowitz and A. Greenspoon for helpful conversations.

## REFERENCES

1. Chang, M. C. and Z. Ran, "The Kodaira dimension of the moduli space of curves of genus 15," *J. Differential Geom.* (in press 1986).

2. ________, "A note on the slope of subvarieties of $\overline{\mathcal{M}}_{15}$ ($6\frac{2}{3} \leq S_{15} \leq 6\frac{3}{4}$)" (preprint).

3. Green, M. B., "Superstrings," *in: Scientific American*, Sept., 1986.

4. Harris, J. and I. Morrison, "Slopes of effective divisors on $\mathcal{M}_g$" (preprint).

5. Harris, J. and L. Tu, "Chern classes of kernel and cokernel bundles," *Invent. Math.* **75**(1984), 467–475.

6. Moore, G., J. Harris, P. Nelson and I. Singer, "Modular forms and the cosmological constant" (preprint).

7. Mumford, D., "Stability of projective varieties," *L'Ens. Math.* **24**(1977), 39–110.

# THE SEVERI PROBLEM: A *Post-Mortem* (?)

Z. Ran*
Department of Mathematics & Comp Sci
University of California
Riverside, CA 92521

Contents:



* Partially supported by NSF.

There is no shortage of surveys on the Severi problem in the literature—see notably the lucid and readable one by Fulton [3], and references therein. Such surveys naturally suffer from the obvious disadvantage of having been written before the problem's ultimate solution for the first time by Harris [5]. I believe however that despite (more likely, because) of this solution this area remains viable and attractive today as a research area. The purpose of this survey, then, is to take full advantage of hindsight in giving an elementary, classically-minded and highly informal account of the problem, some of the classical work on it, as well as an alternative solution of it due to the author [7], concluding with some possible generalizations and open problem.

For including this survey in a String Theory proceedings, I can offer the weak excuse (elaborated somewhat at the end of §1) that an algebraic curve in a Kähler manifold $M$ (such as $\mathbb{P}^2$ ) is automatically a minimal surface, hence forms a "classical particle" in the string theory with spacetime $M$ (no excuse is offered for viewing $\mathbb{P}^2$ as a spacetime). Perhaps more relevant, though, is simply the fact that in the recent past, some methods and results from curve theory have proven useful for string theory, possibly to the point of making curve techniques of general interest to string theorists.

*Acknowledgement.* It is a pleasure to thank S. T. Yau and the Mathematics Department at UCSD for their hospitality during the conference.

## 1. Statement of the problem

Consider the complex projective plane $\mathbb{P}^2$ with homogeneous coordinates $x, y, z$. For any integer $d \geq 1$, we denote by $P_d$ the projectivization of the vector space of homogeneous polynomials of degree $d$ in $x, y, z$. A *plane curve of degree* d is simply the zero-set $C \subset \mathbb{P}^2$ of some nonzero homogeneous polynomial $f$ (if $f$ has multiple factors, the corresponding components of $C$ must be counted with the corresponding multiplicity—we need not worry about such details here). As in the case with most any algebro-geometric situation, there naturally arises the *classification problem*: to classify plane curves of given degree $d$. To get an idea what this might mean, let us have a look at the case $d = 3$, perhaps the first nontrivial one. Here the beginning of the classification is indicated by the following diagram:

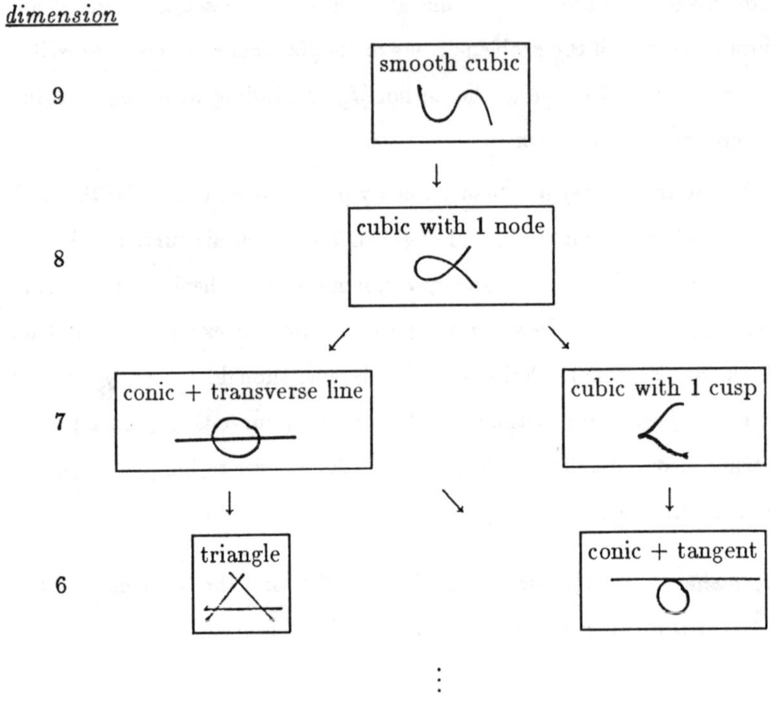


**Fig. 1**

In this diagram each box corresponds to a family of curves parametrized by some irreducible, locally closed subvariety of $P_3$ whose dimension is indicated on the left, and such that the curves in the family, while in general not isomorphic to one another, all "look alike"; in particular they all have irreducible components of the same degrees, and the same number and "type" of singularities such as a node (with local analytic equation $u^2 = v^2$) or cusp (local equation $u^2 = v^3$). The arrows represent (immediate) specialization, i.e., the locus at the end of the arrow is contained (and has codimension 1) in the locus at the beginning. We have, rather arbitrarily, stopped at dimension 6 although a complete classification would not and would be exhaustive in that each curve would belong to (precisely) one box.

This (partial) example suggests rather clearly the form of the general classification problem: namely, this is to describe a *stratification* $P_d = \coprod U_{\alpha,d}$ of the space of curves of degree $d$ by locally closed irreducible algebraic subvarieties, such that the curves in each $U_{\alpha,d}$ look alike in the above sense.

Now it is fairly clear that in this formulation the problem is quite intractable. In fact, even the purely local problem of classifying "singularity types" of plane curves is not completely solved. One is thus led to narrow the problem by restricting the type of singularities allowed. Now the smooth case is of course trivial, and the simplest singularity is a node. So the simplest classification problem for plane curves is that of classifying the *nodal plane curves.*

Let us set up some notation. Denote by $V^{d,g}$ the subset of $P_d$ corresponding to curves with precisely $\delta$ nodes and no other singularities, where $\delta = \frac{(d-1)(d-2)}{2} - g$; and let $V(d,g)$ denote the subset of $V^{d,g}$ corresponding to *irreducible* curves. (Note that $g$ is in fact the (geometric) genus of any curve in $V^{d,g}$, which explains the notation; here the genus of a reducible curve is defined by $g(\bigcup_1^k C_i) = \sum g(C_i) - k + 1$.) Now it is pretty clear that $V^{d,g}$ consists of $V(d,g)$ plus products of other $V(d',g')$'s corresponding to the various ways a curve might split up, so it suffices to deal with $V(d,g)$. The Severi problem asks whether the simplest possible situation holds for $V(d,g)$; namely,

**The Severi problem:** *For* $0 \le g \le \frac{(d-1)(d-2)}{2}$, *is* $V(d,g)$ *irreducible?*

(It is classical—see §2—that in the above range $V(d,g)$ is necessarily nonempty.)

**Theorem 1** (Harris [5]): *The answer to the Severi problem is affirmative for all $d, g$ as above.*

Let us mention an interesting corollary (in fact, a strengthening) of this result.

**Theorem 2:** *Let $R_{d,g}$ denote the space of pairs $(C,f)$ where $C$ is a nonsingular curve of genus $g$ and $f : C \to \mathbb{P}^2$ is a holomorphic map carrying a generator of $H_2(C,\mathbf{Z})$ to $d$ times a generator of $H_2(\mathbb{P}^2,\mathbf{Z})$. Then $R_{d,g}$ is irreducible.*

It is not obvious that Theorem 2 is a consequence of Theorem 1, but this follows from [11] or [1] which show that the set of $(C,f)$ for which $f(C)$ is *not* a nodal curve of degree $d$ is nowhere dense.

Now Theorem 2 may be considered as analogous to some string-theoretic considerations: namely, if $\mathbb{P}^2$ with its Fubini-Study metric is considered as a space-time, then a "classical particle" for the corresponding string theory is just a harmonic map $f : C \to \mathbb{P}^2$ from some Riemann surface. Now it is well known that such an $f$ must be either holomorphic or antiholomorphic. Hence $\bigcup_{d,g} R_{d,g}$ plus its complex conjugate may be identified with the space of all classical particles in $\mathbb{P}^2$.

## 2. Anhang F

The classical work on the Severi problem is usually attributed to Severi's famous "Anhang F," an appendix to his book [9]. It seems to be less generally known, however, that this Anhang F is supposed to be an elaboration of certain paragraphs from the body of Severi's text, and that these make reference to an earlier paper by Enriques. I have not attempted to disentangle just what is due to whom, and shall simply refer to the work of the two by the catch-phrase "Anhang F."

We begin with some quite simple Anhang F facts.

**Proposition 1:** *Let $[C] \in V^{d,g}$; then*

(i) *$V^{d,g}$ is smooth at $[C]$, of dimension $3d+g-1=\frac{(n+1)(n+2)}{2}-\delta-1$;*

(ii) *the tangent space to $V^{d,g}$ at $[C]$ is naturally isomorphic to the space of forms of degree $d$ vanishing at the nodes of $C$, modulo $C$.*

Note that $V^{d,g}$ is, in a local sense, given by "imposing" $\delta$ nodes on a curve, and each node "should" impose one condition, so $\frac{(d+1)(d+2)}{2}-\delta-1$ is the "expected" dimension of $V^{d,g}$. In contemporary deformation-theoretic terminology, one may phrase assertion (i) as saying that nodal plane curves have "unobstructed deformations."

Proposition 1, which incidentally is rather easy to prove, tells us everything about the local structure of $V^{d,g}$. On the other hand, to make any use of it one has to have at least a curve $C$ to start with. But the sad fact, which remains true today, is that for most $d, g$, it is impossible to write down "explicitly" even a *single* curve in $V(d,g)$, say. Thus it is *a priori* unclear, for instance, even whether $V(d,g)$ is nonempty.

To go further, Anhang F introduces a simple but far-reaching idea. Note that, after all, the $V^{d,g}$'s for different $g$'s should be unions of strata of one and the same stratification, so it is natural to expect a relation between the closure of $V^{d,g}$ and the various $V^{d,g'}$, $g'<g$ (namely, $V^{d,g'}\subset\overline{V^{d,g}}$). Then, if one can understand the local structure of $\overline{V^{d,g}}$ along $V^{d,g'}$, one might try to infer from knowledge of $V^{d,g'}$ something about $V^{d,g}$. On the other hand, if one is willing to go far enough down, the $V^{d,g}$'s do eventually become rather harmless; e.g. $V^{d,-d+1}$, the "lowest" one, is just the locus of unions of $d$ lines in general position, about which one knows "everything" (another harmless locus is $V(d,0)$, the irreducible rational curves).

One of Anhang F's main contributions was to show that the local structure of $\overline{V^{d,g}}$ along $V^{d,g'}$ is indeed as nice as possible.

**Proposition 2.2:** *Fix integers $-d+1\le g'<g\le\frac{(d-1)(d-2)}{2}$. Then*

(i) *we have $V^{d,g'}\subset\overline{V^{d,g}}$;*

(ii) *near any* $[C] \in V^{d,g'}$, $\overline{V^{d,g}}$ *is a union of smooth branches whose common intersection is* $V^{d,g'}$;

(iii) *fix* $C \in V^{d,g'}$, *and let* $\Delta$ *be the set of nodes of* $C$; *then there is a natural bijection between the set of subsets* $\Lambda$ *of* $\Delta$ *of cardinality* $g - g'$ *and the set of branches* $B_\Lambda$ *of* $\overline{V^{d,g}}$ *at* $C$; *this correspondence has the property that as* $C$ *deforms to a general curve in the branch* $B_\Lambda$, *the nodes in* $\Lambda$ *smooth out while those in* $\Delta - \Lambda$ *remain nodes;*

(iv) *moreover* $B_\Lambda \subset \overline{V(d,g)}$ *iff* $C \setminus (\Delta - \Lambda)$ *is connected.*

Now this proposition, together with the above-mentioned philosophy concerning its use, already form a fairly powerful combination: for instance, it is now clear that $V(d,g)$ is nonempty whenever $0 \leq g \leq (d-1)(d-2)/2$ (start with $c \in V(d,0)$ (respectively, $V^{d,-d+1}$) and smooth out an arbitrary (respectively, an appropriately chosen) set of $g$ nodes).

The projected application of the above philosophy to the Severi problem is based on the following.

**Proposition 2.3:** $V^{d,-d+1}$ *(respectively,* $V(d,0)$*) is contained in the closure of a unique component of* $V(d,g)$ *for any* $0 \leq g \leq \frac{(d-1)(d-2)}{2}$.

Equivalently, as $C$ moves around $V^{d,-d+1}$ or $V(d,0)$, the monodromy acts transitively on the set of branches of $\overline{V(d,g)}$ at $C$, i.e., by Proposition 2.2, on the set of appropriate (respectively, all) sets of $g$ nodes on $C$. The latter is not hard to prove, either combinatorially for $V^{d,-d+1}$ (cf. [4]) or based on the geometry of secant planes of the rational normal curve (cf. [5]) for $V(d,0)$ (which incidentally also implies the assertion for $V^{d,-d+1}$).

Using Proposition 2.3 and an obvious induction, we are led to the following result, which may be regarded as encapsulating the mathematical content of Anhang F.

**Theorem 2.4:** *The answer to the Severi problem is affirmative provided the following holds: for every component $V$ of $V(d, g)$, the closure $\overline{V}$ meets $V^{d,g'}$ for some $g' < g$.*

In other words, the general member of $V$ can be degenerated to a (possibly reducible) nodal curve with more nodes. Now in the actual Anhang F, Severi does purport to show the existence of such a degeneration. However, the only degeneration that he does in fact produce is one to a "star," i.e., a union of $d$ lines through a point. As we shall demonstrate in the next section, such a degeneration is in general useless.

This then is essentially the way things have stood since Anhang F of 1921 and until Harris' work of 1986, with some exceptions, notably some work by Zariski [11] and, independently, Arbarello and Cornalba [1]. I do not feel competent to comment any further on Harris' work, except to say that it is based on Theorem 2.4 as well as on the aforementioned work of Zariski-Arbarello-Cornalba, and uses curves tangent to higher and higher order to a fixed line to produce the required degeneration.

## 3. A solution

My approach to the Severi problem has as its starting point an idea of Mumford, which has played an essential role in the proof by Harris and Mumford [6] that the moduli space $\mathcal{M}_g$ of curves of genus $g$ is of general type for large $g$. To explain this idea (in a somewhat oversimplified way), fix integers $d, g$, and consider a space

$$H = \{(C, f) : C \text{ nonsingular of genus } g,\ f : C \to \mathbb{P}^1\ d\text{-to-}1\}.$$

We have an obvious map $H \to \mathcal{M}_g$; on the other hand, we have the compactification $\overline{\mathcal{M}_g}$ of $\mathcal{M}_g$ by stable curves. What one needs is a compactification $\overline{H}$ of $H$ to fit in the diagram

$$\begin{array}{ccc} H & \subset & \overline{H} \\ \downarrow & & \downarrow \\ \mathcal{M}_g & \subset & \overline{\mathcal{M}}_g \end{array}.$$

Essentially, the problem is, what should the limit of the map $f$ be, as $C$ degenerates, say to a reducible curve of the form $C_1 \bigcup_p C_2$, i.e., the union of 2 curves joined at a point? Mumford's solution to this problem is simple, elegant and novel: let $\mathbb{P}^1$ itself degenerate to $\mathbb{P}^1_1 \bigcup_q \mathbb{P}^1_2$ (a union of 2 copies of $\mathbb{P}^1$), and $f$ to $(f_1, f_2)$ where $f_i : C_i \to \mathbb{P}^1_i$ and $f_1(p) = f_2(p) = q$.

He then goes on with considerable more work to construct the space $\bar{H}$ (for technical reasons he actually has to work with a somewhat different space than $H$). We remark that Fulton has used this $\overline{H}$ to give a simple proof of the irreducibility of $\mathcal{M}_g$ [4], this being the first purely algebro-geometric proof of this fact. Motivated by Fulton's argument, one is tempted to believe that one could solve the Severi problem by constructing a good compactification of $V(d, g)$, perhaps analogous to $\bar{H}$. In retrospect, however, it seems that partly for reasons having to do with the technical difficulties alluded to above, the latter problem is rather *more* difficult than the Severi problem, and it still remains open. What does turn out to be the case is that all one needs for Severi is a suitable analogue of Mumford's degeneration construction above for maps to $\mathbb{P}^2$ instead of $\mathbb{P}^1$, rather than a full-fledged compactification (the difference between the two is manageable for maps to $\mathbb{P}^1$, but thus far not for $\mathbb{P}^2$).

Over the last couple of years I have been interested in developing the analogue of Mumford's degeneration construction for maps to $\mathbb{P}^r$, $r \geq 2$. The resulting degeneration theory for linear systems turns out to be equally valid for mpas of arbitrary-dimensional varieties to $\mathbb{P}^r$, and has a number of other applications besides the Severi problem. Here however, I will stick to the simplest case, that of maps of curves to $\mathbb{P}^2$.

The first thing to describe is how $\mathbb{P}^2$ can degenerate. This is really quite simple. Let $S(2)$ be the blowup of $\Delta \times \mathbb{P}^2$ at a point $0 \times P$, where $\Delta$ denotes the unit disk, and let $b : S(2) \to \Delta \times \mathbb{P}^2$ be the blowdown map and $\pi : S(2) \to \Delta$ the natural map. Note that for $t \neq 0$, $\pi^{-1}(t) \simeq \mathbb{P}^2$ while $\pi^{-1}(0) =: S(2) = \widetilde{\mathbb{P}}^2_1 \cup \mathbb{P}^2_2$, where $\widetilde{\mathbb{P}}^2_1$ is the blowup of $\mathbb{P}^2$ at $P$ and meets $\mathbb{P}^2_2$ in its exceptional divisor $E$ which is a line on $\mathbb{P}^2_2$. Note for future reference that there is a "dual" blowdown map $b' : S(2) \to \Delta \times \mathbb{P}^2$ which sends $\mathbb{P}^2_2$ isomorphically to $0 \times \mathbb{P}^2$ and each ruling of $\widetilde{\mathbb{P}}^2_1$ to a point. (This realizes $S(2)$ as the blowup of $\Delta \times \mathbb{P}^2$ along $0\times$(a line).)

Now let us put curves in the picture. Suppose we have a family of curves $\{C_t \subset t \times \mathbb{P}^2 : t \in \Delta\}$ of degree $d$, say, and let us consider its limit in $S(2)$ as $t \to 0, t \neq 0$. This limit will in general have the form $C_1 \cup C_2$ where $C_1 \subset \widetilde{\mathbb{P}}^2_1$ is the proper transform of $C_0$, $C_2$ has degree $e = mult_P C_0$ and $C_1 \cap E = C_2 \cap E$. (We will call such a curve $C_1 \cup C_2$ a nice curve of type $(d, e)$ on $S(2)$ and $C_1$ a nice curve of type $(d, e)$ on $\widetilde{\mathbb{P}}^2_1$.) Here $e$ can take any value $0 \leq e \leq d$. In the one extreme $e = 0$, $C_1$ does not pass through $P$ at all; in the other extreme $e = d$, which is in a sense dual to the first, $C_1$ is a star though $P$.

Now note that we could do the following stupid thing: start with a *constant* family $\Delta \times C \subset \Delta \times \mathbb{P}^2$, pull it up to $S(2)$ via the *dual* blowdown $b'$, then push it forward to another copy of $\Delta \times \mathbb{P}^2$ via $b$. We get a family $\{\overline{C}_t \subset t \times \mathbb{P}^2\}$ with all $C_t$, $t \neq 0$ projectively equivalent and $\overline{C}_0$ a star. This shows that a degeneration to a star can be a completely "fake" one, hence such degenerations (e.g. the one constructed by Severi) are, in and of themselves, useless.

From now on we switch for technical reasons from $S(2)$ to $S'(2) = \widetilde{\mathbb{P}}^2_1 \cup \widetilde{\mathbb{P}}^2_2$, the blowup of $S(2)$ at a point on $\mathbb{P}^2_2$, and talk about curves $C_1 \cup C_2$ of type $(d, e, e')$ on $S'(2)$, where $C_1$ has type $(d, e)$ on $\widetilde{\mathbb{P}}^2_1$ and $C_2$ has type $(e, e')$ on $\widetilde{\mathbb{P}}^2_2$ (for $e' = 0$, such curves correspond to curves of type $(d, e)$ on $S(2)$).

Now it is not difficult to set up a complete analogue of the Anhang F theory (Propositions 2.1 and 2.2) for the nice nodal curves on $S'(2)$ and their parameter spaces. And the same can be done for the analogous surface $S'(m) = \widetilde{\mathbb{P}}^2_1 \cup \cdots \cup \widetilde{\mathbb{P}}^2_d$. Note that on $S'(d)$ we have a natural analogue of the "bottom" locus $V^{d,-d+1}$ of $d$ lines in general position: namely, this is the locus $W$ of curves of the form $\bigcup_{1 \le i \le j \le d} M_{ij}$ where $M_{ij}$ for $i < j$ is a ruling on $\widetilde{\mathbb{P}}^2_i$ and $M_{i,i}$ is a line section on $\widetilde{\mathbb{P}}^2_i$.

Note that for any $m' < m$, $S'(m)$ is a degeneration of $S'(m')$, just like $S'(2)$ is a degeneration of $\widetilde{\mathbb{P}}^2 = S'(1)$. Thus we now have a crucial addition to our classical arsenal: not only can we study the nice nodal curves and their families on each $S'(m)$ separately, but we can also study their behavior as $S'(m')$ degenerates to $S'(m)$. In this situation we have an analogue of Proposition 2.2, as well as a natural analogue of Proposition 2.3 for $W$: namely, for any $0 \le g \le \frac{(d-1)(d-2)}{2}$, $W$ is contained in a unique component of the limit of $V(d,g)$ as $\widetilde{\mathbb{P}}^2$ generates to $S'(d)$. (We have identified $V(d,g)$ with a suitable family of curves of type $(d,0)$ on $\widetilde{\mathbb{P}}^2$.)

From this, of course, follows an analogue of Theorem 2.4; namely, that Severi would follow from an appropriate degeneration statement. This statement takes the following form.

**Lemma 3.1.** *Let $V$ be any component of the family of nice nodal curves of genus $g$ and type $(d,e')$ on $\widetilde{\mathbb{P}}^2$ where $e' < d-1$. Then as $\widetilde{\mathbb{P}}^2$ degenerates to $S'(2)$, the limit of $V$ contains a nice nodal curve of type $(d,e,e')$ on $S'(2)$ with $e' < e \le d-1$.*

This lemma is the main lemma of [7]. Without getting into details, we describe the main geometric idea of the proof, which is quite simple. Take $d+e'+2$ general points on $\widetilde{\mathbb{P}}^2$. It is not hard to see that we can find a general curve $C$ in $V$ going through these points. Now specialize these points to $d+1$ general points on $\widetilde{\mathbb{P}}^2_1$, plus $e'+1$ general points on $\widetilde{\mathbb{P}}^2_2$. With some work, one can show that $C$ specializes to a nice nodal curve $C_1 \cup C_2$ on $S'(2)$, which is of type $(d, e, e')$ for some $e' \leq e \leq d$. But if $e = d$ then $C_1$ would just be a sum of $d$ rulings on $\widetilde{\mathbb{P}}^2_1$, hence could not pass through $d+1$ general points. Thus $e < d$. Similarly $e > e'$. QED

Looking back at the proof we have just sketched, we may offer the following philosophical argument for why it "should" be successful: any method of degenerating a plane curve *in* $\mathbb{P}^2$ must necessarily involve looking at curves which are more and more special. As the curve becomes more special, it becomes more difficult to control it and in particular to show that it is still a nodal curve; this is one of the major sources of difficulty in the Severi problem. In our approach however we have specialized or degenerated the ambient $\mathbb{P}^2$ itself to $S(d)$ and simply observed what was happening to the curves on it in the process. In one sense, these curves "specialize" to the curves on $S(d)$; but in another sense, the specialized curves still *are not special at all*: they are quite general in the family on $S(d)$ to which they belong, hence are not difficult to control. Needless to say, this description does gloss over some difficulties. Happily, however, these are manageable.

## 4. Open problems

Once one knows that the Severi varieties $V(d, g)$ are irreducible, one can begin to ask finer questions about them. Perhaps the simplest such question is: what is the degree of $V(d, g)$ (relative to the natural embedding $V(d, g) \subset P_d$). In [8] we give a recursive formula for this degree, but this leaves open

*Problem 1.* Find a closed-form (nonrecursive) formula for the degree of $V(d, g)$.

Such a formula should have some nice applications. Of course in view of [8], Problem 1 reduces to a (albeit rather daunting) purely combinatorial problem; or perhaps a new approach is called for.

*Problem 2.* Construct a good compactification of $V(d, g)$.

Such a compactification would presumably have to encompass the type of degenerations described in §3, which would force it to be fairly complicated.

Once this problem is solved, it would be reasonable to ask about such invariants as the Hodge or Betti numbers of $V(d,g)$ or the Kodaira dimension of the quotient of $V(d,g)$ by $PGL(2)$, in the cases where the map $V(d,g) \to \mathcal{M}_g$ is not dominant (i.e., $d < \frac{2}{3}g + 6$).

Turning to generalizations of the Severi problem, we must first mention the problem of curves with singularities worse than nodes, e.g. cusps. Zariski's famous example (see [10]) of two distinct families of sextics with 6 cusps shows that the naive analogue of the Severi problem is false here.

We turn next to generalizations of the Severi problem to other surfaces. If $S$ is an algebraic surface and $L$ a line bundle on $S$, we may consider the variety $V(L,g)$ of irreducible nodal curves of genus $g$ in the complete linear system $|L|$.

*Problem 3.* Is $V(L,g)$ irreducible?

It seems reasonable to expect results in case $S$ has Kodaira dimension $\leq 0$ and $L$ is, say, very ample. The methods of [7] seem, in principle, capable of tackling this problem, at least in some cases. For instance, for $S = \widetilde{\mathbb{P}}^2$ and $|L|$ the system of curves of type $(d,e)$, an affirmative answer is already contained implicitly in [7].

In a different direction, one might consider irreducibility problems for families of surfaces in $\mathbb{P}^3$ of given degree and (generic) singularity configuration (i.e., degree and genus of the double curve, and number of triple points and pinch points). In yet another direction, the problem of classifying space curves may also be considered a generalization of the Severi problem, and the above degeneration methods may be of use here.

## References


1. Arbarello, E. and M. Cornalba, "Su una proprieta notevole dei morfismi di una curva a moduli generali in uno spazio proiettivo," *Rend. Semin. Mat. Torino* **38**(1980), 87–99.

2. Enriques, F., "Sui moduli d'una classe di superficie e sul teorema d'esistenza per funzioni algebriche di due variabili," *Atti Accad. Sci. Torino* **47**(1912).

3. Fulton, W., "On the irreducibility of $\mathcal{M}_g$" (appendix to [6]).

4. _______, "On nodal curves," *in: Algebraic Geometry: open problems*, LNM #997, Springer-Verlag, New York.

5. Harris, J., "On the Severi problem," *Invent. Math.* **84**(1986), 445–461.

6. Harris, J. and D. Mumford, "On the Kodaira dimension of the moduli space of curves," *Invent. Math.* **67**(1982), 23–86.

7. Ran, Z., "On nodal plane curves," *Invent. Math.* (in press, 1986).

8. _______, "The degree of the Severi variety," preprint.

9. Severi, F., *Vorlesungen über Algebraische Geometrie*, Teubner, Leipzig, 1921.

10. Zariski, O., *Algebraic Surfaces, 2d ed.* Springer-Verlag, Heidelberg, 1971.

11. _______, "Algebraic systems of plane curves," *Amer. J. Math.* **104**(1982), 209–226.

# A NOTE ON THE SLOPE OF SUBVARIETIES OF $\overline{\mathcal{M}}_{15}$ ($6\frac{2}{3} \leq S_{15} \leq 6\frac{3}{4}$)

M. C. Chang*
Department of Mathematics
University of South Carolina
Columbia, SC 29208

Z. Ran*
Department of Mathematics & Comp Sci
University of California
Riverside, CA 92521

ABSTRACT

*It is shown that any effective divisor on* $\overline{\mathcal{M}}_{15}$ *has slope* $\geq 6\frac{2}{3}$.

During the conference, we learned that physicists have an interest in lower bounds on the slopes of effective divisors on the moduli space $\overline{\mathcal{M}}_g$ of stable curves of genus $g$ (see e.g. [5] where such bounds are related to the vanishing of the "string cosmological constant"). Subsequently we realized that for genus 15 such a bound, which cannot be very far from sharp, was in fact implicitly contained in our paper [1]. The purpose of this short note, then, is simply to point out this bound, and in the process to formalize a couple of notions naturally suggested by the proof.

Let $X$ be a $\mathbb{Q}$-factorial projective variety. Recall that this means that for any divisor $D$ on $X$, some multiple $aD$ of $D$, $a > 0$, comes from some line bundle $L$ on $X$, and that it allows us to define an *intersection number* $D \cdot F$ for $D$ a divisor and $F$ a 1-cycle (i.e., a formal linear combination of curves): namely for $F$ a curve define $D \cdot F = \frac{1}{a} L \cdot F = \frac{1}{a} deg(L|_F)$, and extend by linearity. We will say that a 1-cycle $F$ is *geometrically numerically effective* (*g-nef* for short) if the following condition

* Partially supported by NSF.

is satisfied: there exists an irreducible algebraic family $\{F_t : t \in T\}$ of irreducible curves on $X$ with $F_t$ numerically equivalent to some multiple $aF$, $a > 0$, such that $D = \overline{\bigcup_{t\in T} F_t}$ is a divisor with $F \cdot D \geq 0$. Recall that $F$ is said to be *numerically effective* or *nef* if $F \cdot D \geq 0$ for any effective divisor $D$.

The following fact is nearly obvious.

**Proposition 1.** *Any g-nef cycle is nef.*

*Proof:* Using notations as above, let $A$ be an arbitrary effective divisor on $X$ and write $A = dD + \sum b_i B_i$ with $B_i$ prime divisors different from $D$ and $d, b_i \geq 0$. As $(\cup B_i) \cap D \underset{\neq}{\subset} D$, we have $F_t \not\subset \cup B_i$ for general $t$, so that $F_t \cdot B_i \geq 0$ for all $i$; hence

$$aF \cdot A = adF \cdot D + \textstyle\sum b_i B_i \cdot F_t \geq \sum b_i B_i \cdot F_t \geq 0,$$

whence $F$ is nef.

[*Question:* Is the converse true?]

*Remark:* Note, on the other hand, the following property which is stronger than g-nef: that $F$ is numerically equivalent to $\sum a_i F_i$ such that for each $i$, $a_i > 0$ and $F_i$ is an irreducible curve passing through a generic point of $X$.

We now specialize to the case $X = \overline{\mathcal{M}}_g$, which is well known to be $\mathbb{Q}$-factorial. Let $D$ be any effective divisor on $\overline{\mathcal{M}}_g$, and assume $D \sim a\lambda - b\delta$, where $\lambda$ is the Hodge line bundle, $\delta = \sum_{i=0}^{[g/2]} \delta_i$ is the divisor class of the boundary $\Delta$ and $a, b > 0$. When $a, b$ are integers, such divisors $D$ are in 1-1 correspondence with nonzero modular forms of weight $a$, i.e., sections of $a\lambda$, vanishing to order $\geq b$ on $\Delta$, given up to scalars. In physics, such modular forms sometimes arise out of factors of the string integrand (cf. [5]). The *slope* of $D$ (or of the corresponding modular form) is by definition

$$S(D) = \frac{a}{b},$$

and the *slope* (or *big slope*) of $\overline{\mathcal{M}}_g$ is $S_g = \inf_D S(D)$. Dually, define the slope of a nef 1-cycle $F$ as

$$s(F) = \frac{F \cdot \delta}{F \cdot \lambda}$$

provided both numerator and denominator are positive, and the *little slope* of $\overline{\mathcal{M}}_g$ as $s_g = \sup_F s(F)$. This terminology is justified by the following:

**Proposition 2.** *We have* $s_g \le S_g$.

*Proof:* If $S(D)$ and $s(F)$ are both defined, we have

$$0 \le D \cdot F = (a\lambda - b\delta)F = b(S(D)\lambda - \delta) \cdot F = bF \cdot \delta(S(D)\tfrac{\lambda \cdot F}{\delta \cdot F} - 1) = bF \cdot \delta(\tfrac{S(D)}{s(F)} - 1),$$

hence $s(F) \le S(D)$, as claimed.

Specializing to the case $g = 15$, we now come to the main point of this note.

**Theorem 3.** *We have* $6\frac{2}{3} \le s_{15} \le S_{15} \le 6\frac{3}{4}$.

*Proof:* The inequality $S_{15} \le 6\frac{3}{4}$ (more generally, $S_g \le 6 + \frac{12}{g+1}$ for all odd $g$) was established by Harris and Mumford [4] (later extended to all $g$ with $g+1$ composite by Eisenbud and Harris [2]). On the other hand, Corollary 3.1 of [1] exhibits a g-nef, hence nef, curve $F$ in $\overline{\mathcal{M}}_{15}$ of slope $\frac{2120}{318} = 6\frac{2}{3}$, whence $s_{15} \ge 6\frac{2}{3}$.

**Corollary 4.** *On* $\overline{\mathcal{M}}_{15}$ *there are no modular forms of slope* $< 6\frac{2}{3}$.

*Remarks:*

1. In [4], the absence of modular forms of slope 4 for genus $\le 20$ was used to establish the vanishing of the first 20 terms in the perturbation theory series for the cosmological constant of the heterotic string theory.
2. Since any multiple of the canonical divisor $K$ of $\mathcal{M}_{15}$ has slope $6\frac{1}{2}$, the inequality $S_{15} \ge 6\frac{2}{3}$ clearly implies that no multiple $mK$, $m > 0$ is effective, i.e., that $\mathcal{M}_{15}$ has Kodaira dimension $-\infty$, a result established in [1].

3. In [3], it is shown, among other things, that $S_2 = 10$, $S_3 = 9$, $S_5 = 8$. Excepting these values, Theorem 3 yields, to our knowledge, the most accurate estimation to date of any $S_g$.

## REFERENCES


1. Chang, M. C. and Z. Ran, "The Kodaira dimension of the moduli space of curves of genus 15," *J. Differential Geometry* (in press 1986).
2. Eisenbud, D. and J. Harris, "On the Kodaira dimension of the moduli space of curves of genus $\geq 23$" (preprint).
3. Harris, J. and I. Morrison, "Slopes of effective divisors on the moduli space of stable curves" (preprint).
4. Harris, J. and D. Mumford, "On the Kodaira dimension of the moduli space of curves," *Invent. Math.* **67**(1982), 23–86.
5. Moore, G., J. Harris, P. Nelson and I. Singer, "Modular forms and the cosmological constant" (preprint).

# Report on arithmetic intersections

Gerd Faltings
Department of Mathematics
Princeton University
Princeton,NJ 08544

We intend to give an overview of various results on arithmetic intersection-theory. The topic has been started by Néron and Arakelov. By trying to use these methods in the context of the Mordell-conjecture I obtained some results on arithmetic surfaces. These were generalised by A.N.Beilinson, H.Gillet and C.Soulé . For more details the reader should consult their publications ([B],[GS]). This work was partially supported by the NSF, grant DMS-8502316.

1. Arakelov's intersection-theory is an attempt to generalise the classical intersection-theory, on algebraic surfaces, to arithmetic situations. Classically we define the intersection-number of two divisors D and E on a smooth complete surface X by the rule $\langle D,E\rangle$ = number of points (with multiplicity) in $D\cap E$. If X is fibered over a curve B, by $f: X \to B$, we can decompose this as $\langle D,E\rangle = \sum \langle D,E\rangle_b$ , where the some is over all points $b \in B$, and the individual terms $\langle D,E\rangle_b$ measure the number of points in the intersection which map to b.

It is known that algebraic number-fields behave similar to function-fields. So we can hope to have an analogue of the theory above, for curves over a number-field K, which extend to arithmetic surfaces $X \to B = \mathrm{Spec}(\mathcal{O}_K)$ over the ring of integers of K. The points b should correspond to the different places of K, and is rather obvious how to define local multiplicities $\langle D,E\rangle_b$ for the finite places.

It remains to study the infinite ones: Here the local multiplicity $\langle D,E\rangle_b$ should only depend on the analytic structure of the situation, not on the fact that everything can be defined over a number-field. In other words we start with a complex Riemann-surface X, together with two divisors D and E, and want to define an intersection-number $\langle D,E\rangle_\infty$. By bilinearity it suffices to treat the case of divisors of degree 1, given by points $p,q \in X$. By studying the case of non-archimedean local fields we may guess that the intersection multiplicity $\langle D,E\rangle_\infty \in \mathbb{R}$ should grow like $-\log(\mathrm{distance}(p,q))$ if p approaches q. If we define $g(p,q) = -\langle D,E\rangle_\infty$ and and $G(p,q) = \exp(g(p,q))$, it follows that $G(p,q)$ defines a metric of the line-bundle $\mathcal{O}(\Delta)$ ($\Delta$ = diagonal) on $X\times X$, by the rule $G(p,q)$ = norm at $(p,q)$ of the global section $1 \in \Gamma(X\times X,\mathcal{O}(\Delta))$. Up to a constant factor such a metric is determined by its curvature-form $\Omega = (1/i\pi)d'd''(g)$, which can be any closed real 2-form on $X\times X$ which represents the cohomology-class of $\Delta$. The most reasonable choice seems to use a harmonic representative.

For this we need a hermitian metric in X. Assume that X has genus $g>0$, and choose an ON-basis $\omega_1,\ldots,\omega_g$ of the space of holomorphic 1-forms on X. They define a hermitian metric in X, with the volume given by $d\mu = (1/2\pi i g)\sum \omega_j \wedge \overline{\omega}_j$. In this metric a basis of the complex harmonic 2-forms on X×X is given by

$pr_1^*(\omega_j)\wedge pr_2^*(\omega_k)$, $pr_1^*(\overline{\omega}_j)\wedge pr_2^*(\overline{\omega}_k)$, $1\leqslant j,k\leqslant g$,

$pr_1^*(\omega_j)\wedge pr_2^*(\overline{\omega}_k)$, $pr_1^*(\overline{\omega}_j)\wedge pr_2^*(\omega_k)$, $1\leqslant j,k\leqslant g$,

and $pr_1^*(\mu), pr_2^*(d\mu)$.

The class of $\Delta$ is characterised by the fact that its product with the first basis-elements vanishes, while for the others we obtain $\delta_{jk}$, respectively 1. It follows that it is given, up to some $2\pi i$, by

$\sum_j \{pr_1^*(\omega_j)\wedge pr_2^*(\overline{\omega}_j)+pr_1^*(\overline{\omega}_j)\wedge pr_2^*(\omega_j)\} + pr_1^*(\mu) + pr_2^*(\mu)$.

Hence we obtain a function $g(p,q)$ on $X\times X-\Delta$, with $d'd''(g)$ equal to the expression above, multiplied with $(i\pi)$. g is a Green's function for the Laplacian associated to the chosen hermitian metric on X. We can normalise g by requiring that $\int_X g(p,q)d\mu(q) = 1$ for any $p\in X$ (one checks that the right-hand side is harmonic, hence constant). Because of symmetry we also have $g(p,q) = g(q,p)$.

2. In classical algebraic geometry one usually redefines intersection-theory in terms of line-bundles. Then the intersection-number $\langle D,E\rangle$ is given by $\text{degree}(\mathcal{O}(D)|E) = \text{degree}(\mathcal{O}(E)|D)$. This definition has the advantage that it makes sense even if D and E have common components. In our case the analogue would be line-bundles $\mathcal{L}$ on the Riemann-surface X, together with a hermitian metric whose curvature is a constant multiple of our fixed volume $d\mu$. The multiple is determined by the degree of the line-bundle, as the curvature must represent the first Chern-class of $\mathcal{L}$. An example is the line-bundle $\mathcal{O}(p)$, for some $p\in X$, with the hermitian metric given by $G(p,q)$ = norm of 1 at q. By linearity we can extend this to define a hermitian metric on $\mathcal{O}(D)$, for any divisor D on X.

The next problem is to define good hermitian metrics on the cohomology $H^*(X,\mathcal{L})$. There are of course natural choices, i.e. square integration of harmonic forms. However one finds (by computing examples) that these metrics do not allows us to prove the kind of results we want, especially a Riemann-Roch theorem for arithmetic surfaces. It is however possible to define good volume-forms, not on $H^0$ and $H^1$ separately, but on their formal difference. More precisely define $\lambda(V) = \Lambda^{\dim(V)}V$ for any vectorspace V, and $\lambda(R\Gamma(X,\mathcal{L})) = \lambda(H^0(X,\mathcal{L}))\otimes\lambda(H^1(X,\mathcal{L}))^{-1}$. We intend to define a hermitian metric on this one-dimensional vector-space, such that for any $p\in X$ the natural isomorphism (derived from the long exact sequence $0 \to \mathcal{L}(-p) \to \mathcal{L} \to \mathcal{L}[p] \to 0$)

$\lambda(R\Gamma(X,\mathcal{L})) \cong \lambda(R\Gamma(X,\mathcal{L}(-p)))\otimes\mathcal{L}[p]$

is an isometry. Furthermore we require that scaling the metric on $\mathcal{L}$ by a factor $\lambda$ should introduce a factor $\lambda^{\deg(\mathcal{L})+1-g}$ on the metric on $\lambda(R\Gamma(X,\mathcal{L}))$.

It follows easily that this defines a metric on $\lambda(R\Gamma(X,\mathcal{L}))$ for each choice of an isomorphism (which is automatically an isometry up to a constant factor) $\mathcal{L} \cong \mathcal{O}(D)$, and we have to show that the metric is actually independant of this additional choice. For this one reduces easily to the case of line-bundles of degree g-1. Here the metric on $\lambda(R\Gamma(X,\mathcal{L}))$ is actually invariant under scaling, and we have to show that for two divisors D and E, of degree g-1, any isomorphism $\mathcal{O}(D) \cong \mathcal{O}(E)$ induces an isometry $\lambda(R\Gamma(X,\mathcal{O}(D))) \cong \lambda(R\Gamma(X,\mathcal{O}(E)))$.

Both D and E can be written as difference of positive divisors of degree r and $s = r+1-g$, for some big r. Such divisors are parametrised by $X^{r+s}$. On the other hand line-bundles of degree g-1 are parametrised by the jacobian J of X, an abelian variety of dimension g. In J we have a divisor $\Theta$, the theta-divisor, which classifies line-bundles $\mathcal{L}$ of degree g-1 for which $H^0(X,\mathcal{L}) \neq 0$. The vector-spaces $\lambda(R\Gamma(X,\mathcal{L}))$ define a line-bundle on J which is trivial outside $\Theta$, and one shows that it is in fact isomorphic to $\mathcal{O}(-\Theta)$.

Associating to each D the isomorphism-class of $\mathcal{O}(D)$ defines a map $X^{r+s} \to J$. Furthermore the pullback of $\mathcal{O}(-\Theta)$ under this map has a hermitian metric, as it is isomorphic to the line-bundle made up from the $\lambda(R\Gamma(X,\mathcal{O}(D)))$, and for those we have defined metrics. A nice and pleasant computation (see [F]) of the curvature of this metric shows immediately that this 2-form is pullback of the curvature (on J) of the canonical metric on $\mathcal{O}(-\Theta)$. As $X^{r+s}$ is irreducible, the metric itself must be pullback of a constant multiple of the canonical metric. It is immediate that this implies the required invariance under isomorphisms for the metrics on $\lambda(R\Gamma(X,\mathcal{O}(D)))$.

Once we have this key-fact it is very easy to prove a Riemann-Roch theorem for arithmetic surfaces.

3. So far the metrics on $\lambda(R\Gamma(X,\mathcal{L}))$ have been defined only up to a common factor. One can determine this factor by fixing the metric for one $\mathcal{L}$, for example for $\mathcal{L} = \Omega_X$, the sheaf of relative differentials. Here we can use the natural metric on $H^0(X,\Omega_X)$, together with the triviality of $H^1(X,\Omega_X)$, to define a metric. We obtain a metric on $\mathcal{O}(-\Theta)$, which differs from the standard-metric by a factor depending on the curve X. This way we obtain interesting examples of functions on the moduli-space of curves.

4. For higher-dimensional varieties the approach has been generalised by H.Gillet and C.Soulé. They also can define a direct image for product

maps. This involves methods from the index-theorem, and so is rather involved.


| | | |
|---|---|---|
| [B] | A.Beilinson | Height Pairings Between Algebraic Cycles<br>preprint (1985) |
| [F] | G.Faltings | Calculus on Arithmetic Surfaces<br>Ann. of Math. **119** (1984), 387-424 |
| [FW] | G.Faltings, G.Wüstholz | Rational Points<br>Vieweg, Braunschweig 1984 |
| [GS] | H.Gillet, C.Soulé | Intersection sur les variétés d'Arakelov<br>C.R.A.S. Paris **299** (1984), 563-566 |

# A DEFORMATION THEORY FOR COHOMOLOGY OF ANALYTIC VECTOR BUNDLES ON KÄHLER MANIFOLDS, WITH APPLICATIONS

Mark Green* and Robert Lazarsfeld**
Department of Mathematics
University of California
Los Angeles, CA 90024

* Research partially supported by N.S.F. Grant DMS 85-02350.

** Research partially supported by a Sloan Fellowship and N.S.F. Grant DMS 86-03175.

Introduction

In our paper [GL], we developed a deformation theory for the cohomology groups $H^i(M, E)$ for a variable analytic vector bundle $E$ on a fixed Kähler manifold $M$. The crucial ingredient in that theory is the cohomology at the i'th stage of the complex

$$0 \to H^0(M, E) \to H^1(M, E) \to H^2(M, E) \to \dots \to H^m(M, E) \to 0 \qquad (0.1)$$

obtained by cup product with a deformation class

$$\phi \in H^1(M, \mathrm{Hom}(E, E)) .$$

In this paper, we will show that this is a special case of a more general deformation theory for smoothly varying elliptic complexes of differential operators. This theory uses only elementary standard techniques from the theory of elliptic operators (not coincidentally, these are the only ones we know), reassembled in a new algebraic package. Applied to the Dolbeault complex, one recovers our previous results. It should be interesting to see what happens with other families of elliptic complexes.

We have also used this paper as an opportunity to recast our previous work in a $C^\infty$ and complex-analytic framework. For example, we give an analytic proof of the fundamental computation showing that the maps in (0.1) are indeed as described above, which we intend as a friendly gesture to readers who may feel a certain *anomie* in the presence of $\delta$-functors.

It is easiest to explain our basic construction in the context of a variable complex

$$0 \to V^1 \xrightarrow{A^1(t)} V^2 \xrightarrow{A^2(t)} \dots V^{n-1} \xrightarrow{A^{n-1}(t)} V^n \to 0 \qquad (0.2)$$

where the $V^i$ are finite dimensional vector spaces (although in the context we are interested in, the $V^i$ are infinite dimensional). One may think of the $A^i(t)$ as matrices with $C^\infty$ entries with $t$ varying over some open subset $U \subseteq \mathbb{R}^k$, satisfying

$$A^i(t) \circ A^{i-1}(t) = 0 \qquad \text{for all} \quad t \in U,\ i = 1,\ldots,n \quad . \qquad (0.3)$$

Let

$$H_t^i = \frac{\ker A^i(t)}{\operatorname{im} A^{i-1}(t)} \quad .$$

We are interested in studying the <u>cohomological support loci</u>

$$S^i = \{t \in U | H_t^i \neq 0\}$$

$$S_m^i = \{t \in U | \dim H_t^i \geq m\} \quad . \qquad (0.4)$$

Say $0 \in U$ and $v \in T_0(U)$. Then differentiating (0.3) we obtain

$$\frac{\partial A^i}{\partial v}(0) \circ A^{i-1}(0) + A^i(0) \circ \frac{\partial A^{i-1}}{\partial v}(0) = 0$$

and thus the restriction of $\frac{\partial A^i}{\partial v}(0)$ to $\ker A^i(0)$ has image contained in $\ker A^{i+1}(0)$ and it carries $\operatorname{im} A^{i-1}(0)$ to a subspace of $\operatorname{im} A^i(0)$. We thus obtain a linear map

$$d^i(v) : H_0^i \to H_0^{i+1}$$

defined by the commutative diagram

$$\begin{array}{ccc} \ker A^i(0) & \xrightarrow{\frac{\partial A^i}{\partial v}(0)} & \ker A^{i+1}(0) \\ \downarrow & & \downarrow \\ H_0^i & \xrightarrow{d^i(v)} & H_0^{i+1} \end{array} \qquad (0.5)$$

We can differentiate (0.3) twice to obtain the relation

$$\frac{\partial^2 A^i}{\partial v^2}(0) \circ A^{i-1}(0) + 2\,\frac{\partial A^i}{\partial v}(0) \circ \frac{\partial A^{i-1}}{\partial v}(0) + A^i(0) \circ \frac{\partial^2 A^{i-1}}{\partial v^2}(0) = 0$$

which implies that

$$d^i(v) \circ d^{i-1}(v) = 0 \quad .$$

Thus we obtain a complex $D^\bullet(v)$

$$0 \to H_0^1 \xrightarrow{d^1(v)} H_0^2 \xrightarrow{d^2(v)} \cdots \to H_0^{n-1} \xrightarrow{d^{n-1}(v)} H_0^n \to 0 \qquad (0.6)$$

which we call the <u>derivative complex</u> of $V^\bullet$ in the direction $v$.

The cohomology groups

$$H^i(D^\bullet(v)) = \frac{\ker d^i(v)}{\mathrm{im}\, d^{i-1}(v)}$$

play an important role in the deformation theory of $H_t^i$. For example, for any $i \geq 1$ and $m \geq 0$, one has:

If $\dim H^i_{t^\nu} \geq m$ for a sequence of distinct points $t^\nu$ converging to $0$, then $\dim H^i(D^\bullet(v)) \geq m$ for some non-zero $v \in T_0(U)$. (0.7)

This has as an easy corollary that:

If $H^i(D^\bullet(v)) = 0$ for all non-zero $v \in T_0(U)$, then $H_t^i = 0$ for all $t$ in some punctured neighborhood of $0$. (0.8)

If the vector spaces $V^i$ are infinite-dimensional, these results need not hold. In the case of a smoothly varying elliptic complex of differential operators

$$0 \rightarrow C^\infty(M, E^1) \xrightarrow{L_t^1} C^\infty(M, E^2) \xrightarrow{L_t^2} \cdots \xrightarrow{L_t^{n-1}} C^\infty(M, E^n) \rightarrow 0 \qquad (0.9)$$

where $M$ is a compact orientable manifold and the $E^i$ are $C^\infty$ vector bundles on $M$, one can recover (0.7) and (0.8) by using standard elliptic operator techniques. If $M$ and the $E^i$ are given metrics, and $L_t^{i*}$ denotes the adjoint of $L_t^i$, the Laplacians of the complex are

$$\Delta_t^i = L_t^{i*} L_t^i + L_t^{i-1} L_t^{i-1*} \qquad (0.10)$$

and the cohomology spaces

$$H_t^i = \frac{\ker L_t^i}{\operatorname{im} L_t^{i-1}}$$

are isomorphic to the harmonic spaces

$$H_t^i = \ker \Delta_t^i \quad .$$

If $t$ varies in a manifold $X$ and $0 \in X$, $v \in T_0(X)$, then as in the finite dimensional case we obtain a derivative complex $D^\bullet(v)$

$$0 \rightarrow H_0^1 \xrightarrow{d^1(v)} H_0^2 \xrightarrow{d^2(v)} \cdots \xrightarrow{d^{n-1}(v)} H_0^n \rightarrow 0 \qquad (0.11)$$

and (0.7) and (0.8) hold. An interesting feature of the situation is that the first order deformations of $\Delta_0^i$ give no information about the $H_t^i$; it is the second order information on $\Delta_t^i$ contained in the "Laplacian" of the complex (0.11)

$$d^i(v)^* d^i(v) + d^{i-1}(v) d^{i-1}(v)^* \qquad (0.12)$$

which comes into play.

The special case of (0.9) in which we are interested is a variable Dolbeault complex

$$0 \to A^{0,0}(M, E_t) \xrightarrow{\bar{\partial}_t} A^{0,1}(M, E_t) \xrightarrow{\bar{\partial}_t} \cdots \xrightarrow{\bar{\partial}_t} A^{0,n}(M, E_t) \to 0 \qquad (0.13)$$

where $M$ is a compact complex manifold and $E_t$ is an analytic family of analytic vector bundles on $M$ parametrized by $t$ in a complex manifold $X$.

Here

$$H_t^i = H^i(M, E_t) \quad .$$

Associated to such a family there is a deformation map at $0 \in X$,

$$T_0(X) \xrightarrow{\delta} H^1(M, \mathrm{Hom}(E_0, E_0)) \quad . \qquad (0.14)$$

Given $\phi \in H^1(M, \mathrm{Hom}(E_0, E_0))$, there is a natural map $\wedge\phi$ defined by the composition

$$\begin{array}{ccc} H^i(M, E_0) & \xrightarrow{U\phi} & H^{i+1}(M, E_0 \otimes \mathrm{Hom}(E_0, E_0)) \\ & {\scriptstyle \wedge\phi} \searrow & \downarrow \\ & & H^{i+1}(M, E_0) \end{array} \qquad (0.15)$$

where the vertical map is induced by the contraction map

$$E_0 \otimes \mathrm{Hom}(E_0, E_0) \to E_0$$

$$w \otimes \lambda \to \lambda(w) \quad .$$

The essential computation to make is that

$$d^i(v) = \wedge\delta(v) \quad . \qquad (0.16)$$

We give a proof of this in §2, different from the one we gave in [GL]. This done, one is reduced to studying the complex

$$0 \to H^0(E_0) \xrightarrow{\wedge\phi} H^1(E_0) \xrightarrow{\wedge\phi} H^2(E_0) \to \cdots \xrightarrow{\wedge\phi} H^n(E_0) \to 0 \qquad (0.17)$$

where $\phi \in H^1(M, \mathrm{Hom}(E_0, E_0))$. This complex is particularly tractable if $M$ is Kähler and $E_0$ is a unitary flat bundle. In this case, there is a Hodge theory for such bundles. If we take harmonic representatives and conjugate, we obtain the equivalent complex

$$0 \to H^0(E_0^*) \xrightarrow{\wedge\omega} H^0(\Omega_M^1 \otimes E_0^*) \xrightarrow{\wedge\omega} \cdots \xrightarrow{\wedge\omega} H^0(\Omega_M^n \otimes E_0^*) \to 0 \qquad (0.18)$$

where

$$\omega \in H^0(\Omega_M^1 \otimes \mathrm{Hom}(E_0, E_0)) \quad .$$

Let us now restrict attention to topologically trivial line bundles $E_0 \in \mathrm{Pic}^0(M)$. One can show directly that

(a) If $\beta \in H^0(\Omega_M^i \otimes E_0^*)$, $\beta \neq 0$, then $\{\omega \in H^0(\Omega_M^1) \mid \beta \wedge \omega = 0\}$ has codimension $\geq \dim a(M) - i$, where $M \xrightarrow{a} \mathrm{Alb}(M)$ is the Albanese map of $M$. (0.19)

(b) If $i$ is less that the codimension of $Z(\omega) = \{p \in M \mid \omega(p) = 0\}$ in $M$, for some $\omega \in H^0(\Omega_m^1)$, then the complex (0.18) is exact at $H^0(\Omega_M^i \otimes E_0^*)$. (0.20)

Statement (0.19) is elementary, while (0.20)(which in fact holds for $H^j(\Omega_M^i \otimes E_0^*)$ if $i+j < \mathrm{codim}\, Z(\omega)$) follows from a spectral sequence argument in the spirit of Carrell-Lieberman [CL]. Applying results along the lines of (0.7) to the sequence (0.18), one readily obtains from these:

(a) $\{L \in \mathrm{Pic}^0(M) \mid h^i(L) > 0\}$ has codimension $\geq \dim a(M) - i$ in $\mathrm{Pic}^0(M)$ (0.21)

(b) For a generic $L \in \mathrm{Pic}^0(M)$, $H^q(M, \Omega^p(L)) = 0$ if

$$p + q < \max_{\omega \in H^0(\Omega^1_M)} \{\mathrm{codim}\ Z(\omega)\} \tag{0.22}$$

Looking at the case where $M$ is a compact Kähler surface and $L$ is the trivial bundle, the sequence (0.18) is

$$0 \to H^0(\mathcal{O}_M) \xrightarrow{\wedge\omega} H^0(\Omega^1_M) \xrightarrow{\wedge\omega} H^0(\Omega^2_M) \to 0 \quad . \tag{0.23}$$

By the well-known lemma of Castelnuovo on wedge products of two 1-forms (see [B]), (0.23) is exact at the middle term unless $\omega$ pulls back from a smooth curve $C$ of genus $\geq 2$ via an analytic map $M \xrightarrow{B} C$. Such a map is called an irrational pencil of genus $\geq 2$. By (0.7), we obtain:

If $M$ is a compact Kähler surface having no irrational pencil of genus $\geq 2$, the trivial bundle is an isolated point of $\{L \in \mathrm{Pic}^0(M) \mid h^1(L) > 0\}$. (0.24)

This in turn implies a classical conjecture of Enriques on the <u>paracanonical system</u> of such a surface:

On a smooth algebraic surface $M$ with no irrational pencil of genus $\geq 2$, any irreducible effectively parametrized algebraic family of curves $\{C_t\}_{t \in X}$ on $M$ with $C_0 \in |K_m|$ has $\dim X \leq h^{2,0}(M)$. (0.25)

We would like to thank our colleagues S.Y. Cheng and Jim Ralston for answering patiently our questions on perturbations of elliptic operators.

## §1 Deformations of Elliptic Complexes

Let $M$ be a compact oriented $C^\infty$ manifold equipped with a $C^\infty$ volume form. Let $(E^\bullet, L_t^\bullet)$ be a smoothly varying elliptic complex of differential operators on $M$ parametrized by $t$ in a $C^\infty$ manifold $X$, i.e. we have $C^\infty$ Riemannian vector bundles $E^i \to M$ and a sequence of differential operators

$$0 \to C^\infty(M, E^1) \xrightarrow{L^1_t} C^\infty(M, E^2) \xrightarrow{L^2_t} \cdots \xrightarrow{L^{n-1}_t} C^\infty(M, E^n) \to 0 \tag{1.1}$$

satisfying

$$\left.\begin{array}{l} L^i_t \circ L^{i-1}_t = 0 \\ \\ \text{order } L^i_t = k \end{array}\right\} \begin{array}{l} i = 1,\dots n \\ \text{for all } i \text{ and all } t \in X \end{array} . \tag{1.2}$$

Let $L^{i*}_t$ denote the adjoint of $L^i_t$ and let

$$\Delta^i_t = L^{i*}_t L^i_t + L^{i-1}_t L^{i-1*}_t \tag{1.3}$$

be the Laplacians of the complex. For all $t \in X$, we have an orthogonal direct sum decomposition

$$C^\infty(E^i) = H^i_t \oplus \text{im } L^i_t \oplus \text{im } L^{i-1*}_t$$

where

$$H^i_t = \ker \Delta^i_t = \ker L^i_t \cap \ker L^{i-1*}_t .$$

One has

$$H^i_t \simeq \frac{\ker L^i_t}{\text{im } L^{i-1}_t} .$$

We want to consider the sets

$$S^i(E^\bullet, L^\bullet) = \{t \in X \mid H^i_t \neq 0\}$$

$$S^i_m(E^\bullet, L^\bullet) = \{t \in X \,|\, \dim H^i_t \geq m\} \tag{1.4}$$

which we call the *harmonic support loci* or *cohomological support loci* of $(E^\bullet, L^\bullet)$. Let $0 \in X$ and $v \in T_0(X)$. We have the operators

$$C^\infty(M, E^i) \xrightarrow{\frac{\partial L^i}{\partial v}(0)} C^\infty(M, E^{i+1}) \quad .$$

Let $d^i(v)$ be defined by the commutative diagram

$$\begin{array}{ccc} C^\infty(M, E^i) & \xrightarrow{\frac{\partial L^i}{\partial v}(0)} & C^\infty(M, E^{i+1}) \\ \text{inclusion} \uparrow & & \downarrow \text{orthogonal projection} \\ H^i_0 & \xrightarrow{d^i(v)} & H^{i+1}_0 \end{array} \quad . \tag{1.5}$$

Differentiating the relation

$$L^{i+1}_t \circ L^i_t = 0$$

twice yields the equation

$$\frac{\partial^2 L^{i+1}}{\partial v^2}(0) \circ L^i(0) + 2\,\frac{\partial L^{i+1}}{\partial v}(0) \circ \frac{\partial L^i}{\partial v}(0) + L^{i+1}(0) \circ \frac{\partial^2 L^i}{\partial v^2} = 0$$

which implies that

$$d^{i+1}(v) \circ d^i(v) = 0$$

Thus, we have a complex $D^\bullet(v)$

$$0 \to H^1_0 \xrightarrow{d^1(v)} H^2_0 \xrightarrow{d^2(v)} \cdots \xrightarrow{d^{n-1}(v)} H^n_0 \to 0 \tag{1.6}$$

which we call the derivative complex of $(E^{\cdot}, L^{\cdot})$ in the direction v. The cohomology groups

$$H^i(D^{\cdot}(v)) = \frac{\ker d^i(v)}{\operatorname{im} d^{i-1}(v)}$$

play a fundamental role in the first-order deformation theory of $(E^{\cdot}, L^{\cdot})$ as is reflected in the following result:

Theorem 1.7 : If $\dim H^i(D^{\cdot}(v)) \le m$ for all non-zero $v \in T_0(X)$, then for all $t$ in some punctured neighborhood of 0 in X, $\dim H_t^i \le m$.

Corollary 1.8: If $H^i(D^{\cdot}(v)) = 0$ for all non-zero $v \in T_0(X)$, and $0 \in S^i(E^{\cdot}, L^{\cdot})$, then 0 is an isolated point of $S^i(E^{\cdot}, L^{\cdot})$.

Corollary 1.9: If $H^i(D^{\cdot}(v)) = 0$ for some $v \in T_0(X)$, then 0 is not an interior point of $S^i(E^{\cdot}, L^{\cdot})$.

Proof: Theorem $\rightarrow$ Corollaries

Corollary 1.8 is the case $m = 0$ of the Theorem. Corollary 1.9 follows from 1.8 by restricting to a 1-dimensional submanifold of X with tangent vector v having $H^i(D^{\cdot}(v)) = 0$.

Remarks: Before proving (1.7), we would like to indicate how our results fit in with the standard perturbation theory of elliptic operators. One expects a map

$$H_0^i \xrightarrow{\frac{\partial\Delta^i}{\partial v}(0)} H_0^i$$

which turns out to be zero. It is the second derivative

$$H_0^i \xrightarrow{\frac{\partial^2\Delta^i}{\partial v^2}(0)} H_0^i$$

which contributes here; it is given by

$$\frac{\partial^2 \Delta^i}{\partial v^2}(0) = \frac{\partial L^{i*}}{\partial v}(0) \circ \frac{\partial L^i}{\partial v}(0) + \frac{\partial L^{i-1}}{\partial v}(0) \circ \frac{\partial L^{i-1*}}{\partial v}(0) \text{ on } H_0^i$$

which is just the "Laplacian" of the derivative complex of $(E^{\bullet}, L^{\bullet})$. One could invoke standard results from perturbation theory to give our results. There is nothing new analytically in what we do; whatever novelty there is lies in the algebraic form of the answer, which appears quite convenient for applications.

Proof of Theorem 1.7:

Let $t_1, \ldots t_m$ be local coordinates on $X$ centered at $0$. Let $|t|$ denote $\left(\sum_{i=1}^{m} t_i^2\right)^{\frac{1}{2}}$ and let $\| \ \|_l$ denote the l'th Sobolev norms on the $C^{\infty}(E^i)$. Assume there exist distinct $t^{\nu} \in X$ converging to $0$ with

$$\dim H^i_{t^{\nu}} \geq m + 1$$

and assume

$$\dim H^i(D^{\bullet}(v)) \leq m$$

for all non-zero $v \in T_0(X)$.

We would like to consider the orthogonal projection maps

$$H^i_t \xrightarrow{\pi} H^i_0$$

$$H^i_t \xrightarrow{\pi^{\perp}} (H^i_0)^{\perp}$$

As a first step, we show:

For $|t|$ small, there exists a constant $C$ so that

$$\|\pi^{\perp}(\psi)\|_1 \le C|t| \ \|\psi\|_1 \quad \text{for all} \quad \psi \in H_t^i \quad . \tag{1.10}$$

We will adopt the convention of using $C$ to denote a constant which we will not relabel each time we increase it. Since the family $L_t^i$ is continuous in $t$, we have for all $\psi \in C^{\infty}(M, E^i)$,

$$\|L_t^i \psi - L_0^i \psi\|_1 \le C|t| \ \|\psi\|_{1+k} \quad .$$

If $\psi \in H_t^i$, then

$$\|L_0^i \psi\|_1 \le C|t| \ \|\psi\|_{1+k} \quad .$$

Similarly,

$$\|L_0^{i-1*} \psi\|_1 \le C|t| \ \|\psi\|_{1+k} \quad .$$

By ellipticity, there is a constant $C$ so that

$$\|\psi\|_{1+k} \le C(\|L_0^i \psi\| + \|L_0^{i-1*} \psi\|) \quad \text{if} \quad \psi \in (H_0^i)^{\perp} \quad .$$

Since

$$L_0^i \psi = L_0^i \pi^{\perp}(\psi)$$

$$L_0^{i-1*} \psi = L_0^{i-1*} \pi^{\perp}(\psi)$$

we obtain that

$$\|\pi^{\perp}(\psi)\|_{1+k} \le C|t| \ \|\psi\|_{1+k}$$

as desired, proving (1.10).

Now note that since $L_t^i$ is a differentiable family of differential operators, for all $\psi \in C^\infty(M, E^i)$,

$$\|L_t^i \psi - L_0^i\psi - \frac{\partial L^i}{\partial v}(0)\psi\|_l \leq C|t|^2\|\psi\|_{l+k} \quad .$$

If $\psi \in H_t^i$, then

$$\|L_0^i \psi + \frac{\partial L^i}{\partial v}(0)\ \psi\|_l \leq C|t|^2\|\psi\|_{l+k} \quad .$$

We note

$$\frac{\partial L^i}{\partial v}(0)\psi = \frac{\partial L^i}{\partial v}(0)\pi(\psi) + \frac{\partial L^i}{\partial v}(0)\pi^\perp(\psi)$$

and

$$\|\frac{\partial L^i}{\partial v}(0)\pi^\perp(\psi)\|_l \leq C|t|\ \|\pi^\perp\psi\|_{l+k}$$

$$\leq C|t|^2\|\psi\|_{l+k}$$

and thus

$$\|L_0^i \psi + \frac{\partial L^i}{\partial v}(0)\pi(\psi)\|_l \leq C|t|^2\|\psi\|_{l+k} \quad .$$

Taking $l = 0$ and projecting orthogonally on $H_0^i$, we obtain

$$\|d^i(v)(\pi(\psi))\|_0 \leq C|t|^2\|\psi\|_k \quad . \tag{1.11}$$

Similarly,

$$\|d^{i-1}(v)^*(\pi(\psi))\|_0 \leq C|t|^2\|\psi\|_k \quad . \tag{1.12}$$

We now wish to show the following:

Let $v \in T_0(X)$ be a unit vector and

$$H_t^i \xrightarrow{\lambda_v} \ker d^i(v) \cap \ker d^{i-1}(v)^*$$

the orthogonal projection. Then for $t \neq 0$ in a sufficiently small conical neighborhood of $0$ in the direction $v$, $\lambda_v$ is injective. (1.13)

Note: If we identify $T_0(X)$ with a neighborhood of $0$ in $X$, then by a conical neighborhood of $0$ in the direction $v$ we mean $N_{a,b}(v) = \{t \mid |t| < a \text{ and } |\frac{t}{|t|} - v| < b\}$ for positive constants $a$, $b$.

We note that (1.13) proves Theorem 1.7, since every unit vector in $T_0(X)$ belongs to a conical neighborhood satisfying the condition of (1.13). Projecting these neighborhoods onto the unit sphere, we obtain an open cover. Extract a finite subcover using conical neighborhoods for $v_1, \dots v_N$. Choose $\epsilon > 0$ small enough so that these conical neighborhoods cover the ball of radius $\epsilon$ centered at $0$. Then for $0 < |t| < \epsilon$,

$$\dim H_t^i \leq \max_{j = 1,\dots N} \dim (\ker d^i(v_j) \cap \ker d^{i-1}(v_j)^*) \tag{1.14}$$

Since

$$\ker d^i(v_j) \cap \ker d^{i-1}(v_j)^* \simeq H^i(D^i(v_j))$$

the right hand side of (1.14) is $\leq m$, proving the theorem.

It remains to prove (1.13). If $t \in N_{a,b}(v)$ and $\psi \in H_t^i$, assume $\lambda_v(\psi) = 0$. Then

$$\pi(\psi) \in (\ker d^i(v) \cap \ker d^{i-1}(v)^*)^{\perp}$$

and thus, since $H_0^i$ is finite dimensional, there is an estimate

$$\|\pi(\psi)\|_k \le C(\|d^i(\ell v)\pi(\psi))\|_0 + \|d^{i-1}(\ell v)^* \pi(\psi)\|_0) \tag{1.15}$$

Now by (1.11), for a sufficiently small,

$$\|d^i(t)\pi(\psi)\|_0 \le C|t|^2\|\psi\|_k$$

But

$$|\frac{t}{|t|} - v| < 1$$

so

$$\|d^i(t)\pi(\psi) - |t|d^i(v)\pi(\psi)\|_0 < A|t| \; b\|\pi(\psi)\|_k$$

for some constant $A$ so

$$|t| \; \|d^i(\ell v)\pi(\psi)\|_0 \le (A|t|b + C|t|^2)\|\psi\|_k$$

and thus

$$\|d^i(v)\pi(\psi)\|_0 \le (Ab + C|t|)\|\psi\|_k$$

Similarly, we have

$$\|d^{i-1}(v)^*\pi(\psi)\|_0 \le (A'b + C'|t|)\|\psi\|_k$$

So returning to (1.15), we have

$$\|\pi(\psi)\|_k \le ((A + A')b + (C' + C)a)\|\psi\|_k$$

However, by (1.10),

$$\|\pi(\psi)\|_k \ge (1 - C''a)\|\psi\|_k$$

and thus we obtain

$$\|\psi\|_k \le \frac{1}{1 - C''a}((A + A')b + (C + C')a)\|\psi\|_k$$

and this is a contradiction for $a, b$ sufficiently small and $\psi \neq 0$. Thus $\lambda_v$ is injective on $N_{a,b}(v)$, proving (1.13) and thus Theorem 1.7.

## §2 Applications to Deformations of Analytic Vector Bundles

Let $M$ be a compact Kähler manifold and $E_t \to M$ an analytic family of analytic vector bundles parametrized by $t$ in a complex manifold $X$. We let $E$ be the fixed $C^\infty$ vector bundle on $M$ underlying all the $E_t$. Consider the variable Dolbeault complex

$$0 \to A^{0,0}(M, E) \xrightarrow{\bar\partial_t} A^{0,1}(M, E) \xrightarrow{\bar\partial_t} \cdots \xrightarrow{\bar\partial_t} A^{0,n}(M, E) \to 0 \quad . \tag{2.1}$$

After endowing $E$ with a hermitian metric, we are in the situation of §1, with

$$H_t^i \simeq H^i(M, E_t)$$

and

$$S_m^i = \{t \in X | h^i(M, E_t) \ge m\}$$

$$S^i = \{t \in X | H^i(M, E_t) \neq 0\} \quad . \tag{2.2}$$

We have the standard Kodaira-Spencer type map

$$T_0(X) \xrightarrow{\delta} H^1(M, \mathrm{Hom}(E_0, E_0)) \quad . \tag{2.3}$$

The fundamental computation is:

Theorem 2.4: For $0 \in X$, $v \in T_0(X)$, the derivative complex $D^0(v)$ is

$$0 \to H^0(M, E_0) \xrightarrow{\wedge\delta(v)} H^1(M, E_0) \xrightarrow{\wedge\delta(v)} \cdots \xrightarrow{\wedge\delta(v)} H^m(M, E_0) \to 0 \tag{2.5}$$

where $\wedge\delta(V)$ denotes the composition

$$\begin{array}{ccc} H^i(M, E_0) & \xrightarrow{U\delta(v)} & H^{i+1}(M, E_0 \otimes \mathrm{Hom}(E_0, E_0)) \\ & {\scriptstyle \wedge\delta(v)} \searrow & \downarrow \\ & & H^{i+1}(M, E_0) \end{array}$$

where the vertical map is induced by the tautological map of bundles

$$E_0 \otimes \mathrm{Hom}(E_0, E_0) \to E_0$$

$$e \otimes \lambda \to \lambda(e)$$

Remark: When $E$ is a line bundle, $\wedge\delta(v)$ is just wedge product with the Dolbeault representative of $\delta(v)$.

Proof:

We may restrict to the case $\dim X = 1$, with $t$ a local coordinate for $X$ at $0$. Let $r = \mathrm{rank}\, E$. Cover $M$ by polydices $U_\alpha$. We can trivialize $E_t$ analytically over $U_\alpha$ so that the transition functions $g_{\alpha\beta}(t)$ with respect to this cover satisfy

$$g_{\alpha\beta}(t) = \gamma_\alpha(t) g_{\alpha\beta}(0) \gamma_\beta(t)^{-1}$$

where

$$\gamma_\alpha(t): U_\alpha \to GL(r, \mathbb{C})$$

are $C^\infty$ maps, and everything depends smoothly on $t$. This is because our family is trivial in the $C^\infty$ catagory. We may take $\gamma_\alpha(0) = \mathrm{id}$.

Then

$$\frac{\partial g_{\alpha\beta}}{\partial t}(0) = \frac{\partial \gamma_\alpha}{\partial t}(0)\, g_{\alpha\beta}(0) - g_{\alpha\beta}(0)\,\frac{\partial \gamma_\beta}{\partial t}(0)$$

Taking $\bar{\partial}$ of both sides,

$$\bar{\partial}\left(\frac{\partial \gamma_\alpha}{\partial t}(0)\right) g_{\alpha\beta}(0) - g_{\alpha\beta}(0)\bar{\partial}\left(\frac{\partial \gamma_\beta}{\partial t}(0)\right) = 0$$

Thus

$$\bar{\partial}\left(\frac{\partial \gamma_\alpha}{\partial t}(0)\right) = g_{\alpha\beta}(0)\bar{\partial}\left(\frac{\partial \gamma_\beta}{\partial t}(0)\right) g_{\alpha\beta}(0)^{-1}$$

so that the

$$\phi_\alpha = \bar{\partial}\left(\frac{\partial \gamma_{\bar{\alpha}}}{\partial t}(0)\right)$$

patch together to give a global $\bar{\partial}$-closed element $\phi \in A^{0,1}(M, \mathrm{Hom}(E_0, E_0))$. This represents the element of $H^1(M, \mathrm{Hom}(E_0, E_0))$ corresponding to $\delta(\partial/\partial t)$.

Now if

$$\psi \in A^{0,i}(M, E)$$

we can represent $\psi$ by

$$\psi_\alpha \in A^{0,i}(U_\alpha) \times \mathbb{C}^{\bar{r}}$$

satisfying

$$\psi_\alpha = g_{\alpha\beta}(0)\psi_\beta \qquad .$$

To obtain the corresponding element of $A^{0,i}(M, E_t)$, we look at

$$\psi_\alpha(t) = \gamma_\alpha(t)\ \psi_\alpha$$

which satisfies

$$\psi_\alpha(t) = g_{\alpha\beta}(t)\psi_\beta(t)$$

Now

$$\bar{\partial}_t\psi \in A^{0,i+1}(M, E_t)$$

is represented in $U_\alpha$ by

$$\bar{\partial}\psi_\alpha(t) = (\bar{\partial}\gamma_\alpha(t))\psi_\alpha + \gamma_\alpha(t)\bar{\partial}\psi_\alpha$$

which corresponds in terms of $E_0$ to

$$\gamma_\alpha(t)^{-1}\,\bar{\partial}\psi_\alpha(t) = \gamma_\alpha(t)^{-1}(\bar{\partial}\gamma_\alpha(t))\psi_\alpha + \bar{\partial}\psi_\alpha$$

Taking $\frac{\partial}{\partial t}$ of this, we get

$$\frac{\partial}{\partial t}\,(\bar{\partial}_t\psi) = \frac{\partial}{\partial t}\,(\gamma_\alpha(t)^{-1}(\bar{\partial}\gamma_\alpha(t))\psi_\alpha)$$

and thus

$$\begin{aligned}\frac{\partial}{\partial t}\,(\bar{\partial}_t\psi)|_{t=0} &= \frac{\partial}{\partial t}\,(\bar{\partial}\gamma_\alpha(t))\psi_\alpha|_{t=0}\\ &= \bar{\partial}\left(\frac{\partial\gamma_\alpha}{\partial t}\,(0)\right)\,\psi_\alpha\\ &= \delta\left(\frac{\partial}{\partial t}\right)_\alpha \wedge \psi_\alpha\end{aligned}$$

and thus

$$\bar{d}^i\left(\frac{\partial}{\partial t}\right)\psi = \delta\left(\frac{\partial}{\partial t}\right)\wedge\psi$$

which proves Theorem 2.4.

We now look at the case when $E$ is of the form $\Omega_M^p \otimes L$, where

$L \to M$ is a line bundle. If we allow $L_t$ to vary in a component of Pic(M) the derivative complex takes the form

$$0 \to H^{p,0}(L_0) \xrightarrow{\wedge\phi} H^{p,1}(L_0) \xrightarrow{\wedge\phi} \cdots \xrightarrow{\wedge\phi} H^{p,n}(L_0) \to 0 \tag{2.6}$$

where

$$\phi \in H^{0,1}(M) \quad .$$

From the deformation theory developed in §1, it is immediate that:

Theorem 2.7: At a general $L$ in any component of Pic(M), the maps

$$H^{p,q}(M, L) \xrightarrow{\wedge\phi} H^{p,q+1}(M, L) \tag{2.8}$$

are zero for all $p$, $q$ and all $\phi \in H^{0,1}(M)$.

If in fact $L \in \mathrm{Pic}^0(M)$, then $L$ can be given the structure of a unitary flat line bundle, and thus there is a conjugation map for harmonic forms

$$H^{p,q}(M, L) \to H^{q,p}(M, L^*)$$

and conjugating (2.6) we obtain the equivalent complex

$$0 \to H^{0,p}(L^*) \xrightarrow{\wedge\omega} H^{1,p}(L^*) \xrightarrow{\wedge\omega} \cdots \xrightarrow{\wedge\omega} H^{n,p}(L^*) \to 0 \tag{2.9}$$

where

$$\omega \in H^0(M, \Omega^1_M)$$

is the conjugate of the harmonic representative of $\phi$. We thus have:

Theorem 2.10: For a general $L \in \mathrm{Pic}^0(M)$, the maps

$$H^{p,q}(M, L) \xrightarrow{\wedge\omega} H^{p+1,q}(M, L) \qquad (2.11)$$

are zero for all $p, q$ and all holomorphic 1-forms $\omega$. In particular, the maps

$$H^0(M, \Omega_M^p \otimes L) \xrightarrow{\wedge\omega} H^0(M, \Omega_M^{p+1} \otimes L) \qquad (2.12)$$

are zero.

If we let

$$M \xrightarrow{a} \mathrm{Alb}(M)$$

be the Albanese map of $M$, then for a general point $p \in M$,

$$\dim \mathrm{im}\, (H^0(M, \Omega_M^1) \to \Omega_{M,p}^1) = \dim a(M) \quad .$$

Since a non-zero section of $H^0(M, \Omega_M^i \otimes L)$ is non-zero at a general point, an elementary computation shows that (2.12) implies that $H^0(M, \Omega_M^i \otimes L) = 0$ for a general $L \in \mathrm{Pic}^0(M)$ if $i < \dim a(M)$. Using this for $L^*$ and conjugating, we obtain:

Theorem 2.13: For a general $L \in \mathrm{Pic}^0(M)$,

$$H^i(M, L) = 0 \quad \text{for} \quad i < \dim a(M)$$

Remark: This was a conjecture of Beauville and Catanese. One can obtain a bound on the codimension of those $L$ for which $H^i(M, L) \neq 0$, see [GL].

Since for $L \in \mathrm{Pic}^0(M)$,

$$\chi(M, L) = \chi(M, \mathcal{O}_M)$$

we obtain that if $\dim a(M) = \dim M$, then

$$\chi(M, L) = (-1)^n h^n(M, L)$$

and thus:

Corollary 2.14: If $\dim a(M) = \dim M$, then

$$(-1)^{\dim M} \chi(M, \mathcal{O}_M) \geq 0 \qquad .$$

If one takes the case when $M$ is a surface and $L_0$ is the trivial bundle, (2.9) becomes

$$0 \to H^0(M,\mathcal{O}_M) \xrightarrow{\wedge\omega} H^0(M,\Omega_M^1) \xrightarrow{\wedge\omega} H^0(M,\Omega_M^2) \to 0$$

By a classical lemma of Castelnuovo, (see [B]), this is exact at the middle term unless $\omega$ pulls back from a holomorphic curve $C$ of genus $\geq 2$ by an analytic map $M \to C$. By (1.8), we thus have that:

For $M$ an irregular Kähler surface with no irrational pencil of genus $\geq 2$, $\mathcal{O}_M$ is an isolated point of $\{L \in \mathrm{Pic}^0(M) | H^1(M, L) \neq 0\}$. (2.15)

From this, one obtains the following, which was conjectured by Eriques:

Theorem 2.16: Let $M$ be a compact Kähler surface with no irrational pencil of genus $\geq 2$. Then any effectively parametrized irreducible family of curves $\{C_t\}_{t\in X}$ with $C_0 \in |K_M|$ has dimension $\leq h^{2,0}(M)$.

Proof:

Let $X \xrightarrow{\alpha} \mathrm{Pic}^0(M)$ be the map

$$t \to L_t = [C_t] \otimes [C_0]^{-1}$$

Then for $t$ small and $\neq 0$,

$$h^0(L_t) = 0, \quad h^1(L_t) = 0$$

so

$$h^2(L_t) = \chi(L_t) = \chi(\mathcal{O}_M) = 1 - q + p_g$$

But

$$\dim \alpha^{-1}(t) \le h^0([C_t])-1 = h^2(L_t)-1$$

so

$$\dim \alpha^{-1}(t) \le p_g - q$$

and thus

$$\dim X \le p_g$$

which proves Theorem 2.16.

Remarks: Beyond the result given here, one may also obtain (see [GL]) an analogue of the Kodaira-Nakano Vanishing Theorem for a general $L \in \mathrm{Pic}^0(M)$. Assume $M$ possesses a holomorphic 1-form $\omega$ whose zero locus

$$Z(\omega) = \{p \in M | \omega(p) = 0\}$$

has codim $Z(\omega) = k$, where we take $k = \dim M + 1$ if $Z(\omega) = \phi$. Then

$$H^{p,q}(M, L) = 0 \quad \text{for} \quad p + q < k$$

for a general element $L \in \mathrm{Pic}^0(M)$. A simple consequence once again is that

$$(-1)^{\dim M-p} \chi(M, \Omega_M^p) \ge 0 \qquad (2.17)$$

if $M$ possesses a holomorphic 1-form with isolated zeros.

# Bibliography

[B] A. Beauville, Complex Algebraic Surfaces, Cambridge University Press, Cambridge (1983).

[CL] J. Carrell and D. Lieberman, Holomorphic vector fields and Kähler manifolds, Invent. Math. 21 (1973), pp. 303-309.

[GL] M. Green and R. Lazarsfeld, Deformation theory, generic vanishing theorems, and some conjectures of Enriques, Catanese and Beauville, to appear Invent. Math.

# Topology and Geometry in Superstring-Inspired Phenomenology

Brian R. Greene

Lyman Laboratory of Physics
Harvard University
Cambridge, MA 02138

Kelley H. Kirklin and Paul J. Miron

Department of Theoretical Physics
Oxford University
Oxford, England

Abstract

We give a brief review of the interplay between topology, geometry and phenomenology in the forum of superstring-inspired models of low energy particle physics. We first discuss the mathematical methods required for extracting low energy phenomenology from chosen compactifications of the Heterotic superstring. To motivate a choice of vacuum, we then show that there is only *one* example of a complete intersection Calabi-Yau manifold quotiented by a free projectively inhereted group action which gives rise to three generations of elementary particle multiplets. Finally, choosing this manifold for the vacuum state, we explicitly compute the information required for extracting low energy phenomenology.

## 1. Introduction

To those with firm belief in the existence of an underlying unification amongst the manifestly distinct forces at work in the observable universe, the mathematically rich solution offered by superstring theory [1]is quite compelling. Providing a consistent and possibly finite quantum theory of gravitation which also possesses sufficient gauge symmetry to encompass the standard model, is, however, only a necessary condition which any such potential unified theory must satisfy. Sooner or later we are obliged to face the facts. There is a wealth of experimental evidence which must be successfully addressed. First and foremost amongst these is the simple fact that the observable universe has four dimensions, while consistent superstring theories begin life in ten. This presents no major obstacle, however, as these extra dimensions may be assumed to be extremely small, presumably on the order of the Planck length $10^{-33}cm$, and hence have and probably always will escape direct detection. Ultimately, one hopes that an assumption of this sort, in which the gross structure of the vacuum is put in by hand, will be replaced by a dynamical principle dictated by the underlying theory, but for now we settle for this mild imposition of phenomenological viability. Beyond the dimensionality of spacetime, low energy particle physics provides the next, and far harsher, experimental forum. Happily, the standard model based on an underlying $SU(3) \times SU(2) \times U(1)$ gauge symmetry successfully provides a theoretical explanation for all such experimental facts. Our task of meeting the experimental constraints of phenomenology is thus, to a large extent, tantamount to reproducing the standard model at low energies.

To appreciate the value of reproducing the standard model, one should bear in mind that the latter successfully describes experimental evidence only after some nineteen free parameters are carefully adjusted. On the contrary, superstrings provide essentially *no* free adjustable parameters. Rather, the topology and geometry of the underlying ten dimensional spacetime manifold and certain vector bundles built upon it, as we shall see, determine the (tree level) low energy Lagrangian [2][3][4].

In theory, then, once the vacuum configuration is known, the resulting low energy phenomenology may be extracted by the tools of classical topology and geometry. Our present level of understanding, however, restricts but does not uniquely determine the vacuum configuration [5][2]. We are thus led to turn the question around: Are there any consistent vacuum configurations which lead to viable low energy phenomenology?

Using this as a guiding question, we shall spend some time reviewing the relevant mathematical machinery from algebraic geometry and topology which has been successfully applied, by a number of authors, to the extraction of low energy phenomenology, and then we shall illustrate these techniques in a particularly interesting example [6]. Our goal is to provide mathematicians with a brief introduction to the interplay between familiar mathematics and low energy particle physics, while also providing physicists with a review of the requisite mathematical methods and a concrete illustration of their use in an exemplary model.

Specifically, in section 2 we describe compactification - the passage from ten to four dimensions, and show how the cohomology of the six-dimensional compactified component of spacetime,K, determines the resulting low energy particle content. In section 3 we describe the physical consequences of choosing a K for which $\pi_1(K)$ is nonzero. In section 4 we concentrate on the role of discrete symmetries of K and the methods of topologically extracting Yukawa couplings. In section 5 we discuss the surprisingly stringent mathematical constraints which arise from imposing the one further phenomenological requirement that the low energy particle spectrum contain three generations. Finally, in section 6 we illustrate these techniques for a three generation superstring model.

## 2. Compactification

We take as our starting point the Heterotic superstring [7]which finds as its home a ten dimensional universe with an $SO(32)$ or $E_8 \times E_8$ gauge symmetry. It is useful to emphasize that this spacetime dimension and the possible gauge symmetry groups are not merely posited but rather dictated by the quantum consistency of the theory. Our interest will focus on the phenomenologically more attractive $E_8 \times E_8$ theory. As is standard practice, one of these $E_8$ factors will be relegated to a hidden sector, leaving the observable sector with a single $E_8$ gauge symmetry. In the introduction we have stressed that low energy implications are our concern, so we immediately discard all massive string excitations, which are on the order of the Planck mass, and concentrate on the massless sector which fills out a ten dimensional supersymmetric Yang-Mills field theory with $E_8$ gauge symmetry, coupled to N=1 supergravity [5][8]. The field content of such a theory is composed of the supergravity multiplet $(g_{\mu\nu}, \psi_\mu, B_{\mu\nu}, \psi, \phi)$ where these are, in order of presentation, the ten dimensional spin two graviton (the metric), the spin 3/2 gravitino, the antisymmetric tensor field, a spin 1/2 field, and a scalar dilaton field, in addition to the superYang-Mills multiplet $(A^a_M, \psi^a)$ , where these are the gauge bosons ( the gauge connection) and their supersymmetric partners the gauginos, and a is an adjoint $E_8$ index. Ignoring the modes associated with the dependence of K on the uncompactified coordinates, which shall not enter into our discussion, this gives the full ten dimensional field content. Unlike in standard four dimensional phenomenology where a bold model builder may introduce gauge groups and particle multiplets at will, the underlying superstring theory *determines* these (in ten dimensions) essentially uniquely. We should mention that one could certainly have studied the properties of this ten dimensional field theory without recourse to anything related to strings. The role of strings, therefore, from this point of view, is more or less motivational. Hence the often used phrase 'superstring inspired models'. Later on, we shall briefly note some inherently stringy effects.

Obviously, we are far away from anything physically reasonable, so we presently embark on a path towards reality by assuming that the ten dimensional universe is of the form $M_4 \times K$ where the former is a four dimensional maximally symmetric spacetime, and the latter is a six dimensional compact manifold whose radius we assume to be on the order of the Planck length. The specific choice of $M_4$ and $K$ is far from arbitrary. As shown from a couple of viewpoints [5]$M_4$ should be chosen to be Minkowski space, while the favoured choice for K is a (three dimensional) complex Kahler manifold with vanishing first Chern class. Such manifolds have a holonomy group contained within $SU(3)$ and by Yau's proof of the Calabi conjecture [9][10]admit Ricci flat metrics. Our interest shall focus on K for which the holonomy group is precisely $SU(3)$, the so called Calabi-Yau manifolds. Two important properties of Calabi-Yau manifolds [5], which we shall shortly exploit, are that they admit admit a covariantly constant spinor field $\eta_+$ where the + denotes positive chirality, and that, by the vanishing of $c_1(K)$ the canonical bundle is trivial and hence admits a global section, $\omega$.

In addition to the constraints which the theory places on the choice of $K$, anomaly cancellation places topological constraints on the Yang-Mills gauge bundles [2][11]. If $V_1$ and $V_2$ are the two $E_8$ bundles, then they must be holomorphic and must satisfy

$$c_1(V_1) = c_1(V_2) = 0(mod2) \tag{2.1}$$

and

$$ch_2(TK) = ch_2(V_1) + ch_2(V_2) \tag{2.2}$$

where $c_1$ is the first Chern class and $ch_2$ is the second Chern character. In addition to these topological requirements, the preservation of supersymmetry (and the string equations motion to lowest order in perturbation theory) [2]dictate the geometrical constraint

$$g^{i\bar{j}}F_{i\bar{j}} = 0 \tag{2.3}$$

where $F_{i\bar{j}}$ is the curvature 2-form for either of the bundles $V_1$ or $V_2$. A recent theorem of Uhlenbeck and Yau [12]ensures a solution to this equation when $V_1$ and $V_2$ meet the previous constraints, are *stable* and satisfy the integrability condition $[J] \wedge [J] \wedge c_1(V_{1,2})$ integrated over $K$ vanishes, where $[J]$ is the cohomology class of the Kahler form for $K$. We do not have space to discuss the origin or restrictiveness of these constraints here, but rather refer the reader to a more detailed discussion in [13].

We shall shortly turn to an explicit solution of these equations, but first we note the effect which such a general solution has on the observable $E_8$ gauge symmetry. Let us assume that the bundle $V_1$, which we choose to view as an $E_8$ bundle, has a structure group $S$ contained within $E_8$ and that $T$ is the maximal subgroup of $E_8$ which commutes with $S$. Only those gauge fields associated with $T$ remain massless, and hence $T$ is the observable unbroken gauge group. For example if $V_1$ has an $SU(4)$ or $SU(5)$ structure group, then the unifying gauge group $T$ is $SO(10)$ or $SU(5)$, both phenomenologically attractive gauge groups. Examples of such $V_1$ can be found, but, as shown in [14]nonperturbative effects tend to spoil these solutions. By judicious choice of $V_1$, such pitfalls can be avoided, as demonstrated in [15][13]. We shall develop the formalism for general $V_1$ but shall only concentrate on examples with $S = SU(3)$. For this solution, which was originally constructed in [5], we take $V_1$ to be $TK$, the tangent bundle of our chosen Calabi-Yau manifold, which has an $SU(3)$ structure group. The topological constraints stated above are identically met, where we take $V_2$ to be trivial. As $E_6$ is the maximal subgroup of $E_8$ which commutes with $SU(3)$, it plays the role of our unifying group. Although not as phenomenologically economical as $SU(5)$ or $S0(10)$, $E_6$ is phenomenologically viable, so our hope for obtaining a realistic theory is not futile, but depends on further analysis to which we now turn.

Having specified that the vacuum configuration is of the form $M_4 \times K$, with $M_4$ four dimensional Minkowski space and K a Calabi-Yau manifold, with Yang-Mills bundle meeting the stated constraints, our next task is to determine how the ten dimensional supergravity/superYang-Mills multiplets (which are the only fields in the low energy theory) appear to an observer in $M_4$. The basic analysis we use is familiar from Kaluza-Klein theory and was originally applied in this context in [5]and [16]. We shall only indicate the arguments for the Yang-Mills multiplet (in fact, only for the ten dimensional gauginos), as the supergravity multiplet will play no significant role in our low energy phenomenological analysis.

For simplicity, we first consider a ten dimensional left-handed Fermi field $\psi$, which has no gauge interactions. Since this field is massless, we have

$$\not{D}_{10}\psi_{10} = 0 \tag{2.4}$$

. Using the assumed form of the vacuum, we may rewrite this as

$$(\not{D}_4 + \not{D}_6)\psi_4 \otimes \psi_6 = 0 \tag{2.5}$$

where the subscript denotes the domain in which the operator is defined. From this last equation we see that the six dimensional Dirac operator acts like a four dimensional mass term. We may thus express $\psi_{10}$ in a harmonic expansion containing summands of the form (2.5), with the $\psi_6$ taken to be eigenfunctions of $\not{D}_6$ Typically, the magnitude of the eigenvalues of $\not{D}_6$ are on the order of the inverse of the radius of K, and hence extremely large on the scale of observable physics. However, this is not so for the *zero* modes of $\not{D}_6$. Hence, as far as observable physics goes, we may truncate the expansion of $\psi_{10}$ to only include terms of the form $\psi_4 \otimes \psi_6$ where $\psi_6$ satisfies

$$\not{D}_6\psi_6 = 0 \tag{2.6}$$

. We now show that the solutions of (2.6)are determined by the cohomology of K.

To do so, we first note that since K has $SU(3)$ holonomy, we can write the usual tangent spinor space Dirac algebra as

$$\{\gamma^i, \gamma^{\bar{j}}\} = 2g^{i\bar{j}} \tag{2.7}$$

where the barred (unbarred) indices are antiholomorphic (holomorphic) coordinates. We now make use of the covariantly constant spinor $\eta$ on K whose existence we mentioned earlier. We can construct $\eta$ so that it is annihilated by all of the $\gamma^i$ matrices and hence the $\gamma^{\bar{j}}$ act like creation operators acting on a 'Fock' vacuum. Notice that in the case of interest $\psi_{10}$ is of definite chirality so that the chirality of $\psi_4$ is determined by that of $\psi_6$.

Furthermore, since the application of gamma matrices reverses chirality, we may explicitly build chiral matter by constructing $\psi_6$ of the form [5][2](with $y$ being coordinates on $K$)

$$\psi_L = (\alpha(y) + \alpha_{\bar{i}\bar{j}}(y)\gamma^{\overline{ij}})\eta \tag{2.8}$$

and

$$\psi_R = (\alpha_{\bar{i}}(y)\gamma^{\bar{i}} + \alpha_{\overline{ijk}}(y)\gamma^{\overline{ijk}})\eta \tag{2.9}$$

where repeated indices on gamma matrices indicate antisymmetrized products. Now, for the reasons discussed, lets us now specialize to massless matter by imposing the massless Dirac equation. This gives,

$$\not{D}_6\psi_R = (\gamma^i D_i + \gamma^{\bar{i}} D_{\bar{i}}\psi_R = 0 \tag{2.10}$$

(where $D_i$ denotes covariant differentiation) which in turn gives

$$D_{[\bar{i}}\alpha_{\bar{j}]}(y)\gamma^{\overline{ij}}\eta = 0 \tag{2.11}$$

$$D^{\bar{j}}\alpha_{\bar{j}}(y)\eta = 0 \tag{2.12}$$

and

$$D^{\bar{j}}\alpha_{\overline{jkl}}(y)\gamma^{\overline{kl}}\eta = 0 \tag{2.13}$$

, where the brackets denote antisymmetrization. From (2.11)and (2.12)we see that $\alpha_{\bar{j}}dy^{\bar{j}}$ is an harmonic $(0,1)$ form (where a $(p,q)$ form has $p$ holomorphic and $q$ antiholomorphic indices) and from (2.13)we see that $\alpha_{\overline{jkl}}dy^{\bar{j}}dy^{\bar{k}}dy^{\bar{l}}$ is an harmonic $(0,3)$ form. The same analysis for $\psi_L$ shows that it is constructed from harmonic functions (of which, of course there are only the constant functions) and harmonic $(0,2)$ forms. Thus, the massless (and hence observable) fields which descend into the four dimensional effective theory, are determined by the Dolbeault cohomology groups $H^{0,*}_{\bar{\partial}}(K)$. It is this fact that is responsible for the close link between the topology of $K$ and the observable low energy physics.

The above discussion is not quite the physically relevant one since the ten dimensional gauginos carry gauge quantum numbers which have to this point been ignored. Explicitly, the gauginos transform in the adjoint representation of $E_8$, that is the **248**. On passing to $M_4 \times K$, the gauge symmetry, for example in the $V_1 = TK$ scenario, is reduced to $E_6$; accordingly the **248** decomposes under $E_6 \times SU(3)$ as

$$248 = (27,3) + (\overline{27},\overline{3}) + (78,1) + (1,8) \tag{2.14}$$

The (78,1) are the $E_6$ gauginos, the (1,8) are $E_6$ singlet fields which we shall not discuss in this article, while the (27,3) and the $(\overline{27},\overline{3})$ are the matter generations and antigenerations of the theory. It is heartening that the 27 of $E_6$ is a phenomenologically viable representation as it can accomodate all of the particles in a single standard model generation. From this decompostion we see that (for the 27 and $\overline{27}$) that the zero modes which we have been calling $\psi_6$ transform as the 3 and $\overline{3}$ of $SU(3)$. Now, the bundles corresponding to the 3 and $\overline{3}$ of $SU(3)$ are just $TK$ and $T^*K$, and hence the appropriate generalization of the earlier result that the $\psi_6$ should lie in $H^{0,*}_{\overline{\partial}}(K)$, is thus seen to be that $\psi_6 \epsilon H^{0,*}_{\overline{\partial}}(K,TK)$ and $H^{0,*}_{\overline{\partial}}(K,T^*K)$ [2]. (For ease of presentation, we shall sometimes drop the $\overline{\partial}$).

For concreteness we have rooted our discussion in the specific example of $V_1 = TK$, but it is clear that nothing has depended on this specialization. More generally, for an arbitrary holomorphic $V_1$ of structure group $S$ which meets the constraints given earlier, the **248** of gauginos decomposes under $T \times S$ (where $T$ is the maximal subgroup of $E_8$ which commutes with $S$) as

$$248 = \oplus(R_a(T), \rho_a(S)) \tag{2.15}$$

The massless four dimensional fields transforming in the representation $R_a(T)$ arise from harmonic $\mathcal{B}(\rho_a)$ valued forms K, $H^{0,*}(K,\mathcal{B}(\rho_a))$, where $\mathcal{B}(\rho_a)$ is an associated vector bundle to $V_1$, built on the representation $\rho_a$ of $S$ which appears in the decomposition of the **248** under $T \times S$.

In fact, we can push this discussion one step further. As mentioned earlier, four dimensional chirality is determined by the parity of $q$ in $H^{0,q}(K,\mathcal{B}(\rho_a))$; say, for $q$ even the spinor is left handed and for $q$ odd right handed. Then chirality in the four dimensional theory is detected by the difference of the number of left and right handed fermions

$$n_L - n_R = h^{0,0}(K,\mathcal{B}(\rho_a)) + h^{0,2}(K,\mathcal{B}(\rho_a)) - h^{0,1}(K,\mathcal{B}(\rho_a)) - h^{0,3}(K,\mathcal{B}(\rho_a)) \tag{2.16}$$

. The right hand side is nothing but the index of the $\overline{\partial}$ operator twisted by the bundle $\mathcal{B}(\rho_a)$, while the left hand side is the net number of generations transforming in the representation $R_a$ of the four dimensional gauge group $T$. Thus, the net number of generations transforming in the representatio $R_a$ of $T$ is given by

$$index_{\overline{\partial}}\mathcal{B}(\rho_a) \tag{2.17}$$

. Using CPT and the fact that the adjoint is a real representation we can reexpress this in terms of a single chirality spinor taking values in $\mathcal{B}(\rho_a)$ and its dual bundle. In the case of $V_1 = TK$ we note that $H^{0,1}(K, T^*K)$ is isomorphic to $H^{1,1}(K)$, while by contracting with $\omega$, we also have $H^{0,1}(K, TK)$ being isomorphic to $H^{1,2}(K)$. Since $h^{p,q} = h^{q,p}$ on a Kahler manifold (where this symbol denotes the dimension of the corresponding cohomology group), the only independent Hodge numbers besides $h^{0,0} = h^{3,0} = 1$ are $h^{1,1}$ and $h^{1,2}$. Thus, the number of $27's$, $n_{27}$, minus the number of $\overline{27}'s$, $n_{\overline{27}}$, is given by

$$n_{27} - n_{\overline{27}} = -index_{\overline{\partial}}(TK) = h^{2,1} - h^{1,1} \tag{2.18}$$

and hence the familiar result that the net number of generations of $27's$ of $E_6$ (of course, the distinction between a 27 and a $\overline{27}$ is just a matter of convention) is given by half of the (absolute value ) Euler characteristic of $K$. We therefore see that superstring inspired models offer a *topological* explanation for the observed replication of generations of particles with identical quantum numbers. In later sections we shall address the problem of finding Calabi-Yau manifolds $K$ which give rise to the experimentally observed value of three generations; first, however, we turn to a discussion of the role of $\pi_1(K)$ in superstring phenomenology.

## 3. FLUX BREAKING

An important distinction between standard Kaluza-Klein theory and the analysis of concern here is that from the former point of view the gauge symmetry arises from continuous isometries of the compactified manifold, while in the latter, the gauge symmetry, by the demands of quantum consistency, exists from the outset. This is quite important in light of the fact that Calabi-Yau manifolds do not admit any continuous isometries. However, they typically *can* have quite large *discrete* automorphism groups, and these turn out to be very beneficial. One of the most important benefits of these is the flux breaking mechanism [16], which we shall turn to first.

Recall that the situation as it stands is that we have a supersymmetric gauge theory with gauge group $T$ at the compactification scale with some topologically determined field content, and we somehow have to make our way down to the low energy of the standard model. We are not allowed to postulate the existence of Higgs fields to break our symmetry group further, as is the norm in the traditional supersymmetric grand unified theories, so how do we proceed ?

Let us consider some generic Calabi-Yau manifold $K$, and assume, without loss of generality, that it is simply connected *, $\pi_1(K) = 0$. Suppose that this manifold has a discrete symmetry group $\mathcal{D}$ (here, a discrete symmetry is a biholomorphic map from $K$ to itself), and let $G \subseteq \mathcal{D}$ be a subgroup that acts freely on $K$ (i.e. it has no fixed points other than the identity). Then, since $G$ acts holomorphically, the quotient manifold $R \equiv K/G$, is still a Calabi-Yau manifold. However, we now have to go further in order to specify the vacuum configuration [5,16].

From our previous discussion, we know that we need to satisfy (2.1)(2.2)(2.3)for some $V_1$ having a stucture group $S \subset E_8$. For example we may choose $S = SU(3)$ so as to leave an unbroken group $T = E_6$ (the $T$ here should not be confused with the tangent bundle), corresponding to a flat $E_6$ bundle (since we assume that the only non-trivial component of the $E_8$ gauge field is the $SU(3)$ subgroup that we equate with the spin connection). Now, in the case of a non-simply connected manifold, parallel transport in the flat $T$ bundle depends on the homotopy of paths pulled back to $R$. So, to every element of $\pi_1(R) = G$, we should associate some element $t \in T$. However, this embedding must be a homomorphism in order to preserve the homotopy operation of path multiplication. In other words, if $g \longrightarrow U_g$, $(g \in G)$ is a mapping of the homotopy group into the flat $T$ bundle, we require

$$U_g U_{g'} = U_{gg'}. \tag{3.1}$$

We shall denote a generic embedding of $G$ as $\overline{G}$. $\overline{G}$ acts just like holonomy in the Yang-Mills bundle and, together with $S$, determines the vacuum configuration. This is nothing but the statement that flat nontrivial bundles over $R$ are determined by elements in $Hom(\pi_1(R), T)$.

We shall now see the benefits of this procedure. First, we note that $T$ is broken down further to the subgroup that commutes with $\overline{G}$. We can provide a physical reason for this by considering a contour $\gamma_g$ as a representative of some $g \subset G$. Then the image of $g$ under the homomorphism can be represented by the Wilson loop [5][16]

$$U_g = \mathcal{P}exp[-i \int_{\gamma_g} T^a A_m^a(x)\, dx^m] \tag{3.2}$$

where the $A_m^a(x)$ are vacuum $T$ gauge fields with group generators $T^a$, and $\mathcal{P}$ denotes path ordering. Since we have non-contractible loops in $R$, we can get cases in which

* If $K$ were *not* simply connected, then we are always guaranteed the existence of a covering space that is [17]

$U_g \neq 1$, corresponding to flux being 'trapped' by the 'holes' in the manifold i.e. we have a non-trivial gauge configuration with vanishing curvature which acts, heuristically, like an adjoint of Higgs vacuum expectation values.

The second repercussion of flux breaking is in the massless modes arising from the Yang-Mills supermultiplet (modes arising from the metric on $K$ are obviously unaffected by the quotienting of isometries). Recall that these originate as harmonic $(0,q)$ forms taking values in various vector bundles. However, when we pass from $K$ to the quotient manifold $K/G$, we must identify all points which lie on a single $G$-orbit. Thus, only those fields arising from sections which are $G$ invariant survive through to $R$. Explicitly, if $\psi(y), (y \in K)$ is some generic section, we might require that

$$\psi(gy) = \psi(y) \tag{3.3}$$

for any $g \in G$. However, note that $\psi(y)$ also carries some representation $R_a$ of $T$, since it is a vector bundle-valued form, and we should allow the $U_g$ to act on such representations. Thus, the *correct* condition on sections is [16]

$$U_g\psi(gy) = \psi(y) \tag{3.4}$$

or, more succinctly,

$$(G \oplus \overline{G})\psi = \psi \tag{3.5}$$

. Now, note that, in general, the various $T$ components will have different transformation properties under $\overline{G}$, whereas, since $G$ acts on the *whole* vector space $H^{0,q}_{\overline{\partial}}(R,E)$, all components in a fixed $\psi$ transform identically under $G$. Thus, in general, not all the different $T$ components will survive through to $R$. This will lead us to a natural multiplet splitting mechanism for avoiding the presence of dangerous coloured states which might otherwise mediate proton decay. It should be noted that the index theory relation of the last section is independent of all the $U_g$, and so must be obeyed in any multiplet splitting. So, in addition to the the number of complete $T$ generations gauranteed by the index theorem ,we will generally have several incomplete $T$ multiplets which will, of course, form complete representations of the unbroken group that commutes with $\overline{G}$. The index theorem further tells us that, for every such additional component $\psi_a$, we must also have $\bar{\psi}_a$ surviving.

The above argument also implies that the normal group theoretical relations amongst Yukawa couplings (which are quite often unwanted) do not now hold, since the various components of a complete $T$ multiplet will not, in general, have originated from the *same* harmonic form on $K$.

There are two other important effect which occur in passing to $K/G$. The first is that, in the $V_1 = TK$ scenarios, the net number of particle generations is reduced by a factor of the number of elements in G, since

$$\chi\left(\frac{K}{G}\right) = \frac{\chi(K)}{|G|} \tag{3.6}$$

where $|G|$ is the cardinality of $G$. In the more general scenarios, a similar formula holds if the action of G lifts to the bundle $V_1$, since in such a situation, the index (which can be expressed as a *local* integral) will also be reduced by the same factor. The second effect has to do with the residual holomorphic symmetries which exist on the quotient manifold. We turn to this point in the next section.

## 4. DISCRETE SYMMETRIES AND YUKAWA COUPLINGS

In this section we turn first to a brief discussion of the discrete symmetries of Calabi-Yau manifolds and their application to low energy particle physics. Our goal is to is to restrict the allowed couplings using discrete symmetry arguments, since we may then have a natural explanation for the absence of certain lepton and baryon number violating operators. It turns out that we can actually determine the full tree level $27^3$ and $\overline{27}^3$ Yukawa couplings (and possibly also the $1^3$ and $1.27.\overline{27}$ couplings, where the 1 is an $E_6$ singlet superfield) as certain products in the cohomology ring of the manifold of compactification. This is the second topic of this section. In hindsight, one might wonder whether it is worth the effort to investigate the restrictions arising purely from discrete symmetrics. However, utilising discrete symmetries, aside from demonstrating some useful tools, yields restrictions on the higher order *non-renormalisable* effective interactions amongst massless fields, for which methods of exact calculation are not available.

A typical Calabi-Yau manifold, $K_0$ generally has some quite large discrete symmetry group $\mathcal{D}$. These symmetries induce maps within the various bundle-valued cohomology groups which give rise to the four dimensional massless modes. It is thus our task to ascertain the action of the symmetries in these cohomology groups and therefore on the fields in our theory. In section 6 we shall accomplish this for a specific manifold by using the powerful machinery of deformation theory [18]and fixed point theorems [19]. First, however, we shall make some general remarks concerning the discrete symmetries of potential vacuum configurations.

We shall be concerned with the previously mentioned case of quotienting out a simply-connected Calabi-Yau space $K_0$ by a freely acting finite discrete group $G$ of holomorphic isometries to give $K \equiv K_0/G$, where the manifold $K_0$ has some discrete symmetry group $\mathcal{D}$. As an example, consider the $\mathcal{Y}_{(4;5)}$ manifold [5], defined by

$$\sum_{i=1}^{5} z_i^5 = 0 \tag{4.1}$$

in $CP^4$. It is easy to show that $\mathcal{Y}_{(4;5)}$ is a simply-connected three dimensional, Ricci-flat, Kähler manifold with vanishing first Chern class. This manifold has a huge discrete symmetry group of order 75,000, given by permutations of the $\{z_i\}$ and scalings $z_i \rightarrow \alpha^{n_i} z_i$, where $\alpha$ is a fifth root of unity, and $n_i = 0, \ldots, 4$. However, eq.(4.1) is only *one* particular choice of moduli for $\mathcal{Y}_{(4;5)}$. We are free to add order five monomials with arbitrary coefficients to eq.(4.1), and this will, in general, change the discrete symmetry group.

Having determined the discrete symmetry group $\mathcal{D}$ of $K_0$, we now wish to investigate the corresponding group for the quotient manifold $K$ [16]. Consider some element $g \in G$ and $y \in K_0$, so that $y$ and $gy$ lie on the same G-orbit in $K_0$. Then, if $d$ is to generate a symmetry of $K$, $d(gx)$ must be equivalent to $d(x)$ on $K$. In other words, we need $g' \in G$ such that $d(gy) = g'd(y)$, or

$$dgd^{-1} = g' \qquad (g' \in G) \tag{4.2}$$

Thus the discrete symmetries of $d$ that survive are just those in the normaliser of $G$. We shall call such symmetries *exact* symmetries of $K$.

How does flux breaking alter things ? Recall that this involved a choice of homomorphism mapping $G$ into the unbroken group — $E_6$ in this $V_1 = TK$ scenario — or, equivalently, a choice of flat $E_6$ bundle over $K \cong K_0/G$. This embedding we call $\overline{G}$. Note that we should now extend our condition eq.(4.2) for exact symmetries to allow for the presence of $E_6$ gauge transformations associated with elements of $\mathcal{D}$. We find that such gauge transformations only make sense if we can find a *homomorphism* of elements of $\mathcal{D}$ into $E_6$, such that the image of $\mathcal{D}$ satisfies

$$V_{d^{-1}} U_g V_d = U_{g'} \tag{4.3}$$

where $U_g$ is the image in $E_6$ of $g \in \pi_1(K) \cong G$, $V_d$ is the image of $d \in \mathcal{D}$ and $d^{-1}hd = h'$. The manifold $\mathcal{Y}_{(4;5)}$ defined by (4.1) above has a group $D$ of freely acting symmetries isomorphic to $Z_5 \times Z_5$ which is generated by $z_i \to \alpha^{n_i} z_i$ and $z_i \to z_{i+1}$. It is a staightforward exercise to apply the procedure described above to find which $\mathcal{Y}_{(4;5)}$ symmetries descend to symmetries of $Y \cong \mathcal{Y}_{(4;5)}/D$.

The purpose of this procedure is to allow us to restrict the couplings of four dimensional massless fields using the discrete symmetries of the given Calabi-Yau manifold. However, this can only tell us of the *possible* presence of terms in the lagrangian. We now proceed to discuss techniques for the explicit calculation of tree level Yukawa couplings [3][4].

We start with the coupling which arises in the ten dimensional supergravity lagrangian

$$\int d^{10}w\sqrt{-g_{10}}\ \overline{\psi}^A(w)\Gamma^M A^B_M(w)\psi^C(w)f_{ABC} \tag{4.4}$$

where $g_{10}$ is the determinant of the ten dimensional metric, $\psi^A$ and $A^A_M$ are ten dimensional gaugino and vector boson fields carrying gauge indices $A \in E_8$, and $f_{ABC}$ is a tensor selecting the gauge singlet from eq.(4.4). Since we shall be interested in the couplings of matter superfields arising from eq.(4.4), we set $M = \overline{m}$, the anti-holomorphic cotangent index. We now just plug the harmonic expansion into eq.(4.4) — explicitly, we make the substitutions:

$$\begin{aligned} A^A_M(w) &\longrightarrow \phi^{R_a}_j A^{\rho_a}_{\overline{m}j}(y) \\ \psi^A(w) &\longrightarrow \psi^{R_a}_i(x)\psi^{\rho_a}_i(y) \\ \Gamma^M &\longrightarrow \gamma_5 \otimes \gamma^{\overline{m}} \\ g_{10} &\longrightarrow g_6 \otimes g_4 \\ f_{ABC} &\longrightarrow d_{abc}h_{abc} \end{aligned} \tag{4.5}$$

The indices $i, j$ and $k$ count the members of the relevant cohomology groups $H^1(K, \mathcal{B}[\rho_a])$ and the "singlet" tensor $d^{abc}, h^{abc}$ pick out singlets from $E_6$ and $SU(3)$ products.

Upon substituting eq.(4.5) we obtain

$$\sum_{i=1}^{h^1(K,\mathcal{B}[\rho_a])} \sum_{j=1}^{h^1(K,\mathcal{B}[\rho_b])} \sum_{k=1}^{h^1(K,\mathcal{B}[\rho_c])} g_{ijk} \int_{M_4} d^4x\sqrt{-g_4}\ \overline{\psi}^{R_a}_i \phi^{R_b}_j \psi^{R_c}_k d^{abc} \tag{4.6}$$

where

$$g_{ijk} = \int_K d^6y\sqrt{g_6}\psi^{\dagger\rho_a}_i \gamma^{\overline{m}} A^{\rho_b}_{\overline{m}j} \psi^{\rho_c}_k h^{abc} \tag{4.7}$$

and these equations contain all $a, b, c$ such that $R_a \otimes R_b \otimes R_c$ and $\rho_a \otimes \rho_b \otimes \rho_c$ both contain singlets. We thus see that $g_{ijk}$ — an overlap integral of *six* dimensional fields — acts as a coupling for *four* dimensional fields.

As a consequence of the correspondence on Calabi-Yau manifolds between $(0,p)$ forms and spinors of definite chirality, we have the correspondence of fields

$$A^{\rho_a}_{\overline{m}} \longleftrightarrow A^{\rho_a}_{\overline{m}} \gamma^{\overline{m}} \eta_+ \equiv \psi^{\rho_a} \tag{4.8}$$

(where $\eta_+$ is the parallel spinor feld, whose existence we mentioned and have used earlier) Substituting this into eq.(4.7)

$$g_{ijk} = \int_K d^6 y \sqrt{g_6} (\eta_+^\dagger \gamma^{\overline{r}\overline{m}\overline{n}} \eta_+) A^{\rho_a}_{\overline{r}i} A^{\rho_b}_{\overline{m}j} A^{\rho_c}_{\overline{n}k} h^{abc} \tag{4.9}$$

It so happens [3][4]that we may also express the anti-holomorphic three form on $K$ as

$$\begin{aligned} \overline{\omega} &= \eta_+^\dagger \gamma_{\overline{r}\overline{m}\overline{n}} \eta_+ d\overline{z}^{\overline{r}} \wedge d\overline{z}^{\overline{m}} \wedge d\overline{z}^{\overline{n}} \\ &\equiv \omega_{\overline{r}\overline{m}\overline{n}} d\overline{z}^{\overline{r}} \wedge d\overline{z}^{\overline{m}} \wedge d\overline{z}^{\overline{n}} \end{aligned} \tag{4.10}$$

and so

$$\begin{aligned} g_{ijk} &= \int_K d^6 y \sqrt{g_6} \omega^{\overline{r}\overline{m}\overline{n}} A^{\rho_a}_{\overline{r}i} A^{\rho_b}_{\overline{m}j} A^{\rho_c}_{\overline{n}k} h^{abc} \\ &= \int_K \omega \wedge A^{\rho_a}_i \wedge A^{\rho_b}_j \wedge A^{\rho_c}_k . \end{aligned} \tag{4.11}$$

We have so far ignored the possibility that the matrix of kinetic terms is not diagonal. In view of this, we try to calculate the "normalisation matrix" in a similar fashion:

$$\langle A^{\rho_a}_i | A^{\rho_b}_j \rangle = \int_K d^6 y \sqrt{g_6} A^{\rho_a}_{\overline{m}i} A^{*\overline{\rho}_a}_{nj} g^{\overline{m}n} \tag{4.12}$$

where $\overline{\rho}_a$ is the dual representation to $\rho_a$, "*" denotes complex conjugation and, once again, the singlet components of $\rho_a \otimes \overline{\rho}_a$ are projected out. We can write this as

$$\langle A^{\rho_a}_i | A^{\rho_a}_j \rangle = \int_K \omega \wedge A^{\rho_a}_i \wedge {}^* A^{\overline{\rho}_a}_j \tag{4.13}$$

the raised star here meaning Kodaira-Serre duality. This expression is independent of the complex structure (as is clear from eq.(4.12)) but, in general, depends upon the cohomology class of the Kähler form of $K$.

From eq.(4.11) and eq.(4.13) we obtain the normalised couplings for $27^3$ ($\overline{27}^3$) fields by setting $\rho_a = 3$ ($\overline{3}$) of $SU(3)$ and $i = 1,\ldots,h^{2,1}(K)$ ($i = 1,\ldots,h^{1,1}(K)$)‡. As an example consider the $27^3$ couplings.

With the substitution $\rho_a$ =holomorphic tangent index

$$g_{ijk} = \int_K \omega \wedge A_i^m \wedge A_j^n \wedge A_k^p. \tag{4.14}$$

Eq.(4.14) depends on the complex structure of $K$ via $\omega$, but does *not* depend on additional information about the metric, such as the Kähler class. Thus, $g_{ijk}$ can be evaluated knowing *only* the cohomology classes of the $A_i^m$ in $H^1(K,T) \cong H_{\bar{\partial}}^{2,1}(K)$.

For the normalisation of the 27 fields we choose a basis for the $A_i^l$s such that

$$G_{i\overline{m}np} \equiv A_{i\overline{m}}^l \omega_{lnp} \tag{4.15}$$

with the $G_i$ generating integral cohomology in $H^3(K,Z)$. Then, eq.(4.14) becomes

$$\langle G_i | G_j \rangle = \int_K G_i \wedge {}^*G_j \tag{4.16}$$

which is the intersection number on three forms, and happens *not* to depend either on the Kähler form or the metric. In practise, we shall find it simpler to rely on some techniques in deformation theory to calculate the $27^3$ couplings. We shall return to this point in section 6.

We turn now to the calculation of the tree level topological Yukawa couplings amongst the $\overline{27}$ fields which, for the duration of this section, we shall denote by $F_i$. Recall that these are associated with harmonic (1,1) forms on $K$. Using eq.(4.11), it is easy to show that, in the field theory limit, the $\overline{27}^3$ Yukawa couplings are given by the integral

$$g_{ijk} \equiv \int_K F_i \wedge F_j \wedge F_k \tag{4.17}$$

Rather than attack the calculation of this integral from the cohomological viewpoint, we shall find it easier to approach the problem using homology. Since $h^{2,0}(K) = 0$, the $F_i$'s span the second *de Rham* cohomology group of $K$; we choose to expand the $F_i$ in the natural basis supplied by the integral cohomology group $H^2(K,Z)$. By duality pairing, this group is isomorphic to $H_4(K,Z)$, the fourth integral homology group of $K$. This is an

‡ One may also write down the corresponding expression for the $1 \cdot 27 \cdot \overline{27}$ and $1^3$ couplings, although it is not yet known how to go about calculating the integrals.

isomorphism of ring structures of $H^*(K,Z)$ and $H_*(K,Z)$, where the intersection product is the homological equivalent of the wedge in cohomology. The intersection product of two homology classes is the oriented intersection of two surfaces which represent these classes and which meet transversally [19].

Using this isomorphism, we can rewrite eq.(4.17) in terms of the intersection in homology as

$$g_{ijk} \equiv (F_i, F_j, F_k) \tag{4.18}$$

where we use $F_i$ to denote both the (1,1) form and its corresponding element in homology. Note that the Yukawa couplings so defined are topological *invariants*. The normalisation matrix for the kinetic terms in the lagrangian can be found, using eq.(4.13), to be [3][4]

$$N(F_i, F_j) = \frac{3(F_i, J, J)(F_j^*, J, J)}{8(J, J, J)} - \frac{(F_i, F_j^*, J)}{4} \tag{4.19}$$

where $J$ is the Kähler form of $K$, and the basis is chosen to be complex (that is, we write the $F_i$ as elements of $C \times H^2(M,Z)$, and the "$*$" conjugates the coefficients in $C$). It is easy to see that eq.(4.19) depends only on the cohomology class of the Kähler form, and *not* directly on the complex structure of $K^{\ddagger}$.

Thus, we can in theory utilise eqns.(4.12),(4.16),(4.18),(4.19) to calculate the normalised $27^3$ and $\overline{27}^3$ Yukawa couplings amongst the four-dimensional massless fields arising from compactification. It should be noted that these couplings arise as scalar functions on the moduli space of the Calabi-Yau manifold $K$. The normalised $27^3$ couplings depend on the complex structure of $K$, while the $\overline{27}^3$ couplings depend only on the class of the Kähler form. In section 6, we shall perform these calculations for a phenomenologically attractive vacuum state.

## 5. THREE GENERATION CALABI-YAU MANIFOLDS

As we have seen, various mainly phenomenological considerations lead us to favour compactification of the field theory limit of the superstring on three dimensional Calabi-Yau manifolds. It is believed that the number of such manifolds is very large (possibly of

‡ Although the discrete symmetries *do* depend on the complex structure, eq.(4.18) vanishes if it is not invariant under an automorphism of the base manifold, so there is no contradiction.

the order of tens of thousands) though finite [20]. In the absence of knowledge regarding the ultimate dynamics selecting *the* vacuum state, the imposition of further phenomenological constraints might, hopefully, reduce the number of possibilities further. The most obvious such constraint is the number of low energy generations of elementary particles; present experimental evidence makes three generation compactifications empirically favourable. In this section we shall outline a search for "three generation" Calabi-Yau manifolds (i.e. those with $|\chi| = 6$, as we shall only be considering the $V_1 = TK$ scenario). The reader is referred to [21]for more detail regarding this procedure. Ideally, the task we would set ourselves is that of generating *all* possible Calabi-Yau manifolds, and then compiling a complete list of those with Euler characteristic $\pm 6$. This is, unfortunately, an extraordinarily difficult task. In light of this, we shall consider the largest known class: all Calabi-Yau manifolds whose simply-connected covering spaces are realisable as complete intersections in arbitrary products of complex projective spaces. There are Calabi-Yau manifolds known which lie outside this class: for example, the $Z$ manifold [5], which is realised as an orbifold with the singularities blown up to yield a differentiable manifold, and *any* manifold with $\chi > 0$. One further assumption it is necessary to make in order for our search to be feasible, is that the symmetries we use for moving from the covering space to the manifold itself are inherited from symmetries of the product of projective spaces in which we embed the manifold (the "ambient" space). It is highly plausible that the class thus defined is a very large proportion of all Calabi-Yau manifolds.

Let us recall, for a moment, the properties of complete intersections. Consider an algebraic variety $M$ defined through the embedding

$$CP^N \supset \prod_{i=1}^{m} CP^{n_i} \supset H^j \supset M \tag{5.1}$$

where $N$ is chosen suitably large to embed the product of $CP^{n_i}$ (this can always be done), and $H^j$ is defined as the zero locus of a homogeneous polynomial $p_j$ in the coordinates of $CP^{n_i} \times \ldots \times CP^{n_m}$. We assume that $M$ is defined by $c$ such polynomials

$$M \equiv \bigcap_{j=1}^{c} H^j \tag{5.2}$$

$M$ is then a complete intersection, it is automatically Kähler (by inheriting the Kähler form from $CP^N$) and has dimension $\sum n_i - c$. Recall that $M$ contains no singularities if the form $\Omega = dp_1 \wedge \ldots \wedge dp_c$ vanishes nowhere on $M$. We can represent $M$ by the "homogeneity matrix" [22]

$$\begin{array}{c||cccc} n_1 & d_{11} & d_{12} & \ldots & d_{1c} \\ n_2 & d_{21} & d_{22} & \ldots & d_{2c} \\ \vdots & \vdots & \vdots & \ddots & \vdots \\ n_m & d_{m1} & d_{m2} & \ldots & d_{mc} \end{array} \tag{5.3}$$

where $n_i$ is the dimension of the i$^{th}$ complex projective space and $d_{ij}$ is the degree of the j$^{th}$ defining polynomial in the homogeneous coordinates of the i$^{th}$ projective space (we consider only matrices which are not block diagonal, as we are interested in Calabi-Yau manifolds not expressable as products of lower dimensional spaces).

To calculate the Chern classes of $M$ defined by a generic matrix eq.(5.3), recall that

$$c(T(CP^n)) = (1+x)^{n+1} \tag{5.4}$$

and, for a normal bundle $N_j$

$$c(N_j) = 1 + \sum_{i=1}^{m} d_{ij} x_i \tag{5.5}$$

where $x_i$ is the Kähler form on $CP^{n_i}$ normalised such that

$$\int_{CP^{n_1} \times \ldots \times CP^{n_m}} x_1^{n_1} x_2^{n_2} \ldots x_m^{n_m} = 1 \tag{5.6}$$

Then, since

$$\{T(CP^{n_1}) \oplus \ldots \oplus T(CP^{n_m})\}|_M \equiv T(M) \oplus N_1 \oplus \ldots \oplus N_c \tag{5.7}$$

and, for any two vector bundles $E$ and $F$

$$c(E \oplus F) = c(E) \wedge c(F) \tag{5.8}$$

we have the relationship

$$\prod_{i=1}^{m} (1+x_i)^{n_i+1} = c(T(M)) \prod_{j=1}^{c} (1 + \sum_{i=1}^{m} d_{ij} x_i) \tag{5.9}$$

Clearly, the vanishing of the first Chern class is given by

$$\sum_{j=1}^{c} d_{ij} x_i = n_i + 1 \qquad \forall i \tag{5.10}$$

and the Euler characteristic is just

$$\chi(M) = \int_M c_3(T(M)) \tag{5.11}$$

and so, since we know $c_3$ from eq.(5.9), we need only know how to perform the integration. This we do using the "restriction form" $\bigwedge^j c(N_j)$ to allow us to integrate over $CP^{n_1}\times\ldots\times CP^{n_m}$ (this follows naturally if we think of $M$ as a homology class in $CP^{n_1}\times\ldots\times CP^{n_m}$). This gives us the final expression

$$\chi(M) = \left.\frac{\sum_{i=1}^{m}(1+x_i)^{n_i+1}\sum_{i=1}^{c} d_{ij}x_i}{\sum_{j=1}^{c}(1+\sum_{i=1}^{m} d_{ij}x_i)}\right|_{\prod_{i=1}^{m} x_i^{n_i}} \tag{5.12}$$

where the subscript on the right hand side means that we pick out the coefficient of the term $\prod_{i=1}^{m} x_i^{n_i}$. The complexity of this term necessitates the use of a computer.

We are now in a position to write down all possible homogeneity matrices, modulo some topological reduction rules, satisfying eq.(5.10) and to calculate their Euler characteristic using eq.(5.12). Having completed this task, we have generated a large class of generically simply connected Calabi-Yau manifolds. Every Calabi-Yau manifold can be constructed by modding out a freely acting finite group of isometries from a simply connected Calabi-Yau manifold [17]. We are thus led to consider the class of "QCICY", or quotients of complete intersection Calabi-Yau manifolds by finite discrete groups. What constraints apply in this case ?

Clearly, for a "CICY" manifold (complete intersection Calabi-Yau manifold) $K_0$, if $\frac{1}{2}\chi(K_0) = n$, then the cardinality $m = |G|$ of a freely acting group $G$ on $K_0$ must divide $n$ (this happens because $\frac{1}{2}\chi(K_0)$ is the index of the Dolbeault operator taking values in the tangent bundle of $K_0$). Thus, all occurrences in the QCICY class with $K_0$ as covering space must have $\frac{1}{2}\chi(K_0/G) = n/m \in Z$ for some $m \in Z$, and so the only CICY entries which can lead to three generations are those with Euler characteristic divisible by 6.

We can extend this type of argument further. Consider the twisted Dolbeault index eq.(2.38) on a Calabi-Yau manifold taking values in the vector bundle $E$ over $K_0$.

$$index\bar{\partial}_E = \frac{1}{12}\int_{K_0}[6c_3(E) - 6c_1(E)c_2(E) + c_1(E)c_2(T) + 2c_1^3(E)] \tag{5.13}$$

This is integral for a holomorphic vector bundle $E$. If $K_0$ admits a freely acting group $G$ and if $G$ lifts in a well defined fashion to a symmetry of $E$, the index eq.(5.13) will be divisible by the cardinality of $G$, $|G| = m$ (since the integrand in eq.(5.13) is invariant under $G$, while taking the quotient of $K_0$ by $G$ divides the region of integration by $m$). Note that setting $E = T$ ($T$ clearly has invariance under $G$) we retrieve our condition on the divisibility of the Euler characteristic by 6. Holomorphic vector bundles which must be invariant under $G$ are the cotangent bundle $T^*$, Whitney sums of $T^*$ and $T$, products of $T$ and $T^*$ and the trivial bundle $\mathcal{O}$. However, some simple algebra reveals that the indices of such objects yield no new divisibility criteria which any would-be covering space for a three generation manifold must satisfy.

However, we *can* go further by making the assumption alluded to earlier — we insist that $G$ is a projectively inherited, freely acting group. This means that, if $K_0$ is defined as a complete intersection in $CP^{n_1} \times \ldots \times CP^{n_m}$, then $G$ is a finite discrete subgroup of $PGL(n_1 + 1, C) \times \ldots \times PGL(n_m + 1, C)$ and the action of $G$ on $K_0$ is isomorphic to its action on the ambient space*. The first of these requirements means that elements of $G$ are linear transformations on the coordinates of the ambient space, and so, given an explicit set of defining equations for $K_0$, the group $G$ is a symmetry if the polynomials transform in a well defined representation of $G$.

One of the more technical issues in the search is the question of freely acting symmetries lying outside the aforementioned class — for example, permutation of $CP^n$ factors. It is possible to exclude the majority of such symmetries. The only case one is left unsure about is the class of symmetry groups which are not realisable as linear transformations in the ambient space. It is believed that such symmetries are rare — at any rate, a more thorough understanding of their existence would lie exterior to the complete intersection technology upon which our investigation shall draw heavily.

Having defined the area of our search — the class of quotients of complete intersection Calabi-Yau manifolds by projectively inherited, freely acting symmetries, or "PQCICY" — we are able to construct a few more restrictions, based upon index divisibility arguments. This we do by considering the normal bundle to $K_0$ in $CP^{n_1} \times \ldots \times CP^{n_m}$ for $K_0$ defined by the homogeneity matrix eq.(5.3):

$$N = N_1 \oplus \ldots \oplus N_c. \tag{5.14}$$

Here, $N$ is the sum of $c$ line bundles restricted to $K_0$, each one induced from a particular polynomial $p_i$. It is then easy to show that the structure of each of the line bundles $N_i$ is

* Note that this definition depends upon the particular embedding of $K_0$ in the ambient space, in addition to the topological class of $K_0$.

invariant under an *abelian* projectively inherited freely acting symmetry group (see [21]for more details on this point). Thus, with this caveat in mind, we can use the index of each normal bundle

$$index_{\overline{\partial}}(N_l) = \frac{1}{12}\int_{K_0}[c_1(N_l)c_2(T) + 2c_1^3(N_l)] \tag{3.15}$$

to restrict the manifolds further. The index of arbitrary products of the $N_i$ give a further set of constraints, but, after this, no more constraints are possible, or, indeed, necessary.

We have too little space to explain the details of topological equivalence of homogeneity matrices, reduction rules for making topologically equivalent homogeneity matrices out of larger ones, non-abelian freely acting symmetry groups and so on; for these we refer the reader to [21]. We therefore move on to the results of the algorithm. The CICY class is effectively realisable as (at most) 21,215 homogeneity matrices. Searching through these yields only seven cases which are capable of providing covering spaces for three generation manifolds. However, some elementary algebraic geometry reveals that these fall into two topologically distinct cases:

$$R_0 \equiv \begin{matrix} 3 \\ 3 \end{matrix}\left\|\begin{matrix} 3 & 0 & 1 \\ 0 & 3 & 1 \end{matrix}\right. \qquad B \equiv \begin{matrix} 3 \\ 3 \\ 3 \end{matrix}\left\|\begin{matrix} 2 & 0 & 0 & 1 & 1 & 0 \\ 0 & 2 & 0 & 1 & 0 & 1 \\ 0 & 0 & 2 & 0 & 1 & 1 \end{matrix}\right. \tag{5.16}$$

The first of these has $\chi(R_0) = -18$; it is easy to find a freely acting projectively inherited $Z_3$ on $R_0$. The resulting quotient manifold, $R$, first constructed by Yau [23], shall be the basis of the next section. The second manifold has $\chi(B) = -48$; we thus need a freely acting group of order 8. Such groups are: $Z_2 \times Z_2 \times Z_2$; $Z_2 \times Z_4$; $Z_8$; the square dihedral group $D_4$ and the quaternions $Q$. It appears possible to find freely acting order 4 groups, but *not* order 8. There is, at yet, no rigorous proof that a suitable freely acting $G$ does not exist on $B$, but it appears highly unlikely.

We thus come to the conclusion that, of the brobdignagian CICY class, only two manifolds can lead to three generation PQCICY quotients, and, of these, it appears that only one of the quotients can be constructed.

One might wonder whether this result also applies to other generation manifolds. It appears not. There exist simply connected CICY manifolds with every number of generations from zero to fifty-nine, inclusive, with the exception of 1, 3 and 5. However, there *are* potential covering spaces for 1 and 5 generations.

There is one last consideration that we have to confront: what of the four known $|\chi| = 6$ Calabi-Yau manifolds [23][24]? One of these four is the manifold $R$ described above, while the other three are constructed by resolving quotient singularities resulting from modding out projectively inherited, non-freely acting symmetries on CICY manifolds, and so lie outside the class we searched. However, two of these turn out to be diffeomorphically equivalent to $R$ [25], while the last is generically singular [21], and thus ill defined. Details of these proofs would take us too far afield, and so we refer the reader to the literature.

In conclusion, our search has reduced the number of known three generation Calabi-Yau manifolds to one, and has ruled out a huge number of cases from which one might, a priori, have constructed such manifolds. This is a curious and surprising result, and one can only speculate upon what mathematical principle might underlie it.

## 6. A THREE-GENERATION SUPERSTRING MODEL

We now are ready, at last, to present our model and to start investigating the resulting low energy theory [6]. The simply connected covering space of the manifold is defined by the locus of bi-degree (3,0), (0,3) and (1,1) equations in $CP^3 \times CP^3$ [23](a bi-degree $(a, b)$ equation in $CP^3 \times CP^3$ is an equation of degree $a$ in the coordinates of one $CP^3$ and of degree $b$ in the other). Let us call the 3-fold so defined $R_0$. We calculate first the Hodge numbers of $R$, the quotient of $R_0$ by a free $Z_3$ action.

Consider a single cubic surface (i.e. a degree 3 polynomial in $CP^3$) — call it $C$. We use the adjunction formula [19]

$$c(C) = \frac{(1+J)^4}{(1+3J)} \tag{6.1}$$

where $J$ is the Kähler form inherited from the $CP^3$, to obtain $c_2(C) = 3J^2$, and thus

$$\chi(C) = \int_C 3J^2 = 9 \tag{6.2}$$

since $\int_Y J^{n-k} = d_1 \ldots d_k$ for $k$ equations of degrees $\{d_1, \ldots, d_k\}$ in $CP^n$. Since $C$ is connected, $h^{0,0}(C) = 1$; moreover, we can invoke the Lefschetz Theorem [19]which tells us that $h^{1,0}(C) = 0$, and therefore

$$2h^{0,2}(C) + h^{1,1}(C) = 7. \tag{6.3}$$

However, the first Chern class of $C$ is non-vanishing from eq.(6.1), and so $h^{2,0}(C) = 0$. Thus, we have the Hodge diamond for $C$:

$$\begin{array}{ccccccc} & & & 1 & & & \\ & & 0 & & 0 & & \\ C: & 0 & & 7 & & 0 & \\ & & 0 & & 0 & & \\ & & & 1 & & & \end{array} \tag{6.4}$$

It is now a simple procedure to obtain the Hodge numbers of the product of two cubics in $CP^3 \times CP^3$, $C \times C'$, by using the Künneth formula,

$$h^{p,q}(C \times C') = \sum_{p=r+m,\ q=s+n} h^{r,s}(C) h^{m,n}(C') \tag{6.5}$$

and we find the Hodge diamond for $C \times C'$ is:

$$C \times C' : \quad \begin{array}{ccccccc} & & & 1 & & & \\ & & 0 & & 0 & & \\ & 0 & & 14 & & 0 & \\ 0 & & 0 & & 0 & & 0 \\ 0 \quad 0 & & & 51 & & & 0 \quad 0 \\ 0 & & 0 & & 0 & & 0 \\ & 0 & & 14 & & 0 & \\ & & 0 & & 0 & & \\ & & & 1 & & & \end{array} \tag{6.6}$$

To calculate the Hodge numbers of the vanishing locus of a bi-degree (1,1) equation in $C \times C'$ we can once again use the Lefschetz Theorem:

$$h^{p,q}(R_0) \begin{cases} = h^{p,q}(C \times C') & p+q < 3 \\ \leq h^{p,q}(C \times C') & p+q = 3 \end{cases} \tag{6.7}$$

This tells us that $h^{0,0}(R_0) = 1$, $h^{1,0}(R_0) = h^{2,0}(R_0) = 0$, and $h^{1,1}(R_0) = 14$. Now, $R_0$ is a Calabi-Yau manifold, so we know that $h^{3,0}(R_0) = 1$, and so $h^{2,1}(R_0) = 23$. We then have the Hodge diamond for $R_0$:

$$R_0 : \quad \begin{array}{ccccccc} & & & 1 & & & \\ & & 0 & & 0 & & \\ & 0 & & 14 & & 0 & \\ 1 & & 23 & & 23 & & 1 \\ & 0 & & 14 & & 0 & \\ & & 0 & & 0 & & \\ & & & 1 & & & \end{array} \tag{6.8}$$

As it stands, $R_0$ is a simply connected 9 generation Calabi-Yau manifold — to reach the promised three generation model we need a little more work. The extra ingredient is a freely acting order 3 discrete symmetry group $G$. This is generated by:

$$g : [x_0, x_1, x_2, x_3] \times [y_0, y_1, y_2, y_3] \rightarrow [x_0, \alpha^2 x_1, \alpha x_2, \alpha x_3] \times [y_0, \alpha y_1, \alpha^2 y_2, \alpha^2 y_3] \tag{6.9}$$

(where $\alpha = exp(2\pi i/3)$, and the $\{x_i\}$ and $\{y_i\}$ are homogeneous coordinates on the two $CP^3$'s). Clearly, the defining equations for $R_0$ must be invariant under $G$ (up to an overall scaling) in order to obtain an isometry, and one can then easily check that $G$ acts freely on the resulting manifold (with suitably chosen defining equations) $R \equiv R_0/G$, and that it can be chosen to be a non-singular complete intersection (this can be accomplished by ensuring that the "transversality form", defined as $\omega = dp^1 \wedge \ldots \wedge dp^3$, for $p^i$ the 3 defining equations, vanishes nowhere on the manifold). We shall have more to say on the precise form of the defining equations later on. However, we must now re-examine the cohomology in the light of having just taken the quotient of $R_0$ by $G$. Clearly, we must determine which of the cohomology classes of $R_0$ descend to $R$ — i.e. which are invariant under $G$. To do this, we make an excursion into deformation theory on complex manifolds [18].

Consider the most general defining equation for a cubic surface $C$:

$$\sum_{i,j,k=0}^{3} a_{ijk} x_i x_j x_k = 0 \tag{6.10}$$

where the $a_{ijk}$ are 19 complex parameters (as stated earlier, one must check that the manifold so defined is non-singular by examining the transversality form). Now, we may perform a linear transformation on the homogeneous coordinates of the $CP^3$ i.e. a transformation by an element of the projective linear group $PGL(4, C)$, which removes $4^2 - 1 = 15$ of the parameters leaving us with 4. We can choose the transformation to put eq.(6.10) in the form

$$\sum_{i=0}^{3} x_i^3 + a_1 x_0 x_1 x_2 + a_2 x_0 x_1 x_3 + a_3 x_0 x_2 x_3 + a_4 x_1 x_2 x_3 = 0. \tag{6.11}$$

The four monomials with coefficients $a_i$ are the so-called "deformations" of $C$'s complex structure. Now, it so happens that $dim\, H^1(C,T) = 4$ as well — this is no mere coincidence. It is a consequence of the work of Kodaira and Spencer [18]that the space of deformations of an algebraic variety $M$ is (with some exceptions; see for example [26]) isomorphic to the cohomology group $H^1(M,T)$. A useful consequence of this theorem is that, since $H^1(R_0,T) \cong H^{2,1}_{\bar\partial}(R_0)$ for $R_0$ Calabi-Yau, the massless 27 fields are in one-to-one correspondence with deformations of $R_0$'s complex structure. We shall have more to say on this point in due course.

Pressing on to the case of the product of two cubic surfaces, $C \times C'$, we find that we now have 8 deformations after using the $PGL$ freedom on both spaces. Now, when we put in the bi-degree (1,1) hyperplane equation, we have no more freedom left, and thus our "reduced" defining equations for $R_0$ are

$$\begin{gathered}\sum_{i=0}^{3} x_i^3 + a_1x_0x_1x_2 + a_2x_0x_1x_3 + a_3x_0x_2x_3 + a_4x_1x_2x_3 = 0\\ \sum_{i=0}^{3} y_i^3 + b_1y_0y_1y_2 + b_2y_0y_1y_3 + b_3y_0y_2y_3 + b_4y_1y_2y_3 = 0\\ \sum_{i,j=0}^{3} c_{ij}x_iy_j = 0\end{gathered} \tag{6.12}$$

where $c_{00} = 1$ in order to avoid generic singularities when we move to $R_0/G$. Thus, the *total* number of deformations is $4+4+16-1 = 23$ (we can remove one parameter by scaling any one of the monomials in the bi-degree (1,1) equation to 1). Thus, $dim\, H^1(R_0, T) \cong dim\, H^{2,1}_{\bar{\partial}}(R_0) = 23$, agreeing with the results of eq.(6.8).

Now, however, we wish to move to $R \equiv R_0/G$, and so we must look for those deformations that transform trivially under $G$. Using eq.(6.9), we find that the most general "reduced" defining equations for $R$ are

$$\begin{gathered}\sum_{i=0}^{3} x_i^3 + a_1x_0x_1x_2 + a_2x_0x_1x_3 = 0\\ \sum_{i=0}^{3} y_i^3 + b_1y_0y_1y_2 + b_2y_0y_1y_3 = 0\\ x_0y_0 + c_{11}x_1y_1 + c_{22}x_2y_2 + c_{33}x_3y_3 + c_{32}x_3y_2 + c_{23}x_2y_3 = 0\end{gathered} \tag{6.13}$$

giving rise to 9 deformations, and so we have the result $h^{2,1}(R) = 9$. It is now a simple matter to write down the full Hodge diamond for $R$:

$$R: \quad \begin{matrix} & & & 1 & & & \\ & & 0 & & 0 & & \\ & 0 & & 6 & & 0 & \\ 1 & & 9 & & 9 & & 1 \\ & 0 & & 6 & & 0 & \\ & & 0 & & 0 & & \\ & & & 1 & & & \end{matrix} \tag{6.14}$$

Note that eq.(6.13) also implies that the complex structure of the *vacuum*, $R$, is specified by choosing 9 complex numbers. Later in this section, when we find how the discrete symmetries of $R$ act on the massless fields of the theory, we shall demonstrate another way of deducing $h^{1,1}(R) = 6$ using fixed point theorems.

Having established our defining equations, we are now in a position to investigate the discrete symmetries of $R$. Since these depend, as discussed, on the chosen flux breaking scenario, we first describe such a choice.

Recall that the $G$ that we found freely acting on $R_0$, and by which we modded out to get $R$, was a discrete group of order 3 — $Z_3$ (see eq.(6.9)). We thus have to make a choice of embedding, $\overline{G}$ of $Z_3$ in $E_6$, which we represent by it's maximal decomposition into $[SU(3)]^3$. Our choice is the following

$$g \in G \to U_g = \begin{pmatrix} 1 & & \\ & 1 & \\ & & 1 \end{pmatrix} \otimes \begin{pmatrix} \alpha & & \\ & \alpha & \\ & & \alpha \end{pmatrix} \otimes \begin{pmatrix} \alpha & & \\ & \alpha & \\ & & \alpha \end{pmatrix} \in \overline{G} \tag{6.15}$$

embedded in $SU(3)_C \otimes SU(3)_L \otimes SU(3)_R$, where $\alpha$ is a non-trivial cube root of unity. This clearly breaks $E_6$ to $[SU(3)]^3$ *. Consider now the 27 of $E_6$. Under this decomposition

$$27 \longrightarrow (1,3,\bar{3}) \oplus (\bar{3},1,\bar{3}) \oplus (3,3,1) \tag{6.16}$$

and so leptons, which lie in $(1,3,\bar{3})$, are *invariant* under $\overline{G}$ (they transform as $\alpha \cdot \bar{\alpha} = 1$), left-handed antiquarks $Q$ transform as $Q \longrightarrow \alpha^2 Q$ under $\overline{G}$, while left-handed quarks $q$ transform as $q \longrightarrow \alpha q$. From the Hodge diamond of $R$, we see that 9 fields transformas the 27 of $E_6$, and 6 fields transform as the $\overline{27}$ of $E_6$. Each of these multiplets decomposes under $SU(3)^3$ as above. We are now ready to determine the discrete symmetry group of $R$ and its action on the fields in the theory.

Recall from that a simple way to do this is to find symmetries of $R_0$ which belong to the normaliser of $G$. That is, if $d$ is a symmetry of $R_0$, and $g$ a generator of $G$, we must be able to find a $g' \in G$ such that

$$dgd^{-1} = g'. \tag{6.17}$$

However, we first have to make a decision concerning the choice of parameters $a_i, b_i, c_{ij}$ in eq.(6.13), since, in general, different choices will lead to different discrete symmetry groups. Our choice is the following form:

$$\begin{aligned} \sum x_i^3 &= 0 \\ \sum y_i^3 &= 0 \\ (x_0 y_0 + x_1 y_1) + c(x_2 y_2 + x_3 y_3) &= 0 \end{aligned} \tag{6.18}$$

* There are other non-trivial flux breaking patterns leading to $[SU(3)]^3$ and $SU(6) \otimes U(1)$.

where we leave $c$ as an arbitrary constant. We shall motivate this choice in subsequent discussions.

The symmetry group of $R_0$ is easily seen to be the group $\mathcal{D}$ defined by

i $(x_0, y_0) \leftrightarrow (x_1, y_1)$ ; $(x_2, y_2) \leftrightarrow (x_3, y_3)$ (This is enlarged to the full diagonal permutation group if $c = 1$);

ii diagonal scalings of the coordinates by cube roots of unity, $x_i \to \alpha^{n_i} x_i$, and $y_i \to \alpha^{-n_i} y_i$ with $n_0 = 0$;

iii interchange of $x$ and $y$ coordinates, $x_i \leftrightarrow y_i$.

It is now a relatively simple matter to determine which elements of $\mathcal{D}$ are of the form eq.(6.18). They are generated by $g$ and

$$B : [x_0, x_1, x_2, x_3] \times [y_0, y_1, y_2, y_3] \to [x_0, x_1, \alpha x_2, x_3] \times [y_0, y_1, \alpha^2 y_2, y_3] \tag{6.19a}$$

$$C : [x_0, x_1, x_2, x_3] \times [y_0, y_1, y_2, y_3] \to [x_0, x_1, x_3, x_2] \times [y_0, y_1, y_3, y_2] \tag{6.19b}$$

$$D : [x_0, x_1, x_2, x_3] \times [y_0, y_1, y_2, y_3] \to [y_0, y_1, y_2, y_3] \times [x_0, x_1, x_2, x_3] \tag{6.19c}$$

$$P : [x_0, x_1, x_2, x_3] \times [y_0, y_1, y_2, y_3] \to [x_1, x_0, x_2, x_3] \times [y_1, y_0, y_2, y_3] \tag{6.19d}$$

We also define a redundant generator $A$

$$A : [x_0, x_1, x_2, x_3] \times [y_0, y_1, y_2, y_3] \to [\alpha x_0, \alpha x_1, x_2, x_3] \times [\alpha^2 y_0, \alpha^2 y_1, y_2, y_3] \tag{6.19e}$$

which will be of use in future calculations (note that $A = CB^2CB^2$).

Recall that these symmetries can survive flux breaking if we can find a homomorphism of the generator into an $E_6$ element $V_d$ such that

$$V_d^{-1} U_h V_d = U_{h'} \tag{6.20}$$

where $U_h$ is the image in $E_6$ of $h \in \pi_1(R)$, and $d^{-1} h d = h'$. For the symmetries $A$, $B$ and $C$, the trivial homomorphism suffices, since these generators commute with $G$. However, for $D$ and $P$ we need to work a bit harder, but such a homomorphism can be found [6]. Notice that our exact symmetry group contains *all* of the symmetry group $\mathcal{D}$ of $R_0$, and thus there are no pseudosymmetries (symmetries of $R_0$ which do not descend to symmetries of $R$) for this case. However, had we taken $c = 1$ in eq.(6.18), we would have found that the exact symmetry group was the same as for the $c \neq 1$ case, but that the pseudosymmetry group was the group of diagonal cyclic permutations (and their products with exact symmetries).

To determine the action of $\mathcal{D}$ on the fields in the theory we first recall that the deformations of the complex structure of $R_0$ are isomorphic to $H^1(R_0, T) \cong H^{2,1}_{\bar{\partial}}(R_0)$, and so the massless fields transforming as the 27 of $E_6$ can be represented by such objects. We therefore need to find representatives in terms of monomials for the 23 fields on $R_0$ (see eq.(6.8)) which descend to leptons and quarks on $R$. These can be found in eq.(6.12), and in Table 1 we list these monomials and their transformation properties under $G$.

| $G : \lambda_i \longrightarrow \lambda_i$ | $G : q_i \longrightarrow \alpha q_i$ | $G : Q_i \longrightarrow \alpha^2 Q_i$ |
|---|---|---|
| $\lambda_1 = x_0 x_1 x_2$ | $q_1 = x_1 x_2 x_3$ | $Q_1 = x_0 x_2 x_3$ |
| $\lambda_2 = x_0 x_1 x_3$ | $q_2 = y_0 y_2 y_3$ | $Q_2 = y_1 y_2 y_3$ |
| $\lambda_3 = y_0 y_1 y_2$ | $q_3 = x_0 y_1$ | $Q_3 = x_1 y_0$ |
| $\lambda_4 = y_0 y_1 y_3$ | $q_4 = x_1 y_2$ | $Q_4 = x_0 y_2$ |
| $\lambda_5 = x_1 y_1$ | $q_5 = x_1 y_3$ | $Q_5 = x_0 y_3$ |
| $\lambda_6 = x_2 y_2 + x_3 y_3$ | $q_6 = x_2 y_0$ | $Q_6 = x_2 y_1$ |
| $\lambda_7 = x_2 y_2 - x_3 y_3$ | $q_7 = x_3 y_0$ | $Q_7 = x_3 y_1$ |
| $\lambda_8 = x_2 y_3$ | | |
| $\lambda_9 = x_3 y_2$ | | |

**Table 1**

*Representatives of 27 fields in terms of deformations of $R_0$*

We see that the 23 harmonic (2,1) forms on $R_0$ descend to 9 leptonic and 7 quark/anti-quark fields on $R$, with the 9 harmonic (2,1) forms on $R$ being the 9 $G$ invariant lepton fields. It is now a simple matter to derive the transformation properties of the 27 fields under our exact symmetries eq.(6.19) using these representations in terms of complex deformations. These results are displayed in Table 2.

| | | $A$ | $B$ | $C$ | $D \oplus V_d$ | $P \oplus V_p$ |
|---|---|---|---|---|---|---|
| Leptons | $\lambda_1$ | $\alpha^2$ | $\alpha$ | $\to \lambda_2$ | $\to \lambda_3$ | 1 |
| | $\lambda_2$ | $\alpha^2$ | 1 | $\to \lambda_1$ | $\to \lambda_4$ | 1 |
| | $\lambda_3$ | $\alpha$ | $\alpha^2$ | $\to \lambda_4$ | $\to \lambda_1$ | 1 |
| | $\lambda_4$ | $\alpha$ | 1 | $\to \lambda_3$ | $\to \lambda_2$ | 1 |
| | $\lambda_5$ | 1 | 1 | $+1$ | 1 | $\to -\lambda_5 - \lambda_6$ |
| | $\lambda_6$ | 1 | 1 | $+1$ | 1 | 1 |
| | $\lambda_7$ | 1 | 1 | $-1$ | 1 | 1 |
| | $\lambda_8$ | 1 | $\alpha$ | $\to \lambda_9$ | $\to \lambda_9$ | 1 |
| | $\lambda_9$ | 1 | $\alpha^2$ | $\to \lambda_8$ | $\to \lambda_8$ | 1 |
| Left | $q_1$ | $\alpha$ | $\alpha$ | $+1$ | $\to Q_2$ | $\to Q_1$ |
| handed | $q_2$ | $\alpha^2$ | $\alpha^2$ | $+1$ | $\to Q_1$ | $\to Q_2$ |
| quarks | $q_3$ | 1 | 1 | $+1$ | $\to Q_3$ | $\to Q_3$ |
| | $q_4$ | $\alpha$ | $\alpha^2$ | $\to q_5$ | $\to Q_6$ | $\to Q_4$ |
| | $q_5$ | $\alpha$ | 1 | $\to q_4$ | $\to Q_7$ | $\to Q_5$ |
| | $q_6$ | $\alpha^2$ | $\alpha$ | $\to q_7$ | $\to Q_4$ | $\to Q_6$ |
| | $q_7$ | $\alpha^2$ | 1 | $\to q_6$ | $\to Q_5$ | $\to Q_7$ |
| Left | $Q_1$ | $\alpha$ | $\alpha$ | $+1$ | $\to q_2$ | $\to q_1$ |
| handed | $Q_2$ | $\alpha^2$ | $\alpha^2$ | $+1$ | $\to q_1$ | $\to q_2$ |
| anti- | $Q_3$ | 1 | 1 | $+1$ | $\to q_3$ | $\to q_3$ |
| quarks | $Q_4$ | $\alpha$ | $\alpha^2$ | $\to Q_5$ | $\to q_6$ | $\to q_4$ |
| | $Q_5$ | $\alpha$ | 1 | $\to Q_4$ | $\to q_7$ | $\to q_5$ |
| | $Q_6$ | $\alpha^2$ | $\alpha$ | $\to Q_7$ | $\to q_4$ | $\to q_6$ |
| | $Q_7$ | $\alpha^2$ | 1 | $\to Q_6$ | $\to q_5$ | $\to q_7$ |

**Table 2**

*Transformation of 27 fields under exact symmetries*

Having determined the transformation properties of the fields arising in the 27 of $E_6$, we now turn to those arising in the $\overline{27}$. Recall that these are *independent* fields, and are not related in a simple way to the 27's. In particular, we have no simple representation for them in terms of complex deformations. We thus proceed in a different manner which we now describe.

Our technique will be to rely on a very powerful result due to Lefschetz [19]known as the Fixed Point Theorem:

**Fixed Point Theorem.** *Let $f$ be an isometry of a Kähler manifold $M$ of complex dimension $n$. Define the Lefschetz number of $f$ as*

$$L(f) = \sum_{p,q=0}^{n} (-1)^{p+q} Tr_{(p,q)_M}(f) \tag{6.21}$$

*where $Tr_{(p,q)_M}(f)$ denotes the trace in the vector space $H^{p,q}_{\bar{\partial}}(M)$ of the action of $f$ on harmonic (p,q) forms. Then*

$$L(f) = \sum_{\mu_j} \chi(\mu_j) \tag{6.22}$$

*where $\mu_j$ are the sub-manifolds of $M$ left fixed by $f$ and $\chi(\mu_j)$ are their Euler characteristics.*

Note also that

$$L(identity) = \chi(M). \tag{6.23}$$

To apply this theorem on our harmonic (1,1) forms note that, as a consequence of the Lefschetz Hyperplane Theorem, of the 14 members of $H^{1,1}_{\bar{\partial}}(R_0)$, 7 originate on one $CP^3$ cubic and 7 on the other. In other words, we can, for many calculational purposes, decompose

$$H^{1,1}_{\bar{\partial}}(R_0) \cong H^{1,1}_{\bar{\partial}}(C) \oplus H^{1,1}_{\bar{\partial}}(C') \tag{6.24}$$

where $C$ and $C'$ are the two cubic surfaces. In fact, the hyperplane theorem tells us that these forms do not "see" the hyperplane at all. Thus we can just consider the action of symmetries on the (1,1) forms of the cubic surfaces originating from the Hodge diamond of eq.(6.4).

For any $d \epsilon \mathcal{D}$ we split $d$ into two generators $d_x$ and $d_y$ acting on the two cubic surfaces. Then

$$L(d_x) = \chi(\textit{fixed point set of } d_x) = 2 + Tr_{(1,1)_C}(d_x) \tag{6.25}$$

where the second equality follows from eq.(6.4) (recall that the (0,0) and (2,2) forms are necessarily invariant under all isometries). We can write a similar expression for $L(d_y)$. Then we have

$$\begin{aligned} Tr_{(1,1)_{R_0}}(d) &= Tr_{(1,1)_C}(d_x) + Tr_{(1,1)_{C'}}(d_y) \\ &= \chi(\textit{fixed point set of } d_x \textit{ on } C) + \chi(\textit{fixed point set of } d_y \textit{ on } C') - 4 \end{aligned} \tag{6.26}$$

Hence, for each $d$ we can determine the trace of its action on $H^{1,1}$. Using this information, together with the character table for $\mathcal{D}$ and a list of its irreducible represenatation [6]( all easily calculable using standard finite group theory), the action of $\mathcal{D}$ on the $\overline{27}$ is determined. Details may be found in [6]. The final results are contained in Table 3.

| | | $A$ | $B$ | $C$ | $D \oplus V_d$ | $P \oplus V_p$ |
|---|---|---|---|---|---|---|
| Leptons | $\overline{\lambda}_1$ | 1 | 1 | +1 | $\to \overline{\lambda}_2$ | +1 |
| | $\overline{\lambda}_2$ | 1 | 1 | +1 | $\to \overline{\lambda}_1$ | +1 |
| | $\overline{\lambda}_3$ | $\alpha$ | $\alpha$ | −1 | $\to \overline{\lambda}_5$ | −1 |
| | $\overline{\lambda}_4$ | $\alpha$ | $\alpha$ | −1 | $\to \overline{\lambda}_6$ | −1 |
| | $\overline{\lambda}_5$ | $\alpha^2$ | $\alpha^2$ | −1 | $\to \overline{\lambda}_3$ | −1 |
| | $\overline{\lambda}_6$ | $\alpha^2$ | $\alpha^2$ | −1 | $\to \overline{\lambda}_4$ | −1 |
| Left | $\overline{Q}_1$ | 1 | $\alpha$ | $\to \overline{Q}_2$ | $\to \overline{q}_4$ | $\to \overline{q}_1$ |
| handed | $\overline{Q}_2$ | 1 | $\alpha^2$ | $\to \overline{Q}_1$ | $\to \overline{q}_3$ | $\to \overline{q}_2$ |
| quarks | $\overline{Q}_3$ | 1 | $\alpha$ | $\to \overline{Q}_4$ | $\to \overline{q}_2$ | $\to \overline{q}_3$ |
| | $\overline{Q}_4$ | 1 | $\alpha^2$ | $\to \overline{Q}_3$ | $\to \overline{q}_1$ | $\to \overline{q}_4$ |
| Left | $\overline{q}_1$ | 1 | $\alpha$ | $\to \overline{q}_2$ | $\to \overline{Q}_4$ | $\to \overline{Q}_1$ |
| handed | $\overline{q}_2$ | 1 | $\alpha^2$ | $\to \overline{q}_1$ | $\to \overline{Q}_3$ | $\to \overline{Q}_2$ |
| anti- | $\overline{q}_3$ | 1 | $\alpha$ | $\to \overline{q}_4$ | $\to \overline{Q}_2$ | $\to \overline{Q}_3$ |
| quarks | $\overline{q}_4$ | 1 | $\alpha^2$ | $\to \overline{q}_3$ | $\to \overline{Q}_1$ | $\to \overline{Q}_4$ |

**Table 3**

*Transformation of $\overline{27}$ fields under exact symmetries*

There is one important point that we have so far failed to mention. We shall eventually be using our exact symmetries and their action upon the fields in the theory as a way of restricting Yukawa couplings arising in the low energy superpotential. Now, it should be noted that the "physical" quantity of interest here is really the integral $\int d^2\theta W$ over the Grassmannian superfield coordinates, where $W$ is the superpotential. We should therefore check that none of our symmetries leave $W$ invariant, but *not* $\int d^2\theta W$ — such symmetries would be called "R–symmetries". To check this we must investigate how the holomorphic three-form $\omega$ transforms i.e. a symmetry is not an $R$-symmetry if it leaves the holomorphic three-form invariant [16].

To do this, we once again invoke the Lefschetz Fixed Point Theorem, and use our knowledge of how the harmonic (2,1) and (1,1) forms transform. Let $\xi_f$ represent the action of an isometry $f$ on $\omega$ i.e. $f : \omega \longrightarrow \xi_f\omega$. Then, in our usual notation,

$$L(f) = \sum_{\{\mu_j\}} \chi(\mu_j) = 2[1 + Tr_{1,1}(f) - Re(Tr_{2,1}(f))] - 2Re(\xi_f). \tag{6.27}$$

The presence of only the real components of two of the these terms is due to the fact that, for example, $H^{3,0}_{\overline{\partial}}(R_0)$ and $H^{0,3}_{\overline{\partial}}(R_0)$ are related by complex conjugation. Now, if $\xi_f = 1$, $f$ leaves $\omega$ invariant, and so cannot be an R-symmetry. It is straightforward to test whether our symmetries are in fact R-symmetries using this procedure, and the reader can check that *all* of our symmetries leave $\omega$ invariant.

For the remainder of this section, we turn to the explicit calculation of the $27^3$ and $\overline{27}^3$ Yukawa couplings arising from compactification on the manifold $R$. First we discuss the $27^3$ couplings [27], which are more important from the phenomenological viewpoint, since they give rise to the observable low-energy matter fields.

Let $M$ be a Kähler manifold realised as a complete intersection in the ambient space $CP^{n_1} \times ...CP^{n_p}$. Then we have the exact sequence of holomorphic vector bundles

$$0 \longrightarrow T_M \longrightarrow T_{CP^{n_1}\times ...\times CP^{n_p}}|_M \longrightarrow N \longrightarrow 0 \tag{6.28}$$

$N$ is the normal bundle of $M$ in $T_{CP^{n_1}\times ...\times CP^{n_p}}$, and consists of a Whitney sum of projective line bundles restricted to $M$ — one for each polynomial defining $M$ — with first Chern classes given by the degrees of the defining equations [19]. Thus, for our manifold $R_0$, the normal bundle $N$ to $R_0$ in $CP^3 \times CP^3$ is given by the Whitney sum of three line bundles with first Chern classes given by three times the hyperplane class of one $CP^3$ (pulled back to $R_0$), three times the hyperplane class of the other $CP^3$, and the sum of the two hyperplane classes. We call these $\mathcal{O}_{R_0}(3,0), \mathcal{O}_{R_0}(0,3)$ and $\mathcal{O}_{R_0}(1,1)$ respectively, and their global sections are given by all possible polynomials of bi-degree (3,0), (0,3) and (1,1) in $CP^3 \times CP^3$ restricted to $R_0$. Associated to eq.(6.28), we have the long exact sequence in cohomology:

$$0 \to H^0(M,T_M) \to H^0(M,T_{CP^{n_1}\times ...\times CP^{n_p}}|_M) \to H^0(M,N) \to H^1(M,T_M) \to \quad (6.29)$$

Now, if $M$ is a Calabi-Yau complete intersection — call it $R_0$ — $dimH^0(R_0,T) = h^{2,0}(R_0) = 0$. If, further, $H^1(R_0,T_{CP^{n_1}\times ...\times CP^{n_p}}|_{R_0}) = 0$, then eq.(6.29) contains the *short* exact sequence

$$0 \longrightarrow H^0(R_0,T_{CP^{n_1}\times ...\times CP^{n_p}}|_{R_0}) \xrightarrow{\gamma} H^0(R_0,N) \xrightarrow{\delta} H^1(R_0,T) \longrightarrow 0 \qquad (6.30)$$

which forms the basis for deformation theory on complex algebraic varieties [18]. In eq.(6.39), $H^0(R_0,T_{CP^{n_1}\times ...\times CP^{n_p}}|_{R_0})$ is the space of infinitesimal global coordinate transformations of $T_{CP^{n_1}\times ...\times CP^{n_p}}$ restricted to $R_0$ — i.e. just $PGL(n_1+1)\times\ldots\times PGL(n_p+1)$. These transformations clearly leave the complex structure of $R_0$ unaffected. So we can think of elements of $H^0(R_0,N)$ in eq.(6.30) as deforming $R_0$ by adding infinitesimal factors of their polynomials to $R_0$'s defining equations. However, some of the elements of $H^0(R_0,N)$ will *not* be the pullbacks of non-trivial deformations which lie in $H^1(R_0,T)$ — these are contained in the image of the map $\gamma$. To see this, suppose an element of $H^0(R_0,T_{CP^{n_1}\times ...\times CP^{n_p}}|_{R_0})$ is represented by the $PGL$ transformation $\{z_i^A \mapsto c_{iAB} z_i^B \; i = 1,\ldots,p\}$, where $z_i^A$ are coordinates on $CP^{n_i}$. To first order, this acts on the defining polynomials $p^\alpha$ of $R_0$ as

$$p^\alpha \mapsto c_{iAB} z_i^B \frac{\partial p^\alpha}{\partial z_i^A} \quad [\alpha = 1,\ldots,(\sum_{l=1}^{p} n_l - 3)] \qquad (6.31)$$

The complex structure remains unchanged by eq.(6.31), and arbitrary choices for the $c_{iAB}$ identify elements of $H^0(R_0,N)$ in the kernel of the map $\delta$ of eq.(6.30). Alternatively, if we think of elements of $H^0(R_0,N)$ as vectors, each component corresponding to one of $R_0$'s defining equations, the trivial deformations of eq.(6.31) correspond to the ideal $\mathcal{I}$ generated by the vectors $\frac{\partial p^\alpha}{\partial z_i^A}$ for all $i, A$ ($\alpha$ is the vector index). The *non-trivial* complex structure deformations of $R_0$ then correspond to all vectors with correct degree polynomials in each coordinate, modulo the ideal: $H^0(R_0,N)/\mathcal{I}$. Note that the complex structure of $R_0$ enters the calculation via the ideal, as expected.

One point that must be resolved is whether $H^1(R_0, T_{CP^{n_1}\times\ldots\times CP^{n_p}}|_{R_0}) = 0$. It turns out that this condition can, in many cases, be relaxed and, indeed, for our manifold it is *not* zero [28]. However, deformation theory can still be shown to be valid [28]and so the method still works.

We can now begin to see the relevance of this discussion to our calculation. In terms of $H^1(R_0, T)$, the Yukawa couplings are calculated as a cup product in cohomology [3]§:

$$H^1(R_0,T) \times H^1(R_0,T) \times H^1(R_0,T) \xrightarrow{\cup} H^3(R_0, T\otimes T\otimes T) \longrightarrow H^3(R_0, \mathcal{O}) \cong C \quad (6.32)$$

The second step in eq.(6.32) projects on to the singlet in the $3\otimes 3\otimes 3$ of the holonomy $SU(3)$. To represent this cubic product in terms of deformations, we need to lift it to the representations of $H^1(R_0,T)$ in $H^0(R_0,N)/\mathcal{I}$. This lifting is given by [29]

$$\left[\frac{H^0(R_0,N)}{\mathcal{I}}\right]^3 \xrightarrow{\phi} \frac{H^0(R_0, N\otimes N\otimes N)}{\mathcal{I}} \cong H^0(R_0, \Omega^3) \cong C \quad (6.33)$$

where the map, in terms of deformations, is given by

$$(q^\alpha, r^\beta, s^\gamma) \xrightarrow{\phi} q^\alpha r^\beta s^\gamma; \quad q^\alpha, r^\beta, s^\gamma \in \frac{H^0(R_0,N)}{\mathcal{I}}. \quad (6.34)$$

It can be shown (using simple algebra if $R_0$ is defined by one equation) that

$$H^0(R_0, N\otimes N\otimes N)/\mathcal{I}$$

is always one dimensional for Calabi-Yau $R_0$ (a non-trivial point, since for other types of manifolds in other dimensions one would not end up with such a unique projection). So the Yukawa coupling is given by multiplying together the deformation representatives of the 27 fields and modding out the ideal. We can then write this quantity as a complex number times the unique element of $H^0(R_0, N\otimes N\otimes N)/\mathcal{I}$. We should then, in principle, integrate this quantity over $R_0$ to give an overall factor.

One can also apply this technique to the calculation of the normalisation matrix, but for our purposes it will be enough to restrict these using the discrete symmetries alone. The "proper" (i.e. topological) calculation of the normalisations is given elsewhere [30].

| Field | Representation | Normalisation |
|---|---|---|
| $\lambda_1$ | $x_0x_1x_2$ | $N_1$ |
| $\lambda_2$ | $x_0x_1x_3$ | $N_1$ |
| $\lambda_3$ | $y_0y_1y_2$ | $N_1$ |
| $\lambda_4$ | $y_0y_1y_3$ | $N_1$ |
| $\lambda_{5M}$ | $x_1y_1$ | – |
| $\lambda_{6M}$ | $x_2y_2$ | – |
| $\lambda_{7M}$ | $x_3y_3$ | – |
| $\lambda_8$ | $x_2y_3$ | $N_2$ |
| $\lambda_9$ | $x_3y_2$ | $N_2$ |
| $\lambda_{5+}$ | $\sqrt{2}(\lambda_{6M}+\lambda_{7M})$ | $N_3$ |
| $\lambda_{5-}$ | $-(2\lambda_{5M}+c\lambda_{6M}+c\lambda_{7M})$ | $N_4$ |
| $\lambda_7$ | $(\lambda_{6M}-\lambda_{7M})$ | $N_5$ |

**Table 4**

Using these symmetries, we can write down a convenient choice of normalised fields — this is presented in Table 4.

We shall denote the normalisation matrix by

$$N[a,b] \equiv \langle a,b\rangle \tag{6.35}$$

which is constructed as an hermitian inner product on the polynomial representatives $a^\alpha$ and $b^\beta$. Then, using the discrete symmetries one obtains the normalisations denoted in the third column of Table 4. Note that, in the limit $c \to 1$ (the maximally symmetric case), $N_3 = N_4 = N_5$, $N_1 = N_6$, and $N_2 = N_7 = N_8$, leaving three independent normalisations.

| | | |
|---|---|---|
| $q_1$ | $x_1x_2x_3$ | $N_6$ |
| $q_2$ | $y_0y_2y_3$ | $N_6$ |
| $q_3$ | $x_0y_1$ | $N_7$ |
| $q_4$ | $x_1y_2$ | $N_8$ |
| $q_5$ | $x_1y_3$ | $N_8$ |
| $q_6$ | $x_2y_0$ | $N_8$ |
| $q_7$ | $x_3y_0$ | $N_8$ |
| $Q_1$ | $x_0x_2x_3$ | $N_6$ |
| $Q_2$ | $y_1y_2y_3$ | $N_6$ |
| $Q_3$ | $x_1y_0$ | $N_7$ |
| $Q_4$ | $x_0y_2$ | $N_8$ |
| $Q_5$ | $x_0y_3$ | $N_8$ |
| $Q_6$ | $x_2y_1$ | $N_8$ |
| $Q_7$ | $x_3y_1$ | $N_8$ |

**Table 4, continued**

*Polynomial representation for the normalised states for the manifold defined by eq.(6.18).*

This follows from the defining equations

$$\begin{aligned} p^1 &\equiv \frac{1}{3}\sum_A x_A^3 = 0 \\ p^2 &\equiv x_0y_0 + x_1y_1 + c(x_2y_2 + x_3y_3) = 0 \\ p^3 &\equiv \frac{1}{3}\sum_A y_A^3 = 0 \qquad . \end{aligned} \tag{6.36}$$

by squaring the second of these with $c = 1$:

$$\sum_A |x_A y_A|^2 + \sum_{B \neq A} [x_A y_A][\overline{x_A y_B}] = 0 \tag{6.37}$$

and substituting into the normalisation matrix.

Equipped with this information, we can now calculate the Yukawa couplings using the method of ref. [29]. We shall use the "ideal" equivalence

$$Q^{\alpha\beta\gamma} \cong Q^{\alpha\beta\gamma} + X^{(\alpha\beta|A|} \frac{\partial p^{\gamma)}}{\partial x^A} + Y^{(\alpha\beta|A|} \frac{\partial p^{\gamma)}}{\partial y^A} + F_\delta^{\alpha\beta\gamma} p^\delta \tag{6.38}$$

where $Q^{\alpha\beta\gamma}$ is a polynomial resulting from the product of three representatives of the 27 fields, and $X, Y, F$ are arbitrary polynomials of the appropriate degree. Some simple linear algebra reveals that eq.(6.38) allows us to express any $Q$ in terms of, for example, $Q^{123}$. For instance, suppose only $X^{13A}$ were non-zero:

$$\begin{aligned} Q^{113} &\cong Q^{113} + \frac{2}{3} X^{13A} x_A^2 \\ Q^{123} &\cong Q^{123} + \frac{1}{3} X^{13A} c_A y_A \end{aligned} \tag{6.39}$$

($c_A = 1$, $A = 0,1$, $C_A = c$, $A = 2,3$). Then, if $Q^{113} = x_0^3 x_2^3 y_0^3$, for example, choosing $X^{13A} = -\frac{3}{2} x_0^3 x_2 y_0^3$ will give $Q^{113} = 0$, $Q^{123} \cong -\frac{c}{2} x_0^3 y_0^3 x_2 y_2$. In Table 5, we list the factors which take any $Q^{\alpha\beta\gamma}$ to $Q^{123}$.

| Initial monomial | Multiplicative factor taking this to $Q^{123}$ |
|---|---|
| $Q^{222}$ | $\frac{1}{6} \frac{x_A^2}{c_A y_A} \frac{y_B^2}{c_B x_B}$ |
| $Q^{111}$ $(Q^{333})$ | $-\frac{1}{6} \frac{c_A y_A}{x_A^2} \frac{c_B y_B}{x_B^2} \frac{y_C^2}{c_C x_C}$ |
| $Q^{112}$ $(Q^{233})$ | $\frac{1}{2} \frac{c_A y_A}{x_A^2} \frac{y_B^2}{c_B x_B}$ |
| $Q^{113}$ $(Q^{133})$ | $-\frac{1}{2} \frac{c_A y_A}{x_A^2}$ |
| $Q^{122}$ $(Q^{223})$ | $-\frac{1}{2} \frac{y_A^2}{c_A x_A}$ |

**Table 5**

*Multiplicative factors taking monomials to the standard form. For the bracketed entries, the result is obtained by the substitution $x \leftrightarrow y$. The indices $A, B, C$ may be chosen to suit the particular monomial.*

Having reduced all the $Q^{\alpha\beta\gamma}$ to $Q^{123}$, we merely have to relate the different monomials in terms of which $Q^{123}$ appears. This is not difficult. If $\xi_A = x_A y_A$ $(A = 0,1,2,3)$, an application of eq.(6.38) with only $X^{13A}$ and $Y^{12E}$ non-zero yields

$$Q^{123} \cong Q^{123} \frac{\xi_B}{\xi_A} \left[ \frac{c_A^2}{c_B^2} \right] \tag{6.40}$$

for $Q^{123} \in \xi_A^2 \xi_B$. Thus

$$\begin{aligned} \xi_A^3 \xi_B &= \frac{c_A^2}{c_B^2} \xi_A^2 \xi_B^2 \\ \xi_A \xi_B \xi_C^2 &= \frac{c_C^2}{c_B^2} \xi_A \xi_B^2 \xi_C \end{aligned} \tag{6.41}$$

Further identities may be obtained by using $p^2 \equiv 0$. For example

$$\begin{aligned} \xi_1 \xi_2 \xi_3^2 &= -\frac{c}{2}(\xi_2 + \xi_3)\xi_2 \xi_3^2 = -c\xi_2^2 \xi_3^2 \\ \prod_i \xi_i &= \frac{1}{c}(1 + 2c^3)\xi_2^2 \xi_3^2 \\ \xi_{0,1}^4 &= -\xi_0^2 \xi_1^2 - \frac{2}{c}\xi_0^2 \xi_2^2 \\ \xi_{2,3}^4 &= -2c\xi_0^2 \xi_2^2 - \xi_2^2 \xi_3^2 \\ \xi_0^2 \xi_1^2 &= \frac{c(1+2c^3)}{2+c^3}\xi_2^2 \xi_3^2 \\ \xi_0^2 \xi_2^2 &= \frac{3c^2}{2+c^3}\xi_2^2 \xi_3^2 \end{aligned} \tag{6.42}$$

With these results we obtain the various Yukawa couplings, presented in Tables 6,7 and 8.

The normalisation factors should clearly be included in the final lagrangian, but we leave them out of these Tables for the sake of clarity.

Note that these results can lead to a considerable *hierarchy* of couplings if $|c|$ is large or small. Also, in the limit $|c| \to 0$ or $|c| \to \infty$, the manifold degenerates into an orbifold (with 3 or 9 singular points respectively) and certain Yukawa couplings become infinite.

We now turn to the direct calculation of the tree level $\overline{27}^3$ Yukawa couplings for compactification based on $R$ [31]

Consider a generic $\overline{27}^3$ Yukawa coupling on $R_0$. The Yukawa coupling integral of eq.(4.17) on $R$ is non-zero only if it involves two forms from one cubic surface and one from the other — this is obvious from the cohomological expression

$$\int_{R_0} F_i^x \wedge F_j^x \wedge F_k^y = \int_C F_i^x \wedge F_j^x \int_{C'} J^y \wedge F_k^y \tag{6.43}$$

| | |
|---|---|
| $\lambda_1\lambda_4\lambda_9 + \lambda_2\lambda_3\lambda_8$ | $\frac{2c^3+1}{c}$ |
| $\lambda_1^3 + \lambda_2^3 + \lambda_3^3 + \lambda_4^3$ | $\frac{c}{6}\frac{1+2c^3}{2+c^3}$ |
| $\lambda_8^3 + \lambda_9^3$ | $\frac{1}{6c^2}$ |
| $[\lambda_1\lambda_3 + \lambda_2\lambda_4]\lambda_{5+}$ | $\frac{2\sqrt{2}(1+2c^3)}{c(2+c^3)}$ |
| $[\lambda_1\lambda_3 - \lambda_2\lambda_4]\lambda_7$ | $\frac{-2(1+2c^3)(1+c^3)}{c(2+c^3)}$ |
| $\lambda_8\lambda_9\lambda_{5+}$ | $\frac{\sqrt{2}}{3c^2}$ |
| $\lambda_{5-}^2\lambda_{5+}$ | $\frac{4\sqrt{2}}{3}$ |
| $\lambda_{5+}^3$ | $\frac{8\sqrt{2}(1-c^3)}{3c^2(2+c^3)}$ |
| $\lambda_7^2\lambda_{5+}$ | $-\frac{4\sqrt{2}}{3}\frac{(1+2c^3)}{c^2(c^3+2)}$ |

**Table 6**

$\lambda^3$ *couplings in units of* $\xi_2^2\xi_3^2$

where $C$ and $C'$ are the usual cubic surfaces, and superscripts denote forms on $C$ ($x$ coordinates) or $C'$ ($y$ coordinates) — $J^y$ is the Kähler form on $C'$. The homological version of this is

$$(F_i^x, F_j^x, F_k^y) = n(F_k^y, J^y) \tag{6.44}$$

where $n$ is the number of points at which $F_i^x$ and $F_j^x$ intersect on $C$. Thus, the problem reduces to that of computing the intersection pairings on the cubic surface.

Fortunately, the cubic surface in $CP^3$ is a very well understood subject [32]. The key result in this area is the fact that all such non-singular cubic surfaces can be realised as $CP^2$ with six points blown up. These blown up points can be thought of as just $CP^1$'s, and are termed exceptional divisors [19]. They increase the dimension of $H_2(C)$ (we are working in a complex basis now) from one — for the $CP^2$ alone — to seven. This ties in with our previous results on the dimension of $H^2(R_0)$.

Another useful result is the renowned statement that a cubic surface contains precisely 27 lines (or $CP^1$'s) which can be partitioned into the seven homology classes. For example,

| | |
|---|---|
| $\lambda_1 q_2 Q_7, \lambda_2 q_2 Q_6, \lambda_3 q_5 Q_1, \lambda_4 q_4 Q_1 + (q \leftrightarrow Q)$ | $\frac{1}{c}(1+2c^3)$ |
| $\lambda_1 q_3 Q_4, \lambda_2 q_3 Q_5, \lambda_3 q_6 Q_3, \lambda_4 q_7 Q_3 + (q \leftrightarrow Q)$ | $\frac{1+2c^3}{2(2+c^3)}$ |
| $\lambda_1 q_6 Q_6, \lambda_2 q_7 Q_7, \lambda_3 q_4 Q_4, \lambda_4 q_5 Q_5 + (q \leftrightarrow Q)$ | $\frac{c(1+2c^3)}{2(2+c^3)}$ |
| $\lambda_8 q_1 Q_6, \lambda_8 q_5 Q_2, \lambda_9 q_4 Q_2, \lambda_9 q_1 Q_7 + (q \leftrightarrow Q)$ | $\frac{1}{2}$ |
| $\lambda_8 q_4 Q_7, \lambda_9 q_5 Q_6, +(q \leftrightarrow Q)$ | $-\frac{1}{6c}$ |
| $\lambda_{5+} q_1 Q_2 + \lambda_{5+} q_2 Q_1$ | $-2\sqrt{2}$ |
| $\lambda_{5-} q_1 Q_2 - \lambda_{5-} q_2 Q_1$ | $\frac{2(1+c^3)}{c}$ |
| $\lambda_{5+} q_3 Q_3$ | $\frac{-\sqrt{2}(1+2c^3)}{3(2+c^3)}$ |
| $\lambda_{5+}(q_4 Q_6 + q_6 Q_4 + q_5 Q_7 + q_7 Q_5)$ | $\frac{\sqrt{2}(c^3-1)}{3c(2+c^3)}$ |
| $\lambda_{5-}(q_4 Q_6 - q_6 Q_4 + q_5 Q_7 - q_7 Q_5)$ | $-\frac{1}{3}$ |
| $\lambda_7(q_4 Q_6 - q_5 Q_7 + q_6 Q_4 - q_7 Q_5)$ | $\frac{(1+2c^3)}{3c(2+c^3)}$ |

**Table 7**

*$\lambda qQ$ Yukawas (diagonal basis) in units of $\xi_2^2 \xi_3^2$.*

| | |
|---|---|
| $q_1^3, q_2^3, Q_1^3, Q_2^3$ | $\frac{c}{6}$ |
| $q_3^3, Q_3^3$ | $\frac{c(1+2c^3)}{6(2+c^3)}$ |
| $q_i^3, Q_i^3, (i = 4, \ldots, 7)$ | $\frac{c}{2(2+c^3)}$ |
| $q_3 q_4 q_6, q_3 q_5 q_7, Q_3 Q_4 Q_6, Q_3 Q_5 Q_7$ | $-\frac{(1+2c^3)}{6(2+c^3)}$ |
| $q_1 q_2 q_3, Q_1 Q_2 Q_3$ | $\frac{1}{c}(1+2c^3)$ |

**Table 8**

*$q^3$, $Q^3$ quark Yukawas in units of $\xi_2^2 \xi_3^2$*

on $C$, the 27 $CP^1$'s are

$$\begin{aligned} &x_0 + \alpha^j x_1 = x_2 + \alpha^k x_3 = 0 \\ &x_0 + \alpha^j x_2 = x_1 + \alpha^k x_3 = 0 \qquad (0 \le j, k \le 2) \\ &x_0 + \alpha^j x_3 = x_1 + \alpha^k x_2 = 0 \end{aligned} \tag{6.45}$$

The homology class of the Kähler form can be represented by the surface $H^x = \{x_0 = 0\} \in C$ — a torus in $C$. We also need to choose six combinations of the 27 $CP^1$'s to form a basis for $H_2(C)$ — these must be non-intersecting since they are, effectively, the six exceptional divisors. There are, in fact, 72 sets of disjoint lines; however, a discrete automorphism group (the Weyl group of $E_6$ [32]) takes any of these sets into any other. The choice we shall make is

$$\begin{aligned} E_1 &= \{x_0 + x_2 = x_1 + x_3 = 0\} \\ E_2 &= \{x_0 + \alpha x_2 = x_1 + \alpha^2 x_3 = 0\} \\ E_3 &= \{x_0 + \alpha^2 x_2 = x_1 + \alpha x_3 = 0\} \\ E_4 &= \{x_0 + x_3 = x_1 + \alpha x_2 = 0\} \\ E_5 &= \{x_0 + \alpha x_3 = x_1 + x_2 = 0\} \\ E_6 &= \{x_0 + \alpha^2 x_3 = x_1 + \alpha^2 x_2 = 0\} \end{aligned} \tag{6.46}$$

One can compute the necessary intersection pairings in various ways. For instance, using the fact that the self intersection of the Kähler form on $C$ — $(H^x, H^x)$ — is the degree of $N_{H^x|C}$, the normal bundle of $H^x$ in $C$, and the degree is the integral of the first Chern class over $H^x$, we have

$$\begin{aligned} c(N_{H^x|C}) &= \frac{c(C)}{c(H^x)} \\ &= \frac{(1+H^x)^4}{(1+3H^x)} \frac{(1+3H^x)}{(1+H^x)^3} \\ &= 1 + H^x \end{aligned} \tag{6.47}$$

since we described $H^x$ by a hyperplane in $CP^3$. Then, using standard results (see below eq.(6.2)), we obtain

$$(H^x, H^x) = \int_{H^x} c_1(N_{H^x|C}) = 3. \tag{6.48}$$

All of the $E_i^x$ intersect $H^x$ once and are, by definition, non-intersecting, so

$$(E_i^x, H^x) = 1 \tag{6.49a}$$

$$(E_i^x, E_j^x) = 0 \qquad (i \neq j) \tag{6.49b}$$

The self intersection of the $E_i^x$ is given by

$$\begin{aligned} c(N_{E_i^x|C}) &= \frac{c(C)}{c(E_i^x)} \\ &= \frac{(1+H^x)^4}{(1+3H^x)(1+H^x)^2} \\ &= 1 - H^x \end{aligned} \tag{6.50}$$

so that

$$(E_i^x, E_i^x) = -1 \tag{6.51}$$

Of the remaining 21 lines, there is a partition into those which intersect two exceptional divisors, $E_i^x$ and $E_j^x$, and those that intersect all but one, $E_k^x$. The former 15 lines are usually referred to as $F_{ij}$ and the latter 6 as $G_k$. One can write these lines in terms of our homology basis as

$$\begin{aligned} G_i^{x,y} &\sim \frac{1}{3}(2H^{x,y} - \sum_{k=1}^{6} E_k^{x,y}) + E_i^{x,y} \\ F_{ij}^{x,y} &\sim \frac{1}{3}(H^{x,y} + \sum_{k=1}^{6} E_k^{x,y}) - E_i^{x,y} - E_j^{x,y} \end{aligned} \tag{6.52}$$

where $\sim$ indicates equivalence under homology. All these results carry through to the $C'$ case as well.

It is now trivial to calculate the full intersections:

$$\begin{aligned} (H^x, H^x, H^y) &= 9 \\ (E_i^x, H^x, H^y) &= 3 \\ (H^x, H^x, E_i^y) &= 3 \\ (E_i^x, H^x, E_j^y) &= 1 \\ (E_i^x, E_j^x, H^y) &= -3\delta_{ij} \\ (E_i^x, E_j^x, E_k^y) &= -\delta_{ij} \end{aligned} \tag{6.53}$$

plus the corresponding cases with $x \leftrightarrow y$, the others all vanishing.

To relate the homology basis to the physical fields, we examine the action of our discrete symmetries on the lines. For example

$$g : (E_1, E_2, E_3, E_4, E_5, E_6, H) \rightarrow (E_2, E_3, E_1, E_5, E_6, E_4, H) \tag{6.54}$$

so that $H$, $E_1+E_2+E_3$ and $E_4+E_5+E_6$ are invariant under $g$, these being the leptons, and the combinations transforming as $\alpha$ and $\alpha^2$ are the quarks and antiquarks. Similarly

$$\begin{aligned} P &: (E_1,E_2,E_3,E_4,E_5,E_6,H) \to (G_6,G_5,G_4,G_3,G_2,G_1,H) \\ B &: (E_1,E_2,E_3,E_4,E_5,E_6,H) \to (G_3,G_1,G_2,F_{46},F_{45},F_{56},H) \\ C &: (E_1,E_2,E_3,E_4,E_5,E_6,H) \to (G_6,G_4,G_5,G_2,G_3,G_1,H) \end{aligned} \tag{6.55}$$

while $D$ merely exchanges the (1,1) forms from $C$ and $C'$. This now uniquely determines the identification with the physical fields, and the results are tabulated in Table 9.

| Field | Homology representatives |
|---|---|
| $\overline{\lambda}_1$ | $\sqrt{2/3}(H^x+H^y)$ |
| $\overline{\lambda}_2$ | $H^x-H^y$ |
| $\overline{\lambda}_3$ | $\sqrt{2/3}[H^x+\alpha(E_1^x+E_2^x+E_3^x)+\alpha^2(E_4^x+E_5^x+E_6^x)]$ |
| $\overline{\lambda}_4$ | $\sqrt{2/3}[H^y+\alpha^2(E_1^y+E_2^y+E_3^y)+\alpha(E_4^y+E_5^y+E_6^y)]$ |
| $\overline{\lambda}_5$ | $\sqrt{2/3}[H^x+\alpha^2(E_1^x+E_2^x+E_3^x)+\alpha(E_4^x+E_5^x+E_6^x)]$ |
| $\overline{\lambda}_6$ | $\sqrt{2/3}[H^y+\alpha(E_1^y+E_2^y+E_3^y)+\alpha^2(E_4^y+E_5^y+E_6^y)]$ |
| $\overline{q}_1$ | $\sqrt{2}[\alpha^2E_4^x+\alpha E_5^x+E_6^x]$ |
| $\overline{q}_2$ | $\sqrt{2}[E_1^x+\alpha^2E_2^x+\alpha E_3^x]$ |
| $\overline{q}_3$ | $\sqrt{2}[\alpha^2E_4^y+\alpha E_5^y+E_6^y]$ |
| $\overline{q}_4$ | $\sqrt{2}[E_1^y+\alpha E_2^y+\alpha^2E_3^y]$ |
| $\overline{Q}_1$ | $\sqrt{2}[E_1^x+\alpha E_2^x+\alpha^2E_3^x]$ |
| $\overline{Q}_2$ | $\sqrt{2}[\alpha E_4^x+\alpha^2E_5^x+E_6^x]$ |
| $\overline{Q}_3$ | $\sqrt{2}[E_1^y+\alpha^2E_2^y+\alpha E_3^y]$ |
| $\overline{Q}_4$ | $\sqrt{2}[\alpha^2E_4^y+\alpha E_5^y+E_6^y]$ |

**Table 9**

*Representation of $\overline{27}$ fields in homology basis*

We can now write down the allowed couplings:

$$\begin{aligned} g_{ijk} = 2\overline{\lambda}_1^3 - \overline{\lambda}_1\overline{\lambda}_2^2 - \overline{\lambda}_1(\overline{\lambda}_3\overline{\lambda}_5+\overline{\lambda}_4\overline{\lambda}_6+\overline{Q}_1\overline{q}_2+\overline{Q}_2\overline{q}_1+\overline{Q}_3\overline{q}_4+\overline{Q}_4\overline{q}_3) \\ +\sqrt{\frac{3}{2}}\overline{\lambda}_2(\overline{\lambda}_3\overline{\lambda}_5-\overline{\lambda}_4\overline{\lambda}_6+\overline{Q}_1\overline{q}_2+\overline{Q}_2\overline{q}_1-\overline{Q}_3\overline{q}_4-\overline{Q}_4\overline{q}_3) \end{aligned} \tag{6.56}$$

Choosing the homology class of the Kähler form of $R$ to be invariant under our discrete symmetry group i.e. $J \sim a(H^x + H^y)$, makes the normalisation matrix eq.(6.42) proportional to the identity matrix. This tells us that this choice makes the symmetries of our manifold symmetries of the full effective lagrangian, not just of the superpotential.

Note that this procedure has prohibited certain couplings that *would* have been allowed by discrete symmetries alone, namely

$$\begin{aligned}&\sum_{i=1}^{4}(\overline{Q}_i^3 + \overline{q}_i^3)\\ &\overline{\lambda}_1(\overline{Q}_1\overline{q}_4 + \overline{Q}_4\overline{q}_1)\\ &\overline{\lambda}_1(\overline{Q}_2\overline{q}_3 + \overline{Q}_3\overline{q}_2)\\ &\overline{\lambda}_2(\overline{Q}_1\overline{q}_4 + \overline{Q}_4\overline{q}_1)\\ &\overline{\lambda}_2(\overline{Q}_2\overline{q}_3 - \overline{Q}_3\overline{q}_2).\end{aligned} \tag{6.57}$$

As shown in [31], however, nonperturbative instanton effects contribute nonzero values for all of these couplings except the last two.

Using these Yukawa couplings for the renormalizable interactions and the list of discrete symmetries, the low energy phenomenology is now within reach using methods familiar from standard model building. For the case at hand, much analysis of this sort has been carried out [6]with a surprising degree of success. For a particular choice of supersymmetry breaking (and resulting intermediate stages of gauge symmetry breaking), the model essentially gives the particle content of the three generation supersymmetric standard model with sufficiently small mass neutrinos, a reasonable Kobayashi-Maskawa matrix, and an acceptable rate of baryon number violation. Of course, much further work needs to be done, especially in the area of supersymmetry breaking as this has a strong bearing on the resulting low energy phenomenology. It is encouraging, though, that at this early stage superstring phenomenology can so closely reproduce the standard model.

## Acknowledgements

BRG is supported in part by NSF contract PHY-82-15249. KHK acknowledges the support of an NSF graduate fellowship, and PJM acknowledges an SERC studentship.

## References


[1] For a review, see J.H.Schwarz, "Superstrings" ( World Scientific, 1986)

[2] E.Witten, Nucl.Phys. B268 (1986) 79

[3] A. Strominger, E. Witten, Comm.Math.Phys.101(1985)341

[4] A. Strominger, Phys.Rev.Letts.55(1985)2547

[5] P.Candelas,G.Horowitz,A.Strominger and E. Witten, Nucl.Phys. B258(1985) 46

[6] B.R. Greene, K.H. Kirklin, P.J. Miron, G.G. Ross, Phys.Lett.180B(1986)69;Nucl.Phys.B278(198 667,
Oxford preprint 11/87 (1987)

[7] D.Gross, J.Harvey, E.Martinec and R.Rohm, Phys.Rev.Lett. (1985) 55; Nucl.Phys.B256 (1985)253

[8] G.F.Chapline and N.S.Manton, Phys.Lett. 120B(1983)105

[9] E. Calabi, "Algebraic Geometry and Topology", (Princeton,1957) 78

[10] S-T. Yau, Proc. Natl. Acad. Sci. 74(1977)1798

[11] C.Hull and E. Witten, Phys. Letts. 160B (1985) 398

[12] K.Uhlenbeck and S-T. Yau, On the Existence of Hermitian Yang-Mills Connections in Stable Vector Bundles

[13] J.Distler and B.R.Greene, in preparation

[14] M. Dine, N. Seiberg, X-G. Wen, E. Witten, Princeton preprints (1986,87)

[15] J.Distler, to appear in Phys.Letts. B

[16] E.Witten, Nucl.Phys. 258B (1985) 75

[17] J. Cheeger, D. Gromoll, J. Diff. Geom. 6(1971)119 .

[18] K. Kodaira, "Complex manifolds and deformations of complex structure" (Springer-Verlag, 1986)

[19] See, for example, P. Griffiths, J. Harris "Principles of Algebraic Geometry" (Wiley Interscience, 1978)

[20] S-T. Yau, private communication

[21] P.S. Aspinwall, K.H. Kirklin, P.J. Miron, B.R. Greene, Oxford preprint 26/87 (1987)

[22] T. Hubsch, Univ. of Maryland preprint (1986); P. Green, T. Hübsch, Maryland preprints (1987).

[23] S-T. Yau, Argonne Symposium on Anomalies, Geometry and Topology (World Scientific, 1985)

[24] R. Schimmrigk, Texas preprint (1987)

[25] B.R.Greene and K.H.Kirklin, Harvard/Oxford preprint

[26] P.Green and T.Hubsch,Maryland Preprint, (1987)

[27] B.R. Greene, K.H. Kirklin, P.J. Miron, G.G. Ross, "$27^3$ Yukawa Couplings for a Three-Generation Superstring Model", Harvard and Oxford preprint 18/87 (1987)

[28] T.Hubsh (private communication)

[29] P.Candelas, "Yukawa couplings between (2,1) forms", University of Texas preprint UTTG-05-87

[30] B.R. Greene, K.H. Kirklin, P.J. Miron, G.G. Ross, in preparation

[31] J. Distler, B.R. Greene, K.H. Kirklin, P.J. Miron, "Evaluation of $\overline{27}^3$ Yukawa couplings in a three generation superstring model", Oxford and Harvard preprints, 1987.

[32] See, for example, R. Hartshorne, "Algebraic Geometry", (Springer-Verlag)

# Yukawa Couplings Between (2,1)-Forms

P. Candelas
Theory Group
Department of Physics
The University of Texas
Austin, Texas 78712

and

Institute for Theoretical Physics
Ellison Hall
The University of California
Santa Barbara, CA 93106

Supported in part by Robert A. Welch Foundation and NSF Grants PHY 8503890, 8605978, and PHY 8217853 as supplemented by NASA.

Abstract

The compactification of superstrings leads to an effective field theory for which the space-time manifold is the product of a four dimensional Minkowski space with a six dimensional Calabi-Yau space. The particles that are massless in the four dimensional world correspond to differential forms of type (1,1) and of type (2,1) on the Calabi-Yau space. The Yukawa couplings between the families correspond to certain integrals involving three differential forms. For an important class of Calabi-Yau manifolds, which includes the cases for which the manifold may be realized as a complete intersection of polynomial equations in a projective space, the families correspond to (2,1)-forms. The relation between (2,1)-forms and the geometrical deformations of the Calabi-Yau space is explained and it is shown, for those cases for which the manifold may be realized as the complete intersection of polynomial equations in a single projective space or for many cases when the manifold may be realized as the transverse intersection of polynomial equations in a product of projective spaces, that the calculation of the Yukawa coupling reduces to a purely algebraic problem involving the defining polynomials. The generalization of this process is presented for a general Calabi-Yau manifold.

## 1. Introduction

In the low energy regime superstring theory is believed to reduce to an effective ten-dimensional field theory propagating on a spacetime manifold which is the product of a four dimensional Minkowski space with a Calabi-Yau space, i.e., a compact Kähler manifold of vanishing first Chern class and complex dimension three. The massless families and antifamilies correspond to harmonic (1,1) and (2,1) forms of the internal manifold [1]. The Yukawa couplings arise as integrals of triples of harmonic forms[2]. The precise form of the integral depends on whether the three zero modes correspond to (1,1)-forms or (2,1)-forms. If they correspond to three (1,1) forms a,b,c, say, then the coupling is

$$\kappa = \int_M a \wedge b \wedge c. \tag{1.1}$$

As has been pointed out by Strominger this leads, in this case, to an interesting topological interpretation for the coupling. Through deRham cohomology each of the two-forms is associated with a four-surface. In a six (real) dimensional space three four-surfaces meet in some number of points. There exists a basis in which $\kappa(a,b,c)$ is the number of points of intersection of these surfaces. Couplings of this type have been discussed for a number of cases by Strominger [3] and it is not our purpose to dwell further on them here, rather this article is devoted to a discussion of the couplings for the case that the zero modes are related to (2,1)-forms which is the case for many manifolds of interest.

One might reasonably enquire how it is that three 3-forms can be integrated over a six-dimensional manifold. The way in which this comes about depends on the special properties of Calabi-Yau spaces. It is a crucial property of these spaces that there exists a nowhere zero holomorphic 3-form. That is a nowhere zero 3-form

$$\Omega = \frac{1}{3!}\,\Omega_{\mu\nu\rho}(x)dx^{\mu}\wedge dx^{\nu}\wedge dx^{\rho} \tag{1.2}$$

which has $\Omega_{\mu\nu\rho}$ as its only nonzero components[f1] and where the $\Omega_{\mu\nu\rho}$ are moreover holomorphic functions of the coordinates.

Given $\Omega$ it is possible to define for every (2,1)-form

$$\omega = \frac{1}{2}\,\omega_{\mu\nu\bar{\rho}}\,dx^{\mu}\wedge dx^{\nu}\wedge dx^{\bar{\rho}} \tag{1.3}$$

an equivalent object

$$\omega^{\mu} = \frac{1}{2}\,\bar{\Omega}^{\mu\rho\sigma}\omega_{\rho\sigma\bar{\nu}}\,dx^{\bar{\nu}}. \tag{1.4}$$

$\omega^{\mu}$ is a (0,1)-form with a holomorphic vector index. It may be regarded as a (0,1)-form that takes values in the holomorphic tangent bundle T. It is straightforward to show that the original (2,1)-form $\omega_{\rho\sigma\bar{\nu}}$ is harmonic with respect to the exterior derivative d if an only if $\omega^{\mu}$ is harmonic with respect to $\bar{\partial}$, the antiholomorphic part of d. The Yukawa coupling involves three such forms $a^{\mu}$, $b^{\nu}$, $c^{\rho}$, say, and takes the explicit form [2]

$$\kappa = \int \Omega\wedge a^{\mu}\wedge b^{\nu}\wedge c^{\rho}\Omega_{\mu\nu\rho}\ . \tag{1.5}$$

The purpose of this article is to discuss the significance of the coupling in relation to deformations of the complex structure of M and to evaluate it for

---

f1. Notation: We shall use $x^{\mu}$, $\mu = 1,2,3$ to denote the coordinates of M, Latin indices m,n•• will refer to a real coordinate basis and run over six values

a number of cases of interest. We shall see that for the case that M may be realized as a complete intersection of polynomial equations in a projective space, and in a number of other cases, the evaluation of $\kappa$ reduces to a purely algebraic problem.

Much of what follows is material that is already present in the literature, though some is new. The discussion of the deformation of complex structure draws heavily on the classic work of Kodaira and Spencer [4,5]. The relationship between the holomorphic three-form and the defining polynomials was known to the mathematicians but I learnt it from the articles by Witten [6] and Strominger and Witten [7]. I review these relations here in order to make the account reasonably self contained. Witten [6] also showed that the Yukawa couplings are strongly constrained by the pseudosymmetries of the manifold. Finally the fact that statements about the cohomology classes of algebraic varieties translate into algebraic statements is a fact I learnt from conversations with Bott.

The process of evaluating $\kappa$ breaks naturally into two parts. The first is to understand the geometrical meaning of the integral and the second is its explicit evaluation.

## 2. The geometrical meaning of $H_{\bar\partial}^{(0,1)}(M,T)$

Let us suppose initially that M is given as a hypersurface in an (N+3) dimensional projective space $P_{N+3}$ defined as the transverse intersection of N polynomials

$$p^{\alpha}(z) = 0 \quad , \quad \alpha = 1,\cdots,N \tag{2.1}$$

where $z^A$, $A = 1,\cdots,N+4$ are homogeneous coordinates for the projective space and the degrees of the polynomials $p^\alpha$ are so chosen that the hypersurface has vanishing first Chern class. If the hypersurface has vanishing first Chern class then, in virtue of Yau's theorem, we may take it to be endowed with a Ricci-flat metric $g_{mn}$

$$R_{mn}(g) = 0 \tag{2.2}$$

Consider an infinitesimal variation of M which takes us to another nearby Ricci-flat manifold with metric $g_{mn} + h_{mn}$

$$R_{mn}(g+h) = 0 \tag{2.3}$$

Linearization of this equation and imposition of the coordinate condition $\nabla^m h_{mn} = 0$ leads to a differential equation for $h_{mn}$, the Lichnerowicz equation,

$$\Delta_L h_{mn} \equiv h_{mn} + 2R_m{}^r{}_n{}^s h_{rs} = 0 . \tag{2.4}$$

The first point to note is that, owing to the special proprties of Kähler spaces the Lichnerowicz operator $\Delta_L$ does not mix modes of different type i.e., the $h_{\mu\bar{\nu}}$, $h_{\mu\nu}$, $h_{\bar{\mu}\bar{\nu}}$ parts of $h_{mn}$ separately satisfy the equation.

First let us examine the mixed part of $h_{mn}$. Form from $h_{\mu\bar{\nu}}$ a (1,1)-form

$$k = ih_{\mu\bar{\nu}} dx^{\mu} \wedge dx^{\bar{\nu}}. \tag{2.5}$$

It turns out that, in virtue of the fact that $h_{\mu\bar{\nu}}$ satisfies the Lichnerowicz equation, k satisfies the Hodge-de Rham equation

$$(dd^{+} + d^{+}d)k = 0 . \tag{2.6}$$

In other words k is harmonic. For the case we are considering the embedding space consists of a single projective space. For this case it is known that $b_{11} = 1$. The unique (1,1)-form is the Kähler-form

$$J = ig_{\mu\bar{\nu}} dx^{\mu} \wedge dx^{\bar{\nu}} . \tag{2.7}$$

Thus the only metric perturbation of this type is a constant multiple of the metric itself. This perturbation corresponds simply to a change in the overall scale of M.

Of greater interest for our present purposes are the pure modes $h_{\mu\nu}$ and $h_{\bar{\mu}\bar{\nu}}$ which are, of course, related by complex conjugation so that it suffices to consider $h_{\bar{\mu}\bar{\nu}}$. Form from this a (0,1)-form that takes values in T.

$$h^{\mu} = h^{\mu}{}_{\bar{\nu}} dx^{\bar{\nu}} . \tag{2.8}$$

It may be shown that $h_{\bar{\mu}\bar{\nu}}$ satisfies the Lichnerowicz equation if and only if $h^{\mu}$ satisfies the Hodge-de Rham equation corresponding to the operator $\bar{\partial}$

$$(\bar{\partial}\bar{\partial}^{+} + \bar{\partial}^{+}\bar{\partial})h^{\mu} = 0. \tag{2.9}$$

In other words $h_{\bar{\mu}\bar{\nu}}$ satisfies the Lichnerowicz equation if and only if $h^{\mu}$ is harmonic with respect to $\bar{\partial}$. The elements of the cohomology group $H_{\bar{\partial}}^{(0,1)}(M,T)$ are in one-one correspondence with the Ricci-flat perturbations of M. These perturbations are also in one-one correspondence with the deformations of the complex structure of M. It can happen that two manifolds are diffeomorphic, that is they are the same as real manifolds, and yet they are different as complex manifolds because there is no biholomorphic map between them. The simplest example of this phenomenon concerns two tori

$$T_1 = \{(x,y)|(x,y) \simeq (x+1,y) \simeq (x,y+1)\}$$

$$T_2 = \{(\xi,\eta)|(\xi,\eta) \simeq (\xi+1,\eta) \simeq (\xi,\eta+2)\} . \tag{2.10}$$

These manifolds are diffeomorphic as real manifolds since

$$(\xi,\eta) = (x,2y) \tag{2.11}$$

defines a $C^{\infty}$ map between them. If however we set $z = x+iy$ and $\zeta = \xi+i\eta$ then $\zeta$ is not a holomorphic function of z

$$\zeta = \frac{3}{2}z - \frac{1}{2}\bar{z} \tag{2.12}$$

and it is not possible to eliminate the $\bar{z}$.

The various Ricci-flat manifolds that were discussed previously are related in just such a way. They may be smoothly deformed into each other and are in fact diffeomorphic. They are also all Kähler manifolds, being analytic submanifolds of $P_{N+3}$. However they have different complex structures. We know this because a Kähler manifold has a hermitean metric with only mixed components whereas the metric perturbation $h_{mn}$ has pure parts $h_{\mu\nu}$ and $h_{\bar{\mu}\bar{\nu}}$. The complex structures of the two manifolds are different because although the pure part of $h_{mn}$ can be removed by a transformation of coordinates it cannot be removed by a holomorphic coordinate transformation. Under a holomorphic transformation

$$x^\mu \to \xi^\mu = x^\mu + \varepsilon^\mu(x) \tag{2.13}$$

the metric perturbation transforms according to the familiar rule

$$h_{mn} \to \tilde{h}_{mn} = h_{mn} - \varepsilon^r{}_{,m} g_{rn} - \varepsilon^r{}_{,n} g_{mr}, \tag{2.14}$$

if $\varepsilon^\mu$ is holomorphic then $h_{\mu\nu}$ and $h_{\bar{\mu}\bar{\nu}}$ are invariant.

For a space realized as a complete intersection of polynomials $p^\alpha$, $\alpha = 1,\cdots,N$ in $P_{N+3}$ the deformations of the complex structure can be simply realized as deformations of the defining polynomials $p^\alpha$. Consider infinitesimal changes $p^\alpha \to p^\alpha + q^\alpha$ in the defining polynomials

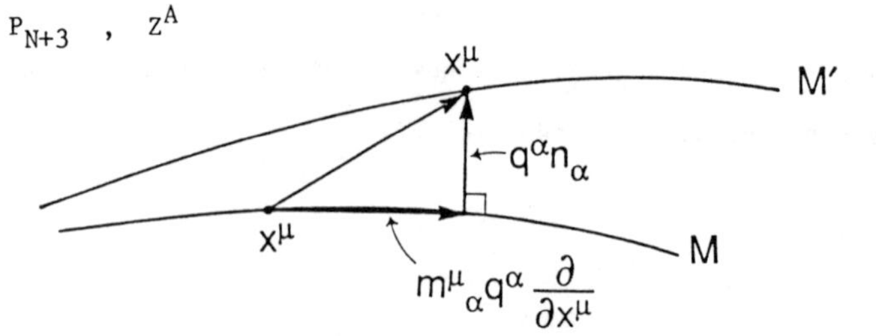


Figure 1

We may study the deformation as in General Relativity where it is common to decompose deformations into lapses and shifts [7]. The situation here is analogous except that there are now several "times" labelled by $\alpha$. On M we choose a holomorphic atlas, i.e., coordinates $x^{\mu}{}_{j}$ defined in patches $U_j$ and we choose a similar atlas for the deformed manifold $M'$ also labelling the points by $x^{\mu}{}_{j}$ in such a way that the point labelled by $x^{\mu}{}_{j}$ on $M'$ approaches that labelled by $x^{\mu}{}_{j}$ on M as the $q^{\alpha} \to 0$. From now on we shall largely suppress the j that labels the different coordinate patches. We join the point labelled by $x^{\mu}$ on the original manifold to that with same label $x^{\mu}$ on the perturbed manifold by a vector and decompose this vector into a normal lapse and a tangential shift which, being linear in the deformation $q^{\alpha}$, takes the form

$$m^{\mu}{}_{\alpha} q^{\alpha} \frac{\partial}{\partial x^{\mu}} .$$

Associated with the embedding of M in the ambient space is an important geometrical tensor the extrinsic curvature or second fundamental form. Since the extrinsic curvature plays a pivotal role in what follows we shall pause briefly to recall how it arises in the study of embedded hypersurfaces.

It proves useful to choose a reciprocal set of basis vectors and one forms

$$e_{\mu} = \frac{\partial}{\partial x^{\mu}} \qquad ; \qquad e^{\mu} = dx^{\mu} + m^{\mu}{}_{\alpha} n^{\alpha}$$

$$n_{\alpha} = \frac{\partial}{\partial p^{\alpha}} - m^{\mu}{}_{\alpha} e_{\mu} \qquad ; \qquad n^{\alpha} = dp^{\alpha}. \qquad (2.15)$$

With respect to this basis the lapse vector in Fig. 2 is $q^{\alpha} n_{\alpha}$.

The extrinsic curvature $\chi_{mn}{}^{a}$ is defined by the relation

$$\hat{\nabla}_m e_n = \nabla_m e_n + \chi_{mn}{}^{a} n_a \tag{2.16}$$

where a "^" is used to distinguish quantities that refer to the embedding space while quantities without the "^" are intrinsic to the surface.

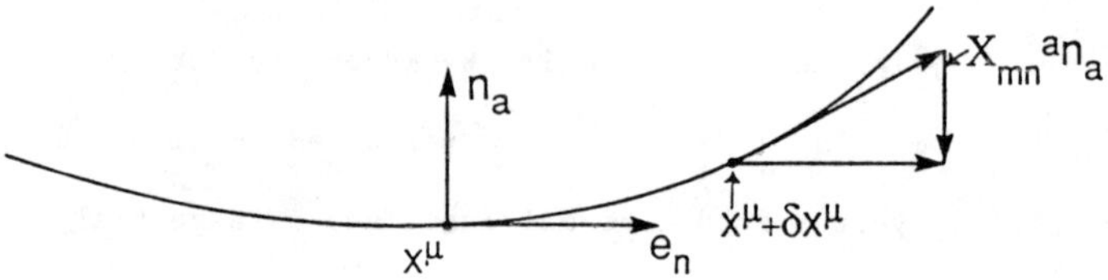


Figure 2

The difference between parallel propagation in the embedding space and parallel propagation in the hypersurface is illustrated by Fig. 3. This difference which is due to the rotation of the tangent vectors $e_n$ measures the curvature of the surface. Alternatively one may concentrate on the rotation of the normal vectors $n_a$. In terms of frame rotation coefficients we have

$$\hat{\nabla}_m n_a = \hat{\Gamma}_{ma}{}^{r} e_r + \hat{\Gamma}_{ma}{}^{b} n_b \tag{2.17}$$

Now

$$\hat{\Gamma}_{mas} = \hat{g}(e_s, \hat{\nabla}_m n_a) = - \hat{g}(\hat{\nabla}_m e_s, n_a) = - \chi_{msa}. \tag{2.18}$$

Thus

$$\hat{\nabla}_m n_a = - \chi_m{}^{r}{}_{a} e_r + \hat{\Gamma}_{ma}{}^{b} n_b \tag{2.19}$$

which provides an alternative definition of $\chi_{mn}{}^{a}$.

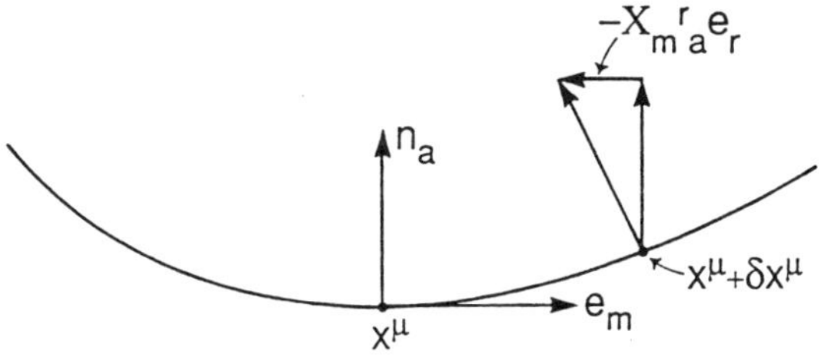


Figure 3. An alternative but equivalent way to measure the curvature of a hypersurface is to parallely propagate the normals $n_a$ from a point $x^\mu$ to a nearby point $x^\mu+\delta x^\mu$. There it will differ from the preexisting normals by an amount proportional to the extrinsic curvature.

It is straightforward to show that for the case we are considering, of a hypersurface embedded in a Kähler manifold by holomorphic equations, that the extrinsic curvature has

$$\chi_{\mu\nu}{}^{\alpha} \quad \text{and} \quad \chi_{\bar{\mu}\bar{\nu}}{}^{\bar{\alpha}} \tag{2.20}$$

as its only nonzero components.

The fact that the extrinsic curvature relates the rotation of the tangent vectors to that of the normal vectors is important in the following. From equation (2.16) we may obtain an explicit expression for the extrinsic curvature in terms of the defining polynomials $p^\alpha$.

$$\chi_{\mu\nu}{}^{\alpha} = \frac{\partial z^i}{\partial x^\mu}\frac{\partial z^j}{\partial x^\nu}\left(\frac{\partial^2 p^\alpha}{\partial z^i \partial z^j} - \hat{\Gamma}_{ij}{}^{k}\frac{\partial p^\alpha}{\partial z^k}\right) \tag{2.21}$$

where in this equation i, j, k run over N+3 values corresponding to choosing one of the $z^A$ to be unity. If we endow $P_{N+3}$ with the Fubini-Study metric

$$\hat{g}_{j\bar{k}} = \frac{1}{\sigma}\left(\delta_{j\bar{k}} - \frac{z_j z_{\bar{k}}}{\sigma}\right) \tag{2.22}$$

with

$$z_j = \delta_{j\bar{k}} z^{\bar{k}} \quad , \quad z_{\bar{k}} = \delta_{j\bar{k}} z^j \tag{2.23}$$

and

$$\sigma = \sum_{A=1}^{N+4} |z^A|^2 \ , \tag{2.24}$$

then it is easy to see that the term involving the Christoffel symbol does not contribute to $\chi_{\mu\nu}{}^{\alpha}$, moreover we may use the fact that one of the z's is a constant to write $\chi_{\mu\nu}{}^{\alpha}$ in the more symmetric form

$$\chi_{\mu\nu}{}^{\alpha} = \frac{\partial z^A}{\partial x^\mu}\frac{\partial z^B}{\partial x^\nu}\frac{\partial^2 p^\alpha}{\partial z^A \partial z^B} \ . \tag{2.25}$$

The extrinsic curvature may also be related, via equation (2.19), to the shift $m^\mu{}_\alpha$

$$\chi_{\bar{\lambda}}{}^{\mu}{}_{\alpha} = \frac{\partial}{\partial x^{\bar{\lambda}}}(m^\mu{}_\alpha) \ . \tag{2.26}$$

This last equation enables us to relate explicitly the deformations $q^\alpha$ of the defining polynomials to the elements of $H_{\bar{\partial}}^{(0,1)}(M,T)$.

Set

$$a^\mu = \chi_{\bar{\lambda}}{}^{\mu}{}_{\alpha} q^\alpha dx^{\bar{\lambda}} \ . \tag{2.27}$$

$a^\mu$ is a (0,1)-form valued in T. Moreover by (2.26) we have

$$a^\mu = \bar{\partial}(m^\mu{}_\alpha q^\alpha) . \qquad (2.28)$$

Since $\bar{\partial}^2 = 0$ we see that $a^\mu$ is closed

$$\bar{\partial} a^\mu = 0. \qquad (2.29)$$

Naively one might suppose that eqn. (2.28) shows that $a^\mu$ is also exact, however this is in general false owing to the fact that our construction is tied to a particular coordinate patch. Since M is compact it cannot be covered by a single patch and there is no reason to suppose that where two patches overlap that the corresponding shifts transform correctly. In other words the shift $m^\mu{}_\alpha$ does not, in general, extend to a non-singular tensor field. On the other hand $\chi_{\mu\nu}{}^\alpha$ is a bona fide geometrical tensor defined everywhere on M since M is assumed smooth. Thus eqn. (2.27) defines $a^\mu$ globally.

It is of interest to enquire under what circumstances $a^\mu$ is exact i.e., when can we write

$$a^\mu = \bar{\partial} v^\mu \qquad (2.30)$$

with

$$v = v^\mu \frac{\partial}{\partial x^\mu} \qquad (2.31)$$

a globally defined holomorphic vector field on M. For this purpose and for later use we introduce some notation:

Set

$$p^{\alpha}{}_{A} = \frac{\partial p^{\alpha}}{\partial z^{A}} , \quad p^{\bar{\beta}} = \overline{(p^{\beta})} \tag{2.32}$$

and

$$k^{\alpha\bar{\beta}} = \sum_{A} p^{\alpha}{}_{A} p^{\bar{\beta}}{}_{\bar{A}} \tag{2.33}$$

$k^{\alpha\bar{\beta}}$ is closely related to the lapse. The perpendicular part of the metric is

$$h^{\alpha\bar{\beta}} = \sigma k^{\alpha\bar{\beta}} . \tag{2.34}$$

We denote by $k_{\alpha\bar{\beta}}$ the matrix inverse to $k^{\alpha\bar{\beta}}$ and use $k^{\alpha\bar{\beta}}$ and $k_{\alpha\bar{\beta}}$ to raise and lower polynomial indices. It is also convenient to denote complex conjugation of the embedding coordinates by raising and lowering indices. For example we set

$$dz_{A} = dz^{\bar{A}} \quad \text{and} \quad \partial^{A} = \frac{\partial}{\partial z^{\bar{A}}} . \tag{2.35}$$

This facilitates a summation convention for the embedding coordinates. With these conventions we have

$$k^{\alpha\bar{\beta}} = p^{\alpha}{}_{A} p^{\bar{\beta}A} = p^{\alpha\bar{A}} p^{\bar{\beta}}{}_{\bar{A}}. \tag{2.36}$$

Returning to the problem at hand, we can show from (2.22), (2.25) and (2.27) that

$$a^{\mu} \frac{\partial}{\partial x^{\mu}} = -\bar{\partial}\left(q^{\alpha} p_{\alpha}{}^{A} \frac{\partial}{\partial z^{A}}\right). \tag{2.37}$$

Here, again, one is tempted to conclude that we have shown $a^{\mu}$ to be exact since the quantity inside the brackets on the right hand side is globally defined on M. This is true, the catch is that

$$q^{\alpha} p_{\alpha}{}^{A} \frac{\partial}{\partial z^{A}} \tag{2.38}$$

is not in general a vector field that is tangent to M.

For a vector field

$$\nu = \nu^{A} \frac{\partial}{\partial z^{A}} \tag{2.39}$$

defined on the embedding space to be tangent to M it is necessary (and sufficient) that the quantities

$$\nu^A \frac{\partial p^\alpha}{\partial z^A} \tag{2.40}$$

vanish with the $p^\alpha$. That is there must exist polynomials $f^\alpha{}_\beta$ of the appropriate degrees (a polynomial of negative degree will be understood to be zero) such that

$$\nu^A \frac{\partial p^\alpha}{\partial z^A} = f^\alpha{}_\beta p^\beta \ . \tag{2.41}$$

Note also that owing to the presence of the $\bar{\partial}$ operator in (2.37) there is a certain freedom in writing

$$a^\mu \frac{\partial}{\partial x^\mu} = - \bar{\partial}\left(\nu^A \frac{\partial}{\partial z^A}\right) \tag{2.42}$$

we can take

$$\nu^A = q^\alpha p_\alpha{}^A - c^A \tag{2.43}$$

where the $c^A(z)$ are any linear holomorphic functions of the $z^B$

$$c^A(z) = c^A{}_B z^B \ . \tag{2.44}$$

(The $c^A$ have to be linear in order to have the same homogeneity degree as the first term on the right-hand side of eqn. (2.43)).

Thus $a^\mu \frac{\partial}{\partial x^\mu}$ is exact if an only if

$$(q^{\alpha}p_{\alpha}{}^{A} - c^{A})\ p^{\gamma}{}_{A} = f^{\gamma}{}_{\delta}p^{\delta} \tag{2.45}$$

for some choice of $c^A$ and $f^{\gamma}{}_{\delta}$. On making use of eqn. (2.36) this condition becomes

$$q^{\alpha} = c^{A}p^{\alpha}{}_{A} + f^{\alpha}{}_{\beta}p^{\beta}\ . \tag{2.46}$$

This condition admits of a simple interpretation. Let us consider the two terms on the right-hand side separately. To understand the first consider making a linear change of coordinates in the embedding space.

$$z^{A} \rightarrow z^{A} + c^{A}(z) = z^{A} + c^{A}{}_{B}z^{B} \tag{2.47}$$

Under such a transformation the defining polynomials transform, to linearized order, in the following way

$$p^{\alpha}(z) \rightarrow p^{\alpha}(z+c) = p^{\alpha}(z) + c^{A}p^{\alpha}{}_{A}\ . \tag{2.48}$$

In other words a deformation $q^{\alpha}$ which is of the form $c^{A}p^{\alpha}{}_{A}$ corresponds merely to a change of coordinates in the embedding space rather than to a change in the shape of M.

The second term on the right-hand side of eqn. (2.46) corresponds to the freedom to describe M either by the equations

$$p^{\alpha} = 0 \tag{2.49}$$

or by the equivalent set, $f^{\alpha}{}_{\beta}$ is assumed small,

$$p^{\alpha} + f^{\alpha}{}_{\beta}p^{\beta} = 0. \tag{2.50}$$

The burden of (2.46) is therefore that $a^{\mu}$ is exact precisely when we would not wish to think of $q^{\alpha}$ as describing a genuine perturbation. Such a $q^{\alpha}$ corresponds merely to a coordinate change together with a rearrangement of the defining polynomials. In fact we would wish to identify

$$q^{\alpha} \simeq q^{\alpha} + c^{A}p^{\alpha}{}_{A} + f^{\alpha}{}_{\beta}p^{\beta} \tag{2.51}$$

for all $c^{A}$ and $f^{\alpha}{}_{\beta}$ since the left and the right-hand sides represent the same physical deformation.

For a given choice of defining polynomials $p^{\alpha}$ let us define sets $I^{\alpha}$ of polynomials which consist precisely of the polynomials of the form (2.46).

$$I^{\alpha} = \{q^{\alpha} |\ q^{\alpha} = c^{A}p^{\alpha}{}_{A} + f^{\alpha}{}_{\beta}p^{\beta}\} \tag{2.52}$$

for all possible holomorphic choices of the $c^{A}$ and $f^{\alpha}{}_{\beta}$. The above discussion has established the following transcription of closed forms, exact forms and cohomology classes into algebraic terms. The closed forms are in one-one correspondence with polynomials $q^{\alpha}$ of the appropriate degrees. Let us denote the set of polynomials of the same degree as $p^{\alpha}$ by $C^{\alpha}$ then a closed form corresponds to a vector of polynomials $q^{\alpha}$ with $q^{\alpha} \varepsilon C^{\alpha}$ for each $\alpha$. The exact forms are in one-one correspondence with the elements of $I^{\alpha}$, i.e., with vectors of polynomials $q^{\alpha} \varepsilon I^{\alpha}$. Finally the cohomology classes correspond to closed forms identified modulo exact forms i.e., to vectors of polynomials subject to the identification (2.51). These are vectors of polynomials $q^{\alpha} \varepsilon P^{\alpha}$ where we define

$$P^{\alpha} = C^{\alpha}/I^{\alpha} . \tag{2.53}$$

This situation is summarized in Table 1.

| | | |
|---|---|---|
| closed forms | $\leftrightarrow$ | $q^{\alpha} \varepsilon C^{\alpha}$ |
| exact forms | $\leftrightarrow$ | $q^{\alpha} \varepsilon I^{\alpha}$ |
| cohomology classes | $\leftrightarrow$ | $q^{\alpha} \varepsilon P^{\alpha}$ |

Table 1.

The correspondence between cohomology classes and polynomials will allow us to transcribe operations on the cohomology classes into algebraic operations on polynomials. Before proceeding to the evaluation of the Yukawa couplings, however, we need to discuss the properties of the holomorphic three-form $\Omega$ which appears explicitly in expression (1.5).

## 3. A Digression on the Holomorphic 3-Form and some Jacobians

Our purpose here is to derive and relate a number of relations between different representations of the holomorphic 3-form $\Omega$ that are of importance in what follows. These relations follow from the fact that, for the class of spaces that we are presently considering, the holomorphic 3-form admits explicit representations in terms of the defining polynomials.

We shall first work in the coordinate patch

$$U_{N+4} = \{z^{N+4} \neq 0\} . \tag{3.1}$$

We set

$$z^{N+4} = \zeta , \tag{3.2}$$

with $\zeta$ an arbitrary nonzero constant, and use the other N+3 z's as coordinates in this patch. We shall use lower case latin indices to range over the values 1,2,···,N+3.

Choose i,j,k to be distinct in the first instance and set

$$\Delta^{ijk} = \varepsilon^{ijkm_1\cdots m_N} \frac{\partial p^1}{\partial z^{m_1}} \frac{\partial p^2}{\partial z^{m_2}} \cdots \frac{\partial p^N}{\partial z^{m_N}}$$

$$= \frac{\partial(z^i, z^j, z^k, p^1, \cdots, p^N)}{\partial(z^1, z^2, \cdots\cdots, z^{N+3})} . \tag{3.3}$$

$\Delta^{ijk}$ is the Jacobian of the transformation between the coordinates $(z^1,\cdots,z^{N+3})$ and another set consisting of $z^i, z^j, z^k$ and the N polynomials $p^\alpha$. We may use $(z^i, z^j, z^k)$ as coordinates for M. The Jacobian between this set and another such set $(z^m, z^n, z^r)$, say, is

$$\frac{\partial(z^i, z^j, z^k)}{\partial(z^m, z^n, z^r)} = \frac{\Delta^{ijk}}{\Delta^{mnr}} \tag{3.4}$$

in virtue of (3.3).

The holomorphic 3-form can be written(f2) in terms of $\Delta^{ijk}$

$$\Omega = \frac{\zeta dz^i_{N+4}\, dz^j_{N+4}\, dz^k_{N+4}}{\Delta^{ijk}_{N+4}} \tag{3.5}$$

where there is no summation over the indices i,j,k and we have appended subscripts to emphasize that what is, in reality, meant by setting $z^{N+4} = \zeta$ is that we choose coordinates

$$z^i_{N+4} = \frac{\zeta z^i}{z^{N+4}} \quad , \quad i = 1,\cdots,N+3. \tag{3.6}$$

Relation (3.5) requires a certain amount of checking.

(i) Note first of all that the homogeneity degree of the right-hand side is

$$N+4 - \sum_{\alpha=1}^{N} \deg(\alpha) \tag{3.7}$$

where $\deg(\alpha)$ denotes the degree of $p^\alpha$. Since $z^A$ and $\lambda z^A$ represent the same point in $P_{N+3}$ the homogeneity degree of $\Omega$ must be zero for this representation to be sensible. Thus we must have

$$\sum_{\alpha} \deg(\alpha) = N+4 \tag{3.8}$$

this is the condition for M to have vanishing first Chern class.

f2. In the following we shall frequently omit the "$\wedge$" symbol when exterior multiplication is clear from the context.

(ii) We must check that the right-hand side of (3.5) is independent of the choice of i,j,k. This is easy since we have

$$\frac{dz^i\, dz^j\, dz^k}{\Delta^{ijk}} = \frac{1}{\Delta^{ijk}} \frac{\partial(z^i,z^j,z^k)}{\partial(z^m,z^n,z^r)}\, dz^m\, dz^n\, dz^r \tag{3.9}$$

$$= \frac{dz^m\, dz^n\, dz^r}{\Delta^{mnr}}$$

in virtue of (3.4).

(iii) We must check that the representation is independent of our choice of $z^{N+4}$ as the constant coordinate. To this end we transform to coordinates $z^m{}_1$, say, $m = 2,3,\ldots,N+4$ obtained by taking $z^1 = \eta$. On the set $U_1 \cap U_{N+4}$ the transformation between the two sets of coordinates takes the form

$$z^1_{N+4} = \frac{\eta\zeta}{z^{N+4}_1}\ , \qquad z^i_{N+4} = \frac{\zeta z^i{}_1}{z^{N+4}_1}\ , \quad i = 2,\cdots,N+3\ . \tag{3.10}$$

In virtue of (ii) above it is sufficient to show that, for example

$$\frac{\zeta dz^1_{N+4}\, dz^2_{N+4}\, dz^3_{N+4}}{\Delta^{123}_{N+4}} = -\frac{\eta\, dz^{N+4}_1\, dz^2_1\, dz^3_1}{\Delta^{(N+4)23}_1}\ . \tag{3.11}$$

(The sign change is due to the permutation that arises in going from (N+4, 1, 2, 3,$\cdots$,N+3) to (1, N+4, 2, 3,$\cdots$,N+3)). Given (3.10) eqn. (3.10) follows straightforwardly once account is taken of the fact that it is implicit that on the left-hand side of (3.11) the arguments of the polynomials $p^\alpha$ in $\Delta^{123}_{N+4}$ are the $z^i_{N+4}$ while on the right-hand side the arguments of the $p^\alpha$ are the $z^i_1$.

(iv) Finally we must check that, as defined, $\Omega$ is never singular. For a singularity to occur within $U_{N+4}$, say, all the $\Delta^{ijk}_{N+4}$ would have to vanish simultaneously. This however would imply the vanishing of $dp^1dp^2\cdots dp^N$ which cannot happen if M is smooth.

There is a very beautiful construction of $\Omega$ due to Bott [8] which illuminates the properties we have described. There is a natural "form" which may be constructed from the homogeneous coordinates $z^A$:

$$\mu = \frac{(-1)^{N+3}}{(N+3)!}\epsilon_{A_1A_2\cdots A_{N+4}} z^{A_1}\, dz^{A_2}\cdots dz^{A_{N+4}}. \tag{3.12}$$

Note that in coordinates with $z^{N+4}$ = const. this is just

$$z^{N+4}\, dz^1\, dz^2 \ldots dz^{N+3}.$$

$\mu$ is not well defined on $P_{N+3}$ since under a scaling $z^A \to \lambda z^A$ it is not invariant, it scales as

$$\mu \to \lambda^{N+4}\mu\ . \tag{3.13}$$

This difficulty can be cured by dividing $\mu$ by the product of the N polynomials $p^\alpha$. The (N+3)-form

$$\nu = \frac{\mu}{p^1p^2\cdots p^N} \tag{3.14}$$

is invariant under $z^A \to \lambda z^A$ in virtue of (3.8). $\nu$ is well-defined on $P_{N+3}$ but has poles on the hypersurfaces $p^\alpha = 0$. Let us integrate $\nu$ around a contour

$$\Gamma_N = \gamma_1 \times \gamma_2 \times \cdots \times \gamma_N \tag{3.15}$$

which is the cartesian product of the N one-dimensional curves $|p^\alpha| = \delta$, with $\delta$ assumed small which wind around the hypersurfaces $p^\alpha = 0$.

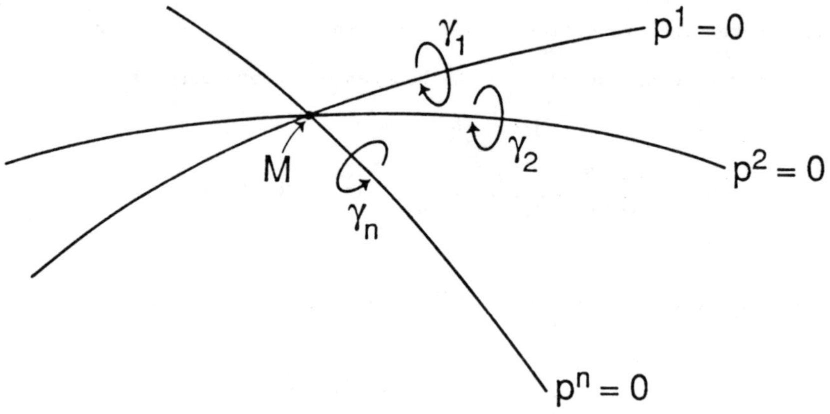


Figure 4. The contour $\Gamma_N$

Consider the integral (the limit $\delta \to 0$ is understood)

$$\int_{\Gamma_N} \frac{\mu}{p^1 p^2 \cdots p^N} = \int_{\Gamma_N} \frac{z^{N+4} dz^1 \cdots dz^{N+3}}{p^1 \cdots p^N}$$

$$= \int_{\Gamma_N} \frac{z^{N+4}}{p^1 \cdots\cdots p^N} \frac{\partial(z^1, z^2, \cdots\cdots\cdots z^{N+3})}{\partial(z^i, z^j, z^k, p^1, \cdots, p^N)}$$

$$= \int_{\Gamma_N} \frac{z^{N+4} dz^i dz^j dz^k}{\Delta^{ijk}} \frac{dp^1 dp^2 \cdots dp^N}{p^1 p^2 \cdots p^N}$$

$$= (2\pi i)^N \Omega. \tag{3.16}$$

The second line follows by changing variables of integration from the N+3 z's to any three distinct z's and the N polynomials $p^\alpha$, and the fourth by integrating out the $p^\alpha$ in the limit $\delta \to 0$.

## 4. Computation of the Yukawa Coupling

We return now to the task of computing the coupling

$$\kappa(a,b,c) = \int_M \Omega \wedge a^\mu \wedge b^\nu \wedge c^\rho \Omega_{\mu\nu\rho} \tag{4.1}$$

between three elements of the group $H_{\bar\partial}^{(0,1)}(M,T)$.

We have seen that each of the elements of this group corresponds to a polynomial defined modulo $I^{\alpha}$. Let us denote the polynomials corresponding to $a^{\mu}$, $b^{\nu}$, $c^{\rho}$ by $q^{\alpha}$, $r^{\beta}$, $s^{\gamma}$ respectively.

We have the explicit relation

$$\begin{aligned} a^{\mu} &= q^{\alpha} \chi_{\bar{\lambda}}{}^{\mu}{}_{\alpha} dx^{\bar{\lambda}} \\ &= - q_{\bar{\alpha}} \frac{\partial z^{\bar{J}}}{\partial x^{\bar{\rho}}} g^{\mu\bar{\rho}} \, dp^{\bar{\alpha}}{}_{\bar{J}} \end{aligned} \tag{4.2}$$

the second relation following from (2.25).

We need to form the (0,3)-form

$$\bar{\omega} = a^{\mu} \wedge b^{\nu} \wedge c^{\rho} \, \Omega_{\mu\nu\rho}, \tag{4.3}$$

a little algebra which makes use of the relations of the previous section leads to the expression

$$\bar{\omega} = - \frac{\sigma}{|z^{N+4}|^2} q_{\bar{\alpha}} r_{\bar{\beta}} s_{\bar{\gamma}} \, z^{N+4} \, k^{1/2} \, \Delta^{\bar{I}\bar{J}\bar{K}} \, dp^{\bar{\alpha}}{}_{\bar{I}} \, dp^{\bar{\beta}}{}_{\bar{J}} \, dp^{\bar{\gamma}}{}_{\bar{K}} \tag{4.4}$$

where

$$k = \det(k_{mn}) = \det{}^2(k_{\alpha\bar{\beta}}). \tag{4.5}$$

We know on general grounds that $\bar{\omega}$ is covariant under changes of coordinates so we may multiply this equation by $\frac{|z^{N+4}|^2}{\sigma}$ and add similar equations, one for each coordinate patch, since

$$\frac{1}{\sigma}\sum_A |z^A|^2 = 1 \tag{4.6}$$

we arrive at the important expression

$$\bar{\omega} = q^{\alpha} r^{\beta} s^{\gamma} e_{ABC}\, \bar{\partial} p_{\alpha}{}^{A} \bar{\partial} p_{\beta}{}^{B} \bar{\partial} p_{\gamma}{}^{C} \tag{4.7}$$

where

$$e_{ABC} = \frac{1}{N!}\,\varepsilon_{ABCDM_1\ldots M_N}\, z^D \varepsilon^{\sigma_1\ldots\sigma_N} p_{\sigma_1}{}^{M_1}\ldots p_{\sigma_N}{}^{M_N} \tag{4.8}$$

This expression has very nice properties but is, perhaps, somewhat beset by indices. In order to appreciate what is going on let us specialize temporarily to the simplest case $P_4(5)$ of the quintic in $P_4$ for which there is only one polynomial so that we may omit the $\alpha$ indices provided we remember that

$$k^{11} = \frac{1}{k_{11}} = \bar{p}^A p_A. \tag{4.9}$$

To save a little writing we write

$$\bar{p}^A p_A = \|p\|^2 \quad \text{and} \quad \pi^A = \frac{\bar{p}^A}{\|p\|^2} \tag{4.10}$$

so that

$$\pi^A p_A = 1. \tag{4.11}$$

then expression (4.7) becomes

$$\bar{\omega} = qrs\, \varepsilon_{ABCDE} z^A \pi^B \bar{\partial}\pi^C \bar{\partial}\pi^D \bar{\partial}\pi^E \tag{4.12}$$

Now quite apart from other considerations $\bar{\omega}$ is a (0,3)-form which is closed in virtue of the fact that $a^\mu$, $b^\nu$ and $c^\rho$ are. For a Calabi-Yau manifold there is precisely one harmonic (0,3)-form, $\bar{\Omega}$. Thus we may write

$$\bar{\omega} = \kappa\bar{\Omega} + \bar{\partial}\bar{\nu} \tag{4.13}$$

with $\bar{\kappa}$ a constant that is determined by ($a^\mu$, $b^\nu$, $c^\rho$) or equivalently by (q, r, s) and $\bar{\nu}$ is a (0,2)-form. The exact part $\bar{\partial}\bar{\nu}$ does not contribute to the integral. Hence

$$\int \Omega \wedge \bar{\omega} = \kappa \tag{4.14}$$

which identifies the constant appearing in (4.12) as the value of the coupling.

If for some choice of q, r and s, $\bar{\omega}$ is exact then the corresponding $\kappa$ is zero. We wish to enquire, therefore, when the expression (4.12) is exact. We shall now show that $\bar{\omega}$ is exact if the product qrs is of the form $E^A p_A$ with the $E^A$ any eleventh order polynomials. This requires an identity. Consider the quantity

$$\varepsilon_{AFCDE} z^A \bar{\partial}\pi^C \bar{\partial}\pi^D \bar{\partial}\pi^E, \tag{4.15}$$

the permutation symbol has F as a free index and is contracted on the other four indices with the tensor

$$z^A \bar{\partial}\pi^C \bar{\partial}\pi^D \bar{\partial}\pi^E. \tag{4.16}$$

The latter tensor is orthogonal to $p_A$ on all its indices, either because

$$z^A p_A = 5p = 0 \quad \text{on } M \tag{4.17}$$

or because

$$p_C \bar{\partial}\pi^C = \bar{\partial}(p_C \pi^C) = 0. \tag{4.18}$$

It follows that the free index F of the quantity (4.15) must be parallel to $p_F$. Thus we obtain the identity

$$\varepsilon_{AFCDE} z^A \bar{\partial}\pi^C \bar{\partial}\pi^D \bar{\partial}\pi^E = p_F \varepsilon_{ABCDE} z^A \pi^B \bar{\partial}\pi^C \bar{\partial}\pi^D \bar{\partial}\pi^E \tag{4.19}$$

Now our immediate aim is to show that $\bar{\omega}$ is $\bar{\partial}$-exact if qrs $\varepsilon I$. Suppose that qrs = $E^F p_F$ then

$$\bar{\omega} = E^F p_F \epsilon_{ABCDE} z^A \pi^B \bar{\partial}\pi^C \bar{\partial}\pi^D \bar{\partial}\pi^E$$

$$= E^B \epsilon_{ABCDE} z^A \bar{\partial}\pi^C \bar{\partial}\pi^D \bar{\partial}\pi^E$$

$$= \bar{\partial}(E^B \epsilon_{ABCDE} z^A \pi^C \bar{\partial}\pi^D \bar{\partial}\pi^E) \tag{4.20}$$

where the second line follows in virtue of the identity (4.19). We knew previously that each of the polynomials q, r, s is defined modulo I. We have learnt that the same is true of their product. The possible cohomology classes of the product qrs are the elements of

$$P_{15} = C_{15}/I . \tag{4.21}$$

It is natural at this point to enquire how many of those there are. The answer comes as a surprise on first aquaintance. There is precisely one. This is easy to see explicitly for the simplest example for which

$$p = (z^1)^5 + (z^2)^5 + (z^3)^5 + (z^4)^5 + (z^5)^5 \tag{4.22}$$

for this case I is generated by the monomials $(z^A)^4$ and it is clear that there is precisely one polynomial of fifteenth order than can be written down without writing any $z^A$ to the fourth power. It is

$$20^5 (z^1)^3(z^2)^3(z^3)^3(z^4)^3(z^5)^3 \tag{4.23}$$

the numerical factor being inserted for later convenience. In other words

$$\dim P_{15} = 1. \tag{4.24}$$

This is clear for this special case but is true more generally. Being an integer dim $P_{15}$ does not change under a sufficiently small variation of p and, in fact, does not change as p varies so long as the manifold defined by p is nonsingular. Let us denote the unique fifteenth order polynomial that is not in P by Q. We may take Q to be given by the determinant of the matrix of second derivatives p

$$Q = \frac{1}{5!}\varepsilon^{A_1\cdots A_5}\varepsilon^{B_1\cdots B_5}\frac{\partial^2 p}{\partial z^{A_1}\partial z^{B_1}}\cdots\frac{\partial^2 p}{\partial z^{A_5}\partial z^{B_5}}\,. \tag{4.25}$$

This is a more general prescription for the form of Q than (4.22).

It is shown in appendix A that with the product qrs replaced by Q the value of the integral (4.13) is unity. The burden of these remarks is that there is a unique way of writing

$$qrs = \kappa Q + E^A p_A \tag{4.26}$$

with $\kappa$ a constant which we identify as the Yukawa coupling. Thus the problem of computing the integral (4.15) has been reduced to a purely algebraic process. One multiplies the three polynomials q,r,s and writes the product as a multiple of Q the unique fifteenth order polynomial not in I plus a remainder that is in I . This leads to a unique expression and the coefficient of Q is the Yukawa coupling. For the simplest case with p given by (4.23) the procedure is very simple. One multiplies the polynomials q,r,s and the coefficient of the term

$$20^5\,(z^1)^3(z^2)^3(z^3)^3(z^4)^3(z^5)^3$$

in the product is the Yukawa coupling.

Before returning from the simplest case of only one polynomial to the general case. We would like to remark on the identifications that have been made between forms and polynomials. These are summarized in Table 2.

| Cohomology group | Space of polynomials |
|---|---|
| $H^{(3,0)}$ | $P_0$ |
| $H^{(2,1)}$ | $P_5$ |
| $H^{(1,2)}$ | $P_{10}$ |
| $H^{(0,3)}$ | $P_{15}$ |

Table 2. The correspondence between cohomology classes and sets of polynomials for the manifold $P_4(5)$. Here $P_k$ denotes the set of homogeneous polynomials of degree k identified modulo the ideal I.

The only entry that has not been explicitly discussed is the correspondence between (1,2)-forms and $P_{10}$. This is provided by the observation that any (1,2)-form can be expressed as a sum of the terms of the form

$$\Omega_{\mu\nu\rho} dx^\mu a^\nu b^\rho .$$

It follows that tenth order polynomials u(z), defined modulo I, can be identified with (1,2) forms $\nu^{(1,2)}$ via the correspondence

$$\begin{aligned} \nu^{(1,2)} &= \frac{u}{\|p\|^4} \Omega_{\mu\nu\rho} \chi^\nu{}_{\bar\sigma} \chi^\rho{}_{\bar\tau} dx^\mu dx^{\bar\sigma} dx^{\bar\tau} \\ &= - u\varepsilon_{ABCDE} z^A \pi^B dz^C \bar\partial\pi^D \bar\partial\pi^E \end{aligned} \tag{4.27}$$

it is straightforward to show that $\nu^{(1,2)}$ is exact if $u\varepsilon I$.

Suppose now that there are N polynomials $p^{\alpha}$. We shall first enquire as to when the (0,3)-form (4.7) is exact and then we shall discuss the generalization of Table 2.

The assertion is that

$$\bar{\omega} = q^{\alpha}r^{\beta}s^{\gamma}\epsilon_{ABC}\bar{\partial}p_{\alpha}{}^{A}\bar{\partial}p_{\beta}{}^{B}\bar{\partial}p_{\gamma}{}^{C} \tag{4.28}$$

is exact whenever the tensor of polynomials $q^{(\alpha}r^{\beta}s^{\gamma)}$ can be written in the form

$$c^{A(\alpha\beta}p^{\gamma)}{}_{A} + f^{\alpha\beta\gamma}{}_{\delta}p^{\delta}$$

for some choice of polynomials $c^{A\alpha\beta}$ and $f^{\alpha\beta\gamma}{}_{\delta}$. We shall denote the set of all tensors of polynomials whose components $q^{\alpha\beta\gamma}$ are of degree $\deg(\alpha) + \deg(\beta) + \deg(\gamma)$ and are of this form by $I^{\alpha\beta\gamma}$. The assertion is that $\bar{\omega}$ is exact if $q^{(\alpha}r^{\beta}s^{\gamma)} \epsilon I^{\alpha\beta\gamma}$.

The substantiation of the assertion is analogous to that for the case of one polynomial. Note first that

$$k_{F}{}^{C} = p^{\gamma}{}_{F}p_{\gamma}{}^{C} \tag{4.29}$$

is a tensor that projects normal to M:

$$k_{F}{}^{C}\, p^{\alpha}{}_{C} = p^{\alpha}{}_{F} \quad , \quad k_{F}{}^{C}\, \frac{\partial z^{F}}{\partial x^{\mu}}\, dx^{\mu} = 0 \; . \tag{4.30}$$

Now if in expression (4.29) $q^{\alpha}r^{\beta}s^{\gamma}$ is replaced by $c^{F\alpha\beta}p^{\gamma}{}_{F}$ then, in virtue of the observation that

$$p^{\gamma}{}_{F}\bar{\partial}p_{\gamma}{}^{C} = \bar{\partial}k_{F}{}^{C} \tag{4.31}$$

we find that

$$\begin{aligned}\bar{\omega} &= C^{F\alpha\beta}\, e_{ABC}\, \bar{\partial}p_{\alpha}{}^{A}\bar{\partial}p_{\beta}{}^{B}\bar{\partial}k_{F}{}^{C} \\ &= \bar{\partial}\left\{C^{F\alpha\beta}k_{F}{}^{C}e_{ABC}\bar{\partial}p_{\alpha}{}^{A}\bar{\partial}p_{\beta}{}^{B}\right\} \\ &- \frac{1}{(N-1)!}C^{F\alpha\beta}k_{F}{}^{C}\varepsilon_{ABCDM_1\cdots M_N}\,\varepsilon^{\sigma_1\cdots\sigma_N}\,p_{\sigma_1}{}^{M_1}\cdots p_{\sigma_{N-1}}{}^{M_{N-1}}z^{D}\bar{\partial}p_{\sigma_N}{}^{M_N}\,\bar{\partial}p_{\alpha}{}^{A}\,\bar{\partial}p_{\beta}{}^{B}\end{aligned} \tag{4.32}$$

The second term vanishes, this is seen most easily from the fact that the permutation symbol is projected orthogonal to $p_{\sigma}{}^{A}$ on the five indices $(A,B,C,D,M_N)$. Since the $p_{\sigma}{}^{A}$, $\sigma = 1,\cdots,N$ are a set of N linearly independent vectors that span an N-dimensional space projecting the permutation symbol orthogonal to the $p_{\sigma}{}^{A}$ on more than four indices must give zero.

It is compelling that there is the following correspondence between spaces of tensors of polynomials and harmonic 3-forms:

| Cohomology group | Space of polynomials | | | Identification |
|---|---|---|---|---|
| $H^{(3,0)}$ | $1$ | | | |
| $H^{(2,1)}$ | $p^{\alpha}$ | $=$ | $C^{\alpha}/I^{\alpha}$ | $C^{A}p^{\alpha}{}_{A} + f^{\alpha}{}_{\beta}p^{\beta}$ |
| $H^{(1,2)}$ | $p^{\alpha\beta}$ | $=$ | $C^{\alpha\beta}/I^{\alpha\beta}$ | $C^{A(\alpha}p^{\beta)}{}_{A} + f^{\alpha\beta}{}_{\gamma}p^{\gamma}$ |
| $H^{(0,3)}$ | $p^{\alpha\beta\gamma}$ | $=$ | $C^{\alpha\beta\gamma}/I^{\alpha\beta\gamma}$ | $C^{A(\alpha\beta}p^{\gamma)}{}_{A} + f^{\alpha\beta\gamma}{}_{\delta}p^{\delta}$ |

Table 3. The correspondence between cohomology classes and spaces of polynomials

Since there is precisely one harmonic (0,3)-form there is precisely one polynomial $Q^{\alpha\beta\gamma}$ which is not in $I^{\alpha\beta\gamma}$. Thus the rule for computing the Yukawa coupling is to take the quantity $q^{(\alpha}{}_r{}^{\beta}{}_s{}^{\gamma)}$ and to express it as a multiple of $Q^{\alpha\beta\gamma}$ plus a remainder that lies in $I^{\alpha\beta\gamma}$

$$q^{(\alpha}{}_r{}^{\beta}{}_s{}^{\gamma)} \quad = \quad \kappa Q^{\alpha\beta\gamma} + C^{A(\alpha\beta}{}_p{}^{\gamma)}{}_A + f^{\alpha\beta\gamma}{}_\delta p^\delta \tag{4.33}$$

It is straightforward, though tedious, to check these assertions by calculating the dimensions of $P^\alpha$, $P^{\alpha\beta}$ and $P^{\alpha\beta\gamma}$ by computing directly the difference between the number of independent polynomials of the relevant degree and the number of independent polynomials in the corresponding tensor ideal. However it is simpler to defer the substantiation of Table 3 to the next section where it will be subsumed in an argument valid for all Calabi-Yau spaces.

Finally we list representative forms for each of the cohomology classes of rank 3 and give an explicit expression for $Q^{\alpha\beta\gamma}$. These expressions are useful, when M has discrete symmetries, since they permit us to relate the transformation properties of the polynomials to those of the differential forms

| Differential form | Representation |
|---|---|
| $\Omega$ | $e_{ABC}\, dz^A\, dz^B\, dz^C$ |
| $\omega^{(2,1)}$ | $q^\gamma\, e_{ABC}\, dz^A\, dz^B\, \bar\partial p_\gamma{}^C$ |
| $\omega^{(1,2)}$ | $q^{\beta\gamma}\, e_{ABC}\, dz^A\, \bar\partial p_\beta{}^B\, \bar\partial p_\gamma{}^C$ |
| $\bar\Omega$ | $q^{\alpha\beta\gamma}\, e_{ABC}\, \bar\partial p_\alpha{}^A\, \bar\partial p_\beta{}^B \bar\partial p_\gamma{}^C$ |

Table 4. Explicit expressions for representatives of third rank cohomology groups. The polynomials $q^\gamma$, $q^{\beta\gamma}$, $q^{\alpha\beta\gamma}$ belong to $P^\gamma$, $P^{\beta\gamma}$, $P^{\alpha\beta\gamma}$ respectively.

Finally we record an explicit representative for $\bar{\Omega}$:

$$Q^{\alpha\beta\gamma} = p^{\alpha}{}_{;A_1B_1} p^{\beta}{}_{;A_2B_2} p^{\gamma}{}_{;A_3B_3} e^{A_1A_2A_3A_4}{}_{;B_4} e^{B_1B_2B_3B_4}{}_{;A_4} \tag{4.34}$$

where

$$e^{ABCD} = \frac{1}{N!}\varepsilon^{ABCDM_1\cdots M_N}\varepsilon_{\sigma_1\cdots\sigma_N} p^{\sigma_1}{}_{M_1}\cdots p^{\sigma_N}{}_{M_N}. \tag{4.35}$$

Note that $e^{ABCD}$ is a polynomial in the $z^A$ which is of degree 4 in virtue of (3.8). The expression we have given for $Q^{\alpha\beta\gamma}$ is, in fact, the only polynomial of the appropriate degree that can be constructed from the natural ingredients $p^{\alpha}$ and $e^{ABCD}$.

The considerations of this section are suggestive of more general arguments applicable to all Calabi-Yau spaces. It is to this subject that we now turn.

## 5. The General Case

Heretofore we have discussed the calculations of the Yukawa couplings for manifolds realized by giving N polynomial equations in a single projective space $P_{N+3}$. There are interesting generalizations of this class, among which are spaces realized as complete intersections of polynomials in products of projective spaces such as the interesting space given by Yau which is realized by giving the three polynomial equations

$$\sum_{A=1}^{4} x_A^3 = 0 \quad , \quad \sum_{A=1}^{4} y_A^3 = 0 \tag{5.1}$$

$$\sum_{A=1}^{4} x_A y_A = 0$$

in $P_3 \times P_3$. Our analysis generalizes in a straightforward way to cover these examples but rather than repeat it it is simpler to pass to the general case.

The essential observation is that the Yukawa couplings are intimately related to the deformations of the complex structure of the manifold. Fortunately the theory of the deformation of complex structures has been well developed by Kodaira and Spencer and later writers [4,5]. The complex structure of a complex manifold depends on a certain number, say m, of complex parameters.[f3]

In our previous examples m is equal to the number of coefficients in the defining polynomials once account is taken of the freedom to remove some of the coefficients by choice of coordinates.

Following Kodaira [5] we shall denote the parameters on which M depends by $t^\alpha$, $\alpha = 1,\cdots,m$ (technically, these are the effective parameters). This is a somewhat different use of the index $\alpha$ from our previous use, which we hope will not confuse the reader. To emphasize the dependence of the manifold on the parameters we shall sometimes write $M_t$ for M. The parameter space is denoted by B, in general this is itself a complex manifold but since we shall be concerned with the infinitesimal deformations of a given manifold $M_0$, corresponding to $t^\alpha = 0$, say, it is sufficient to take B to be a neighborhood of the origin in $C^m$.

f3. For a general complex manifold $m \leq b_{21}$. Since there might exist solutions to the linearized perturbation equations which are not the linearization of some perturbation. For Calabi-Yau manifolds however it has been shown by Tian [9] that all the perturbations are unobstructed so that $m = b_{21}$.

The fiber bundle

$$\mathscr{M} = \{(M_t,t)|t\varepsilon B\} \tag{5.2}$$

with projection $M_t \to t$ is a complex manifold of dimension m + 3 in which the given manifold M is the embedded submanifold specified by the m holomorphic equations

$$t^{\alpha}(y^A) = 0 \tag{5.3}$$

with the $y^A$, A=1,•••,m+3, coordinates for . We may proceed in close analogy with our previous treatment. As before we may choose basis vectors and basis forms adapted to the deformations

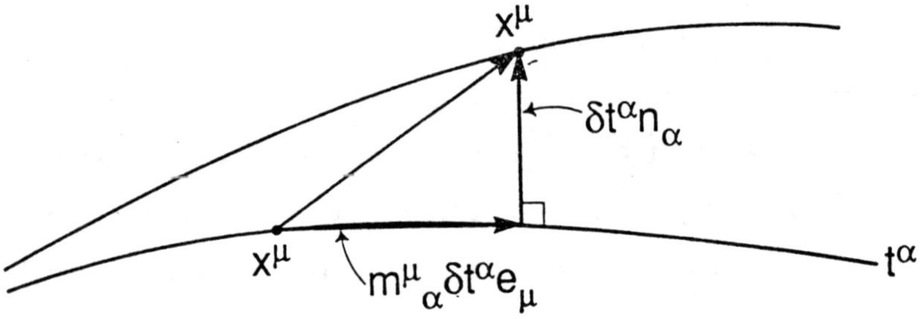


Figure 4. The displacement vector, which joins the point labelled by $x^{\mu}$ in $M_t$ to the point with the same coordinate labels in $M_{t+\delta t}$, can again be decomposed into a normal lapse and a tangential shift.

$$e_\mu = \frac{\partial}{\partial x^\mu} \quad ; \quad e^\mu = dx^\mu + m^\mu{}_\alpha n^\alpha$$

$$n_\alpha = \frac{\partial}{\partial t^\alpha} - m^\mu{}_\alpha \frac{\partial}{\partial x^\mu} \quad ; \quad n^\alpha = dt^\alpha \tag{5.4}$$

There is one difference between this and the previous case which is that we may not assume that the ambient manifold $\mathcal{M}$ is Kähler. However since it is a complex manifold so we may take it to be endowed with a hermitean metric and the hermitean connection [10]. The previous analysis goes through almost unchanged except that now the extrinsic curvature

$$\chi^\alpha_{\mu\nu} = \frac{\partial y^A}{\partial x} \frac{\partial y^B}{\partial x^\nu} \left(\frac{\partial^2 t^\alpha}{\partial y^A \partial y^B} - \Gamma^C_{AB} \frac{\partial t^\alpha}{\partial y^C}\right) \tag{5.5}$$

is not now necessarily symmetric in A and B since the hermitean connection has torsion and is not symmetric in its lower indices unless $\mathcal{M}$ is in fact Kähler. This makes no difference to the analysis since the symmetry of $\chi_{\mu\nu}{}^\alpha$ was never used. We must however take care to represent the elements of $H^{(0,1)}_{\bar\partial}(M,T)$ as in eqn. (2.27) with the $dx^{\bar\nu}$ contracted on the first index. The elements of $H^{(0,1)}_{\bar\partial}(M,T)$ are now represented in the form

$$a^\mu = q^\alpha \chi_{\bar\nu}{}^\mu{}_\alpha dx^{\bar\nu} \tag{5.6}$$

with the $q^\alpha$ corresponding to variations in the parameters $t^\alpha$. More precisely let us suppose that the equations

$$t^\alpha(y^A,u) = 0 \tag{5.7}$$

depend holomorphically on a parameter u. Then with a deformation of M we associate a vector

$$q^{\alpha}(y^A) = \frac{\partial t^{\alpha}}{\partial u}(y^A,0). \tag{5.8}$$

We are again interested in the question as to when the form (5.6) is exact. A short calculation reveals that we may write

$$a^{\mu} \frac{\partial}{\partial x^{\mu}} = \bar{\partial}\left(q^{\alpha} t_{\alpha}{}^{D} \frac{\partial}{\partial y^{D}}\right) \tag{5.9}$$

where we employ a notation similar to that of Sect. 2. The $\bar{\partial}$-operator refers to

$$t^{\alpha}{}_{A} = \frac{\partial t^{\alpha}}{\partial y^{A}} \tag{5.10}$$

and the α-indices are raised and lowered by means of the normal metric

$$h^{\alpha\bar{\beta}} = t^{\alpha}{}_{A} t^{\bar{\beta}A}. \tag{5.11}$$

$a^{\mu}$ is exact if and only if

$$(q^{\alpha} t_{\alpha}{}^{D} - c^{D}) \frac{\partial t^{\gamma}}{\partial y^{D}} = 0 \tag{5.12}$$

on $t^{\beta} = 0$ for some holomorphic $c^D(y)$ i.e., if and only if

$$q^{\alpha} = c^{D}\, t^{\alpha}{}_{D} + f^{\alpha}{}_{\beta}\, t^{\beta} \tag{5.13}$$

for some holomorphic $c^D(y)$ and $f^{\alpha}{}_{\beta}(y)$.

Let us return now to a consideration of the product of three elements of $H_{\bar{\partial}}^{(0,1)}(M,T)$

$$\bar{\omega} = a^{\mu}{}_{\wedge} b^{\nu}{}_{\wedge} c^{\rho} \Omega_{\mu\nu\rho} \tag{5.14}$$

and set

$$\bar{\omega}^{\mu\nu\rho} = a^{[\mu}{}_{\wedge} b^{\nu}{}_{\wedge} c^{\rho]} \tag{5.15}$$

and

$$\|\Omega\|^2 = \frac{1}{3!} \Omega_{\mu\nu\rho} \bar{\Omega}^{\mu\nu\rho} \tag{5.16}$$

so that

$$\bar{\omega} = \bar{\omega}^{\mu\nu\rho} \Omega_{\mu\nu\rho} \tag{5.17}$$

and

$$\bar{\omega}^{\mu\nu\rho} = \frac{1}{3! \|\Omega\|^2} \bar{\Omega}^{\mu\nu\rho} \bar{\omega}. \tag{5.18}$$

Although it might seem at first sight improbable the quantity

$$\frac{\bar{\Omega}^{\mu\nu\rho}}{\|\Omega\|^2}$$

is a holomorphic tensor field

$$\bar{\partial}\left(\frac{\bar{\Omega}^{\mu\nu\rho}}{\|\Omega\|^2}\right) = 0. \tag{5.19}$$

The most explicit way to see this is to recall that $\Omega$ is holomorphic and nowhere zero. Thus in local coordinates

$$\Omega_{\mu\nu\rho} = f(x) \varepsilon_{\mu\nu\rho} \tag{5.20}$$

with f(x) holomorphic and nowhere zero. It follows that

$$\frac{\bar{\Omega}^{\mu\nu\rho}}{\|\Omega\|^2} = \frac{1}{f(x)}\varepsilon^{\mu\nu\rho}. \tag{5.21}$$

Since f is nonzero we verify (5.19). With this observation in hand it is strightforward to check that, given two forms $\bar{\omega}$ and $\bar{\omega}^{\mu\nu\rho}$ related by (5.17) that $\bar{\omega}$ is closed if and only if $\bar{\omega}^{\mu\nu\rho}$ is and $\bar{\omega}$ is exact if and only if $\bar{\omega}^{\mu\nu\rho}$ is.

We enquire first when $\bar{\omega}^{\mu\nu\rho}$ is exact. If $a^\mu$, $b^\nu$, $c^\rho$ correspond to perturbations $q^\alpha$, $r^\beta$, $s^\gamma$ then

$$\bar{\omega}^{\mu\nu\rho} = q^\alpha r^\beta s^\gamma \, \chi_{\bar{\kappa}}{}^{\mu}{}_{\alpha} \, \chi_{\bar{\lambda}}{}^{\nu}{}_{\beta} \, \chi_{\bar{\sigma}}{}^{\rho}{}_{\gamma} \, dx^{\bar{\kappa}} \, dx^{\bar{\lambda}} \, dx^{\bar{\sigma}} \, . \tag{5.22}$$

If we replace the product $q^\alpha r^\beta s^\gamma$ by any holomorphic tensor $q^{\alpha\beta\gamma}$ of the appropriate degree then the corresponding (0,3)-form $\bar{\omega}^{\mu\nu\rho}$, and hence $\bar{\omega}$, is still closed. By an analysis that is precisely analogous to the discussion of the deformations $q^\alpha$ we find the corresponding (0,3)-form is exact if and only if

$$q^{\alpha\beta\gamma} = C^{A(\alpha\beta}_{A} t^{\gamma)} + f^{\alpha\beta\gamma}(y,t) \tag{5.23}$$

with the $f^{\alpha\beta\gamma}$ holomorphic functions that vanish with $t^\alpha$. In a similar way it is possible to relate (1,2)-forms to symmetric tensors $q^{\alpha\beta}$ subject to an appropriate tensor ideal. These results are summarized in Table (5).

| Type of form | Representative | Tensor ideal |
|---|---|---|
| (3,0) | 1 | |
| (2,1) | $q^{\alpha}$ | $c^{A}t^{\alpha}{}_{A} + f^{\alpha}$ |
| (1,2) | $q^{\alpha\beta}$ | $c^{A(\alpha}t^{\beta)}{}_{A} + f^{\alpha\beta}$ |
| (0,3) | $q^{\alpha\beta\gamma}$ | $c^{A(\alpha\beta}t^{\gamma)}{}_{A} + f^{\alpha\beta\gamma}$ |

Table 5. The correspondence between cohomology groups and spaces of functions for a general Calabi-Yau space.

We know on general grounds that there is precisely one cohomology class of (0,3)-forms. Thus there exists a tensor $Q^{\alpha\beta\gamma}$, say, which represents $\bar{\Omega}$ which is such that an arbitrary tensor $q^{\alpha\beta\gamma}(y)$ is equal to some multiple of $Q^{\alpha\beta\gamma}(y)$ modulo the tensor ideal. We may take $Q^{\alpha\beta\gamma}$ to be given by the expression

$$Q^{\alpha\beta\gamma} = \chi_{\mu_1\nu_1}{}^{\alpha}\,\chi_{\mu_2\nu_2}{}^{\beta}\,\chi_{\mu_3\nu_3}{}^{\gamma}\,\frac{\bar{\Omega}^{\mu_1\mu_2\mu_3}\bar{\Omega}^{\nu_1\nu_2\nu_3}}{\|\Omega\|^4}\,. \tag{5.24}$$

$Q^{\alpha\beta\gamma}$ is holomorphic in virtue of our earlier remarks concerning the quantity $\bar{\Omega}^{\mu\nu\rho}/\|\Omega\|^2$. We may check that, as defined, $Q^{\alpha\beta\gamma}$ does not correspond to an exact form i.e., it is not in the tensor ideal, by showing that the integral

$$\int_M \Omega\wedge\bar{\omega}$$

with $\bar{\omega}$ the corresponding (0,3)-form, is nonzero. This is straightforward

$$\bar{\omega} = Q^{\alpha\beta\gamma} \chi_{\bar{\mu}_1}{}^{\nu_1}{}_{\alpha} \chi_{\bar{\mu}_2}{}^{\nu_2}{}_{\beta} \chi_{\bar{\mu}_3}{}^{\nu_3}{}_{\gamma} \Omega_{\nu_1\nu_2\nu_3} dx^{\bar{\mu}_1} dx^{\bar{\mu}_2} dx^{\bar{\mu}_3} \tag{5.25}$$

and we have the elementary identity

$$dx^{\bar{\mu}_1} dx^{\bar{\mu}_2} dx^{\bar{\mu}_3} = \frac{\Omega^{\bar{\mu}_1\bar{\mu}_2\bar{\mu}_3}}{\|\Omega\|^2} \bar{\Omega} \, . \tag{5.26}$$

Thus

$$\bar{\omega} = Q^{\alpha\beta\gamma} \bar{Q}_{\alpha\beta\gamma} \|\Omega\|^2 \bar{\Omega} \tag{5.27}$$

moreover since

$$\Omega_\wedge \bar{\Omega} = \|\Omega\|^2 g^{1/2} d^6x \tag{5.28}$$

we find that

$$\int_M \Omega_\wedge \bar{\omega} = \int_M Q^{\alpha\beta\gamma} \bar{Q}_{\alpha\beta\gamma} \|\Omega\|^4 g^{1/2} d^6x \, . \tag{5.29}$$

The right-hand side of this equality is the integral of a positive real quantity and so cannot vanish.

Finally, the relation that determines the Yukawa coupling is

$$q^{(\alpha_r \beta_s \gamma)} = \kappa Q^{\alpha\beta\gamma} + C^{A(\alpha\beta_t\gamma)}{}_A + f^{\alpha\beta\gamma} \, . \tag{5.30}$$

## 6. Symmetries and Pseudosymmetries

Of particular importance for application to the compactification of superstring theories are manifolds M endowed with discrete isometry groups. We shall not attempt a detailed discussion of the implications of such symmetries but will instead make some observations for the case that M may be realized as the transverse intersection of polynomials $p^{\alpha}$ in $P_{N+3}$. Even in this case we have very little to add to the detailed discussion of Witten [6] who first pointed out that the symmetries and pseudosymmetries of M greatly restrict the possible structure of the couplings $\kappa(a,b,c)$. Indeed for the case of $P_4(5)$ with the simplest quintic (4.24). Witten showed purely on the basis of the pseudosymmetries of the manifold that the coupling between the quintics q, r and s vanishes unless the product qrs contains the term

$$(z^1)^3(z^2)^3(z^3)^3(z^4)^3(z^5)^3 .$$

The ratios of nonzero couplings $\kappa(q,r,s)$ i.e., those for which the product qrs does contain the above term are not determined purely on symmetry grounds. Nevertheless it is surprising that it is possible to go so far purely on the basis of symmetries especially since the procedures we have developed in this article has a priori nothing to do with symmetries. The simple but essential observation is that the polynomial $Q^{\alpha\beta\gamma}$ which represents $\bar{\Omega}$ is covariant under the symmetry transformation, a consequence of this is that noncovariant polynomials are in the tensor ideal $I^{\alpha\beta\gamma}$ and hence correspond to exact forms. More precisely suppose that M has a group $\Delta$ of discrete isometries that are induced by a group of $\hat{\Delta}$ of isometries of the embedding space. Then for each $g\epsilon\hat{\Delta}$ there are unitary matrices $L^A{}_B$ and $\ell^\alpha{}_\beta$ such that the action of g is represented by

$$z^A \to L^A{}_B z^B \tag{6.1}$$

and

$$p^\alpha(Lz) = \ell^\alpha{}_\beta p^\beta(z) \tag{6.2}$$

($\ell^\alpha{}_\beta$ must be unitary to preserve the normal metric $k^{\alpha\bar{\beta}}$). For the simplest cases $\ell^\alpha{}_\beta$ is just the identity $\delta^\alpha{}_\beta$ and for these cases we wish to show that for the purposes of computing the coupling it suffices to consider the symmetric part $(q^{(\alpha_r\beta_s\gamma)})_0$ of the product $q^{(\alpha_r\beta_s\gamma)}$. For an arbitrary polynomial $\phi$ we may make a decomposition into a symmetric and a non symmetric part by writing

$$\phi(z) = \phi_0(z) + \phi_1(z) \tag{6.3}$$

with

$$\phi_0(z) = \frac{1}{|\hat{\Delta}|} \sum_{g\epsilon\hat{\Delta}} \phi(Lz) \tag{6.4}$$

where $|\hat{\Delta}|$ denotes the order of $\hat{\Delta}$. Now it is easy to see from the explicit representation (4.35) that $Q^{\alpha\beta\gamma}$ is invariant i.e., that

$$Q^{\alpha\beta\gamma} = Q_0^{\alpha\beta\gamma} \tag{6.5}$$

it is also clear that if a polynomial $R^{\alpha\beta\gamma}(z)$ is in the ideal then so is $R^{\alpha\beta\gamma}(Lz)$ for each $g\varepsilon\hat{\Delta}$. Hence if for some polynomial $q^{\alpha\beta\gamma}$ we may write

$$q^{\alpha\beta\gamma} = \kappa Q^{\alpha\beta\gamma} + R^{\alpha\beta\gamma} \tag{6.6}$$

with $R^{\alpha\beta\gamma}$ a remainder in the tensor ideal then

$$q_0^{\alpha\beta\gamma} = \kappa Q^{\alpha\beta\gamma} + R_0^{\alpha\beta\gamma} \tag{6.7}$$

so the symmetric part of $q^{\alpha\beta\gamma}$ leads to the same value of the coupling. In particular the non-symmetric part $q_1^{\alpha\beta\gamma}$ is in the tensor ideal.

Finally we note that representation (4.35) also affords a simple transformation rule for the holomorphic 3-form, relaxing the assumption that $\ell^{\alpha}{}_{\beta}$ is the identity, we have

$$\Omega \rightarrow \overline{\det(\ell^{\alpha}{}_{\beta})}\ \Omega \tag{6.8}$$

if $\det(\ell^{\alpha}{}_{\beta})$ is not unity then the low energy theory has an R-symmetry.

## 7. The Normalization Matrix

The elements of the cohomology group $H_{\bar{\partial}}^{(0,1)}(M,T)$ correspond to the families of light particles in the low energy effective theory. An important question of normalization arises. The kinetic terms of the low energy effective Lagrangian will not be diagonal and canonically normalized if a basis for $H_{\bar{\partial}}^{(0,1)}(M,T)$ is chosen arbitrarily. In order to diagonalize the kinetic terms one must diagonalize the matrix

$$N(a^{\mu},b^{\nu}) = \int a \wedge \bar{b} \qquad (7.1)$$

where a and b are the (2,1)-forms corresponding to $a^{\mu}$ and $b^{\nu}$. The computation of this matrix is an important problem about which little is known. In terms of deformations $q^{\alpha},r^{\beta}$ we have

$$N(q,r) \propto \int q^{\alpha} r^{\bar{\beta}} \chi^{\mu\nu}{}_{\alpha} \chi_{\mu\nu\bar{\beta}} \, .$$

For the special cases of manifolds endowed with large symmetry or pseudosymmetry groups much information about the normalization maxtrix is available on the basis of symmetry alone. However even for the simplest manifolds it does not seem to be known how to determine $N(q^{\alpha},r^{\bar{\beta}})$ completely short of numerical computation.

The reason that the evaluation of $N(q^{\alpha},r^{\beta})$ is difficult seems to be that whereas, for the case of a complete intersection of polynomials in $P_{N+3}$ say, the coupling $\kappa(q^{\alpha},r^{\beta},s^{\gamma})$ are polynomial functions of the coefficients of the defining polynomial $p^{\alpha}$ this is not true of the normalization matrix. It seems that $N(q^{\alpha},r^{\beta})$ depends transcendentally[f4] on the coefficients of $p^{\alpha}$.

There is an alternative and intriguing way to pose the problem of calculating the normalization matrix it is the following. Consider for definiteness the

f4. I am grateful to P. Griffiths for this observation.

manifold $P_4(5)$ corresponding to a quintic in $P_4$. We know from our previous discussion that (2,1)-forms are in correspondence with quintic polynomials in $P_5$ and (1,2)-forms are in correspondence with tenth order polynomials in $P_{10}$. Thus every quintic q defines a (2,1)-form and hence, by complex conjugation, also a (1,2)-form. The question is to which tenth order polynomial does q correspond?

Acknowledgements

It is a pleasure to acknowledge fruitful conversations with R. Bott, P. Griffiths and R. O. Wells, Jr. The author wishes to thank the Santa Barbara Institute for Theoretical Physics where part of this work was performed.

## Appendix: Direct Evaluation of the Coupling for $P_4(5)$

The purpose of this appendix is to record the fact that it is possible, at least for the case of $P_4(5)$, to evaluate the coupling $\kappa(q,r,s)$ directly. Ultimately this is what one must do if one wishes to fix its overall normalization.

We begin by setting

$$\bar{\omega} = a^{\mu} b^{\nu} c^{\rho} \Omega_{\mu\nu\rho} \tag{A1}$$

as in Sect.5 and extending the domain of integration to the whole of $P_4$ by means of a δ-function

$$\kappa = -\frac{1}{2i} \int_{P_4} \Omega\bar{\omega}\, dp\, d\bar{p}\, \delta^{(2)}(p) , \tag{A2}$$

where $\delta^{(2)}(p)$ means $\delta(\mathrm{Re}p)\delta(\mathrm{Im}p)$. To proceed we observe that

$$\delta^{(2)}(p)d\bar{p} = \frac{1}{\pi}\bar{\partial}\left(\frac{1}{p}\right) \tag{A3}$$

and also that

$$\begin{aligned} \Omega dp &= \frac{1}{4!}\epsilon_{ABCDE}z^{A}dz^{B}dz^{C}dz^{D}dz^{E} \\ &= z^{5}dz^{1}dz^{2}dz^{3}dz^{4} \\ &= z^{5}d^{4}z \end{aligned} \tag{A4}$$

which follow from (3.5). The second equality holding in the coordinate patch $U_5$ hence

$$\kappa = \frac{1}{2\pi i.4!}\int_{P_4}\bar{\omega}\bar{\partial}\left(\frac{1}{p}\right)z^{5}d^{4}z \ . \tag{A5}$$

The next maneuver is to integrate by parts. From eqn (4.4) we have the expression, valid in $U_5$,

$$\frac{|z^{5}|^{2}\bar{\omega}}{\sigma} = z^{5}\frac{qrs}{\|p\|^{8}}\,\epsilon^{\bar{I}\bar{J}\bar{K}\bar{L}}\,\bar{p}_{\bar{I}}d\bar{p}_{\bar{J}}d\bar{p}_{\bar{K}}d\bar{p}_{\bar{L}} \tag{A6}$$

where the indices have the range 1,..., 4. It follows that

$$\bar{\partial}\left(\frac{|z^{5}|^{2}\bar{\omega}}{\sigma}\right) = \frac{qrs}{\|p\|^{10}}\,z^{5}p_{5}\,\bar{\tau} \tag{A7}$$

with

$$\tau = \epsilon^{ABCDE}\,p_{A}dp_{B}dp_{C}dp_{D}dp_{E} \tag{A8}$$

by adding four similar expressions we find

$$\bar{\partial}\bar{\omega} = \frac{5pqrs}{\|p\|^{10}}\,\bar{\tau} \,. \qquad \text{(A9)}$$

Notice that there is an explicit factor of p in this expression which cancels the $\frac{1}{p}$ in the integrand. So far we have

$$\kappa = -\frac{5}{2\pi i}\int_{P_4} \frac{qrs}{\|p\|^{10}}\, z^5 d^4z\,\bar{\tau} \,. \qquad \text{(A10)}$$

It is possible to transform the integral to an integral over the nine sphere $S^9$. This is, of course, a consequence of the fact that $S^9$ is the Hopf fibration of $P_4$ but it may be understood in less sophisticated language as a consequence of using the scaling freedom associated with the homogenous coordinates to enforce the coordinate condition $\sigma = 1$ rather than the more usual condition of taking one of the $z^A$ to be a constant. To effect the transformation insert unity in the guise of

$$1 = \frac{i}{2\pi}\int d\eta\, d\bar{\eta}\, \frac{|z^5|^2}{|\eta|^4}\, \delta\Big(\sigma - \frac{|z^5|^2}{|\eta|^2}\Big) \qquad \text{(A11)}$$

into the integral and make the change of variable

$$z^k \to \frac{z^k z^5}{\eta} \quad , \quad k = 1,\ldots,4 \qquad \text{(A12)}$$

noting that if a function $F(z^k,z^5)$ is homogeneous of degree m then

$$F\Big(\frac{z^k z^5}{\eta}\,,\, z^5\Big) = \Big(\frac{z^5}{\eta}\Big)^m F(z^k,\eta) \,. \qquad \text{(A13)}$$

Finally rename the variable $\eta \rightarrow z^5$. We find

$$\kappa = \frac{-5}{(2\pi)^2} \int_{C_5} d^5z \; \bar{\tau} \; dz^{\bar{5}} \; \frac{qrs}{\|p\|^{10} z^{\bar{5}}} \; \delta(\sigma - 1) \tag{A14}$$

where $d^5z = dz^1dz^2dz^3dz^4dz^5$. We wish now to eliminate the specific reference to $z^{\bar{5}}$ in the integrand. To this end note that

$$\tau \frac{dz^5}{z^5} = \varepsilon^{ABCDE} \, p_A dp_B dp_C dp_D dp_E \frac{1}{z^5} \frac{\partial z^5}{\partial p_F} dp_F$$

$$= \varepsilon^{ABCDE} \, \varepsilon_{BCDEF} \, p_A \frac{1}{z^5} \frac{\partial z^5}{\partial p_F} d^5p \tag{A15}$$

$$= 4! \frac{1}{z^5} p_A \frac{\partial z^5}{\partial p_A} d^5p$$

$$= 3! d^5p$$

where $d^5p = dp_1dp_2dp_3dp_4dp_5$ and the last equality follows from the fact that each $z^A$ is a function homogeneous of degree 1/4 of the $p_B$. There is a further identity which is useful and which may be established analogously

$$\tau \frac{\partial \sigma}{\sigma} = 3 \; d^5p. \tag{A16}$$

Thus

$$\tau \frac{dz^5}{z^5} \doteq \tau \frac{\partial \sigma}{\sigma} \, . \tag{A17}$$

This identity permits us to perform the σ-integration in (A14)

$$\kappa = \frac{-5}{(2\pi)^2} \int_{S_9} \frac{qrs}{\| p \|^{10}} d^5 z \,\bar{\tau} \,. \tag{A18}$$

We have brought the integral to a standard form to which we may apply the Bochner-Martinelli theorem [11]. This leads to a drastic simplication

$$\kappa = \frac{-5}{(2\pi)^2} \int_{\Gamma_5} \frac{qrs d^5 z}{p_1 p_2 p_3 p_4 p_5} \tag{A19}$$

where the contour of integration $\Gamma_5 = \gamma_1 \times \gamma_2 \times .. \times \gamma_5$ is the direct product of five 1-dimensional contours that wind around the five curves $p_A = 0$.

We may now recover our previous results in virtue of Cauchy's theorem. First note that we may take the product qrs to be defined modulo I since a shift $qrs \to qrs + E^A p_A$ has no effect on the integral. For each term of the point $E^A p_A$ that is in the ideal the factor $p_A$ cancels one of the poles in the denominator the corresponding $\gamma_A$ is contractible to a point. Hence the contribution vanishes. Thus the only integral that needs to be computed is

$$\int_{\Gamma_5} \frac{Q d^5 z}{p_1 p_2 p_3 p_4 p_5} = \int_{\Gamma_5} \frac{d^5 p}{p_1 p_2 p_3 p_4 p_5} = (2\pi i)^5 \tag{A20}$$

where the first equality is a consequence of the fact that

$$Q = \det\left(\frac{\partial^2 p}{\partial z^A \partial z^B}\right) = \frac{\partial(p_1, p_2, \ldots, p_5)}{\partial(z^1, z^2, \ldots, z^5)} \tag{A21}$$

is the Jacobian of the transformation from the $p_A$ to the $z^A$.

References


1. P. Candelas, G. Horowitz, A. Strominger and E. Witten, Nucl. Phys. B258(1985) 46.

2. A. Strominger and E. Witten, Commun. Math. Phys. 101, 341 (1985).

3. A. Strominger, Phys. Rev. Lett. 55 (1985) 2547, and in "Unified String Theories", M. Green and D. Gross Eds, World Scientific, 1986.

4. K. Kodaira and D. C. Spencer, Ann. Math. 67 (1958) 328, Ann. Math 67 (1958) 403, Ann. Math. 71 (1960) 43.

5. K. Kodaira, Complex Manifolds and Deformations of Complex Structure, Springer-Verlag 1985.

6. E. Witten, Nucl. Phys. B258 (1985) 75.

7. R. Arnowitt, S. Deser and C. W. Misner, "The Dynamics of General Relativity" in Gravitation: An Introduction to Current Research, L. Witten Ed. Wiley, New York 1962.

8. M. F. Atiyah, R. Bott and L. Garding, Acta Mathematica 131, 145 (1973).

9. G. Tian, "Smoothness of the Universal Deformation Space of Compact Calabi-Yau Manifolds and its Peterson-Weil Metric" preprint, Department o Mathematics, University of California, San Diego.

10. S.Kobayashi and K. Nomizu, Foundations of Differential Geometry, Interscience, New York, 1963.

11. P. Griffiths and J. Harris, Principles of Algebraic Geometry, Interscience, New York, 1978.

# Three dimensional algebraic manifolds with $C_1 = 0$ and $\chi = -6$

G. Tian* and S. T. Yau**
Department of Mathematics
University of California, San Diego
La Jolla, CA 92093

Algebraic threefolds were used in [CHSW] to study low energy phenomenological questions. It was desirable to have manifolds with Euler number equal to +6 or -6. While a lot of efforts were tried by many people, the only known examples were constructed by us in [ Y]. They are quotient of some manifolds as complete intersections in the product of projective spaces. In [ Y], we constructed three examples. It turns out the K. Kirland and B. Green pointed out to us that the first two examples are diffeomorphic to each other. (The fundamental group of the second example should have been $Z_3$ and $h^{1,1}$, $h^{2,1}$ of the first should have been 6, 9 respectively.) It is not clear whether the complex structure of the first example can be deformed to the one in the second example. It turns out that there are singularities in the third example. These singularities can be resolved to give rise to an example with zero Euler number.

In this paper, we demonstrate a method of constructing another three dimensional algebraic manifold with $C_1 = 0$ and Euler number -6 from a given one. Such a method was used in the past few years in studying algebraic threefolds. The manifolds constructed in such a way are all birational to each other. Conversely, algebraic threefolds with $C_1 = 0$ which are

*Supported by Alfred P. Sloan Foundation Fellowship for doctoral dissertation.
**Research supported in part by NSF grant #DMS84-09447 and ONR contract #N00014-85-K-0367.

birational to each other, are obtained by such constructions. Roughly speaking, we take a rational curve in the threefold and replace it by another rational curve in a different way. Applying the construction to our previous example, we find new algebraic manifolds with $C_1 = 0$ and $\chi = -6$. They have the same cohomology ring but different differentiable structure because the first Pontragin class is different for these manifolds. It would be interesting to know whether these are the only algebraic threefolds with $C_1 = 0$ and $\chi = -6$.

§1. The construction

Let $X$ be a three dimensional manifold with $C_1 = 0$, then for any smooth rational curve $C \subset X$, the normal bundle $N_{C|X}$ is a holomorphic rank 2 bundle and $C_1(N_{C|X}) = -C_1(C)$. Adapting the usual notation, $K_C = \theta_C(-2)$, i.e. $C_1(C) = 2$. By Grotherdieck's theorem [GH], $N_{C|X} = \theta_C(m_1) \oplus \theta_C(m_2)$. Since $C_1(N_{C|X}) = -C_1(C) = -2$, $m_1 + m_2 = -2$. In this paper, we only consider the case that $m_1 = m_2 = -1$. Then $N^*_{C|X} = \theta_C(1) \oplus \theta_C(1)$. As a rational curve, $C$ is covered by two open subsets $U_0, U_1 \subset X$, and in each $U_i$, we have local coordinates $(z_i, u_i, v_i)$ such that $U_i \cap C = \{U_i = V_i = 0\}$ and $z_{i|C}$ is the local coordinate of $U_i \cap C$. Since $C \cong CP^1$, we can choose $z_1 = \frac{1}{z_0}$ on $U_0 \cap U_1$. Furthermore,

$$\begin{cases} u_1 = u_0 f_{11}(u_0, v_0, z_0) + v_0 f_{12}(u_0, v_0, z_0) \\ v_1 = u_0 f_{21}(u_0, v_0, z_0) + v_0 f_{22}(u_0, v_0, z_0) \end{cases} \quad \text{on } U_0 \cap U_1 . \qquad //$$

Over $U_i$, $(du_i, dv_i)$ defines a frame for $N^*_{C|X}$. Then the transition matrix is given by

$$\begin{pmatrix} f_{11}(0,0,z_0), & f_{12}(0,0,z_0) \\ f_{21}(0,0,z_0), & f_{22}(0,0,z_0) \end{pmatrix}^{-1} \quad \text{on } U_0 \cap U_1 \cap C .$$

Since $N^*_{C|X} \cong \theta(1) \oplus \theta(1)$, there are holomorphic matrices

$$\begin{pmatrix} g_{11}(z_0), g_{12}(z_0) \\ g_{21}(z_0), g_{22}(z_0) \end{pmatrix}, \begin{pmatrix} h_{11}(z_1), h_{12}(z_1) \\ h_{21}(z_1), h_{22}(z_1) \end{pmatrix} \text{ on } U_0 \cap C,\ U_1 \cap C,$$

respectively, such that

$$\begin{pmatrix} h_{11}(z_1), h_{12}(z_1) \\ h_{21}(z_1), h_{22}(z_1) \end{pmatrix} \begin{pmatrix} f_{11}(0,0,z_0), f_{12}(0,0,z_0) \\ f_{21}(0,0,z_0), f_{22}(0,0,z_0) \end{pmatrix} \begin{pmatrix} g_{11}(z_0), g_{12}(z_0) \\ g_{21}(z_0), g_{22}(z_0) \end{pmatrix} = \begin{pmatrix} z_0 & 0 \\ 0 & z_0 \end{pmatrix}$$

and

$$\det \begin{pmatrix} h_{11}(z_1), h_{12}(z_1) \\ h_{21}(z_1), h_{22}(z_1) \end{pmatrix} \neq 0 \text{ in } U_1 \cap C, \quad \det \begin{pmatrix} g_{11}(z_0), g_{12}(z_0) \\ g_{21}(z_0), g_{22}(z_0) \end{pmatrix} \neq 0 \text{ in } U_0 \cap C.$$

The following transformations

$$(u_0', v_0') \to (g_{11}u_0' + g_{12}v_0', g_{21}u_0' + g_{22}v_0') = (u_0, v_0)$$

$$(u_1, v_1) \to (h_{11}u_1 + h_{12}v_1, h_{21}u_1 + h_{22}v_1) = (u_1', v_1')$$

are invertible when $U_0$ and $U_1$ are small enough. With respect to the new frames $(du_0', dv_0')$ and $(du_1', dv_1')$, the transition matrix of $N^*_{C|X}$ is given by

$$\begin{pmatrix} \frac{1}{z_0} & 0 \\ 0 & \frac{1}{z_0} \end{pmatrix}.$$

For simplicity, we write $u_0 = u_0'$, $v_0 = v_0'$, $u_1 = u_1'$, $v_1 = v_1'$. Then on $U_0 \cap U_1$,

$$f_{11}(0,0,z_0) = f_{22}(0,0,z_0) = z_0$$

$$f_{12}(0,0,z_0) = f_{21}(0,0,z_0) = 0 .$$

By shrinking $U_0$ and $U_1$ more and using the method of power series, one can find new coordinates $u_0', v_0'$ on $U_0$ and $u_1', v_1'$ on $U_1$, respectively, such that

$$\begin{aligned} u_1' &= z_0 u_0' \\ v_1' &= z_0 v_0' \end{aligned} \qquad \text{on } U_0 \cap U_1 \quad .$$

We still write $u_1 = u_1'$, $v_1 = v_1'$; $u_0 = u_0'$, $v_0 = v_0'$.

Let $\varphi : U_0 \to C^4$ be defined by $\varphi(z_0, u_0, v_0) = (u_0, v_0, z_0u_0, z_0v_0) = (x, y, z, w)$. Then the image of $\varphi$ lies in $xw = yz$.

On $U_0 \cap U_1$, $\varphi$ can be written as $\varphi(z_1, u_1, v_1) = (z_1u_1, z_1v_1, u_1, v_1)$. Hence $\varphi$ can be defined over $U_1$ and image $\varphi$ is a neighborhood of the origin in $\{xw = yz\}$. Obviously, $\varphi(C) = 0$ and $\varphi : (U_0 \cup U_1) \setminus C \to C^4 \setminus \{0\}$ is one-to-one.

Now, we can blow down $X$ to a 3-fold $\tilde{X}$ with an ordinary singular point, i.e. $\tilde{X} = (X \setminus C) \cup_\varphi \varphi(U)$, the 3-fold produced by identifying $U \setminus C$ with $\varphi(U) \setminus \{0\}$ through $\varphi$, where $U = U_0 \cup U_1 \subset X$. There is a natural projection $\pi: X \to \tilde{X}$. One can also blow $X$ down along several disjoint smooth rational curves with normal bundle isomorphic to $\theta_{CP^1}(-1) \oplus \theta_{CP^1}(-1)$.

An immediate question is the algebraicity of $\tilde{X}$. If $\tilde{X} \subset CP^m$ for some $m$, the pull-back of the hyperplane line bundle of $CP^m$ to $X$ by $\pi$ is very ample outside $C_1, \ldots, C_N$ and every global section is constant along each $C_i$. This pulled-back line bundle has no base points on $X$. Conversely, if we can find a line bundle on $X$ with the above mentioned properties, then one can prove that $\tilde{X}$ is algebraic, since each global section of this line bundle can be pushed down to be a global section of $\tilde{X}$.

Clearly, $X$ can be regarded as a resolution of $\tilde{X}$. The blowing up of $C^4$ at the origin induces a resolution $\hat{X}$ of $\tilde{X}$ with exceptional curve $CP^1 \times CP^1$, since the singular point of $X$ is defined by $xw = yz$ near the origin. Considering that the exceptional curve of $X$ over $\tilde{X}$ is a one dimensional curve $C \cong CP^1$, we call $X$ the small resolution of $\tilde{X}$. Now $\tilde{X}$ has another small resolution $X'$ which can be described as follows:

$$X = (X \setminus C) \cup_{\varphi} \varphi(U), \quad \varphi(U) = \{xw = yz\} \cap \{\text{a neighborhood of } 0 \text{ in } C^4\},$$

where $U = U_0 \cup U_1$ and $\varphi: U_0 \to C^4, (z_0, u_0, v_0) \to (u_0, v_0, z_0u_0, z_0v_0)$

$$\varphi: U_1 \to C^4, (z_1, u_1, v_1) \to (z_1u_1, z_1v_1, u_1, v_1)$$

Now take two copies of $C^3$ and define

$$V_0 \cong C^3, \; V_1 \cong C^3,$$

$$\psi_0 : V_0 \to C^4, \; (z_0, u_0, v_0) \to (u_0, z_0u_0, v_0, z_0v_0)$$

$$\psi_1 : V_1 \to C^4, \; (z_1, u_1, v_1) \to (z_1u_1, u_1, z_1v_1, v_1)$$

Identifying $(z_0, u_0, v_0)$ with $(z_1, u_1, v_1)$ when $z_0z_1 \neq 0$ by $z_1 = 1/z_0$, $u_1 = z_0u_0$, $v_1 = z_0v_0$, then $\psi_1(z_1, u_1, v_1) = \psi_0(z_0, u_0, v_0)$. Under the identification, $V_0 \cup V_1$ becomes the total space of $\theta_{CP^1}(-1) \oplus \theta_{CP^1}(-1)$ and $\psi_0, \psi_1$ induce a map $\psi: \theta_{CP^1}(-1) \oplus \theta_{CP^1}(-1) \to C^4$ which contracts the zero section $CP^1$ to the origin. Put $V = \psi^{-1}(\varphi(U))$, and $X' = (X \setminus C) \cup_{\psi^{-1} \circ \varphi} V$, we obtain a smooth 3-fold $X'$, which is another (small) resolution of $\tilde{X}$ with exceptional curve $\cong CP^1$.

In the literature, $X'$ is called the flop of $X$ along $C$. One can also flop $X$ along several disjoint rational curves.

## §2. The projectivity of the new manifold

Next, we should consider the projectivity of $X'$. Suppose that $X$ is an algebraic 3-fold. For simplicity, we suppose that $X'$ is the flop of $X$ along one rational curve, although the arguments can be used in more general cases.

Lemma 1: $b_1(X) = 0$.

Proof: If $b_1(X)$ is not zero, then by Bogomolov's theorem [B], there is a finite covering $Y$ of $X$ such that $Y \cong W \times T^k$, $k = \frac{1}{2} b_1(X)$, $W$ is simply-connected. $\tau : Y \to X$ is the projection, then $\tau^{-1}(C)$ is a disjoint union of rational curves. For any $C_i \subset \tau^{-1}(C)$, connected, $N_{C_i|Y} \cong \theta_{CP^1}(-1) \oplus \theta_{CP^1}(-1)$. On the other hand, since $T^k$ cannot contain any rational curve, $C_i \subset W$. So $N_{C_i|Y} = \theta_{C_i}(m) \oplus \theta_{C_i}$. A contradiction. Hence, $b_1(X)$ is zero. //

Corollary: $H^2(X, \theta_X) = 0$.

Proof: By Riemann-Roch theorem,

$$h^0(X, \theta_X) - h^1(X, \theta_X) + h^2(X, \theta_X) - h^3(X, \theta_X) = \frac{C_1(X) \cdot C_2(X)}{24}.$$

Since $C_1(X) = 0$, $h^3(X, \theta_X) = 1$. Hence, $h^2(X, \theta_X) = h^1(X, \theta_X) = 0$, by lemma 1. //

Lemma 2: $H^2(X', \theta_{X'}) \cong H^2(X, \theta_X) = 0$.

Proof: Note first that $X'$ and $X$ are bimeromorphic and there is a biholomorphism $\phi: X \setminus C \to X' \setminus C'$ where $C'$ is the exceptional curve over $\tilde{X}$. By the definition of Čech cohomology [GH], $H^2(X', \theta_{X'}) = H^2(X' \setminus C', \theta_{X'}) = H^2(X \setminus C, \theta_X) = H^2(X, \theta_X) = 0$. //

Proposition 1: If $\tilde{X}$ is projective (i.e. algebraic), then so is $X'$.

Proof: By Poincare duality, there is a differential 2-form $\Omega$ on $X'$ such that

$$\int_{C'} \Omega > 0 .$$

We may also assume that $\Omega \in H^2(X', Z)$.

$$\Omega = \Omega^{(2,0)} + \Omega^{(1,1)} + \Omega^{(0,2)}, \quad \Omega^{(2,0)} = \overline{\Omega^{(0,2)}}, \quad \Omega^{(1,1)} = \overline{\Omega^{(1,1)}} .$$

$$0 = d\Omega = \partial\Omega^{(2,0)} + (\bar{\partial}\Omega^{(2,0)} + d\Omega^{(1,1)} + \partial\Omega^{(0,2)} + \bar{\partial}\Omega^{(0,2)}$$

$$\therefore \quad \bar{\partial}\Omega^{(0,2)} = 0, \quad \text{i.e.} \quad \Omega^{(0,2)} \in H^2(X', \Omega_{X'}) \cong H^{0,2}_{\bar{\partial}}(X', C).$$

By lemma 2, $\Omega^{(0,2)} = \bar{\partial}\alpha^{(0,1)}$.

Put $\Omega' = \Omega - d\alpha$, then $\int_{C'} \Omega' > 0$ and $\Omega'$ is (1, 1)-form. We simply replace $\Omega'$ by $\Omega$. As $\Omega|_{C'} \in H^2(C', Z) \cong H^2(CP^1, Z) = Z$, there exists a function $f \in C^2(C', R)$, s.t. $\Omega|_{C'} + \partial\bar{\partial}f > 0$. Extending $f$ to $X'$, we define $\Omega' = \Omega + \partial\bar{\partial}f$. Then $\Omega'$ is a closed (1, 1)-form with $\Omega'|_{C'} > 0$.

Now since $\tilde{X}$ is projective, i.e. $\tilde{X} \hookrightarrow CP^N$, the pull-back of the Fubini-Study form on $CP^N$ to $X'$ is a semi-positive (1, 1)-closed form $\Omega''$ and $\Omega''$ only degenerates along tangent directions of $C'$. Let $m$ be a large integer, then $\Omega' + m\Omega''$ is a positive (1, 1)-closed form on $X'$, and $\Omega' + m\Omega'' \in H^2(X', Z)$. This implies that $X'$ is projective by Kodaira's embedding theorem.

## §3. The homology of the new manifold

By standard arguments of chasing homological exact sequences, one can prove

<u>Lemma 3</u>: The natural projection $\pi : X \to \tilde{X}$ induces isomorphisms $\pi_* : H_i(X, Z) \to H_i(\tilde{X}, Z)$ for $i \neq 2$. $\pi_* : H_2(X, Z) \to H_2(\tilde{X}, Z)$ is surjective and vanishes the homological class, represented by $C$. Here $Z$ is the ring of the integer.

Lemma 4: (i) $\chi(X) = \chi(X')$.

(ii) $K_X = 0$ is equivalent to $K_{X'} = 0$.

Proof: The first follows from the previous lemma. The second follows from the fact that $X \setminus C$ are $X' \setminus C'$ are biholomorphic and $\operatorname{codim}(C, X) = 2 = \operatorname{codim}(C', X')$.

§4. The diffeomorphic type of the new manifold

Now we come to decide the diffeomorphic type of $X'$, the flop of $X$. By C. T. C. Wall [W], we should look at the trilinear form on $H(X', Z)$ and the linear homomorphism $P_1: H_4(X', Z) \to Z$, induced by the second Chern class $C_2(X')$. Most likely, the trilinear form of $X'$ is equivalent to that of $X$, at least, we can provide an example, in which $X$ and $X'$ have equivalent trilinear forms. Therefore, what remains, which can distinguish the differential type of $X'$ from $X$, is the effect of $C_2(X')$ on $H_4(X', Z)$. The homology group $H_4(X', Z)$ is equal to $H_4(X, Z)$ via the projections from $X'$, $X$ to the singular variety $\tilde{X}$. In the following, we give a criterion to distinguish $X'$ from $X$.

Lemma 4: Let $S \subset X$ be a smooth surface containing the exceptional curve $C$. Let $\pi: X \to \tilde{X}$ and $\pi': X' \to \tilde{X}$ be the natural projections. Then $S$ is the C blowing-up of $\pi(S)$ at the double point of $\tilde{X}$, and there is a smooth surface $S'$ in $X'$ such that $\pi'(S') = \pi(S)$ and $\pi'|_{S'}: S' \to S$ is a biholomorphism. Moreover, $S'$ is transversal to the exceptional curve $C'$.

Proof: Let $N_{C|S}$ be the normal bundle of C in S, then it is a subbundle of $N_{C|X}$. Since $N_{C|X} \cong \theta_C(-1) \oplus \theta_C(-1)$, $N_{C|S} \cong \theta_C(-1)$. Then we can choose the following covering of C in X, $\{U_0, (z_0, u_0, v_0)\}$, $\{U_1, (z_1, u_1, v_1)\}$ such that $z_1 = 1/z_0$, $u_1 = z_0 u_0$, $v_1 = z_0 v_0$ on $U_0 \cap U_1$ and $\{U_0 \cap S, (z_0, u_0)\}$, $\{U_1 \cap X, (z_1, u_1)\}$ is a coordinate covering of C in S. By the definition of $\tilde{X}$, $\tilde{X} = (X \setminus C) \cup_{\varphi} \varphi(U)$ where $\varphi(U)$ is a neighborhood of the origin in $\{(x, y, z, w) \in C^4 | xw = yz\}$,

$$\varphi\big|_{U_0} : U_0 \to C^4, (z_0, u_0, v_0) \to (u_0, v_0, z_0 u_0, z_0 v_0)$$

and

$$\varphi\big|_{U_1} : U_1 \to C^4, (z_1, u_1, v_1) \to (z_1 u_1, z_1 v_1, u_1, v_1) \quad .$$

Hence $\pi(S) = (S \setminus C) \cup_{\varphi} \{(x, 0, z, 0) | x, z \text{ small}\}$. Note that $\pi(S)$ is smooth and S is the blowing up of $\pi(S)$ at $(0, 0, 0, 0) \in \pi(S)$. Clearly,

$$X' = (X \setminus C) \cup_{\psi^{-1} \circ \varphi} V, \text{ where } V = V_0 \cup V_1, \ \psi: V_0 \subset C^3 \to C^4$$

with

$$\psi(z_0, u_0, v_0) = (u_0, z_0 u_0, v_0, z_0 v_0)$$

and

$$\psi : V_1 \subset C^3 \to C^4$$

with

$$\psi(z_1, u_1, v_1) = (z_1 u, u_1, z_1 u_1, v_1) \quad .$$

Here

$$z_1 = 1/z_0 \ , \ u_1 = z_0 u_0 , \quad v_1 = z_0 v_0 \quad \text{on } V_0 \cap V_1 \ .$$

Define $S' = (S \setminus C) \cup_{\psi} \{(x, 0, z, 0) | x, z \text{ small}\}$, then $S'$ is what we need. //

Note that lemma 4 has the following converse. If $S \subset X$ is transversal to $C$, then one can find a smooth $S'$, containing $C'$, such that $\pi'(S') = \pi(S)$, and $S'$ is a blowing up of $\pi(S) \cong S$. The proof is the same as above.

<u>Lemma 5</u>: For any smooth surface $Y \subset X$, $C_2(X)\,[Y] = C_2(Y) - C_1^2(Y)$. In particular, $C_2(X)\,[S] = C_2(X') \cdot S' + 2$ for $S, S'$ in lemma 4.

<u>Proof</u>: One has the exact sequence

$$0 \to TY \to TX|_Y \to N_{Y|X} \to 0 \quad .$$

Hence,

$$C_1(X)|_Y = C_1(N_{Y|X}) + C_1(Y)$$

$$C_2(X) \cdot Y = C_2(X)|_Y = C_2(Y) + C_1(Y) \cdot C_1(N_{Y|X}) \cdot$$

Now, $C_1(X) = 0$, so $C_2(X) \cdot Y = C_2(Y) - C_1^2(Y)$.

By lemma 4, $S$ is the blowing-up of $S'$ at one point. Hence $K_S = K_{S'} + C$, where $C$ is the exceptional curve of $S$ over $S'$. Therefore $C_1^2(S) = C_1^2(S') - 1$.

On the other hand, $C_2(S) = \chi(S) = \chi(S') + 1 = C_2(S') + 1$. Therefore

$$C_2(X) \cdot S = C_2(X') \cdot S' + 2 \; . \qquad //$$

Choose a basis $(Y_1, \ldots, Y_r)$ in $H_4(X, Z)$ where $r = \dim H_4(X, Z)$. Then the r-typle of integers $(n_1, \ldots, n_r)$, $n_i = C_2(S) \cdot Y_i$ determines $C_2(X)$.

Proposition 2: Suppose that $H_4(X, Z)$ has a basis $(Y_1, Y_2, \ldots, Y_r)$ such that each $Y_i$ is a smooth surface in $X$ and $C \subset Y_1$. If the maximal common divisor $\langle n_1, \ldots, n_r \rangle$ of $n_1, \ldots, n_r$ is greater than 2, then $X'$ and $X$ are differentially different.

Proof: Let

$$\begin{array}{ccc} X & & X' \\ & {\scriptstyle \pi}\searrow \quad \swarrow{\scriptstyle \pi'} & \\ & \tilde{X} & \end{array}$$

By lemma 4, there are $Y_1', \ldots, Y_r' \subset X'$ such that $\pi'(Y_i') = \pi(Y_i)$ and $Y_1$ is the blowing-up of $Y_1'$ at one point.

We have proved that $\pi_* : H_4(X, Z) \to H_4(\tilde{X}, Z)$ and $\pi_*' : H_4(X', Z) \to H_4(\tilde{X}, Z)$ are isomorphisms. It follows that $Y_1', \ldots, Y_r'$ is a basis of $H_4(X', Z)$. Let $n_i' = C_2(X') \cdot Y_i'$, then by the assumption and lemma 4, $n_1' = n_1 - 2$; $n_i' = n_i + 2k_i$ for $i \geq 2$, where $k_i$ are integers, $k_i = 0$ if $Y_i \cap C = \phi$.

Since $C_1(X) = 0$ and $C_2(X') = 0$, $C_2(X)$ and $C_2(X')$ are Pontragin classes of $X$, $X'$, respectively. $f: X \to X'$ is a diffeomorphism, then $C_2(X) = f^* C_2(X')$, and

$$(n_1', n_2', \ldots, n_r') = (C_2(X) \cdot f_*^{-1}(Y_1'), \ldots, C_2(X) \cdot f_*^{-1}(Y_r'))$$

$$= (n_1, n_2, \ldots, n_r) A ,$$

$A$ is a $r \times r$ integer matrix, i.e.

$$(n_1, n_2, \ldots, n_r)(A - id) = -(2, k_2, \ldots, k_r) .$$

This contradicts the assumption that $\langle n_1,\ldots,n_r\rangle > 2$. Therefore, $X$, $X'$ are not diffeomorphic to each other.

Remark: The proof here holds for more general situations. For example, one can drop the assumption of the smoothness of $Y_i$ for $2 \le i \le r$ in Proposition 2.

§5. The example

Example: Consider the algebraic 3-fold $X$ constructed in [Y]. Recall that $X = M/\sigma$, where $M = \Sigma_6^1 \times \Sigma_6^2 \cap H$, $\Sigma_6^1 = \{x_0^3+x_1^3+x_2^3+x_3^3 = 0\}$ in $CP^3$, $\Sigma_6^2 = \{y_0^3+y_1^3+y_2^3 = 0\}$ in $CP^3$, $H = \{x_0y_0+x_1y_1+x_2y_2+x_3y_3 = 0\}$ is the hyperplane in $CP^3 \times CP^3$, and $\sigma: [x_0,x_1,x_2,x_3]\times[y_0,y_1,y_2,y_3] \to [x_1,x_2,x_0,\omega x_3]\times[y_1,y_2,y_0,\omega^2 y_3]$, $\omega = \exp\left(\frac{2\pi\sqrt{-1}}{3}\right)$.

The cubic surface $\Sigma_6^1$ has 27 lines:

$\ell_1^1$: $x_0 = -x_1$, $x_2 = -x_3$; $\ell_4^1$: $x_0 = -\omega x_1$, $x_2 = -x_3$;

$\ell_2^1$: $x_0 = -x_1$, $x_2 = -\omega x_3$; $\ell_5^1$: $x_0 = -\omega x_1$, $x_2 = -\omega x_3$;

$\ell_3^1$: $x_0 = -x_1$, $x_2 = -\omega^2 x_3$; $\ell_6^1$: $x_0 = -\omega x_1$, $x_2 = -\omega^2 x_3$;

. . . . . .

Similarly, $\Sigma_6^2$ has 27 lines $\ell_1^2,\ldots,\ell_{27}^2$.

Define $\sigma_1$: $CP^3 \to CP^3$, $[x_0,x_1,x_2,x_3] \to [x_1,x_2,x_0,\omega x_3]$,

$\sigma_2$: $CP^3 \to CP^3$, $[y_0,y_1,y_2,y_3] \to [y_1,y_2,y_0,\omega^2 y_3]$,

then
$$\sigma = \sigma_1 \times \sigma_2 .$$

Since $H_4(X,Z) = H_4(M,Z)^\sigma$, the $\sigma$-invariant homology of $M$, it has a basis consisting of

$$E_{11} = (\ell_1^1 + \sigma_1(\ell_1^1) + \sigma_1^2(\ell_1^1)) \times \Sigma_6^2 \cap H ,$$

$$E_{12} = (\ell_5^1 + \sigma_1(\ell_5^1) + \sigma_1^2(\ell_5^1)) \times \Sigma_6^2 \cap H ,$$

$$A_1 = \frac{1}{3}(H_1 - \ell_1^1 - \sigma_1(\ell_1^1) - \sigma_1^2(\ell_1^1) - \ell_5^1 - \sigma_1(\ell_5^1) - \sigma_1^2(\ell_5^1)) \times \Sigma_6^2 \cap H ,$$

$$E_{21} = \Sigma_6^1 \times (\ell_1^2 + \sigma_2(\ell_1^2) + \sigma_2^2(\ell_1^2)) \cap H ,$$

$$E_{22} = \Sigma_6^1 \times (\ell_5^2 + \sigma_2(\ell_5^2) + \sigma_2^2(\ell_5^2)) \cap H ,$$

$$A_2 = \frac{1}{3} \Sigma_6^1 \times (H_2 - \ell_1^2 - \sigma_2(\ell_1^2) - \sigma_2^2(\ell_1^2) - \ell_5^2 - \sigma_2(\ell_5^2) - \sigma_2^2(\ell_5^2)) \cap H ,$$

where $H_1, H_2$ are generic hyperplane sections of $\Sigma_6^1$, $\Sigma_6^2$, respectively. One can easily check that each $E_{ij}$ consists of three rational elliptic surfaces, i.e. $C_2(E_{ij}) = 12 \times 3$.

$$H_1 \times \Sigma_6^2 \cap H \cong \{[x_0, x_1, x_2, 0] \times [y_0, y_1, y_2, y_3] \,|\, x_0^3 + x_1^3 + x_2^3 = 0 , \; y_0^3 + y_1^3 + y_2^3 + y_3^3 = 0, \; x_0y_0 + x_1y_1 + x_2y_2 = 0 \} .$$

This is an elliptic surface with Euler number 36. Hence, by lemma 5,

$$\begin{aligned} C_2(M) \cdot A_1 &= \frac{1}{3} C_2(M) \cdot (H_1 \times \Sigma_6^2 \cap H) - \frac{1}{3} C_2(M) \cdot (E_{11} + E_{12}) \\ &= \frac{1}{3} C_2(H_1 \times \Sigma_6^2 \cap H) - \frac{1}{3}(C_2(E_{11}) + C_2(E_{12})) \\ &= -12 \end{aligned}$$

$$C_2(M) \cdot E_{ij} = C_2(E_{ij}) = 36$$

$C_2(M) \cdot A_2 = -12$ by symmetry.

Clearly, $E_{11}, E_{12}, E_{21}, E_{22}, A_1, A_2$ induce a basis of $H_4(X, Z)$. Call it $\{Y_1, \ldots, Y_6\}$. Then

$$C_2(X) \cdot Y_i = 12 \quad \text{for } i = 1, 2, 3, 4,$$

$$C_2(X) \cdot Y_i = -4 \quad \text{for } i = 5, 6.$$

Hence, $\langle C_2(X) \cdot Y_1, \ldots, C_2(X) \cdot Y_6 \rangle = 4$. It is easy that each $Y_i$ can be represented by a smooth surface.

The curve $C_1 = \ell_1^1 \times [1, 1, -1, -1]$ is a line in $M$ whose normal bundle is isomorphic to $\theta_{CP^1}(-1) \oplus \theta_{CP^1}(-1)$. The curves

$$C_2 = \sigma(C_1) = \sigma_1(\ell_1^1) \times [1, -1, 1, -\omega^2]$$

$$C_3 = \sigma^2(C_1) = \sigma_1^2(\ell_1^1) \times [-1, 1, 1, -\omega]$$

are disjoint from $C_1$. They contract down to a rational curve $C$ on $X$ with the normal bundle $\theta_C(-1) \oplus \theta_C(-1)$. Hence by section 1, we obtain a flop $X'$ of $X$ along $C$. Hence by the above computation and proposition 2, $X'$ is differentially different from $X$. We also know that the Euler number $\chi(X') = -6$, $K_{X'} = 0$. Actually, $X'$ has the same homology groups as $X$.

It remains to prove that $X'$ is algebraic. By proposition 1 and the discussion in section 1, it suffices to find a line bundle $L$ on $X$ such that $L$ has no base point on $X$, very ample outside $C$ and each global section of $L$ is constant along $C$. The lines $\ell_1^1$, $\sigma_1(\ell_1^1)$, $\sigma_1^2(\ell_1^1)$ are three rational curves on $\Sigma_6^1$ with self-intersection number $-1$. Blowing down these rational curves, one gets an algebraic surface $\Sigma_3$, i.e. $CP^2 \# 3CP^2$.

Let $L^1 = \pi^*(-K_{\Sigma_3}) \otimes (-K_{\Sigma_6^2})\big|_M$ , and the required line bundle L is induced from $L^1$, i.e. $L = L^1 \otimes \sigma^* L^1 \otimes \sigma^{*2} L^1$. It is easy to check that such a L has all properties stated above. Therefore, we get a different algebraic 3-fold $X'$ with $\chi(X') = -6$, and $C_1(X') = 0$.

Remarks: (i) There are 54 rational curves on X with the normal bundle isomorphic to $\theta_{CP^1}(-1) \oplus \theta_{CP^1}(-1)$.

(ii) By using the basis given above, one can prove that the trilinear form on $H^2(X, Z)$, induced from intersection form of X, is the same as that on $H^2(X', Z)$.

References

[CHSW] P. Candelas, G. T. Horowitz, A. Strominger, E. Witten, Superstring Phenomenology. Symp. on Anomalies, Geometry, Topology (1985). Edited by W. A. Bardeen et cl. World Scientific.

[B] F. A. Bogomolov, Kähler manifolds with trivial canonical class. Izv. Akad. Nauk SSSR Ser. Mat. 38 (1974), 11-21.

[GH] P. Griffiths, J. Harris. Principles of Algebraic Geometry. A Wiley-Interscience Publication. John Wiley & Sons, New York.

[W] C. T. C. Wall. Classification Problems in Differential Topology. V. On certain 6-manifolds. Inv. Math. (1986), 355-374.

[Y] S. T. Yau. Compact three dimensional Kähler manifolds with zero Ricci curvature. Symp. on Anomalies, Geometry, Topology (1985). Edited by W. A. Bardeen et al. World Scientific.

# Hermitian-Yang-Mills Connection on Non-Kähler Manifolds

Jun Li and Shing Tung Yau

Department of Mathematics

University of California, San Diego, La Jolla, CA 92093

In [3], K. Uhlenbeck and the second author demonstrated the existence of Hermitian-Yang-Mills connection on a stable holomorphic vector bundle over a compact Kähler manifold. Some physicists have expressed an interest in establishing a similar theorem for Hermitian non-Kähler manifolds. There is a possibility that such an existence theorem would be useful for the study of non-Kähler complex manifolds. Notice that in the last ten years, a large class of non-Kähler complex manifolds has been constructed. They may even play a role in studying algebraic manifolds. The major difficulty in studying these manifolds is the lack of analytic tools. Hopefully, Hermitian-Yang-Mills connection would provide such a tool for the vector bundle.

First of all, we have to define the concept of stability which was the necessary condition for the existence of a Hermitian-Yang-Mills connection.

Given any compact complex manifold $M^n$, we can attach groups in the following way. For each (p, q), we associate the group $A^{p,q}(M)$ to be the quotient of the group $\{\omega \mid \partial\bar{\partial}\omega = 0 \text{ and } \omega \text{ is a } (p,q) \text{ form}\}$ by the

*Research supported in part by NSF grant #DMS84-09447 and ONR contract #N00014-85-K-0367.

subgroup $\{\partial\alpha + \bar{\partial}\beta \mid \alpha$ is a (p-1, q) form and $\beta$ is a (p, q-1) form$\}$. It is not difficult to prove that the dimension of $A^{p,q}$ is finite. There are two multiplications in the ring $\oplus A^{p,q}$, one is given by $\alpha \cdot \beta = \alpha \wedge \partial\beta$ and the other is given by $\alpha \cdot \beta = \alpha \wedge \bar{\partial}\beta$.

There is another smaller group $\tilde{A}^{p,q}$ defined by the quotient of $\{\omega \mid \omega$ is a (p, q) form with $\partial\omega = \bar{\partial}\alpha$ and $\bar{\partial}\omega = \partial\beta$ for some (p+1, q-1) form $\alpha$ and (p-1, q+1) form $\beta\}$ by the subgroup $\{\partial\alpha + \bar{\partial}\beta \mid \alpha$ is a (p-1, q) form and $\beta$ is a (p, q-1) form$\}$. Clearly, $\tilde{A}^{p,q}$ is a subgroup of $A^{p,q}$.

Using an argument of Sullivan, one can show that for each p, there is positive (p, p) form $\omega$ in $\tilde{A}^{p,p}$.

The following proposition tells us that there are a lot of $\partial\bar{\partial}$ closed (n-1, n-1) forms.

Proposition. Let M be a compact complex manifold of dimension n, then for any positive (n-1, n-1) form $\Omega$, there is a positive function f so that $f\Omega \in A^{n-1,n-1}$.

Proof. The proof is essentially due to P. Ganduchon [1], if one notices that any positive (n-1, n-1) form $\Omega$ can be written as $\Omega = \omega^{n-1}$ with $\omega$ positive (1, 1) form (simply because $\Omega$ gives a conformal structure on $T^*M$).

Let us now define the concept of stability for a holomorphic vector bundle V over M. It will depend on an element $\Omega$ in the positive cone of the group $A^{n-1,n-1}$. We say that the bundle V is stable iff for all coherent subsheaf $\mathfrak{F}$ of V,

$$\frac{1}{\operatorname{rank} \mathfrak{F}} \int_M c_1(\mathfrak{F}) \wedge \Omega < \frac{1}{\operatorname{rank} V} \int_M c_1(V) \wedge \Omega$$

Here $c_1(\mathfrak{F})$ and $c_1(V)$ are represented by the curvature of some Hermitian metric on $\mathfrak{F}$ and $V$, respectively. Note that they are defined up to $\partial\bar{\partial}f$ for some function $f$.

In the remaining of this paper, we will mainly follow the notation of the paper [3].

For $(M, g)$, we always mean a compact complex manifold with Hermitian metric $g$ and $(V, h)$, holomorphic Hermitian vector bundle over $M$ of rank $r$, $\omega$ the corresponding Hermitian form. We denote

$$D = \partial + \bar{\partial}$$

its holomorphic and anti-holomotphic parts.

In a local holomorphic frame $\{e_\alpha\}$ of $V$, the connection $D$ is given by

$$D = d + h^{-1}(\partial h)$$

and the curvature

$$F = \bar{\partial}(h^{-1} \partial h).$$

Fix $g$, we can define an operation $tr_g$: $\Gamma(\operatorname{End} V \otimes T^{(1,1)}M) \to$ $\to \Gamma(\operatorname{End} V)$. We also adopt the following convention if no confusion is aroused.

$$\begin{cases} F_{j\bar{k}} = (F^{\beta}_{\alpha j\bar{k}})_{1 \leq \alpha,\, \beta \leq r} \\ H = tr_g F - \mu_\omega \cdot id = \sum_{j,k} g^{j\bar{k}} F_{j\bar{k}} - \mu_\omega \cdot id \,. \end{cases} \tag{1}$$

here

$$\mu_\omega = \frac{1}{\mathrm{Vol}(M)} \frac{\deg_\omega V}{\mathrm{rank}\ V} = \frac{1}{\mathrm{Vol}(M)} \frac{\int_M c_1(V) \wedge * \omega}{\mathrm{rank}\ V} \tag{2}$$

Definition 1. A holomorphic vector bundle $V$ over $(M, g)$ is Hermitian-Yang-Mills if there exists a Hermitian metric $h$ for which the Hermitian curvature $F$ satisfies

$$H = \mathrm{tr}_g F - \mu_\omega \cdot \mathrm{id} = 0 \ .$$

Here we state our main theorem.

Theorem 1. A stable holomorphic vector bundle over a compact manifold $(M, g)$ admits a unique Hermitian-Yang-Mills connection.

The proof of the theorem is parallel to the proof in [3].

Fix a Hermitian metric $h_0$ on $E$, then any other metric $h_1$ can be written as

$$\langle s, t \rangle_{h_1} = \langle hs, t \rangle_{h_0} \tag{3}$$

or in local coordinate

$$h_1 = h_0 h \ . \tag{3'}$$

Here $h \in \Gamma(\mathrm{End}\ E)$ is positive definite, symmetric w.r.t. $h_0$. The corresponding Hermitian connection $D^0$ and $D^1$ are related by

$$D^1 = D^0 + h^{-1} \partial^0 h$$

hence

$$F^1 = D^1 \circ D^1 = F^0 + D^0[h^{-1}\partial h] + (h^{-1}\partial^0 h) \wedge (h^{-1}\partial^0 h)$$

The tensor H,

$$H - H_0 = \mathrm{tr}_g\ \bar\partial(h^{-1}\partial^0 h) = -g^{ij}\,\bar\partial_i(h^{-1}\partial^0_j h) \tag{4}$$

(In the following, we simply write $\bar\partial\partial$ instead of $g^{ji}\bar\partial_j\partial_i$)

We try to find h by considering the perturbed equation

$$H + \varepsilon \ln h = H_0 - \bar\partial(h^{-1}\partial^0 h) + \varepsilon \ln h = 0 \tag{5}$$

Since $\partial\bar\partial(\omega^{n-1}) = 0$, an easy argument shows that we can normalize $h_0$ by a conformal factor $e^{\varphi}$ so that $\mathrm{tr}_E H_0 = 0$.

To solve (5) for large $\varepsilon$, we use continuity argument to solve the following one-parameter family of equations.

$$L_{\varepsilon,\sigma}(h) = H_0 - \bar\partial(h^{-1}\partial^0 h) + \varepsilon \ln h - \sigma h^{-\frac{1}{2}} H_0 h^{\frac{1}{2}} = 0 \tag{6}$$

with $0 \le \sigma \le 1$.

To make the operator $L_{\varepsilon,\sigma}$ behave well, we had better consider the operator $\tilde L_{\varepsilon,\sigma}$.

$$\tilde L_{\varepsilon,\sigma}(h) = h L_{\varepsilon,\sigma}(h) \tag{6'}$$

Proposition 2. The operation $\tilde L_{\varepsilon,\sigma}$ is an operator from sections of self-adjoint endomorphism of V into itself.

Proof. It is easy to show that both

$$H = H_0 - \bar{\partial}(h^{-1}\partial^0 h) ,$$

$$\ln h ,$$

and $$h^{-\frac{1}{2}} H_0 h^{\frac{1}{2}}$$

are symmetric w.r.t. $\langle \; , \; \rangle_{h_1}$ .

So $\tilde{L}_{\epsilon,\sigma} = h L_{\epsilon,\sigma}$ is symmetric w.r.t. $\langle \; , \; \rangle_{h_0}$ . #

Let $T_\epsilon = \{\sigma \in [0,1] \mid L_{\epsilon,\sigma}(h) = 0$ has a solution$\}$. Then $\sigma = 1 \in T_\epsilon$ , and all we have to do is to show $T_\epsilon$ is closed and open.

First implication of equation (6) is

Proposition 3. Suppose h solves (6), under the normalization $\operatorname{tr} H_0 = 0$, $\det h = 1$.

Proof. Take a trace of (6), one can see

$$\bar{\partial}\partial(\operatorname{tr} \ln h) - \epsilon(\operatorname{tr} \ln h) = 0 .$$

Maximal principle tells us $\operatorname{tr}(\ln h) = 0$, i.e. $\det h = 1$. #

To show the openness of $T_\epsilon$, it is sufficient to show the linearized operator $\delta\tilde{L}_{\epsilon,\sigma}$ is surjective.

Proposition 4. If $h$ solves (6) and $m = \max|\ln h|$ then

$$\int_X \langle h^{-1}\delta\tilde{L}_{\epsilon,\sigma}(h)\phi,\ \phi h^{-1}\rangle * 1$$

$$\geq \left[-e^{2m}\max\left|\bar{\partial}\left(\frac{\omega^{n-1}}{(n-1)!}\right)\right| + \frac{\epsilon m}{e^{m}-1} - 2\sigma e^{2m}\max|H_0|\right]\int_X |\phi\cdot h^{-1}| * 1 . \qquad (7)$$

Proof. Since

$$\delta\tilde{L}_{\epsilon,\sigma}(h)\phi = -h\bar{\partial}(h^{-1}\partial^0(\phi\cdot h^{-1})h)$$

$$+ \ \epsilon h\,\delta(\ln h) - \sigma h\,\delta(h^{-\frac{1}{2}}H_0\,h^{\frac{1}{2}}) , \qquad (8)$$

so

$$-\int_X \langle\partial(h^{-1}\partial^0(\phi\cdot h^{-1})h),\ \phi\cdot h^{-1}\rangle \frac{\omega^n}{n!}$$

$$= \int_X \mathrm{tr}_g\langle h^{-1}\partial^0(\phi\cdot h^{-1})h,\ \partial^0(\phi\cdot h^{-1})\rangle \frac{\omega^{n-1}}{(n-1)!}$$

$$- \int_X \mathrm{tr}_g\langle h^{-1}\partial^0(\phi\cdot h^{-1})h,\ \phi\cdot h^{-1}\rangle\, \bar{\partial}\left(\frac{\omega^{n-1}}{(n-1)!}\right)$$

$$\geq e^{-2m}\int_X |\partial^0(\phi\cdot h^{-1})|^2 \frac{\omega^n}{n!} - \left(e^{-2m}\int_X |\partial^0(\phi\cdot h^{-1})|^2 \frac{\omega^n}{n!}\right.$$

$$\left. - e^{2m}\max\left|\bar{\partial}\frac{\omega^{n-1}}{(n-1)!}\right| \int |h^{-1}\cdot\phi|^2 \frac{\omega^n}{n!}\right)$$

$$= -e^{2m}\max\left|\bar{\partial}\frac{\omega^{n-1}}{(n-1)!}\right| \int |\phi\cdot h^{-1}|^2 * 1 .$$

The rest are quite a standard calculation (see [3]). #

Lemma 5. Suppose

$$H_0 - \bar{\partial}(h^{-1}\partial^0 h) + \varepsilon \ln h - h^{-\frac{1}{2}} \tilde{H}_0 h^{\frac{1}{2}} = 0 ,$$

and $\ln h = u$, then

$$|u|^{\frac{1}{3}} |H_0 - \tilde{H}_0| \geq -\bar{\partial}\partial |u|^{\frac{4}{3}} + \varepsilon |u|^{\frac{4}{3}}$$

and

$$m = \max_x |u| \leq \varepsilon^{-1} \max |H_0 - \tilde{H}_0| . \tag{9}$$

Proof. Without loss of generality, we can assume on an open set with full measure,

$$u = \sum \lambda_\alpha e_\alpha \otimes e_\alpha^* .$$

Let $\partial_i e_\alpha = A_{i\alpha}^\beta e_\beta$, then

$$\partial^0 h = \sum e^{\lambda\alpha} (\partial_i \lambda_\alpha) e_\alpha \otimes e_\alpha^* + \sum (e^{\lambda_\alpha} - e^{\lambda_\beta}) A_{i\alpha}^\beta e_\beta \otimes e_\alpha^* ,$$

$$\partial^0 u = \sum \frac{1}{3} \lambda_\alpha^{-\frac{2}{3}} (\partial_i \lambda_\alpha) e_\alpha \otimes e_\alpha^* + \sum (\lambda_\alpha^{\frac{1}{3}} - \lambda_\beta^{\frac{1}{3}}) A_{i\alpha}^\beta e_\beta \otimes e_\alpha^* \tag{10}$$

So

$$\langle \partial^0 u^{\frac{1}{3}}, h^{-1}\partial^0 h \rangle = \sum \frac{1}{3} \lambda_\alpha^{-\frac{2}{3}} |\partial_i \lambda_\alpha|^2$$

$$+ \sum \frac{1}{3} \lambda_\alpha^{-\frac{2}{3}} |A_{i\alpha}^\beta|^2 (e^{\lambda_\alpha - \lambda_\beta} - 1)(\lambda_\alpha^{\frac{1}{3}} - \lambda_\beta^{\frac{1}{3}}) \geq 0 .$$

On the other hand,

$$\langle u^{\frac{1}{3}}, -\bar{\partial}(h^{-1}\partial^0 h) + \epsilon\, \ln h\rangle$$

$$= -\bar{\partial}\langle u^{\frac{1}{3}}, h^{-1}\partial^0 h\rangle + \langle \partial^0 u^{\frac{1}{3}}, h^{-1}\partial^0 h\rangle + \epsilon\langle u^{\frac{1}{3}}, \ln h\rangle$$

$$\geq -\frac{3}{4}\operatorname{tr}_g \partial\bar{\partial}|u|^{\frac{4}{3}} + \epsilon|u|^{\frac{4}{3}} ,$$

then follows (8) and applying maximum principle, (9).

By using the Harnack inequality of non-divergence equation to (8) [2],

Corollary. If $\int |u|^2 \leq m^2$, then

$$\max|u| \leq c(m + \max|H_0 - \tilde{H}_0|) \tag{11}$$

Based on Lemma 5, we can control $m$ by making $\epsilon$ large, in other words, when $\epsilon$ is large ,

$$-e^{2m}\max\left|\bar{\partial}\frac{\omega^{n-1}}{(n-1)!}\right| + \frac{em}{e^m - 1} - 2\sigma e^{2m}\max|H_0| > 0 .$$

This shows that $\delta\tilde{L}_{\epsilon,\sigma}(h)$ is invertible when $\epsilon$ is large. The standard argument tells us $T_\epsilon$ is open when $\epsilon$ is large.

Now we are going to show $T_\epsilon$ is closed when $\epsilon$ is large. Since we have the estimate that $m = \max|\ln h|$ is small when $\epsilon$ is large, the same argument as in [3] shows:

Proposition 6. If

$$H_0 - \bar{\partial}(h^{-1}\bar{\partial}^0 h) + \epsilon \, \ell n \, h - h^{-\frac{1}{2}} \tilde{H}_0 h^{\frac{1}{2}} = 0$$

and $m = \max |\ell n \, h|$ is small, then

$$\|h\|_{c^k(M)} \le C(m, \|H_0\|_{c^k(m)}, \|\tilde{H}_0\|_{c^k(m)}) \; . \qquad \#$$

This estimate guarantees the closedness of $T_\epsilon$. In particular, we know equation (6), $L_{\epsilon,\sigma}(h) = 0$, always has a solution when $\epsilon$ is large, especially, $L_\epsilon(h) = 0$ has a solution for $\epsilon$ large.

Now, we turn our attention to the solution of

$$L_\epsilon(h) = L_{\epsilon,0}(h) = 0 \qquad (12)$$

or

$$\tilde{L}_\epsilon(h) = \tilde{L}_{\epsilon,0}(h) = 0 \; .$$

By previous argument, the set $T = \{\epsilon \ge 0 \,|\, L_\epsilon(h) = 0$ has one solution $\}$ is non-empty. A simple calculation shows that the symbols of the linearized operator

$$\delta \tilde{L}_\epsilon(h)$$

and its adjoint

$$(\delta \tilde{L}_\epsilon(h))^*$$

are the same.

By Atiyah-Singer index theorem, Index$(\delta L_\epsilon(h)) = 0$. So $\delta \tilde{L}_\epsilon(h)$ is surjective iff it is injective.

Proposition 7. Suppose $\tilde{L}_\varepsilon(h) = 0$ and $\delta\tilde{L}_\varepsilon(h)(\phi) + \alpha h \,\ell n\, h = 0$, then for $\Psi = h^{-\frac{1}{2}}(\phi)\, h^{-\frac{1}{2}}$,

$$-\mathrm{tr}_g\, \partial\bar{\partial}\,|\Psi|^2 + \varepsilon\,|\Psi|^2 \le -2\alpha\,\langle \ell n\, h, \Psi\rangle \quad . \tag{13}$$

Proof. See [3].

In particular, when $\alpha = 0$ and $\Psi$ satisfies (13), $\Psi = 0$, or $\Phi = 0$, which says $\delta\tilde{L}_\varepsilon(h)$ is injective, so is surjective.

Now since $T$ is open, we differentiate the equation $L_\varepsilon(h) = 0$ with respect to $\varepsilon$ and obtain

$$\delta L_\varepsilon(h)(\phi) + \ell n\, h = 0,$$

here $\phi = \delta h$.

Applying the Harnack inequality [2] to (13),

Lemma 8. If $\delta L_\varepsilon(h)(\delta h) + \ell n\, h = 0$ and $m = \max|\ell n\, h|$, then

$$\max|\delta h| \le c(m) \; .$$

Following the same argument in [3], we can deduce

Theorem 2. There exists a solution to the equation $L_\varepsilon(h) = 0$ for all $\varepsilon > 0$. Moreover, if $\int_M |\ell n\, h_\varepsilon|^2$ stays bounded for $\varepsilon > 0$, then there exists a solution in an interval in $(-\varepsilon_0, \infty)$ for some $\varepsilon_0 > 0$.

Remark. If one goes over the proof carefully, one should notice that we have never used the condition $\partial\bar{\partial}(\omega^{n-1}) = 0$. So Theorem 2 holds for arbitrary Hermitian metric on $M$.

As done by the second author, one can construct a coherent subsheaf $F$ in the following way.

Proposition 9. Suppose that $\overline{\lim_{\epsilon}} \int_M |\ln h_\epsilon|^2 = \infty$, then there exists a subsequence $\epsilon_l \to 0$ and $\rho_l \to 0$ such that

$$h_l = \rho_\ell h_{\epsilon_\ell} \leq I ,$$

$$h_\ell \to h_\infty \neq 0 \quad \text{in } L^\infty$$

$$h_\ell \to h_\infty \quad \text{weakly in } L_1^2$$

Furthermore, $\pi = I - \lim_{\sigma \to 0} h_\infty^\sigma$ represents a weakly holomorphic subbundle, or $F = \pi(E)$ is a holomorphic subsheaf.

Proof. See [3].

To see if the resulting sheaf violates the stability condition of $E$, we calculate

$$c_1(F, h_0) \cup \frac{\omega^{n-1}}{(n-1)!}$$

$$= \frac{i}{2\pi} \int \mathrm{tr}_v(F_0 \pi) \frac{\omega^{n-1}}{(n-1)!} + \frac{i}{2\pi} \int \mathrm{tr}_v((\pi^\perp \partial^0 \pi)^* \wedge (\pi^\perp \partial^0 \pi)) \frac{\omega^{n-1}}{(n-1)!}$$

$$= \frac{1}{2\pi} \int \mathrm{tr}_v((H_0 + \mu I)\pi) - \frac{1}{2\pi} \int |\partial^0 \pi|^2 \qquad (14)$$

but

$$\int \mathrm{tr}(H_0 \pi) = \lim \int \mathrm{tr}(H_0 h_\ell^\sigma)$$

$$= \lim \int \mathrm{tr}(H_\ell h_\ell^\sigma + \bar\partial(h_t^{-1} \partial h_t) h_t^\sigma)$$

$$= \lim \left[ \int \mathrm{tr}(\epsilon(\ln h_t) h_t^\sigma) + \int \mathrm{tr}(h_t^{-1} \partial h_t \wedge \bar\partial h_\ell^\sigma) \frac{\omega^{n-1}}{(n-1)!} \right]$$

$$+ \int \mathrm{tr}(h^{-1} \partial h \cdot h_t^\sigma) \wedge \frac{\partial \omega^{n-1}}{(n-1)!}$$

A simple calculation shows

$$\lim \int \mathrm{tr}(h_t^{-1} \partial h_t \wedge \bar\partial h_\ell^\sigma) \frac{\omega^{n-1}}{(n-1)!} \geq \int |\partial^0 \pi|^2$$

and

$$\mathrm{tr}(\ln h_t \cdot h_t^\sigma) \geq 0 \quad \text{since} \quad \mathrm{tr} \ln h_t = 0$$

$$\int \mathrm{tr}(h_t^{-1} \partial h_t \cdot h_t^\sigma) \frac{\bar\partial \omega^{n-1}}{(n-1)!} = \int \mathrm{tr}(\partial(h_t^\sigma)) \frac{\partial \omega^{n-1}}{(n-1)!} = 0 \ .$$

So

$$\int \mathrm{tr}(H_0 \pi) \geq \int |\partial^0 \pi|^2 ,$$

that is

$$\mu_\omega(F) \geq \mu_\omega(E) \ .$$

This inequality violates the stability condition of V. In other words, we can find the desired metric $h$ such that

$$\mathrm{tr}_g F = \mu(V)\,\mathrm{id}$$

that completes the proof of the main Theorem 1.

References

1. Trudinger, N. S.: Local estimates for subsolution and supersolutions of general second order elliptic quasilinear equations. Invent. Math. 61, 67-79 (1980).

2. Gauduchon, P. Fibrés hermitiens a endomorphisme de Ricci non négatif. Bull. Soc. Math. France, 105, 113-140 (1977). Comm. Pure Appl. Math. (1986).

3. Uhlenbeck, K. and S. T. Yau. On the existence of Hermitian-Yang-Mills Connections in Stable Vector Bundles.

# Existence of Kähler-Einstein metrics on complete Kähler manifolds and their applications to algebraic geometry

by

G. Tian and S. T. Yau*

Department of Mathematics

University of California, San Diego, La Jolla, CA 92093

*Research supported in part by NSF grant #DMS84-09557 and ONR contract #N00014-85-K-0367.

It is more than ten years ago [ Ya2 ] that the problem of the existence of Kähler-Einstein metrics on compact Kähler manifolds is solved when the first Chern class is either negative or zero. Almost immediately, Cheng and the second author [ CY1 ] [ CY2 ] understood how to generalize these theorems to non-compact Kähler manifolds. We shall focus our discussion on those canonical Kähler-Einstein manifolds with negative Ricci curvature.

In [CY1], Cheng and the second author gave a geometric criterion for the existence of complete Kähler-Einstein metrics. Besides some technical assumptions (which will be removed in this paper), the criterion is rather satisfactory. In order to enlarge the class of Kähler manifolds which admit canonical Kähler-Einstein metrics, it was introduced in [ CY1 ] the concept of almost complete Kähler-Einstein metrics with negative Ricci curvature. They are Einstein metrics which can be approximated by complete Kähler metrics whose Ricci tensor is bounded from below approximately by the negative of the metric. It can be proved that such metrics are unique and hence reflects a lot about the complex structure of the manifold. They can be constructed for a large category of Kähler manifolds. As far as we know, these metrics can automatically be complete. In fact, Mok and the second author [MY] did demonstrate this for bounded domains in Stein manifolds.

In constructing Kähler-Einstein metrics with negative Ricci curvature, one can reduce a lot of estimates to local consideration. In other words, with enough initial estimates, the regularity at a point can be studied by the concept of quasi-local coordinates introduced in [CY2]. It was used to study quasi-projective manifolds and manifolds with singularities. A very preliminary case was announced in [Ya4] and a much more detailed work was described by the second author in a series of seminars in the Institute for Advanced Study in 1981. Cheng and the second author wrote up some of the work which was published in [CY2]. It was mysterious that while the full manuscript was submitted in [CY2], only half of it appeared.

In this paper, we carried out much more detailed analysis than the previous work. We believe that we are close to finding a necessary and sufficient condition for a quasi-projective manifold to admit a complete Kähler-Einstein metric with finite volume. Note that we expect complete Kähler-Einstein manifolds with finite volume to be quasi-projective.

Given a compact manifold $\overline{M}$ with a divisor D, we study the question of putting a canonical Kähler-Einstein metric on $M = \overline{M} \setminus D$. A simple condition we find is that $K_{\overline{M}} + [D]$ is numerically effective, $(K_{\overline{M}} + D)^n \neq 0$ and the sections of multiples of $K_{\overline{M}} + [D]$ gives an immersion of M. Then we study the case where we allow the metric to have singularities arised locally as the orbit space of some finite group. On the process we also find generalization of Chern number inequalities for these manifolds and in particular, characterization of orbit spaces of a discrete group acting on the complex ball. Theorem 2.1, Theorem 3.1 and Theorem 3.2 will be considered as the main theorems of this paper.

We should mention that some of the rather special cases were studied recently by R. Kobayashi [Ko1] and Tsuji [Ts] who also used the methods developed in [CY1] [CY2].

§1. Kähler-Einstein metrics on complex manifolds

Suppose that M is an n-dimensional complex manifold and S is a divisor in M. By a S-structure on M we mean a finite open covering $\{U_\ell\}_{1\le \ell \le L}$ of S in M such that for each $\ell$, there exists a branched covering $\pi_\ell : \tilde{U}_\ell \to U_\ell$ where $\tilde{U}_\ell$ is a smooth manifold and the branched locus in $S \cap U_\ell$. If $U_\ell \cap U_{\ell'} \neq \phi$, we assume that there is a holomorphic covering map $\psi_{\ell'\ell}$ from $\pi_\ell^{-1}(U_\ell \cap U_{\ell'})$ into $\pi_{\ell'}^{-1}(U_\ell \cap U_{\ell'})$ which commutes with the branched coverings $\pi_\ell$, $\pi_{\ell'}$, i.e. $\pi_{\ell'} \circ \psi_{\ell'\ell} = \pi_\ell$. We shall use the abbreviation $\{\pi_\ell, \tilde{U}_\ell, \psi_{\ell'\ell}\}_{1\le \ell \le L}$ to denote this S-structure on M. Let N be another complex manifold with a divisor T and a T-structure $\{\pi_k', \tilde{V}_k, \psi_{k'k}'\}_{1\le k\le K}$. A holomorphic map f from M into N is said to be compatible with $\{\pi_\ell, \tilde{U}_\ell, \psi_{\ell'\ell}\}$ and $\{\pi_k', \tilde{V}_k, \psi_{k'k}\}$ if the following is true: (i) for each k, there are liftings of $f\big|_{f^{-1}(V_k)\setminus S}$ at points of $f^{-1}(T)\setminus S$ and liftings of $f\big|_{f^{-1}(V_k)\cap S_\ell}$ such that $f = \pi_k' \circ f_{kx}$ on $B_x$ where $B_x$ is a small neighborhood of $x \in f^{-1}(T)\setminus S$ and $f_{kx}$ is the inclusion map of $B_x$ into $\tilde{V}_k$. If $f^{-1}(V_k) \cap S_\ell \neq \phi$, then there is a map $f_{k\ell} : \pi_\ell^{-1}(f^{-1}(V_k)\cap S_\ell) \to \tilde{V}_k$ such that $\pi_k' \circ f_{k\ell} = f \circ \pi_\ell$ on $\pi_\ell^{-1}(f^{-1}(V_k)\cap S_\ell)$, (ii) there are the following compatible relations, $f_{k,x} = \psi_{kk'}' \circ f_{k'x'}$ on $B_x \cap B_{x'}$; $f_{k,x} = f_{k\ell} \circ \pi_\ell$ on $\pi^{-1}(B_x) \cap \pi_\ell^{-1}(f^{-1}(V_k)\cap S_\ell)$ and $\psi_{kk'} \circ f_{k'\ell'} = f_{k\ell} \circ \psi_{\ell\ell'}$ on $\pi_{\ell'}^{-1}(f^{-1}(V_{k'}) \cap S_{\ell'}) \cap \psi_{\ell\ell'}^{-1} \pi_\ell^{-1}(f^{-1}(V_k)\cap S_\ell) = \pi_{\ell'}^{-1}(f^{-1}(V_k \cap V_{k'}) \cap S_\ell \cap S_{\ell'})$. We call $(M, \{\pi_\ell, \tilde{U}_\ell, \psi_{\ell'\ell}\})$ and $(N, \{\pi_k', \tilde{V}_k, \psi_{k'k}'\})$ biholomorphic if f has an inverse which is compatible with corresponding structures on M and N.

Clearly, if $M$ has another S-structure $\{\pi'_k, \tilde{V}_k, \psi_{k'k}\}$, then two structures are biholomorphically equivalent if the identity on $M$ is compatible with these two structures.

Definition 1.1: A Kähler metric $g$ on $M \setminus S$ is said to be compatible with $\{\pi_\ell, \tilde{U}_\ell, \psi_{\ell'\ell}\}$ iff for each $\ell$, $\pi_\ell^* g$ can be extended smoothly to $U$ as a Kähler metric. Such a metric is said to be complete if $g$ is a complete Kähler metric on $M \setminus \bigcup_{\ell=1}^{L} U_\ell$ with boundary $\partial\left(\bigcup_{\ell=1}^{L} U_\ell\right)$ and the extension of each $\pi_\ell^* g$ is a complete metric on $\tilde{U}_\ell$ with boundary $\pi_\ell^{-1}(\partial U_\ell)$.

Remark. A typical example of our $M$ comes from the resolution of a V-manifold $M_o$ with quotient singularities. In this case, $S$ is the exceptional divisor over the singularities in $M_o$, and $\{\tilde{U}_\ell\}_{1 \le \ell \le L}$ are just obtained by blowing up local neighborhoods of singularities in $M_o$.

The purpose of this section is to formulate and construct a Kähler-Einstein metric compatible with $\{\pi_\ell, U_\ell, \psi_{\ell'\ell}\}$ on $M \setminus S$ under certain assumptions. We shall reduce the problem of finding Kähler-Einstein metric to the solution of certain complex Monge-Ampère equations. Let $(Z^1, \ldots, Z^n)$ be a local coordinate in $M \setminus S$. We denote the Kähler metric $g$ by $\sum_{i,j} g_{i\bar{j}}\, dZ^i d\bar{Z}^j$ in this coordinate. Then its Ricci curvature tensor $R_{i\bar{j}}$ is given by

$$R_{i\bar{j}} = -\frac{\partial^2}{\partial Z_i\, \partial \bar{Z}_j} (\log \det(g_{i\bar{j}})) \tag{1.1}$$

Both the Kahler form $\omega_g = \sum_{i,j} g_{i\bar{j}}\, dZ^j \wedge d\bar{Z}^j$ and the Ricci form $\mathrm{Ric}(g) = \sum_{i,j} R_{i\bar{j}}\, dZ^i \wedge d\bar{Z}^j$ are globally defined on $M \setminus S$. We assume that they are cohomologous on $M \setminus S$, i.e. for a certain smooth function $f$ on $M \setminus S$,

$$\mathrm{Ric}(g) + \omega_g = \partial\bar{\partial} f \quad \text{on } M \setminus S \tag{1.2}$$

Consider the following complex Monge-Ampère equation

$$\begin{cases} (\omega_g + \partial\bar{\partial}\varphi)^n = e^{f+\varphi}\,\omega_g^n \\ \omega_g + \partial\bar{\partial}\varphi > 0 \end{cases} \quad \text{on } M \setminus S \tag{1.3}$$

where $\omega_g^n = \overbrace{\omega_g \wedge \cdots \wedge \omega_g}^{n}$ is the volume form of $g$. In local coordinate $(Z^1, \ldots, Z^n)$ in $M \setminus S$, (1.3) can be written as

$$\begin{cases} \det(g_{i\bar{j}} + \varphi_{i\bar{j}}) = e^{f+\varphi} \det(g_{i\bar{j}}) \\ (g_{i\bar{j}} + \varphi_{i\bar{j}}) > 0 \end{cases} \tag{1.4}$$

If $\varphi$ is a solution of (1.3), $\omega_g + \partial\bar{\partial}\varphi$ defines a new Kähler metric on $M \setminus S$. By (1.1) and (1.2), one easily checks that this new metric is Kähler-Einstein on $M \setminus S$. Moreover, such a new metric is compatible with $\{\pi_\ell, \tilde{U}_\ell, \psi_{\ell'\ell}\}$ if for each $\ell$, $\pi_\ell^*\varphi$ can be smoothly extended to $\tilde{U}_\ell$ with $\pi_\ell^*\omega_g + \partial\bar{\partial}(\pi^*\varphi) > 0$ on $\tilde{U}_\ell$.

Before we state our main theorem in this section, we need the following definitions.

Definition 1.2: Let $V$ be a ball in $\mathbb{C}^n$ and $\phi$ be a holomorphic map from $V$ into $M$, compatible with the structure $\{\pi_\ell, \tilde{U}_\ell, \psi_{\ell'\ell}\}$. $\phi$ is called a quasi-coordinate map if either $\phi(V) \cap S = \phi$ and $\phi$ is of maximal rank everywhere, or $\phi(V) \subset U_\ell$ for certain $\ell$ and the lifting $\phi_\ell : V \to \tilde{U}_\ell$ is of maximal rank everywhere, and satisfy the equation $\phi = \pi_\ell \circ \phi_\ell$. In both cases, $(V, \phi)$ is called a local quasi-coordinate of $M$ with S-structure $\{\pi_\ell, \tilde{U}_\ell, \psi_{\ell'\ell}\}$.

Definition 1.3: Let $M$ have a S-structure $\{\pi_\ell, \tilde{U}_\ell, \psi_{\ell'\ell}\}$ and $g$ be a compatible $C^\infty$-Kähler metric on $M \setminus S$. Then we say that $(M, g)$ has bounded geometry of order $m+\beta$, where $m \in Z$, $\beta \in [0,1)$, if the following conditions hold: There exists a system of quasi-coordinates $\mathcal{V} = \{(V_\alpha, v_\alpha^i)\}$ such that:

(i) $\bigcup_\alpha$ (Image of $V_\alpha$) $= M$ and every $x \in M$ is the image of the center of some $V_\alpha$.

(ii) There are positive numbers $\epsilon$ and $\delta$ independent of $\alpha$ such that $\epsilon \le$ radius of $V_\alpha \le \delta$.

(iii) There exist constants $c$ and $\mathfrak{C}_k$ $(k = 0, 1, \ldots, m)$ such that $0 < c^{-1}(\delta_{ij}) \le (g_{\alpha i\bar{j}}) \le c(\delta_{ij})$ and

$$\left| \partial^{|p|+|q|} g_{\alpha i\bar{j}} / \partial v_\alpha^p \, \partial \bar{v}_\alpha^q \right|_{C^\beta(V_\alpha)} \le \mathfrak{C}_{|p|+|q|} \tag{1.4}$$

for all multi-indices $p, q$ with $|p|+|q| \le m$, where $g|_{V_\alpha} = \sum_{i,j} g_{\alpha i\bar{j}} \, dv_\alpha^i d\bar{v}_\alpha^j$ or $\pi_\ell^* g|_{V_\alpha} = \sum_{i,j} g_{\alpha i\bar{j}} \, dv_\alpha^i d\bar{v}_\alpha^j$. Here $|\cdot|_{C^\beta(V_\alpha)}$ denotes the standard $\beta$-Hölder norm on $V_\alpha$.

Remark. Clearly, if $(M, \{\pi_\ell, \tilde{U}_\ell, \psi_{\ell'\ell}\})$ has bounded geometry of order 3, the curvature tensor of the Kähler metric g and the covariant derivatives of the Ricci curvature are bounded on $M \setminus S$. Later on we shall prove a weak converse, i.e. the boundedness of the curvatures and the covariant derivatives implies that M has bounded geometry of order $2+\beta$ for some $\beta \in [0, 1)$.

Definition 1.4: Let $\{\pi_\ell, \tilde{U}_\ell, \psi_{\ell'\ell}\}$ be a S-structure on M. A Kähler-Einstein metric g on $M \setminus S$, compatible with $\{\pi_\ell, \tilde{U}_\ell, \psi_{\ell'\ell}\}$, is said to be almost-complete if there is a sequence of complete Kähler metrics $g_i$ on $M \setminus S$, compatible with $\{\pi_\ell, \tilde{U}_\ell, \psi_{\ell'\ell}\}$, such that

(i) $Ric(g_i) \geq -t_i \omega_{g_i}$, $\lim t_i = 1$.

(ii) g is the pointwise $C^\infty$-limit of $\{g_i\}$ on $M \setminus S$.

The concept of almost-completeness first appeared in [CY1] when they constructed Kähler-Einstein metrics on weakly pseudoconvex domains. The definition here is different from that in [CY1] but both definitions assure the uniqueness of almost-complete Kähler-Einstein metrics in the corresponding category.

Theorem 1.1. Assume that $(M, g, \{\pi_\ell, \tilde{U}_\ell, \psi_{\ell'\ell}\})$ have bounded geometry of order $2+\beta$ with $\beta \in (0, 1)$. Here $\{\pi_\ell, \tilde{U}_\ell, \psi_{\ell'\ell}\}$ is a S-structure on M where $S \subset M$ is a divisor. Let $Ric(g) + \omega_g = \partial\bar{\partial} f$ on $M \setminus S$ for certain non-positive function $f \in C^\infty(M \setminus S, R) \cap C^0(M, R)$. Suppose that for any $\varepsilon > 0$, there is a compact subset $K_\varepsilon$ of M such that $Ric(g) + \partial\bar{\partial} f_\varepsilon \leq \varepsilon\, \omega_g$ in

$(M \setminus S) \setminus K_\varepsilon$ for a certain bounded smooth function $f_\varepsilon$ on M. Then M admits a unique almost-complete Kähler-Einstein metric compatible with $\{\pi_\ell, \tilde{U}_\ell, \psi_{\ell'\ell}\}$.

The rest of this section is devoted to the proof of this theorem. First we prove the uniqueness. We shall need a maximum principle and a generalized Schwarz lemma for volume forms.

Lemma 1.1 (Maximum principles). Let $\{\pi_\ell, \tilde{U}_\ell, \psi_{\ell'\ell}\}$ be a S-structure on complex manifold M with a Kähler metric g compatible with this structure, $\varphi \in C^\infty(M \setminus S, R)$ and for each $\ell$, $\pi_\ell^* \varphi$ can be extended smoothly to $\tilde{U}$. We will call such a $\varphi$ compatible with $\{\pi_\ell, \tilde{U}_\ell, \psi_{\ell'\ell}\}$. Then we have

(i) if the sectional curvature tensor of g is bounded from below and $\sup_{M\setminus S} \varphi < +\infty$, then there is a sequence $\{x_i\} \subset M \setminus S$ such that

$\lim \varphi(x_i) = \sup \varphi$, $\lim |\nabla\varphi(x_i)|_g = 0$ and

$$\overline{\lim}\, \text{Hessian}(\varphi)(x_i) \le 0 \tag{1.5}$$

where the Hessian is taken with respect to g.

(ii) If the Ricci curvature of g is bounded from below and $\varphi$ is a nonnegative function satisfying

$$\Delta\varphi \ge C_1\varphi^\alpha - C_2\varphi - C_3 \quad \text{on } M \setminus S \tag{1.6}$$

where $\alpha > 1$, $C_1 > 0$, $C_2, C_3 \ge 0$ are constants. Then

$$\sup_{M\setminus S} \varphi \le \max\left\{1, \left(\frac{C_2 + C_3}{C_1}\right)^{1/\alpha}\right\}.$$

Proof: (i) The corresponding result on a complete Riemannian manifold is proved by using supporting functions (see [CY1] ). But we don't have such supporting functions here, the trouble is that the metric $g$ is no longer complete along $S$ and $\inf_{x \in S} \sup\{\delta \mid B_{\pi_\ell^* g}(x, \delta) \subset \tilde{U}_\ell\}$ could be zero. The latter happens when the infinity of $M$ is a cusp. We modify the argument here.

Take a fixed point $x_o \in M \setminus S$, let $d(x) = d(x_o, x)$ be the distance function. By our assumption, $\varphi$ is continuous in $M$. Hence for $\epsilon > 0$, $\sup_{M \setminus S} (\varphi(x) - \epsilon d(x))$ can be attained at a point $x_\epsilon \in M$. Let $\gamma$ be the minimal geodesic in $M$ connecting $x_o$ and $x_\epsilon$. The existence of such a $\gamma$ can be shown along the same lines as we did for complete Riemannian manifolds. The argument we use is based on the following observation: For all $\tilde{x} \in \gamma$, let $d_{\tilde{x}}(x) = d(\tilde{x}, x)$, then $\sup_{M \setminus S} (\varphi(x) - \epsilon d_{\tilde{x}}(x)) = \varphi(x_\epsilon) - \epsilon d_{\tilde{x}}(x_\epsilon)$. This is simply the corollary of the triangle inequality and $d(x, x_\epsilon) = d(\tilde{x}, x_\epsilon) + d(\tilde{x}, x)$.

The point $x_\epsilon$ is either in $M \setminus S$ or in one of $U_\ell$. The proof for both cases are the same except that we will work in $\tilde{U}_\ell$ with metric $\pi_\ell^* g$ instead of $g$ in the second case. So we may assume that $x_\epsilon \in M \setminus S$.

Choose $\tilde{x} \in \gamma$ within the cut-locus of $x_\epsilon$. Then as $\varphi(x_\epsilon) - d_{\tilde{x}}(x_\epsilon) = \sup_{M \setminus S} (\varphi(x) - \epsilon d_{\tilde{x}}(x))$, we have

$$|\nabla \varphi(x_\epsilon)| \leq \epsilon \, |\nabla d_{\tilde{x}}(x_\epsilon)| = \epsilon \tag{1.7}$$

$$\text{Hessian}\, \varphi(x_\epsilon) \leq \epsilon \; \text{Hessian}(d_{\tilde{x}}(x_\epsilon)) \tag{1.8}$$

$$\varphi(x_\epsilon) \geq \varphi(x) - \epsilon d_{x_o}(x) \quad \text{for } x \in M \setminus S \tag{1.9}$$

By the well-known Hessian comparison theorem and the fact that the sectional curvature of g is bounded below on $M \setminus S$.

$$\text{Hessian}(d_{\tilde{x}}(x_\varepsilon)) \leq C \text{ for a constant } C \text{ independent of } \varepsilon \tag{1.10}$$

Now (i) follows from (1.7) - (1.10) by taking a sequence of $\{\varepsilon_i\}$ with $\lim \varepsilon_i = 0$.

(ii) First we note that a similar proof as (i) shows that for any smooth function $\psi$, compatible with $\{\pi_\ell, \tilde{U}_\ell, \psi_{\ell'\ell}\}$ with $\sup \psi < +\infty$, there is a sequence $\{x_i\} \subset M \setminus S$, such that $\lim \psi(x_i) = \sup_{M \setminus S} \psi$, $\lim |\nabla\psi(x_i)| = 0$, $\overline{\lim}\, \Delta\psi(x_i) \leq 0$. Hence by (1.6), we only need to show that $\varphi$ is bounded. If not, $\sup_{M \setminus S} \varphi = +\infty$. Define $v(x) = \frac{-1}{1+\varphi(x)}$, then

$$\Delta v \geq \left(C_1 \frac{\varphi^\alpha}{1+\varphi} - 2|\nabla v|^2\right)\frac{1}{-v} - \frac{C_2\varphi + C_3}{(1+\varphi)^2} \tag{1.11}$$

Since $\sup_{M \setminus S} v = 0$, there is a sequence $\{x_i\}$ such that $\lim v(x_i) = 0$, $\lim |\Delta v(x_i)|$ and $\overline{\lim}\, \Delta v(x_i) \leq 0$. But this violates (1.11) and so $\varphi$ must be bounded. //

Lemma 1.2 (Schwarz lemma): Let M and N be n-dimensional complex manifolds equipped with S-structure $\{\pi_\ell, \tilde{U}_\ell, \psi_{\ell'\ell}\}$ and T-structure $\{\pi_k, \tilde{V}_k, \psi'_{k'k}\}$ respectively. Let g be a complete Kähler metric on $M \setminus S$ compatible with $\{\pi_\ell, \tilde{U}_\ell, \psi_{\ell'\ell}\}$ such that its Ricci curvature is bounded from below on $M \setminus S$ and its scalar curvature $R_g \geq -K_1$ and let h be a Kähler metric on $N \setminus T$ compatible with $\{\pi'_k, \tilde{V}_k, \psi'_{k'k}\}$ such that the Ricci form is negative definite and

$R_h \le -K_2$. Suppose $\phi: M \to N$ is a nontrivial holomorphic map compatible with S-structure on M and T-structure on N. Then $K_1 > 0$ and

$$\sup_{M\setminus S} (\phi^* \omega_h^n/\omega_g^n) \le (K_1/K_2)^n \tag{1.12}$$

Proof: The proof is the same as that in [YA3] for Kähler manifolds. We sketch the proof here. Let $u = \phi^* \omega_h^n/\omega_g^n$ be the volume ratio. By the assumption that $\phi$ is compatible with two structures on M, N, $u \in C^\infty(M \setminus S, R) \cap C^0(M, R)$ and each $\pi_\ell^* u$ can be extended smoothly to $\tilde{U}_\ell$. For each $x \in M\setminus S$, either $\phi(x) \in N\setminus T$ or $\phi(x) \in \tilde{V}_k$. In the first case, a computation as in [YA3] shows that

$$\Delta u \ge u(-K_1 + K_2 u^{1/n}) = K_2 u^{1+(1/n)} - K_1 u \tag{1.13}$$

In the second case, $\phi = \pi_k' \circ \phi_k$ for certain $\phi_k$ from a neighborhood of x into $\tilde{V}_k$. Then

$$u = \phi^* \omega_h^n/\omega_g^n = \phi_k^* \circ \pi_k'^* \omega_h^n/\omega_g^n$$

$$= \left(\phi_k^* \omega^n_{\pi_k'^* h} / \omega_g^n\right)$$

Hence we still have the inequality (1.13) and the lemma follows from Lemma 1.1(ii). //

We are now ready to prove the uniqueness statement in Theorem 1.1. Suppose that we have two almost-complete Kähler-Einstein metrics $g_1$ and $g_2$, both compatible with $\{\pi_\ell, \tilde{U}_\ell, \psi_{\ell'\ell}\}$. By definition 1.4, there are two sequences of complete Kahler metrics $g_{1i}$ and $g_{2i}$ on $M\setminus S$, compatible with $\{\pi_\ell, \tilde{U}_\ell, \psi_{\ell'\ell}\}$, satisfying: $Ric(g_{1i}) \geq -t_i$; $Ric(g_{2i}) \geq -t_i$ and $\lim t_i = 1$, $g_1 = \lim g_{1i}$, $g_2 = \lim g_{2i}$. By Lemma 1.2,

$$\sup_{M\setminus S}\left(\omega^n_{g_1}/\omega^n_{g_{2i}}\right) \leq t_i^n \quad \text{for all } i \tag{1.13}$$

$$\sup_{M\setminus S}\left(\omega^n_{g_2}/\omega^n_{g_{1i}}\right) \leq t_i^n \quad \text{for all } i \tag{1.14}$$

Letting $i \to \infty$, we obtain $\omega^n_{g_1} = \omega^n_{g_2}$ on $M\setminus S$. Since $g_1$, $g_2$ are Kahler-Einstein, $\omega_{g_1} = -Ric(g_1) = \partial\bar{\partial}\log(\omega^n_{g_1}) = \partial\bar{\partial}\log(\omega^n_{g_2}) = -Ric(g_2) = \omega_{g_2}$, i.e. $g_1 \equiv g_2$. The uniqueness is proved.

Now we prove the existence part of Theorem 1.1. The strategy is to solve equation (1.3) for the required metric by continuity method. We set up the following equations on $M\setminus S$,

$$\begin{cases} (\omega_g + \partial\bar{\partial}\varphi)^n = e^{f+t\varphi}\,\omega^n_g \\ \omega_g + \partial\bar{\partial}\varphi > 0 \end{cases} \quad \text{on } M\setminus S \tag{1.15$_t$}$$

where $t \geq 1$. The advantage of using this family of equations is that whenever $\varphi$ is a solution of $(1.15)_t$, the Kähler metric $g_t$ with associated Kähler form $\omega_g + \partial\bar{\partial}\varphi$ has its Ricci curvature bounded from below by

$-t(\omega_g + \partial\bar{\partial}\varphi)$. In fact,

$$\begin{aligned}\mathrm{Ric}(g_t) &= -\partial\bar{\partial} \log \det(g_{i\bar{j}} + \varphi_{i\bar{j}}) \\ &= -\partial\bar{\partial} \log \det(g_{i\bar{j}}) - \partial\bar{\partial} f - t\partial\bar{\partial}\varphi \\ &= -t\omega_{g_t} + (t-1)\omega_g > -t\omega_{g_t}\end{aligned}$$

This lower boundedness of the Ricci curvature will enable us to apply the generalized Schwarz lemma.

Define $I = \{t > 1 \mid$ for any $t' \geq t$. $(1.15)_t$ has a solution $\varphi_t$ such that the associated Kähler metric $g_{t'}$ is compatible with $\{\pi_\ell, \tilde{U}_\ell, \psi_{\ell'\ell}\}$ and uniformly equivalent to $g$ for $t' \in [t, +\infty)\}$.

We will first prove that $I$ is nonempty, open and closed in $(1, +\infty)$ and then prove the limit of $\{g_t\}$ exists as $t$ goes to 1.

We say a function $\varphi \in C^{j,\beta}(M \setminus S, R)$ is compatible with $\{\pi_\ell, \tilde{U}_\ell, \psi_{\ell'\ell}\}$ if for each $\ell$, $\pi_\ell^* \phi$ can be extended to a smooth function in $C^{j,\beta}(\tilde{U}_\ell, R)$, $j = 0, 1, 2$. Denote by $C_c^{j,\beta}(M \setminus S, R)$ the space of all $C^{j,\beta}$-functions on $M \setminus S$ compatible with $\{\pi_\ell, \tilde{U}_\ell, \psi_{\ell'\ell}\}$. We will introduce a norm on this space. Fix a quasi-coordinate system $\mathcal{V} = \{(V_\alpha, v_\alpha^i)\}$ with (i) - (iii) stated in Definition 1.3 for $m = 2$. Define the norm $\|\cdot\|_{j,\beta}$ on $C_c^{j,\beta}(M \setminus S, R)$ by

$$\|u\|_{j,\beta} = \sup_{\alpha \in \mathcal{V}} \{|\phi_\alpha^* u|_{C^{j,\beta}(V_\alpha)}\}, \quad j = 0, 1, 2 \tag{1.16}$$

where $|\cdot|_{C^{j,\beta}(V_\alpha)}$ is the standard $C^{j,\beta}$-norm on the Euclidean open subset $V_\alpha$ with coordinates $(v_\alpha^i)$, and $(V_\alpha, \phi_\alpha)$ is the local quasi-coordinate. Put $C_b^{j,\beta}(M\setminus S, R) = \{u \in C_c^{j,\beta}(M\setminus S, R) \mid \|u\|_{j,\beta} < +\infty\}$. Then $C_b^{j,\beta}(M\setminus S, R)$ is a Banach space with the norm $\|\cdot\|_{j,\beta}$. Given another system of quasi-coordinates, we can define another norm $\|\cdot\|'_{j,\beta}$ on $C_c^{j,\beta}(M\setminus S, R)$. By (ii), (iii), in Definition 1.3, one can directly check that $\|\cdot\|_{j,\beta}$ and $\|\cdot\|'_{j,\beta}$ are equivalent.

Now consider the equations

$$\begin{cases} ((1-s)\omega_g - s(\mathrm{Ric}(g) + \partial\bar\partial\psi))^n = e^{\psi}\omega_g^n & \text{on } M\setminus S \\ ((1-s)\omega_g - s(\mathrm{Ric}(g) + \partial\bar\partial\psi)) > 0 & \end{cases} \qquad (1.17)_s$$

where $s < 1$. The transformation from $\varphi$ to $f + t\varphi = \psi$ shows that $(1.17)_s$ is equivalent to $(1.15)_t$ for $s = \frac{1}{t} < 1$.

<u>Lemma 1.3</u>. If $(1.17)_{s_o}$ has a solution $\psi_{s_o} \in C_b^{2,\beta}(M\setminus S, R)$ with the associated Kähler metric equivalent to $g$, then $(1.17)_s$ also has a solution $\psi_s \in C_b^{2,\beta}(M\setminus S, R)$ for $s$ sufficiently close to $s_o$ such that the associated Kähler metrics are uniformly equivalent to $g$.

<u>Remark</u>. For a solution $\psi_s$ of $(1.17)_s$, the associated Kähler metric $g_t = g_{1/s}$ can be written as $\sum_{i,j}((1-s)g_{i\bar j} - s(R_{i\bar j} + \psi_{si\bar j})\,dZ^i d\bar Z^j$ which is compatible with $\{\pi_\ell, \tilde U_\ell, \psi_{\ell'\ell}\}$. Although the metric $g_t$ is only $C^\beta$ it is actually $C^\infty$ since $(1.17)_s$ is equivalent to $(1.15)_{1/s}$ and the regularities on the $3^{rd}$ derivatives or higher are purely local.

Proof of Lemma 1.3: It is based on the application of the Implicit Function Theorem. Define an operator $A: [0,1) \times C_b^{2,\beta}(M \setminus S, R) \to C_b^{o,\beta}(M \setminus S, R)$ by

$$A(s, \psi) = \frac{((1-s)\omega_g - s(\mathrm{Ric}(g) + \partial\bar{\partial}\psi))^n}{e^{\psi}\,\omega_g^n} \quad \text{for } (s,\psi) \qquad (1.18)$$

In local quasi-coordinate $(v_\alpha^1, \ldots, v_\alpha^n)$. $\omega_g = \sum_{i,j} g_{i\bar{j}}\, dv_\alpha^i \wedge d\overline{v_\alpha^j}$, $\mathrm{Ric}(g) = \sum_{i,j} R_{i\bar{j}}\, dv_\alpha^i \wedge d\overline{v_\alpha^j}$,

$$A(s,\psi) = \frac{\det((1-s)g_{i\bar{j}} - s(R_{i\bar{j}} + \psi_{i\bar{j}}))}{\det(g_{i\bar{j}})}\, e^{-\psi} \qquad (1.19)$$

where $\psi_{i\bar{j}} = \dfrac{\partial^2\psi}{\partial v_\alpha^i\, \partial \bar{v}_\alpha^j}$.

From this explicit expression, one easily sees that $A$ is a smooth nonlinear operator.

Clearly, $(1.17)_s$ is solvable in $C_b^{2,\beta}(M\ S, R)$ iff the equation $A(s,\psi) \equiv 1$ has a solution in $C_b^{2,\beta}(M \setminus S, R)$. Now, $A(s_o, \psi_{s_o}) \equiv 1$. Hence, the lemma will follow from the Implicit Function Theorem once we prove that the partial derivative $L = D_\psi A(s_o, \psi_{s_o})$ of $A$ is invertible at $(s_o, \psi_o)$. By definition, $L = C_b^{2,\beta}(M \setminus S, R) \to C_b^{o,\beta}(M \setminus S, R)$ is defined by

$$L(\varphi) = \Delta_{s_o}\varphi - \varphi \quad \text{for } \varphi \in C_b^{2,\beta}(M \setminus S, R)$$

where $\Delta_{s_o}$ is the Laplacian of the metric $g_{s_o}$, i.e. $\Delta_{s_o}\varphi = g_{s_o}^{i\bar{j}}\varphi_{i\bar{j}}$ in local quasi-coordinate $(v_\alpha^1, \ldots, v_\alpha^n)$. The injectivity of $L$ is trivially true. For any $\theta \in C_b^{o,\beta}(M \setminus S, R)$, we will solve the equation $L\varphi = \theta$ for

$\varphi \in C_b^{2,\beta}(M \setminus S, R)$. Choose a sequence of positive numbers $\{R_k\}$ with $\lim R_k = +\infty$. Let $B_g(x_o, R_k) = B_k$ be the geodesic ball with center at a fixed point $x_o$ and radius $R_k$ with respect to $g$. Then we can solve the Dirichlet problem

$$\begin{cases} L\varphi = \theta & \text{in } B_g(x_o, R_k) \\ \varphi = 0 & \text{on } \partial B_g(x_o, R_k) \end{cases} \tag{1.20}$$

and obtain a smooth $\varphi_k$, compatible with $\{\pi_\ell, \tilde{U}_\ell, \psi_{\ell'\ell}\}$. The maximum principle implies that

$$\sup_{B_k} |\varphi_k| \le \sup_{M\setminus S} |\theta| . \tag{1.21}$$

We can find another system of quasi-coordinates $\mathcal{V}' = \{(V'_\alpha, v^i_\alpha)\}$ such that for each $\alpha$, $V'_\alpha \subset\subset V_\alpha$ and $\bigcup_\alpha V'_\alpha = M$. It is clear that $\mathcal{V}'$ satisfies (i), (ii), (iii) in Definition 1.3 for $m = 2$. By the standard Schauder estimates (see [GT]) and (1.19), (1.20), (1.21), (ii), (iii) in Definition 1.3 and $\psi_{s_o} \in C_b^{2,\beta}(M \setminus S, R)$, we conclude

$$|\phi^*_\alpha \varphi_k|_{C^{2,\beta}(V'_\alpha)} \le C \tag{1.22}$$

where $C$ is a constant independent of $k$, $\alpha$. It follows that a subsequence of $(\varphi_k)$ converges to a $\varphi \in C_c^{2,\beta}(M\setminus S, R)$ with $L\varphi = \theta$ and

$$|\phi^*_\alpha \varphi|_{C^{2,\beta}(V'_\alpha)} \le C \quad \text{for all } \alpha \tag{1.23}$$

Hence $\varphi \in C_b^{2,\beta}(M \setminus S, R)$ and $L$ is invertible. //

When $s = 0$, $\varphi \equiv 0$ is obviously a solution of $(1.17)_s$ in $C_b^{2,\beta}(M\setminus S, R)$. Therefore by Lemma 1.3, I is nonempty.

Let $t_o$ be in I, then for $s_o = 1/t_o < 1$, $(1.17)_{s_o}$ has a solution $\psi_{s_o}$ such that the associated Kähler metric $g_{t_o}$ is compatible with $\{\pi_\ell, \tilde{U}_\ell, \psi_{\ell'\ell}\}$ and is equivalent to $g$. Hence, $f + t_o\varphi_{t_o} = \psi_{s_o}$ is bounded.

Lemma 1.4: $\psi_{s_o} = f + t_o\varphi_{t_o} \in C_b^{2,\beta}(M \setminus S, R)$.

Proof: The problem is completely local and we can work in a quasi-coordinate $(V, v^i) = (V_\alpha, v_\alpha^i)$. In such a coordinate,

$$g = \sum_{i,j} g_{i\bar{j}}\, dv^i d\bar{v}^j , \qquad g_{t_o} = \sum_{i,j} (g_{i\bar{j}} + \varphi_{t_o i\bar{j}})\, dv^i d\bar{v}^j$$

It follows from the equivalence of $g$ and $g_{t_o}$ that $(\varphi_{t_o i\bar{j}})$ is bounded. By assumption,

$$R(g) + n + t_o \Delta_{g_o} \varphi = \Delta_g (f + t_o\varphi) \text{ for } \varphi = \varphi_{t_o} \tag{1.24}$$

Let $V' = V'_\alpha$ be the subset of $V$ used in (1.22), and $V'' \subset\subset V$ be a ball such that $V' \subset\subset V''$. By the Schauder estimate and the boundedness of $f + t_o\varphi$, there is a constant $C_1$ independent of $\alpha$ such that

$$|\psi_{s_o}|_{C^{1,\beta}(V'')} = |f + t_o\varphi|_{C^{1,\beta}(V'')} \le C_1 \tag{1.25}$$

Let $\theta = s_o \log \det(g_{i\bar{j}}) + \psi_{s_o}$, then

$$\det(g_{i\bar{j}}(1 - s_o) + \theta_{i\bar{j}}) = e^{\theta}(\det(g_{i\bar{j}}))^{1-s_o} \tag{1.26}$$

Differentiating both sides of (1.26), we obtain

$$\Delta_{s_o}\theta_k + g_{s_o}^{i\bar{j}} \frac{\partial g_{i\bar{j}}}{\partial v^k}(1 - s_o) = \theta_k + (1 - s_o)g^{i\bar{j}} \frac{\partial g_{i\bar{j}}}{\partial v^k} \tag{1.27}$$

By (1.25) and the assumptions of Theorem 1.1, $\Delta_{s_o}\theta_k = F$ on $V''$ for a function $F \in C_b^{o,\beta}(V'')$. The standard Schauder estimates show that there is a constant $C_3$ independent of $\alpha$, such that

$$|\theta_k|_{C^{1,\beta}(V')} \le C_3$$

i.e.,

$$|\theta|_{C^{2,\beta}(V')} \le C_3 \tag{1.28}$$

The lemma then follows. //

Lemma 1.3 and 1.4 imply that I is open in $(1, +\infty)$. Next, we prove that I is closed in $(1, +\infty)$. Suppose that there is a sequence $\{t_i\} \subset I$ with $\lim t_i = t > 1$. May assume that $t_1 > t_2 > \cdots > t$. Let $\varphi_i$ be the solution of $(1.15)_{t_i}$.

<u>Lemma 1.5</u>: There are constants $C_4$, $C_5$, $C_6$, depending only on the metric $g$ ($C_4$ may depend on $t$) such that

$$|f + t_i\varphi_i|_{C^o(M\setminus S)} \le C_4 \tag{1.29}$$

$$f + t_i\varphi_i \le C_5 \quad \text{on } M\setminus S \tag{1.30}$$

and

$$\varphi_i \ge -C_6 \quad \text{on } M\setminus S \quad . \tag{1.31}$$

Proof: Note that in Lemma 1.2, the metric $h$ can be replaced by a volume form $\mathrm{Vol}_N$ which is compatible with $\{\pi'_k, \tilde{V}_k, \psi_{k'k}\}$ such that $\partial\bar{\partial}\log(\mathrm{Vol}_N) > 0$ and $(\partial\bar{\partial}\log(\mathrm{Vol}_N))^n \geq K_2^n/n^n(\mathrm{Vol}_N)$. In such a situation, the conclusion will be:

$$\sup_{M\setminus S} (\phi^* \mathrm{Vol}_N/\omega_g^n) \leq (K_1/K_2)^n \tag{1.32}$$

Now put $\mathrm{Vol}_N = e^f \omega_g^n$ and $N = M$. Since $\partial\bar{\partial}\log(\mathrm{Vol}_N) = \partial\bar{\partial}f + \partial\bar{\partial}\log\det(g_{ij}) = \partial\bar{\partial}f - \mathrm{Ric}(g) = \omega_g > 0$ and $\mathrm{Ric}(g_i) \geq -t_i \geq -t_1$, (1.31) follows from (1.32) and $(1.15)_{t_i}$.

Put $\psi_i = f + t_i\varphi_i$, then, applying Lemma 1.1(i) to equation $(1.17)_{s_i}$ for $\psi_i$ and $s_i = 1/t_i$, we get (1.30).

By the assumptions, there is a compact subset $K$ and a bounded function $f_{\varepsilon'}$ for $\varepsilon' = \frac{1}{2}(1 - \frac{1}{t})$, such that $\mathrm{Ric}(g) + \partial\bar{\partial}f_{\varepsilon'} \leq \varepsilon'\omega_g$ in $(M\setminus S)\setminus K$. We may rewrite $(1.17)_{s_i}$ to obtain

$$\left(\left(1 - \frac{1}{t_i}\right)\omega_g - \mathrm{Ric}(g) + \partial\bar{\partial}\theta_i \bar{V}\, \partial\bar{\partial}f_{\varepsilon'}\right)^n = e^{-f_{\varepsilon'} + \theta_i}\omega_g^n \tag{$(1.33)_{s_i}$}$$

where $\theta_i = \psi_i + f_{\varepsilon'}$.

If $\inf_{M\setminus S} \psi_i < -t_i C_6 + \inf_K (-f_{\varepsilon'} + f)$, then applying Lemma 1.1 (i) to the equation $(1.33)_{s_i}$ outside $K$, we have

$$\lim_j e^{-f_{\varepsilon'} + \theta_i}(x_j) \geq \frac{1}{2^n}\left(1 - \frac{1}{t_i}\right)^n \tag{$(1.34)_{s_i}$}$$

where $\lim_j \theta_i(x_j) = \inf_{M\setminus S} \theta_i$. Therefore

$$\lim \psi_i \geq -n\log\left(\frac{t_i - 1}{2t_i}\right) \tag{$(1.35)_i$}$$

Since $f_{\varepsilon'}$ is bounded, (1.29) is proved. //

Remark. We only need a lower bound of $f_{\varepsilon'}$ in the above proof. Hence, the assumptions of Theorem 1.1 can be weakened.

Lemma 1.6: There are three constants $C_7$, $C_8$, $C_9$, depending on g ($C_9$ may depend on t) such that for any solution $\varphi = \varphi_k$,

$$n + \Delta\varphi \leq C_7 e^{C_8 \varphi} \tag{1.36}$$

$$n + \Delta\varphi \leq C_9 \tag{1.37}$$

where $\Delta = \Delta_g$ is the Laplacian with respect to g.

Proof: By the same computations as in [Y2], we obtain in the orthonormal frame of g,

$$\Delta_k (e^{-C\varphi}(n+\Delta\varphi)) \geq e^{-C\varphi}\Big[\Delta f + t_k \Delta\varphi + \frac{1}{2}\sum_{i\neq j} R_{i\bar{i}j\bar{j}}(g) \frac{(\varphi_{j\bar{j}} - \varphi_{i\bar{i}})^2}{(1+\varphi_{i\bar{i}})(1+\varphi_{j\bar{j}})}$$
$$+ (n+\Delta\varphi)(-C\Delta_k\varphi)\Big] \quad \text{on } M\setminus S \tag{1.38}$$

and

$$\Delta_k\left(e^{-C(f+t_k\varphi+f_{\varepsilon'})}(n+\Delta\varphi)\right)$$
$$\geq e^{-C(f+t_k\varphi+f_{\varepsilon'})}\Big[\Delta f + t_k\Delta\varphi + \frac{1}{2}\sum_{i\neq j} R_{i\bar{i}j\bar{j}}(g)\frac{(\varphi_{j\bar{j}} - \varphi_{i\bar{i}})^2}{(1+\varphi_{i\bar{i}})(1+\varphi_{j\bar{j}})}$$
$$- C(n+\Delta\varphi)(\Delta_k f + t_k \Delta_k \varphi + \Delta_k f_{\varepsilon'})\Big] \quad \text{on } M\setminus S \tag{1.39}$$

where $\Delta_k$ is the Laplacian of $g_k$ and $f_\varepsilon$ appears in the proof of Lemma 1.5.

Choose

$$C_8 \geq -\inf_{M\setminus S}\inf_{i\neq j} R_{i\bar{i}j\bar{j}}(g) + 1$$

and

$$C_9' \geq \frac{1}{2(t-1)}\left(-\inf_{M\setminus S}\inf_{i\neq j} R_{i\bar{i}j\bar{j}}(g) + 1\right)$$

then by (1.31), we can find constants $C_{10}, C_{11}, C_{12}, C_{13}$, such that

$$\Delta_k\left(e^{-C_8\varphi}(n+\Delta\varphi)\right) \geq -C_{10} - C_{11}(n+\Delta\varphi)e^{-C_8\varphi} + e^{-C_8\varphi}(n+\Delta\varphi)\sum_i \frac{1}{1+\varphi_{i\bar{i}}} \quad \text{on } M\setminus S, \tag{1.40}$$

$$\Delta_k\left(e^{-C_9'(f+t_k\varphi+f_{\varepsilon'})}(n+\Delta\varphi)\right) \geq \left(-C_{12} - C_{13}(n+\Delta\varphi) + (n+\Delta\varphi)\sum_i \frac{1}{1+\varphi_{i\bar{i}}}\right) e^{-C_9'(f+t_k\varphi+f_{\varepsilon'})} \quad \text{on } (M\setminus S)\setminus K, \tag{1.41}$$

$$\sum_i \frac{1}{1+\varphi_{i\bar{i}}} \geq \left(\frac{n+\Delta\varphi}{\det(g_{i\bar{j}}+\varphi_{i\bar{j}})}\right)^{\frac{1}{n-1}} = \left(e^{-(f+t_k\varphi)}(n+\Delta\varphi)\right)^{\frac{1}{n-1}} \tag{1.42}$$

Hence,

$$\Delta_k\left(e^{-C_8\varphi}(n+\Delta\varphi)\right) \geq -C_{10} - C_{11}e^{-C_8\varphi}(n+\Delta\varphi) + e^{-\frac{1}{n-1}f + \frac{C_8-t_k}{n-1}\varphi}\left(e^{-C_8\varphi}(n+\Delta\varphi)\right)^{\frac{n}{n-1}} \quad \text{on } M\setminus S,$$

and

$$\Delta_k\left(e^{-C_9'(f+t_k\varphi+f_{\varepsilon'})}(n+\Delta\varphi)\right) \geq (-C_{12}-C_{13}(n+\Delta\varphi))\, e^{-C_9'(f+t_k\varphi+f_{\varepsilon'})}$$

$$+\, e^{\frac{C_9'}{n-1}f_{\varepsilon'}+\frac{C_9'-1}{n-1}\varphi}\left(e^{-C_9'(f+t_k\varphi+f_{\varepsilon'})}(n+\Delta\varphi)\right)^{\frac{n}{n-1}} \quad \text{on } (M\setminus S)\setminus K\,.$$

Since $f_{\varepsilon'}$ and $f + t_k\varphi$ are bounded, $f$ bounded from above and $\varphi$ is uniformly bounded from below, this lemma follows from Lemma 1.1 (ii) and the above inequalities. This proves Lemma 1.6.

By Lemma 1.6, all compatible Kähler metrics $g_i = g_{t_i}$ are uniformly equivalent to the Kähler metric $g$. Then $\psi_i = f + t\varphi_i \in C_b^{2,\beta}(M\setminus S, R)$ by the proof of Lemma 1.4 and moreover $\|\psi_i\|_{2,\beta}$ are uniformly bounded. The estimates on the third and higher derivatives of $\varphi_i$ can be localized Hence we have proved that a subsequence of $\{\varphi_i\}$ converges to $\varphi$ such that the associated Kähler metric $g_t$ is compatible with $\{\pi_\ell, \tilde{U}_\ell, \psi_{\ell'\ell}\}$ and is equivalent to $g$. Thus, $I$ is closed in $(1, +\infty)$.

Finally, take a sequence $\{t_i\}$ in $I$ with $\lim t_i = 1$. By (1.36) and the same arguments as above, there is a subsequence of $\{\varphi_i\}$ which converges pointwisely to a compatible $\varphi$ solving (1.3). Obviously, the associated Kähler metric is an almost-complete Kähler-Einstein metric compatible with $\{\pi_\ell, \tilde{U}_\ell, \psi_{\ell'\ell}\}$. Thus, Theorem 1.1 is proved.

Remark. We can generalize Theorem 1.1 to the case when $\underline{M}$ is a variety with a divisor S and has only quotient singularities away from S. Under similar conditions as in Theorem 1.1, we can show that there is a unique almost-complete Kähler-Einstein metric on $M \setminus S \cup \{\text{singularities}\}$ which is compatible with the given S-structure and the pull-back of the metric to local uniformization of singularities are extendable.

## §2. On Quasi-Projective Varieties

A quasi-projective variety is a Zariski open subset in a projective manifold. In this section, we always denote by $X$ a smooth projective manifold of dimension $n$ and by $D$, $C$ divisors in $X$. Put $M_D = X \setminus D$. A divisor $D$ is said to be a divisor with normal crossings if each irreducible component of $D$ is smooth and the irreducible components of $D$ intersect transversely. We will give here some numerical criteria on $X$ and $D$ such that $M_D$ admits an almost-complete Kähler-Einstein metric. The tool is Theorem 1.1 in the previous section.

Let $D$, $C$ be divisors in $X$ and $D+C$ be a divisor with normal crossings. This notation is set up so that $D$ will represent the infinity of $M_D$ and $C$ will give a C-structure on $M_D$. Write $D = \sum_{i=1}^{L} D_i$, $C = \sum_{j=1}^{N} C_j$, where $D_i$, $C_j$ are irreducible components of $D$ and $C$, respectively. With $C$ we associate a sequence of positive rational numbers $\frac{m_1 - 1}{m_1}, \ldots, \frac{m_N - 1}{m_N}$.

For $x \in \bigcap_{i=1}^{\ell} \left\{ C_{j_i} \bigcup_{j \neq i} C_j,\ 1 \leq i_1 \leq \cdots \leq i_\ell \leq N \right\}$, we can choose a coordinate $(Z_1, \ldots, Z_n)$ at $x$ such that $\{Z_{j_i} = 0\}$ locally defines $C_{j_i}$ for $i = 1, 2, \ldots, \ell$, $x = (0, \ldots, 0)$ in this coordinate. Define a neighborhood $V_x$ of $x$ to be $\{|Z_j| < \delta$ for $j \neq j_i$; $|Z_{j_i}| < \delta^{m_{j_i}}$ for $i = 1, 2, \ldots, \ell\}$, where $\delta$ is sufficiently small. Then there is a natural branched covering $\tilde{V}_x$ over $V_x$, where $\tilde{V}_x = \{(\omega_1, \ldots, \omega_n) \in \mathbb{C}^n \mid |\omega_i| < \delta\}$ and the branch map $\pi_x : \tilde{V}_x \to V_x$ is defined by $\pi_x(\omega_1, \ldots, \omega_n) = (Z_1, \ldots, Z_n)$, $Z_j = \omega_j$ for $j \neq j_i$; $Z_{j_i} = \omega_{j_i}^{m_{j_i}}$ for $i = 1, 2, \ldots, \ell$. For $\delta$ small enough, one can easily show that if $V_x \cap V_{x'} \neq 0$, there is a holomorphic covering map $\psi_{xx'}$ from $\pi_{x'}^{-1}(V_x \cap V_{x'})$ into $\pi_{x'}^{-1}(V_x \cap V_{x'})$ which commutes with the branched mappings $\pi_x$, $\pi_{x'}$ and the identity on $V_x \cap V_{x'}$. We can take a finite set of points $x_1, \ldots, x_{L_c}$, such that $\{\pi_\ell = \pi_{x_\ell}, \tilde{U}_\ell = \tilde{V}_{x_\ell}, \psi_{\ell\ell} = \psi_{x_\ell, x_\ell}\}$ is a C-structure on $M_D$.

Lemma 2.1: Suppose that there are rational numbers $\mu_1, \ldots, \mu_L \in [0,1]$ such that

$$K_X + \sum_{i=1}^{L} \mu_i D_i + \sum_{j=1}^{N} \frac{m_j - 1}{m_j} C_j > 0 \text{ on } X \tag{2.1}$$

where $K_X$ is the canonical line bundle on X and $m_1, \ldots, m_N$ are the same as above. Then there is a complete Kähler metric g on $M_D \setminus C$, compatible with the above C-structure $\{\pi_\ell, \tilde{U}_\ell, \psi_{\ell'\ell}\}$ such that $(M_D, g, \{\pi_\ell, \tilde{U}_\ell, \psi_{\ell'\ell}\})$ has bounded geometry of order $\infty$.

Remark: The inequality in (2.1) means that

$$K_X + \sum_{i=1}^{L} \mu_i D_i + \sum_{j=1}^{N} \frac{m_j - 1}{m_j} C_j$$

is ample on X. See [H] for the definition of ampleness. Also in the case $C = C_1$, there is an $m_1$-branched covering Y of X with branched locus C, then (2.1) means that

$$K_Y + \sum_{i=1}^{L} \mu_i D_i > 0 \text{ on } Y.$$

Proof: Let $S_1, \ldots, S_L$ be the defining sections of $D_1, \ldots, D_L$ and $t_1, \ldots, t_N$ be the defining sections of $C_1, \ldots, C_N$. Let $\|\cdot\|_{D_i}$ and $\|\cdot\|_{C_j}$ be Hermitian metrics defined on line bundles $[D_i]$ and $[C_j]$, respectively. Let $C_1(D_i)$ and $C_1(C_j)$ be Chern forms associated with $\|\cdot\|_{D_j}$ and $\|\cdot\|_{C_j}$, respectively. Then on $M_D \setminus C$ (see [GH] for reference),

$$C_1(D_i) = -\partial\bar\partial \log(\varepsilon \|S_i\|_{D_i}) \tag{2.2}$$

$$C_1(C_j) = -\partial\bar\partial \log(\varepsilon \|t_j\|_{C_j}) \tag{2.3}$$

where $\varepsilon > 0$ is to be chosen.

On the other hand, for any volume from $V_X$ on X, the Chern form of the canonical line bundle $K_X$ can be written as

$$C_1(K_X) = \partial\bar\partial \log V_X \tag{2.4}$$

By the assumptions (2.1) - (2.4), we can find a proper volume form $V_X$ such that

$$\partial\bar\partial \log \frac{V_X}{\left(\prod_{i=1}^{L} \varepsilon\|S_i\|_{D_i}^{2\mu_i}\right)\cdot\left(\prod_{i=1}^{N} \varepsilon\|t_j\|_{C_j}^{\frac{2m_j-2}{m_j}}\right)} = C_1(K_X) + \sum_{i=1}^{L} \mu_i C_1(D_i)$$

$$+ \sum_{j=1}^{N} \frac{m_j - 1}{m_j} C_1(C_j) > 0 \text{ on } X \tag{2.5}$$

Now define a Kähler metric g by

$$\omega_g = \partial\bar{\partial} \log \frac{V_X}{\left(\prod_{i=1}^{L} \epsilon \| S_i \|_{D_i}^{2\mu_i} \left(-\log \epsilon \| S_i \|_{D_i}^{2}\right)^2\right)\left(\prod_{j=1}^{N} \epsilon \| t_j \|_{C_j}^{\frac{2m_j-2}{m_j}} m^{n+1}\left(1 - \epsilon \| t_j \|_{C_j}^{\frac{2}{m_j}}\right)^{n+1}\right)}$$

$$= \partial\bar{\partial} \log \frac{V_X}{\left(\prod_{i=1}^{L} \epsilon \| S_i \|_{D_i}^{2\mu_i}\right)\left(\prod_{j=1}^{N} \epsilon \| t_j \|_{C_j}^{\frac{2m_j-2}{m_j}}\right)}$$

$$+ \sum_{i=1}^{L} \left( \frac{2C_1(D_i)}{\log(\epsilon \| S_i \|_{D_i}^2)} + \frac{\partial \log \epsilon \| S_i \|_{D_i}^2 \wedge \bar{\partial} \log \epsilon \| S_i \|_{D_i}^2}{(-\log \epsilon \| S_i \|_{D_i}^2)^2} \right)$$

$$+ (n+1) \sum_{j=1}^{N} \left( \frac{\epsilon \partial\bar{\partial} \left( \| t_j \|_{C_j}^{2/m_j} \right)}{\left(1 - \epsilon \| t_j \|_{C_j}^{2/m_j}\right)} + \frac{\epsilon^2 \partial \| t_j \|_{C_j}^{2/m_j} \wedge \bar{\partial} \| t_j \|_{C_j}^{2/m_j}}{\left(1 - \epsilon \| t_j \|_{C_j}^{2/m_j}\right)^2} \right) \qquad (2.6)$$

Since $\log(\epsilon \| S_i \|_{D_i}^2)$ goes to negative infinity as $\epsilon$ goes to zero, it follows from (2.5) and a direct computation that $\omega_g$ is positive definite if $\epsilon$ is small enough. Hence it defines a Kahler metric on $M_D \setminus C$, and is clearly compatible with the above C-structure. Since $-\partial\bar{\partial} \log(-\log \epsilon \| S_i \|_{D_i}^2)^2$ is equivalent to the Poincaré metric on the punctured disc along normal directions of $D_i$, the Kähler metric $g$ is complete. It remains to prove that $(M_D, g, \{\pi_\ell, U_\ell, \psi_\ell\}_\ell)$ has bounded geometry of order $\infty$. We only need to consider this in a neighborhood of D. Pick up a point $x_o$ in D, for simplicity, we assume that $x_o \in D_1 \cap \cdots \cap D_\ell \setminus \left( \bigcup_{i=\ell+1}^{L} D_i \cup \bigcup_{j=1}^{N} C_j \right)$. The proofs for other cases are basically the same.

Take a local coordinate $(Z_1, \ldots, Z_n)$ of X at $x_o$, such that $x_o = (0, \ldots, 0)$ and $D_i = \{Z_i = 0\}$ for $i = 1, 2, \ldots, \ell$. We identify a neighborhood of $x_o$ in X with the polydisc $P = \{|Z_i| < \delta$ for $i = 1, 2, \ldots, n\}$, where $\delta > 0$ only depends on X, D, C. We will find a covering of $P \cap M_D$ by quasi-coordinates such that the pull-backs of g to these quasi-coordinates satisfy (i), (ii), (iii) stated in Definition 1.3. We need the following two lemmas, whose proofs are elementary.

Lemma 2.2: Let $\Delta^*$ be the punctured disc $\{0 < |Z| < \delta\}$ in complex plane $\mathbb{C}$, $\Delta$ be the disc $\{|\omega| < \delta\}$, then the map: $\Delta \to \Delta^*$, $w \to Z = \delta \exp\left(\frac{\omega + \delta}{\omega - \delta} + i\theta\right)$ is the universal covering map of $\Delta^*$, where $\theta$ is any number. The fundamental domain over $\Delta^* \setminus \{te^{i\theta} \,|\, 0 < t < \delta\}$ is $\{\omega \in \Delta \,|\, 0 < \delta \operatorname{Im}(\omega) < \pi |\delta - \omega|^2\}$.

Lemma 2.3: For $\eta \in (0, 1)$, define a map $\Phi_\eta : \Delta \to \Delta$, $\phi_\eta(\omega) = \delta \frac{\omega - \delta\eta}{\delta - \eta\omega}$. This is the automorphism of $\Delta$ mapping the point $\delta\eta$ in $\Delta$ to the origin. Then $\{w \,|\, 0 < \delta \operatorname{Im}(w) < \pi |\delta - w|^2, \ |w|^2 + (-\log r)|1 - w|^2 < \delta^2\} \subset \bigcup_{0<\eta<1} \phi_\eta^{-1}(\Delta_{\frac{1}{2}})$, where $\Delta_{\frac{1}{2}} = \{|w| < \frac{1}{2}\delta\}$ and r is chosen so that

$$\frac{\pi^2 \delta^2}{(-\log r)\left(\delta - \frac{\pi}{\sqrt{-\log r}}\right)^2} \le \frac{1}{\delta - 1} \quad \text{and} \quad -\log r > \frac{4}{\delta^2} .$$

Now $P \cap M_D = \{0 < |Z_i| < \delta$ for $i = 1, 2, \ldots, \ell$; $|Z_i| < \delta$ for $i \geq \ell+1\} = (\Delta^*)^\ell \times \Delta^{n-\ell}$. Let $\Delta_r^* = \{0 < |Z| < r\delta\} \subset \Delta^*$, then $(\Delta_r^*)^\ell \times \Delta_{\frac{1}{2}}^{n-\ell} = (\Delta_r^\ell \times \Delta_{\frac{1}{2}}^{n-\ell}) \cap M_D$ and $(\Delta_r^*)^\ell \times \Delta_{\frac{1}{2}}^{n-\ell}$ is covered by union

$$\bigcup_{\substack{0<\eta_i<1 \\ \theta_i \in [0, 2\pi) \\ i=1,2,\ldots,\ell}} \psi_{\theta_1} \circ \phi_{\eta_1}^{-1}(\Delta_{\frac{1}{2}}) \times \cdots \times \psi_{\theta_\ell} \circ \phi_{\eta_\ell}^{-1}(\Delta_{\frac{1}{2}}) \times \Delta_{\frac{1}{2}}^{n-\ell},$$

where $\psi_\theta : w \to Z = \exp\left(\frac{w+\delta}{w-\delta} + i\theta\right)$ is defined in Lemma 2.2.

Set $F = F_{(\eta,\theta)}$ to be the immersion $(\psi_{\theta_1} \cdot \phi_{\eta_1}^{-1}, \ldots, \psi_{\phi_\ell} \cdot \phi_{\eta_\ell}^{-1}, \mathrm{id}, \ldots, \mathrm{id})$ from $P$ into $M_D$. Let $(v_1, \ldots, v_n)$ be the coordinate of $P$, $(Z_1, \ldots, Z_n)$ be the local coordinate at $x_o = (0, \ldots, 0)$ in $X$, then

$$F_{(\eta,\theta)}(v_1, \ldots, v_n) = \left(\exp\left(\frac{(1+\eta_1)(v_1+\delta)}{(1-\eta_1)(v_1-\delta)} + i\theta_1\right), \ldots, \right.$$

$$\left.\exp\left(\frac{(1+\eta_\ell)(v_\ell+\delta)}{(1-\eta_\ell)(v_\ell-\delta)} + i\theta_\ell\right)\right), v_{\ell+1}, \ldots, v_n \tag{2.7}$$

for $(v_1, \ldots, v_n) \in \Delta^n$. By a direct computation,

$$\frac{\partial}{\partial v_n} = -\left\{2(1+\eta_i)\exp\left(\frac{(1+\eta_i)(v_i+\delta)}{(1-\eta_i)(v_i-\delta)} + i\theta_i\right) \frac{\partial}{\partial Z_i}\right\} \Big/ (1-\eta_i)(v_i-\delta)^2 \tag{2.8}$$

for $i = 1, 2, \ldots, \ell$.

$$\log|Z_i|^2 = 2\,\frac{(1+\eta_i)(|v_i|^2 - \delta^2)}{(1-\eta_i)\,|v_i - \delta|^2} \quad \text{for } i = 1, 2, \ldots, \ell \tag{2.9}$$

In $P \cap M_D$, write $\omega_g = \omega' + \omega''$, where

$$\omega'' = \sum_{i=1}^{\ell} \frac{\partial \log \epsilon \|S_i\|^2_{D_i} \wedge \bar{\partial} \log \epsilon \|S_i\|^2_{D_i}}{(-\log \epsilon \|S_i\|^2_{D_i})^2}$$

and

$$\omega' = \omega_g - \omega'' = \sum_{i,j=1}^{N} b_{i\bar{j}}\, dZ_i \wedge d\bar{Z}_j$$

Then $\omega'$ is a positive (1, 1)-form on P. By the choice of $(Z_1, \ldots, Z_n)$, $\epsilon \|S_i\|^2_{D_i} = h_i |Z_i|^2$ for certain positive smooth function $h_i$ on P, $i = 1, 2, \ldots, \ell$. Now substituting (2.8), (2.9) into $F^* \omega_g$ on $P \cap M_D$, we obtain

$$F^*\omega_g(v_1, \ldots, v_n) = 4 \sum_{i,j \le \ell} b_{i\bar{j}}(F(v_1, \ldots, v_n)) \frac{(1+\eta_i)(1+\eta_j)}{(1-\eta_i)(1-\eta_j)} \frac{1}{(v_i-1)^2(\bar{v}_j-1)^2}$$

$$\cdot \exp\left(\frac{(1+\eta_i)(v_i+1)}{(1-\eta_i)(v_i-1)} + \sqrt{-i}\theta_i + \frac{(1+\eta_j)(\bar{v}_j+1)}{(1-\eta_j)(\bar{v}_j-1)} - \sqrt{-i}\theta_j\right) dv_i \wedge d\bar{v}_j$$

$$+ \sum_{i,j \ge \ell+1} b_{i\bar{j}}(F(v_1, \ldots, v_n))\, dv_i \wedge d\bar{v}_j$$

$$- 4\,\mathrm{Re}\left(\sum_{\substack{i \le \ell \\ j \ge \ell+1}} b_{i\bar{j}}(F(v_1, \ldots, v_n))\right) \frac{1+\eta_i}{1-\eta_i} \frac{1}{(v_i-1)^2} \exp\left(\frac{(1+\eta_i)(v_i+1)}{(1-\eta_i)(v_i-1)} + \sqrt{-i}\theta_i\right) dv_i \wedge d\bar{v}_j$$

$$\sum_{k=1}^{\ell} \frac{\partial(2(1+\eta_k)(|v_k|^2-\delta) + (1-\eta_k)|v_k-\delta|^2 \log h_k) \wedge \bar{\partial}(2(1+\eta_k)(|v_k|^2-\delta) + (1-\eta_k)|v_k-\delta|^2 \log h_k}{(2(1+\eta_k)(|v_k|^2-\delta) + (1-\eta_k)|v_k-\delta|^2 \log h_k)^2}$$

(2.10)

From expression (2.10) and the fact that $\lim_{x\to\infty} x^p \exp(-x) = 0$ for every real number p, we see that in $(\Delta_{\frac{3}{4}})^{\ell} \times \Delta^{n-\ell}$, $F^* \omega_g$ is equivalent to the standard Euclidean metric and there are constants $\mathcal{C}_k$ $(k = 0, 1, \ldots)$ such that

$$|\partial^{|p|+|q|} g_{i\bar{j}} / \partial v^p \partial \bar{v}^q| \le \mathcal{C}_{|p|+|q|} \tag{2.11}$$

for all multi-indices p, q, where $g_{i\bar{j}}$ is the coefficient of $dv_i \wedge d\bar{v}_j$ in (2.10). For each $x \in (\Delta_{\frac{1}{2}})^n$, let $B_{\frac{1}{4}\delta}(x)$ be the Euclidean ball with center at x and radius $\frac{1}{4}\delta$, then $B_{\frac{1}{4}\delta}(x) \subset \Delta_{\frac{3}{4}}^{n-\ell} \times \Delta^{n-\ell}$ and $(B_{\frac{1}{4}\delta}(x), F|_{B_{\frac{1}{4}\delta}(x)})$ is a quasi-coordinate. Since $\bigcup_{\eta,\theta} F_{(\eta,\theta)}(\Delta_{\frac{1}{2}}^n)$ covers $U_{x_o} \cap M_D$, where $U_{x_o}$ is a neighborhood of $x_o$ in X, and $x_o \in D$ is arbitrary, we have proved that $(M_D, g, \{\pi_\ell, \tilde{U}_\ell, \psi_{\ell'\ell}\})$ has bounded geometry of order $\infty$. //

Remark: For convenience, we assume from now on $\varepsilon = 1$.

Lemma 2.4: Let g be the Kähler metric on $M_D \setminus C$ defined in Lemma 2.1, then

$$\log\left\{\frac{\omega_g^n \left(\prod_{i=1}^{L} \|S_i\|_{D_i}^2 (-\log \|S_i\|_{D_i}^2)^2\right) \cdot \left(\prod_{j=1}^{N} \|t_j\|_{C_j}^{\frac{2m_j-2}{m_j}}\right)}{V_X}\right\}$$

is a bounded smooth function on $M_D \setminus C$, compatible with the C-structure $\{\pi_\ell, \tilde{U}_\ell, \psi_{\ell'\ell}\}$.

Proof: It is easy to see that this function $\varphi$ is compatible with $\{\pi_\ell, \tilde{U}_\ell, \psi_{\ell'\ell}\}$. So it suffices to prove the boundedness of $\varphi$ at the infinity $D$ of $M_D$. For simplicity, we only consider the boundedness of $\varphi$ in a neighborhood of a point $x_o \in D$ C. Let $(\Delta^n_{\frac{1}{2}}, F_{(\eta,\theta)})$ be given in the proof of Lemma 2.1, and $(v_1,\ldots,v_n)$ be the coordinate in $\Delta^n_{\frac{1}{2}}$, $(Z_1,\ldots,Z_n) = F_{(\eta,\theta)}(v_1,\ldots,v_n)$. Then

$$F^*\varphi = \log\det(g_{i\bar{j}}) + \log\left(\frac{dv_1\wedge\cdots\wedge dv_n\wedge d\bar{v}_1\wedge\cdots\wedge d\bar{v}_n}{V_X}\prod_{i=1}^{\ell}\|S_i\|^2_{D_i}(-\log\|S_i\|^2_{D_i})^2\right)\circ F$$

$$+ F^*\log\left(\prod_{i\geq \ell+1}\|S_i\|^2_{D_i}(-\log\|S_i\|^2_{D_i})^2 \cdot \prod_{j=1}^{N}\|t_j\|^{\frac{2m_j-2}{m_j}}\right) = \varphi_1+\varphi_2+\varphi_3 .$$

Obviously, $\varphi_3$ is bounded at $x_o$. By (2.11), $\varphi_1$ is bounded in $\Delta^n_{\frac{1}{2}}$ independent of $(\eta,\theta)$. Write $V_X = f\, dZ_1\wedge\cdots\wedge dZ_n\wedge d\bar{Z}_1\wedge\cdots\wedge d\bar{Z}_n$ where $f$ is a positive function near $x_o$. Then by (2.8),

$$\frac{dv_1\wedge dv_2\wedge\cdots\wedge dv_n\wedge d\bar{v}_p\wedge\cdots\wedge d\bar{v}_n}{f\,dZ_1\wedge\cdots\wedge dZ_n\wedge d\bar{v}_1\wedge\cdots\wedge d\bar{v}_n} = \frac{1}{f}\prod_{i=1}^{\ell}\frac{(1-\eta_i)^2|v_i-\delta|^4}{4(1+\eta_i)^2}$$

$$\cdot \exp\left(-2\,\mathrm{Re}\left(\frac{1+\eta_i}{1-\eta_i}\,\frac{v_i+\delta}{v_i-\delta} + \sqrt{+}\,\theta_i\right)\right)$$

$$= \frac{1}{f}\prod_{i=1}^{\ell}\frac{(1-\eta_i)^2\,|v_i-\delta_i|^2}{4(1+\eta_i)^2}\exp\left(-2\,\frac{(1+\eta_i)(|v_i|^2-\delta^2)}{(1-\eta_i)|v_i-\delta|^2}\right) \tag{2.12}$$

By (2.9),

$$\prod_{i=1}^{\ell} \|s_i\|^2_{D_i}(-\log \|s_i\|^2_{D_i})^2 = \prod_{i=1}^{\ell} h_i \exp\left(2\frac{(1+\eta_i)(|v_i|^2-\delta)}{1-\eta_i}\frac{}{|v_i-\delta|^2}\right)$$

$$\cdot\left(-\log h_i - 2\frac{(1+\eta_i)(|v_i|^2-\delta^2)}{(1-\eta_i)\,|v_i-\delta|^2}\right)^2 \tag{2.13}$$

Combining (2.12) and (2.13),

$$\varphi_2 = \frac{1}{4^\ell f}\prod_{i=1}^{\ell} h_i\left(-\frac{(1-\eta_i)}{(1+\eta_i)}|v_i-\delta|^2 \log h_i - 2(|v_i|^2-\delta^2)\right)$$

is bounded in $\Delta^n_{\frac{1}{2}}$ independent of $(\eta,\theta)$. The lemma is proved.

Lemma 2.5:

$$1 = -\log\left(\frac{\omega_g^n \prod_{i=1}^{L}\|s_i\|_{D_i}^{2\mu_i}\left(-\log\|s_i\|_2^2\right)^2 \cdot \prod_{j=1}^{N}\|t_j\|_{C_j}^{\frac{2m_j-2}{m_j}} m^{n+1}\left(1-\|t_j\|_{C_j}^{\frac{1}{m_j}}\right)^{n+1}}{V_X}\right).$$

then $\mathrm{Ric}(g)+\omega_g = \partial\bar\partial f$ and $f$ is a smooth function on $M_D\setminus C$, compatible with $\{\pi_\ell, \tilde U_\ell, \psi_{\ell'\ell}\}$ and bounded from above.

Proof: The first statement follows from the definitions of $g$ and $\omega_g$. The proof for the second statement is the same as that for Lemma 2.4. //

Remark: By adding a constant, we may assume that $f$ is non-positive.

Lemma 2.6: If the line bundle $E = K_X + \sum_{i=1}^{L} D_i + \sum_{j=1}^{N} \frac{m_j - 1}{m_j} C_j$ is numerical effective, i.e. $E \cdot S \geq 0$ for every curve $S \subset X$, then for each $\varepsilon > 0$, there exists a bounded, compatible smooth function $f_\varepsilon$ on $M_D \setminus C$ such that $\mathrm{Ric}(g) + \partial\bar{\partial} f_\varepsilon \leq \varepsilon \omega_g$ in $M_D \setminus C \cup K_\varepsilon$, where $K_\varepsilon$ is a compact subset of $M_D$.

Proof: Let $H$ be an ample line bundle on $X$, then for any $\delta > 0$, $E + \delta H$ is ample on $X$. Let $C_1(H)$ be a positive (1,1)-form on $X$, representing the first Chern class of $H$. Then $C_1(H)$ is dominated by $\omega_g$ on $M_D \setminus C$. By Lemma 2.4, it suffices to find a bounded smooth function $f'_\varepsilon$ on $X$ such that

$$-\partial\bar{\partial} \log \left( \frac{V_X}{\prod_{i=1}^{L} \|S_i\|^2_{D_i} \left(-\log \|S_i\|^2_{D_i}\right)^2 \prod_{j=1}^{N} \|t_j\|_{C_j}^{\frac{2m_j - 2}{m_j}}} \right) + \partial\bar{\partial} f'_\varepsilon \leq \varepsilon \omega_g$$

in $M_D \setminus C$ near $D$. Since $E + \delta H > 0$, there is a $\tilde{f}_\delta$ on $X$ such that $C_1(E) + \delta C_1(H) + \partial\bar{\partial}\tilde{f}_\delta > 0$ on $X$.

For $1 \leq i_1 < i_2 < \cdots < i_k \leq L$, $k \leq n$, define $D_{i_1 \ldots i_k} = \bigcap_{j=1}^{k} D_{i_j}$, it is a smooth subvariety in $X$ of codimension $k$. Put $D^{[k]} = \bigcup_{i \leq i_1 < \cdots < i_k \leq L} D_{i_1 \ldots i_k}$. We will construct $f'_\varepsilon$ by induction on $k$.

It is clear that in a neighborhood of $D^{[n]}$,

$$-\partial\bar{\partial} \log \left( \frac{V_X}{\prod_{i=1}^{L} \|S_i\|^2_{D_i} \left(-\log \|S_i\|^2_{D_i}\right)^2 \|t_j\|_{C_j}^{\frac{2m_j - 2}{m_j}}} \right) < 0$$

so we take $f^{[n]} \equiv 0$. Suppose now that we have found a bounded smooth $f^{[k+1]}$ on $X$ such that

$$-\partial\bar{\partial}\log\left(\frac{V_X}{\prod_{i=1}^{L}\|S_i\|_{D_i}^2\left(-\log\|S_i\|_{D_i}^2\right)^2\prod_{j=1}^{N}\|t_j\|_{C_j}^{\frac{2m_j-2}{m_j}}}\right)$$

$$+\partial\bar{\partial}f^{[k+1]} \le \frac{(n-k-1)}{n}\,\varepsilon\,\omega_g \qquad (2.14)$$

in a neighborhood $V_{k+1}$ of $D^{[k+1]}$. We need to find a bounded smooth function $f^{[k]}$ on $X$ satisfying (2.14) with $k+1$ replaced by $k$ in a neighborhood $V_k$ of $D^{[k]}$.

Introduce two cut-off functions $p_\lambda(t)$, $q_\lambda(t)$: $p_\lambda(t) \equiv 1$ for $t \le \lambda$ and $p_\lambda(t) \equiv 0$ for $t \ge 4\lambda$, $|p_\lambda'(t)| \le 4/\lambda$, $|p_\lambda''(t)| \le 4/\lambda^2$; $q_\lambda'(t) = p_\lambda(t)$, $q_\lambda(4\lambda) = 0$. Here, $\lambda$ will be defined later. Define ($\delta$ will be chosen later)

$$f^{[k]}(x) = f^{[k+1]}p_\lambda\left(\sum_{\substack{D_j\cap D_{i_1\ldots i_k}\neq\emptyset\\ j\neq i_1,\ldots,i_k}}\log(-\log\|S_j\|_{D_j}^2)^2(x)\right)\left(-\tilde{f}_\delta(x)\right.$$

$$\left.-\sum_{D_j\cap D_{i_1\ldots i_k}\neq\emptyset}\log(-\log\|S_j\|_{D_j}^2)^2(x)\right)$$

$$-\,q_\lambda\left(\sum_{\substack{D_j\cap D_{i_1\ldots i_k}\neq\emptyset\\ j\neq i_1,\ldots,i_k}}\log(-\log\|S_j\|_{D_j}^2)^2(x)\right)\quad\text{for } x\in D_{i_1\ldots i_k}$$

(2.15)

Then $f^{[k]} \equiv f^{[K+1]}$ on $D^{[k+1]}$, and at $x \in D_{i_1 \ldots i_k}$,

$$-\partial\bar{\partial} \log \left( \frac{V_X}{\prod_{i=1}^{L} \|S_i\|^2_{D_i} (-\log \|S_i\|^2_{D_i})^2 \prod_{j=1}^{N} \|t_j\|_{C_j}^{\frac{2m_j-2}{m_j}}} \right) + \partial\bar{\partial} f^{[k]}$$

$$= (1-p_\lambda) \left( -\partial\bar{\partial} \log \left( \frac{V_X}{\prod_{i=1}^{L} \|S_i\|^2_{D_i} (-\log \|S_i\|^2_{D_i})^2 \prod_{j=1}^{N} \|t_j\|_{C_j}^{\frac{2m_j-2}{m_j}}} \right) + \partial\bar{\partial} f^{[k+1]} \right)$$

$$- p_\lambda \left( \partial\bar{\partial} \log \frac{V_X}{\prod_{i=1}^{L} \|S_i\|^2_{D_i} \prod_{j=1}^{N} \|t_j\|_{C_j}^{\frac{2m_j-2}{m_j}}} + \partial\bar{\partial} \tilde{f}_\delta \right)$$

$$- p_\lambda \, \partial\bar{\partial} \log \left( \frac{1}{\prod_{j=1}^{k} (-\log \|S_{i_j}\|^2_{D_{i_j}})^2} \right) - p_\lambda \, \partial\bar{\partial} \log \left( \frac{1}{\prod_{\substack{D_j \cap D_{i_1 \ldots i_k} \\ j \neq i_1, \ldots, i_k}} (-\log \|S_j\|^2_{D_j})^2} \right)$$

$$- p' \, \partial\bar{\partial} \left( \sum_{\substack{D_j \cap D_{i_1 \ldots i_k} \neq \phi \\ j \neq i_1, \ldots, i_k}} \log(-\log \|S_j\|^2_{D_j})^2 \right) \cdot$$

$$\cdot \left( \tilde{f}_\delta + \sum_{D_j \cap D_{i_1 \ldots i_k} \neq \phi} \log(-\log \|S_j\|^2_{D_j})^2 \right)$$

$$- 2\mathrm{Re} \left( p' \partial \left( \tilde{f}_\delta + \sum_{D_j \cap D_{i_1 \ldots i_k} \neq \phi} \log(-\log \|S_j\|^2_{D_j})^2 \right) \right.$$

$$\left. \wedge \bar{\partial} \left( \sum_{\substack{D_j \cap D_{i_1 \ldots i_k} \neq \phi \\ j \neq i_1, \ldots, i_k}} \log(-\log \|S_j\|^2_{D_j})^2 \right) \right) - p'' \left( f_\delta + \sum_{D_j \cap D_{i_1 \ldots i_k} \neq \phi} \log(-\log \|S_j\|^2_{D_j})^2 \right)$$

$$\cdot\ \partial\left(\sum_{\substack{D_j\cap D_{i_1\cdots i_k}\neq\phi\\ j=i_1,\ldots,i_k}}\log(-\log\|S_j\|^2_{D_j})^2\right)\wedge\bar\partial\left(\sum_{\substack{D_j\cap D_{i_1\cdots i_k}\neq\phi\\ j\neq i_1,\ldots,i_k}}\log(-\log\|S_j\|^2_{D_j})^2\right)$$

$$-\ q_\lambda'\ \partial\bar\partial\left(\sum_{\substack{D_j\cap D_{i_1\cdots i_k}\neq\phi\\ j\neq i_1,\ldots,i_k}}\log(-\log\|S_j\|^2_{D_j})^2)\right)$$

$$-\ q_\lambda''\,\partial\left(\sum_{\substack{D_j\cap D_{i_1\cdots i_k}\neq\phi\\ j\neq i_1,\ldots,i_k}}\log(-\log\|S_j\|^2_{D_j})^2\right)\wedge\bar\partial\left(\sum_{\substack{D_j\cap D_{i_1\cdots i_k}\neq\phi\\ j\neq i_1,\ldots,i_k}}\log(-\log\|S_j\|^2_{D_j})^2\right)$$

$$\leq\ (1-p_\lambda)\,\frac{n-k-1}{n}\,\varepsilon\omega_g+\delta p_\lambda C_1(H)+p_\lambda\,\partial\bar\partial\left(\sum_{j=1}^{k}\log(-\log\|S_{i_j}\|^2_{D_{i_j}})^2\right)$$

$$+\ \frac{C}{\lambda}\,\omega_g\quad\text{for certain constant}\quad C>0\,. \tag{2.16}$$

Here we have used the facts that $\partial\bar\partial\left(\sum_{\substack{D_j\cap D_{i_1\cdots i_k}\neq\phi\\ j\neq i_1,\ldots,i_k}}\log(-\log\|S_j\|^2_{D_j})^2\right)$ and

$$\partial\left(\sum_{\substack{D_j\cap D_{i_1\cdots i_k}\neq\phi\\ j\neq i_1,\ldots,i_k}}\log(-\log\|S_j\|^2_{D_j})^2\right)\wedge\bar\partial\left(\sum_{\substack{D_j\cap D_{i_1\cdots i_k}\neq\phi\\ j\neq i_1,\ldots,i_k}}\log(-\log\|S_j\|^2_{D_j})^2\right)$$

are dominated by $\omega_g$.

It follows from (2.16) that for $\lambda$ sufficiently large, $\delta$ sufficiently small,

$$-\partial\bar{\partial}\log\left(\frac{V_X}{\prod_{i=1}^{L}\|S_i\|_{D_i}^2(-\log\|S_i\|_{D_i}^2)^2\prod_{j=1}^{N}\|t_j\|_{C_j}^{\frac{2m_j-2}{m_j}}}\right)+\partial\bar{\partial}f^{[k]}\le\frac{n-k}{n}\epsilon\,\omega_g$$

in a neighborhood $V_k$ of $D^{[k]}$. By induction, we take $f'_\epsilon = f^{[1]}$ and we have proved the lemma.

We recall that a numerical effective divisor $E$ on $X$ is said to be big if $E^n > 0$ and $E$ is said to be ample modulo $D$ if for every effective reduced curve $S$ on $X$ not contained in $D$, $E\cdot S > 0$ holds.

Lemma 2.7. Let

$$E = K_X + \sum_{i=1}^{L}\mu_i D_i + \sum_{j=1}^{N}\frac{m_j-1}{m_j}C_j,\quad \mu_i \in (0,1] \text{ and } F = E + \sum_{i=1}^{L}(1-\mu_i)D_i\,.$$

Then the following are equivalent,

(i) $E > 0$ on $X$ and $F$ is nef. (i.e. numbered effective),

(ii) $F$ is nef., big and ample modulo $D$.

Proof: (i) $\Longrightarrow$ (ii). It suffices to show that $F$ is big. Assume that $\mu_i = p_i/q_i$ are rational numbers. Let $q = q_1\ldots q_L$, then

$$h^o(X,qkF) = h^o\ X,qkE + kq\sum_{i=1}^{L}\frac{q_i-p_i}{q_i}D_i\ \ \ge h^o(X,qkE)$$

since

$$kq \sum_{i=1}^{L} \frac{q_i - p_i}{q_i} D_i$$

is effective. Hence

$$\lim_{k \to +\infty} \left( \frac{h^o(X, qkF)}{k^n} \right) > 0 \quad .$$

By Riemann-Roch, $F^n > 0$.

(ii) $\Longrightarrow$ (i). It follows from a result of Kawamata [Ka]. //

Remark: By [Ka], the above (ii) is equivalent to that for a sufficiently large $\nu$, $\phi_\nu$: $X \to CP^m$ $(m > n)$ is a well-defined morphism and is a biholomorphism from $X \setminus D$ onto its image, where $\phi_\nu$ is the associated morphism with linear system $H^o(X, F)$.

Theorem 2.1: Let

$$K_X + \sum_{i=1}^{L} D_i + \sum_{j=1}^{N} \frac{m_j - 1}{m_j} C_i$$

be nef., big and ample modulo D. Then there exists a unique almost-complete Kähler-Einstein metric on $X \setminus D \cup C$, compatible with the natural C-structure (defined at the beginning of this section).

Proof: This follows from Lemma 2.7, Lemma 2.1, 2.6 and Theorem 1.1. //

An important case is that $C = \phi$. In this case, we have

<u>Corollary 2.1</u>: Let $K_X + \sum_{i=1}^{L} D_i$ be nef., big and ample modulo D. Then $M_D = X \setminus D$ admits a unique almost-complete Kähler-Einstein metric.

<u>Remarks</u>: The case $n=2$ was studied by S. Y. Cheng and S. T. Yau [CY] (see also R. Kobayashi [Ko1], [Ko2]). A special case of Corollary 2.1 is also considered by Tsuji [Ts2]. He uses local branch coverings to construct a Kähler-Einstein metric on $X \setminus D$ in case D is smooth and $K_X + D|_D$ is trivial. His proof for the completeness of the metric apparently is incomplete.

By the same proof and the remark at the end of section 1, we can prove the following generalization of Theorem 2.1. For simplicity of the description, we suppose that $C = \phi$. Recall that a V-manifold is a variety covered by a family of quotients of balls by finite groups.

<u>Theorem 2.2</u>: Let X be a normal V-manifold, and $X_r$ be its regular part, D be a compact divisor in $X_r$ with normal crossings. Suppose that $K_X + D$ be nef., big and ample modulo D. Then there exists a unique almost-complete Kähler-Einstein metric on $X_r \setminus D$ such that near each singular point $x_o$ of X, the pull-back of the metric to the local uniformization of $x_o$ is extendable to the whole ball.

## §3. A Numerical Characterization of Ball Quotients

As in the last section, X denotes a projective manifold, $D, C \subset X$ divisors such that $D + C$ has normal crossings and

$$K_X + \sum_{i=1}^{L} D_i + \sum_{j=1}^{N} \frac{m_j - 1}{m_j} C_j$$

is nef., big and ample modulo D, where $D_i$, $C_i$ are irreducible components of D and C, respectively, $m_j$ integers. In 1976, the second author proved the following by using Kähler-Einstein metrics he constructed on X with $K_X$ ample

$$(-1)^n c_1^n(X) \le (-1)^n \frac{2(n+1)}{n} c_1^{n-2}(X)\, c_2(X) \tag{3.1}$$

and the equality holds iff X is a compact unramified quotient of the unit ball in $C^n$. Later, S.Y. Cheng and S.T. Yau [CY2] proved a generalized inequality of (3.1) on a complex V-surface, i.e., a 2-dimensional variety with some isolated quotient singularities and if the equality holds, then the surface is a ball quotient. In particular, they proved the inequality for algebraic surfaces of general type. This was also proved by [Mi2] by algebraic means. Based on the arguments of [CY2], [Ko1] and [Ko2] also proved some special cases. In case $C = \phi$, there is the logarithmic version of (3.1) on $X \setminus D$,

$$c_1^n(\Omega_X'(\log D)) \le \frac{2(n+1)}{n} c^{n-2}(\Omega_X(\log D))\, c_2(\Omega_X'(\log D)) \tag{3.2}$$

This was first studied by Cheng-Yau and more recently by Tsuji.

In this section, we will prove a generalized inequality of (3.2) and give a numberical characterization of ball quotients. First we consider the special case $C = \phi$ and we will prove by using our Kähler-Einstein metric in the last section that the equality in (3.2) holds iff $X \setminus D$ is the quotient of the unit ball in $\mathbb{C}^n$.

Lemma 3.1: Let $\omega_E$ be the Kähler form of the almost-complete Kähler-Einstein metric on $M_D = X \setminus D$ constructed in Theorem 2.1. Then

$$\int_{M_D} \omega_E^n \le (K_X + D)^n \tag{3.3}$$

where

$$(K_X + D)^n = \int_X C_1(K_X + D)^n .$$

Proof: By the proof of Theorem 1.1 and 2.1, $\omega_E$ is the pointwise $C^\infty$-limit of a sequence of Kähler forms $\omega$

$$\omega_\nu = \frac{1}{\nu}\omega_g - \frac{\nu-1}{\nu}\mathrm{Ric}(g) + \partial\bar\partial\psi_\nu \quad \text{with} \quad \int_{M_D} \omega_\nu^n < +\infty ,$$

where $\psi_\nu$ are bounded functions on $M_D$, $g$ is the Kähler metric constructed in Lemma 2.1. Note that all Kähler metrics $g_\nu$ associated to $\omega_\nu$ are complete and $\mathrm{Ric}(g_\nu) \ge -\frac{\nu}{\nu-1}$.

By Lemma 2.4, we may write

$$\omega_\nu = \frac{1}{\nu}\left(C_1(K_X) + \sum_{i=1}^{L} \mu_i C_1(D_i)\right) + \frac{\nu-1}{\nu}\left(C_1(K_X) + \sum_{i=1}^{L} C_1(D_i)\right)$$

$$+ \partial\bar\partial\left(-\sum_{i=1}^{L} \log(-\log \|S_i\|_{D_i}^2)^2 + f_\nu\right) \tag{3.4}$$

where $f_i$ are bounded, $\|\cdot\|_{D_i}$ is the Hermitian metric of $D_i$, $\{S_i = 0\} = D_i$ and the Chern forms are computed in terms of $\|\cdot\|_{D_i}$ and the volume form $V_X$ on X.

First we evaluate each integral $\int_{M_D} \omega_\nu^n$. Let $p_\lambda$ be the cut-off function used in the proof of Lemma 2.6.

$$p_\lambda(t) \equiv 1 \text{ for } t \le \lambda;\ p_\lambda(t) \equiv 0 \text{ for } t \ge 4\lambda;\ |p'_\lambda| \le \frac{4}{\lambda};\ |p''_\lambda|^2 \le \frac{4}{\lambda^2} .$$

Put

$$\omega_{\nu,\lambda} = C_1(K_X) + \sum_{i=1}^{L} \left(\frac{\nu-1}{\nu} + \frac{\mu_i}{\nu}\right) C_1(D_i)$$

$$\partial\bar{\partial}\left(p_\lambda\left(\sum_{i=1}^{L} \log(-\log \|S_i\|^2_{D_i})^2\right) \cdot \left(-\sum_{i=1}^{L} \log(-\log \|S_i\|^2_{D_i})^2 + f_\nu\right)\right),$$

then $\omega_{\nu,\lambda}(x) = \omega_\nu(x)$ for those $x$ with

$$\sum_{i=1}^{L} \log(-\log \|S_i\|^2_{D_i}(x))^2 \le \lambda .$$

Obviously,

$$\int_{M_D} \omega^n_{\nu,\lambda} = \left(K + \sum_{i=1}^{n} \left(\frac{\nu-1}{\nu} + \frac{\mu_i}{\nu}\right) D_i\right)^n \tag{3.5}$$

$$\omega_\nu - \omega_{\nu,\lambda} = (1-p_\lambda)\,\partial\bar{\partial}(-\log(-\log\|S_i\|^2_{D_i})^2 + f_\nu) - 2\mathrm{Re}\ p'_\lambda\,\partial\left(\sum_{i=1}^{L}\log(-\log\|S_i\|^2_{D_i})^2\right)$$

$$\wedge\,\bar{\partial}(-\log(-\log\|S_i\|^2_{D_i})^2 + f_\nu) - p''_\lambda \cdot (-\log(-\log\|S_i\|^2_{D_i})^2 + f_\nu)\,\partial\left(\sum_{i=1}^{L}\log(-\log\|S_i\|^2_{D_i})^2\right)$$

$$\wedge\,\bar{\partial}(\log(-\log\|S_i\|^2_{D_i})^2) - p'_\lambda \cdot (-\log(-\log\|S_i\|^2_{D_i})^2 + f_\nu)\,\partial\bar{\partial}\left(\sum_{i=1}^{L}\log(-\log\|S_i\|^2_{D_i})^2\right) .$$

Hence $\omega_\nu - \omega_{\nu,\lambda}$ is dominated by $(1-p_\lambda)\omega_\nu + \frac{1}{\lambda}\omega_\nu + \lambda p'_\lambda \omega_\lambda$, since $\omega_\nu$ dominates both forms $\partial\bar\partial h$ and $\partial h \wedge \bar\partial h$ for $h = \sum_{i=1}^{L} \log(-\log \|S_i\|^2_{D_i})^2$. Thus, it follows from (3.5),

$$\int_{M_D} \omega_\nu^n = \left(K + \sum_{i=1}^{n} \left(\frac{\nu-1}{\nu} + \frac{\mu_i}{\nu}\right) D_i\right)^n + O\left(\int_{M_D} (1-p_\lambda)\omega_\nu^n + \frac{1}{\lambda}\omega_\nu^n + \lambda p'_\lambda \omega_\nu^n\right)$$

$$= \left(K + \sum_{i=1}^{n} \left(\frac{\nu-1}{\nu} + \frac{\mu_i}{\nu}\right) D_i\right)^n + O\left(\int_{\sum_{i=1}^{L} \log(-\log \|S_i\|^2_{D_i})^2 > \lambda} \omega_\nu^n + \frac{1}{\lambda}\int_{M_D} \omega_\nu^n\right)$$

Since $\int_{M_D} \omega_\nu^n < +\infty$, letting $\lambda$ go to infinity, we have

$$\int_{M_D} \omega_\nu^n = \left(K + \sum_{i=1}^{n} \left(\frac{\nu-1}{\nu} + \frac{\mu_i}{\nu}\right) D_i\right)^n \tag{3.6}$$

By Fatou's lemma, $\int_{M_D} \omega_E^n \le (K+D)^n$. //

Remark: The above proof also shows that for any effective curve $S \subset X$, $\int_{S \cap M_D} \omega_E \le (K+D)\cdot S$, i.e. as a closed current, $C_1(K+D) - \omega_E$ is numerically positive.

Recall from a result of Kawamata [Ka] that for each positive integer m, $K + \frac{m-1}{m} D$ is ample. Hence by Lemma 2.1, for proper Hermitian metrics $\|\cdot\|_{D_i}$ for $[D_i]$, there is a Kähler metric $g_m$ with its Kahler form

$$\omega_{g_m} = \partial\bar{\partial} \log\left(\frac{V_X}{\prod_{i=1}^{L} \|S_i\|_{D_i}^{\frac{2m-2}{m}} m^{n+1}(1 - \|S_i\|_{D_i}^{2/m})^{n+1}}\right)$$

compatible with a D-structure $\{\pi_\ell^m, \tilde{U}_\ell^m, \psi_{\ell'\ell}^m\}$ defined at the beginning of section 2 with $C_j$ replaced by $D_i$ and $m_j$ by $m$. Now applying Theorem 1.1 to $(M_D, g_m, \{\pi_\ell^m, \tilde{U}_\ell^m, \psi_{\ell'\ell}^m\})$, we obtain a Kähler-Einstein metric $\omega_{m,E}$ on $M_D$, compatible with the above D-structure.

Lemma 3.2:

$$\int_{M_D} \omega_{m,E}^n = \left(K + \frac{m-1}{m} D\right)^n \tag{3.7}$$

Proof: $\omega_{m,E} - \omega_{g_m} = \partial\bar{\partial}\psi_m$ for a bounded smooth function $\psi_m$, compatible with the D-structure. Hence,

$$\omega_{m,E} = C_1(K_X) + \frac{m-1}{m} C_1(D) + \partial\bar{\partial}\left(\psi_m - \sum_{i=1}^{L} \log(1 - \|S_i\|_{D_i}^{2/m})^{n+1}\right)$$

$$= C_1(K_X) + \frac{m-1}{m} C_1(D) + \partial\bar{\partial} f_m$$

where $f_m$ is obviously bounded and compatible with $\{\pi_\ell^m, \tilde{U}_\ell^m, \psi_{\ell'\ell}^m\}$. Then by using partitions of unity, one can easily deduce

$$\int_{M_D} \omega_{m,E}^n = \int_X \left(C_1(K_X) + \frac{m-1}{m} C_1(D)\right)^n = \left(C_1(K_X) + \frac{m-1}{m} D_1(D)\right)^n$$

This proves (3.7).

Proposition 3.1: The sequence of the above Kähler-Einstein metrics $\omega_{m,E}$ converges to $\omega_E$, the almost-complete Kähler-Einstein metric constructed Theorem 2.1. In particular, $\omega^n_{m,E} \le \omega^n_E$ on $M_D = X \setminus D$.

Proof: By Schwarz's lemma (Lemma 1.2) and the almost-completeness of $\omega_E$,

$$\omega^n_{m,E} \le \omega^n_E \quad \text{on } M_D \tag{3.8}$$

By the definitions of $g_m$, $g_{m'}$, $(\omega^n_{g_m}/\omega^n_{g_{m'}})(x) \to +\infty$ as $x$ goes to D. Then it follows that $(\omega^n_{m,E}/\omega^n_{m',E})(x) \to +\infty$ as $x \to D$ and Lemma 1.2 can be applied to conclude that

$$\omega^n_{m',E} \le \omega^n_{m,E} \quad \text{on } M_D \text{ for } m' < m \tag{3.9}$$

Let $\varphi_m = \log(\omega^n_{m,E}/\omega^n_{p,E})$. Then when $p \gg 0$, $\varphi_m \ge 0$ and

$$(\omega_{p,E} + \partial\bar{\partial}\varphi_m)^n = e^{\varphi_m}\omega^n_{p,E} \tag{3.10}$$

Based on the same proof of Lemma 1.5, it can be shown that there is a constant $C > 0$ such that

$$\omega_{m,E} \le C\, e^{C\varphi_m}\omega^n_{p,E} \tag{3.11}$$

By (3.8), $\varphi_m \le \log(\omega^n_E/\omega^n_{p,E})$ for $m \ge p$. Hence it follows that $\{\omega_{m,E}\}$ converge to a Kähler-Einstein metric $\omega'_E$ on $M_D$, and

$$(\omega'_E)^n \le \omega^n_E \tag{3.12}$$

From (3.9) and Lemma 3.2,

$$(K_X + D)^n = \lim \int_{M_D} \omega^n_{m,E} = \int_{M_D} (\omega'_E)^n \leq \int_{M_D} \omega^n_E \qquad (3.13)$$

So Lemma 3.1 implies that $(\omega'_E)^n \equiv \omega^n_E$. Hence $\omega'_E \equiv \omega_E$ as they are Kähler-Einstein. //

Let us now prove (3.2) (see also [Ts2]). Let $C_{i,m}$ be the i-th Chern form of the singular Hermitian metric on the cotangent bundle $\Omega_X$ defined by the Kähler-Einstein form $\omega_{m,E}$. For every partition $I = (i_1, \ldots, i_k)$ of n, one can check that integral

$$C_{I,m} = \int_X C_{i_1,m}, \ldots, C_{i_k,m} = \int_{M_D} C_{i_1,m}, \ldots, C_{i_k,m} \qquad (3.14)$$

exists. Let $V_m$ be the multivalued vector bundle on X defined as follows. Choose a coordinate covering $\{\Delta_\alpha (Z'_\alpha, \ldots, Z_\alpha)\}$ of X such that either $\Delta_\alpha \cap D = \emptyset$ or $\Delta_\alpha \cap D_{i_1} \cap \cdots \cap D_{i_k} \neq \emptyset$, $\Delta_\emptyset \cap D_i = \emptyset$ for $i \neq i_j$, and $\{Z^i_\alpha = 0\}$ locally defines $D_i$ for $i = 1, 2, \ldots, k$. Then define

$$V_m|\Delta_\alpha = \begin{cases} O_X(dZ'_\alpha, \ldots, dZ^n_\alpha) & \text{in the first case} \\ O_X\left((Z^1_\alpha)^{-\frac{m-1}{m}} dZ^1_\alpha, \ldots, (Z^k_\alpha)^{-\frac{m-1}{m}} dZ^k_\alpha, dZ^{k+1}_\alpha, \ldots, dZ^n_\alpha\right) & \text{in the second case} \end{cases}$$

If we define a branched covering $\pi_{\alpha,m}: \tilde{\Delta}_\alpha \subset C^n \to \Delta_\alpha$ in the second case by $\pi_\alpha(\omega'_\alpha, \ldots, \omega^n_\alpha) = ((Z'_\alpha)^m, \ldots, (Z^k_\alpha)^m, \ldots, Z^n_\alpha)$, then $\pi^*_{\alpha,m} V_m$ is nothing but the cotangent bundle of $\tilde{\Delta}_\alpha$.

We have the virtual exact sequence

$$o \to V_m \to \Omega'_X(\log D) \to \oplus_i O_{\frac{1}{m}D_i} \to o \tag{3.15}$$

One can compute the number $C_{I,m}$ in terms of (3.15). In particular,

$$C_{(1,\ldots,1),m} = (C_1(\Omega_X(\log D)) - \frac{1}{m} C_1(D))^n \tag{3.16}$$

$$C_{(1,\ldots,1,2),m} = (C_1(\Omega'_X(\log D)) - \frac{1}{m} C_1(D))^{n-1}(C_2(\Omega'_X(\log D)) - C_2 \oplus_i O_{\frac{1}{m}D_i})$$

$$- (C_1(\Omega'_X(\log D)) - \frac{1}{m} C_1(D)) \cdot \frac{1}{m} C_1(D)) \tag{3.17}$$

Hence,

$$\frac{2(n+1)}{n} \int_X C^{n-2}_{1,m} C_{2,m} - \int_X C^n_{1,m} \to \frac{2(n+1)}{n} C_1^{n-2}(\Omega'_X(\log D))\, C_2(\Omega'_X(\log D))$$

$$- C_1^n(\Omega'_X(\log D))$$

as $m$ goes to infinity. Since $\omega_{m,E}$ is Kähler-Einstein, a computation as in [Ya1] shows

$$\frac{2(n+1)}{n} \int_X C^{n-2}_{1,m} C_{2,m} - \int_X C^n_{1,m} = C(n) \int_{M_D} \|T_m\|^2\, \omega^n_{m,E} \tag{3.18}$$

where $T_m$ is the tensor which measures the deviation from being constant holomorphic sectional curvature and $C(n)$ is a constant depending only on n. Note $T_m \equiv 0$ implies that $\omega_{m,E}$ is locally isometric to the standard complete metric on the unit ball in $C^n$. Now let m go to infinity, it follows from (3.18) and Fatou's lemma,

$$\frac{2(n+1)}{n} C_1^{n-2}(\Omega'_X(\log D))\, C_2(\Omega'_X(\log D)) - C_1^n(\Omega'_X(\log D)) \geq C(n) \int_{M_D} \|T\|^2 \omega_E^n \tag{3.19}$$

Therefore, the equality in (3.2) holds iff $T \equiv 0$, i.e. $(M_D, g_E)$ is locally isometric to the open unit ball in $C^n$.

Theorem 3.1. Let X be a projective manifold, $D \subset X$ a divisor with normal crossings. Suppose that $K_X + D$ is nef., big and ample modulo D. Then (3.2) is true and the equality holds iff $M_D$ is a unramified quotient of the unit ball in $C^n$.

Proof: It suffices to prove that if the equality holds in (3.2), then $M_D$ is a unramified quotient of the unit ball $B^n$ in $C^n$, i.e. the universal covering $\tilde{M}_D$ of $M_D$ is just $B^n$. Let $\tilde{\omega}_E$ be the pull-back to $\tilde{M}_D$ of the Kähler-Einstein metric $\omega_E$. Then that the equality holds in (3.2) implies that $\tilde{\omega}_E$ has constant Ricci curvature and constant holomorphic sectional curvature. There is a developing map $\phi: \tilde{M}_D \to B^n$. Such a $\phi$ is locally immersive. We need to prove that $\phi$ is isomorphic. We will use a trick in [T]. Take an arbitrary totally geodesic complex curve $S \subset B^n$, $\phi^{-1}(S)$ is a totally geodesic complex curve in $\tilde{M}_D$. Since such curves spread over $B^n$, it suffices to prove that for each $S$, $\phi\Big|_{\phi^{-1}(S)} : \phi^{-1}(S) \to S$ is isomorphic.

By the construction of $\omega_E$ in Theorem 2.1 there is a family of smooth functions $\{\varphi_t\}$, $\lim t = 1$, such that $\omega_t = \omega_{g_t} = \omega_g + \partial\bar{\partial}\varphi_t$ for a complete Kähler metric on $M_D$ and

$$\omega_t^n = (\omega_g + \partial\bar{\partial}\varphi_t)^n = e^{f+t\varphi_t}\omega_g^n \tag{3.20}$$

where $g$ is the Kähler metric in Lemma 2.1, $\partial\bar{\partial}f = \mathrm{Ric}(g) + \omega_g$, and $\omega_E$ is the $C^\infty$-limit of $\omega_g + \partial\bar{\partial}\varphi_t$ as $t$ goes to 1. Rewrite (3.20) in terms of $\omega_E$,

$$\omega_t^n = (\omega_E + \partial\bar{\partial}(\varphi_t - \varphi))^n = e^{t\varphi_t - \varphi}\omega_E^n \tag{3.21}$$

where $\varphi = \lim \varphi_t$. Then

$$\tilde{\omega}_t^n = (\tilde{\omega}_E + \partial\bar{\partial}(\tilde{\varphi}_t - \tilde{\varphi}))^n = e^{t\tilde{\varphi}_t - \tilde{\varphi}}\tilde{\omega}_E^n \tag{3.22}$$

where $\tilde{\varphi}_t$, $\tilde{\varphi}$ are pull-backs of $\varphi_t$, $\varphi$ to $\tilde{M}_D$. Since $\tilde{\omega}_E$ has constant Ricci curvature and constant holomorphic sectional curvature and $\mathrm{Ric}(g_t) \geq -t$, it follows from Schwarz lemmas on volume forms (Lemma 1.2) and on distances [Y] that there is a constant $\tilde{C}_1 > 0$, independent of $t$, such that

$$t\tilde{\varphi}_t - \tilde{\varphi}_t \geq -\tilde{C}_1 \tag{3.23}$$

$$\tilde{\omega}_E \leq \tilde{C}_1\tilde{\omega}_t \tag{3.24}$$

By (3.22), (3.24), there is a constant $\tilde{C}_2$ such that

$$\tilde{\omega}_t \leq \tilde{C}_2 e^{t\tilde{\varphi}_t - \tilde{\varphi}} \omega_E \tag{3.25}$$

Hence we have a family of Kähler metrics $g_{t,S} = e^{t\tilde{\varphi}_t - \tilde{\varphi}} \omega_E$ on $\phi^{-1}(S)$. Each $g_{t,S}$ is complete as $\tilde{\omega}_t$ is complete. A computation shows that the Gaussian curvature $K_{t,S}$ of $g_{t,S}$ is $e^{-(t\tilde{\varphi}_t - \tilde{\varphi})}(\Delta_E(-t\tilde{\varphi}_t + \tilde{\varphi}) - 1)$, where the Laplacian $\Delta_E$ is taken with respect to the restriction of $\tilde{\omega}_E$ to $\phi^{-1}(S)$.

$$K_{t,S} = -te^{-(t\tilde{\varphi}_t - \tilde{\varphi})} \Delta_E(\tilde{\varphi}_t - \varphi) - (t-1)e^{-t\tilde{\varphi}_t + \tilde{\varphi}} \Delta_E\varphi - e^{-t\tilde{\varphi}_t + \tilde{\varphi}} \tag{3.26}$$

As $\omega_E\big|_{\phi^{-1}(S)} = \omega_g\big|_{\phi^{-1}(S)} + \partial\bar{\partial}\varphi$, $\Delta_E\varphi \leq 1$, i.e. $\Delta_E\varphi \geq -1$. Hence by (3.23), (3.25), there is a constant $C_3$, independent of $t$, such that

$$K_{t,S} \geq -\tilde{C}_3 \tag{3.27}$$

By uniformization theory of Riemann surfaces, $\phi^{-1}(S)$ is covered by the unit disc in $C^1$ and $\phi^{-1}(S)$ has a complete Kähler metric $\tilde{\omega}_{E,S}$ with constant Gaussian curvature $-1$. By (3.27) and Schwarz's lemma (Lemma 1.2), $\tilde{\omega}_{E,S}$ is uniformly dominated by $g_{t,S}$. As $t$ goes to 1, $g_{t,S}$ converges to $\tilde{g}_E$ on $\phi^{-1}(S)$ and $\tilde{\omega}_{E,S}$ is dominated by $\tilde{\omega}_E\big|_{\phi^{-1}(S)}$. It follows that $\tilde{\omega}_E\big|_{\phi^{-1}(S)}$ is complete and therefore $\phi\big|_{\phi^{-1}(S)} : \phi^{-1}(S) \to S$ must be isomorphic. //

The above proof actually shows that $\omega_E$ is complete on $M_D$. We can apply the argument in the proof to the general case, e.g. we can show the following proposition.

Proposition 3.1: If the Kähler-Einstein metric $\omega_E$, constructed in Theorem 1.1, has negative holomorphic sectional curvature, then $\omega_E$ is complete.

We shall discuss the completeness of $\omega_E$ in Theorem 1.1 in a future paper.

Now let us generalize the inequality (3.2) and Theorem 3.1 to the case $C \neq \phi$. Since the proof is the same as before, we just give the result without proof.

Theorem 3.2: Let $X$ be a projective manifold, $D, C \subset X$ divisors and $D + C$ have normal crossings. Suppose that $K_X + D + \sum_{j=1}^{N} \frac{m_j - 1}{m_u} C_j$ is nef., big and ample modulo $D$, where $C_j$ are irreducible components of $C$. Then

$$\left(c_1(\Omega'_X(\log D + C)) - \sum_{j=1}^{N} \frac{1}{m_j} c_1(C_j)\right)^n \le \frac{2(n+1)}{n} \left(c_1(\Omega'_X(\log D + C)) - \sum_{j=1}^{N} \frac{1}{m_j} c_1(C_j)\right)^{n-2} (c_2(\Omega'_X(\log D + C)) - c_1(\Omega'_X(\log(D + C))) \cdot \sum_{j=1}^{N} \frac{1}{m_j} c_1(C_j) + \sum_{j=1}^{N} \frac{1}{m_j^2} c_1^2(C_j)) \qquad (3.28)$$

and the equality holds iff $X \setminus D = M_D$ is a branched quotient of the unit ball in $C^n$ with branches along $C$.

References

[CY1] Cheng, S. Y. and Yau, S. T., On the existence of a complete Kähler metric on non-compact complex manifolds and the regularity of Fifferman's equation, Comm. Pure Appl. Math. 33 (1980) 507-544.

[CY2] Cheng, S. Y. and Yau, S. T., Inequality between Chern numbers of singular Kähler surfaces and characterization of orbit space of discrete group of SU(2, 1), Contemporary Mathematics 49 (1986) 31-43.

[GT] Gilbarg, D. and Trudinger, N., Elliptic partial differentiation equations of second order. Springer-Verlag, 1977.

[H] Hartshorne, R, Algebraic geometry, Springer-Verlag, 1977.

[Ka] Kawamata, Y., Pluricanonical systems on minimal algebraic varieties, Inv. Math. 79 (1985) 567-588.

[Ko1] Kobayashi, R., Einstein-Kähler V-metrics on open Satake V-surfaces with isolated quotient singularities, Math. Ann. 272 (1985) 385-398.

[Ko2] Kobayashi, R., Einstein-Kähler metric on open algebraic surface of general type, Tohoku Math. J. (1985) 43-77.

[Mi1] Miyaoka, Y., On the Chern numbers of surfaces of general type , Math. Am. 42 (1977) 225-237.

[Mi2] Miyaoka, Y., The maximal number of quotient singularities on surfaces with given numerical invariants, Math. Am. 268 (1984) 159-171.

[MY] Mok, N. and Yau, S. T., Completeness of the Kähler-Einstein metric on bounded domains and the characterization of domains of holomorphy by curvature conditions, Proceedings of Symposia in Pure Mathematics 39 (1983) 41-59.

[Sa] Sakai, F., Semi-stable curves on algebraic surfaces and logarithmic pluricanonical maps, Math. Am. 254 (1980) 89-120.

[Ts] Tsuji, H., A numerical characterization of ball quotients, to appear.

[Ya1] Yau, S. T., Calabi's conjecture and some new results in algebraic geometry, Proc. Natl. Acad. Sci. USA 74 (1977) 1798-1799.

[Ya2] Yaù, S. T., On the Ricci curvature of a compact Kähler manifold and the complex Monge-Ampère equation I. Comm. Pure Appl. Math. 31 (1978) 339-411.

[Ya3] Yau, S. T., A general Schwarz lemma for Kähler manifolds, Amer. J. Math. 100 (1978) 197-203.

[Ya4] Yau, S. T., Métriques de Kähler-Einstein sur les variétés ouvertes, in "Premiere Classe de Chern et courbure de Ricci: Preuve de la conjecture de Calabi, Séminaire Palaisean, France (1978) 163-167.

# Smoothness of the Universal Deformation Space of Compact Calabi-Yau Manifolds and Its Peterson-Weil Metric

Gang Tian

Department of Mathematics
University of California, San Diego
La Jolla, CA 92093

1. Introduction. Bogomolov [B] proves that the universal deformation space of a compact Kählerian-Hamiltonian manifold X with $C_1 = 0$ is locally isomorphic to an open set in $H^1(X,T_X)$, where $T_X$ is the tangent sheaf of X. Later on, he also claimed, in one of his preprints, that the same is true for projective manifolds with $C^1 = 0$, i.e. Calabi-Yau manifolds. His proof seems to be too complicated. Here we first give a different and much simpler proof of that fact (Theorem 1). We only need to assume that X is Kähler. The proof here should be able to be applied to more general cases, for instance, the proof works for those non-Kähler manifolds with trivial canonical bundle and the weight n spectral sequence of the Hodge filtration degenerating at $E_1$. This fact was also pointed out to us by Beauville. It works for any Hamiltonian manifold. In section 5, we recall the definitions of period domains and period maps, in particular, the period domain $D = G/H$ and the map $\Omega$, introduced by P. Griffiths in [G], where G is a real Lie group, H is a compact subgroup, and $\Omega$ maps the universal polarized deformation space $\mathcal{M}$ of a Calabi-Yau manifold X into D. In section 7, we prove that on $\mathcal{M}$, the Peterson-Weil metric is just the pull-back by $\Omega$ of the G-invariant Chern class of certain G-homogeneous complex line bundle on D (Theorem 2).

It implies immediately the Local Torelli theorem and that the Kähler class of the Peterson-Weil metric is an integral cohomolgical class on the moduli space of polarized complex structures on the Calabi-Yau manifold $X$, if the moduli space exists. Another corollary of Theorem 2 is that the Peterson-Weil metric is the pull-back by $\Omega$ of a G-invariant metric on $D$. By a result of Griffiths and Schmid [GS], it follows that the holomorphic sectional curvatures of the Peterson-Weil metric are negative and bounded away from zero (Theorem 3).

In one recent preprint of Todorov, he independently produces results of similar nature.

2. <u>Preliminary</u>. From now on, we suppose that $X$ is a compact Kahler manifold, $\Gamma(X, \Omega^{p,q}(T_X))$ is the set of global (p, q)-forms with coefficients in the tangent sheaf $T_X$. $[\ ,\ ]$ denotes the Lie bracket on $\Gamma(X, \Omega^{p,q}(T_X))$.

By the deformation theory of Kodaira-Spencer [KNS], a new complex structure over $X$ is given by $\omega(t) \in \Gamma(X, \Omega^{0,1}(T_X))$ such that

$$(*) \qquad \bar{\partial}\omega(t) + \frac{1}{2}[\omega(t), \omega(t)] = 0, \quad \omega(0) = 0 .$$

where $t$ is the parameter, one can consider this as a point in the universal deformation space of $X$.

Using the Taylor expansion, $\omega(t) = \omega_1 t + \omega_2 t^2 + \cdots$ , $\omega_i \in \Gamma(X, \Omega^{0,1}(T_X))$, then (*) becomes

$$(**) \qquad \begin{aligned} &\omega_1 \in H^1(X, T_X) \\ &\bar{\partial}\omega_N + \frac{1}{2}\sum_{i=1}^{N-1}[\omega_i, \omega_{N-i}] = 0 \quad (N \geq 2) \end{aligned}$$

Our proof is actually to solve (**) inductively. Then a standard argument shows that the Taylor series $\omega(t)$ is convergent and solves (*). Define the natural isomorphism $i_q : \Gamma(X, \Omega^{0,q}(T_X)) \to \Gamma(X, \Omega^{n-1,q}(K_X^{-1}))$ as follows, where $K_X$ is the canonical sheaf of $X$. For $\phi \in \Gamma(X, \Omega^{0,q}(T_X))$, locally,

$$\phi = \sum_{\substack{i, J \\ |J| = q}} f_{\bar{J}}^{i} \frac{\partial}{\partial z^i} d\bar{z}^J ,$$

define

$$i_q(\phi) = dz^1 \wedge \ldots \wedge dz^n \left( \sum_{\substack{i, J \\ |J| = q}} g_{\bar{J}}^{i} \frac{\partial}{\partial z^i} d\bar{z}^J \right)$$

$$= \sum_{\substack{i, \bar{J} \\ |J| = q}} (-1)^{i-1} g_{\bar{J}}^{i} = dz^1 \wedge \ldots \wedge d\hat{z}^i \wedge \ldots \wedge dz^n \wedge d\bar{z}^J .$$

where $d\hat{z}^i$ means that the term $dz^i$ is omitted.

It is easy to see that $i_q$ is well-defined and isomorphic. By $i_q$, one can define a Lie bracket on $\bigoplus_{q \geq 0} \Gamma(X, \Omega^{n-1,q}(K_X^{-1}))$, i.e., $\forall$

$$\phi_1, \phi_2 \in \Gamma\left(X, \Omega^{n-1,q_1}(K_X^{-1})\right) \quad \text{or} \quad \Gamma\left(X, \Omega^{n-1,q_2}(K_X^{-1})\right),$$

$$[\phi_1, \phi_2] = i_{q_1+q_2}\left[i_{q_1}^{-1}(\phi_1), i_{q_2}^{-1}(\phi_2)\right] \in \Gamma\left(X, \Omega^{n-1,q_1+q_2}(K_X^{-1})\right) .$$

Replacing $\omega_i$ by $i_1(\omega_i)$ in (**), one may assume that

$$\omega_i \in \Gamma(X, \Omega^{n-1,1}(K_X^{-1}))$$

in the sequence (**) of equations.

3. A basic identity. Choose a hermitian metric of $K_X^{-1}$ on X, for example, the induced one from the Kähler metric on X. Then we have a compatible connection D. We denote by $\partial$ and $\bar{\partial}$ the (1, 0)-part and (0, 1)-part of D. The operator $\bar{\partial}$ coincides with the usual $\bar{\partial}$ defined by the complex structure of X.

Lemma 3.1. $\omega_1, \omega_2 \in \Gamma(X, \Omega^{n-1,1}(K_X^{-1}))$ then,

$$[\omega_1, \omega_2] = \partial(i^{-1}(\omega_1) \lrcorner\, \omega_2) - \#(\partial\omega_1) \wedge \omega_2 + \omega_1 \wedge \#(\partial\omega_2)$$

where $\lrcorner$ stands for the inner product, and

$$\#(\partial\omega_\ell) = \sum_{i,j} D_i \phi^\ell_{1\ldots\hat{i}\ldots n\bar{j}} d\bar{z}^j$$

for

$$\omega_\ell = \sum_i (-1)^{i-1} \cdot \phi^\ell_{1\ldots\hat{i}\ldots n\bar{j}} dz^1 \wedge \ldots \wedge d\hat{z}^i \wedge \ldots \wedge dz^n \wedge d\bar{z}^j , \quad \ell = 1, 2.$$

Proof. It suffices to prove it locally. For a given point $p \in X$, choose local coordinates such that at p,

$$D_k \phi_{I\bar{J}} = \frac{\partial \phi_{I\bar{J}}}{\partial z^k} .$$

Write

$$\omega_1 = \sum_{i,j} (-1)^{i-1} g_{12\ldots\hat{i}\ldots n\bar{j}} dz^1 \wedge \ldots \wedge d\hat{z}^i \wedge \ldots \wedge dz^n \wedge d\bar{z}^j$$

$$\omega_2 = \sum_{i,j} (-1)^{i-1} h_{12\ldots\hat{i}\ldots n\bar{j}} dz^1 \wedge \ldots \wedge d\hat{z}^i \wedge \ldots \wedge dz^n \wedge d\bar{z}^j$$

then

$$i_1^{-1}(\omega_1) = \sum_i g_{1\ldots\hat{i}\ldots n\bar{j}} \frac{\partial}{\partial z^i} d\bar{z}^j$$

$$i_1^{-1}(\omega_2) = \sum_i h_{1\ldots\hat{i}\ldots n\bar{j}} \frac{\partial}{\partial z^i} d\bar{z}^j$$

Hence,

$$[\omega_1, \omega_2] = i_2[i_1^{-1}(\omega_1),\ i_1^{-1}(\omega_2)] =$$

$$= i_2\left(\sum_{i,k} g_{1\ldots\hat{i}\ldots n\bar{j}} \frac{\partial h_{1\ldots\hat{k}\ldots n\bar{\ell}}}{\partial z^i} \frac{\partial}{\partial z^k} d\bar{z}^j \wedge d\bar{z}^\ell - \right.$$

$$\left. - h_{1\ldots\hat{i}\ldots n\bar{\ell}} \frac{\partial g_{1\ldots\hat{k}\ldots n\bar{j}}}{\partial z^i} \frac{\partial}{\partial z^k} d\bar{z}^j \wedge d\bar{z}^\ell\right)$$

$$= \sum_k (-1)^{k-1}\left(\sum_i g_{1\ldots\hat{i}\ldots n\bar{j}} \frac{\partial h_{1\ldots\hat{i}\ldots n\bar{\ell}}}{\partial z^i}\right.$$

$$\left. - h_{1\ldots\hat{i}\ldots n\bar{\ell}} \frac{\partial g_{1\ldots\hat{k}\ldots n\bar{j}}}{\partial z^i}\right) dz^1 \wedge \ldots \wedge d\hat{z}^k \wedge \ldots \wedge dz^n \wedge d\bar{z}^j \wedge d\bar{z}^\ell .$$

Now $\psi = i_1^{-1}(\omega_1) \lrcorner\, \omega_2$

$$= \left(\sum_i g_{1\ldots\hat{i}\ldots n\bar{j}} \frac{\partial}{\partial z^i} d\bar{z}^j\right) \lrcorner \left(\sum_k (-1)^{k-1} h_{1\ldots\hat{k}\ldots n\bar{\ell}}\, dz^1 \wedge \ldots \wedge d\hat{z}^k \wedge \ldots \wedge dz^n \wedge d\bar{z}^\ell\right)$$

$$= \sum_{i<k} (-1)^{k+i} g_{1\ldots\hat{i}\ldots n\bar{j}} h_{1\ldots\hat{k}\ldots n\bar{\ell}}\, dz^1 \wedge \ldots d\hat{z}^i \wedge \ldots \wedge d\hat{z}^k \wedge \ldots \wedge dz^n \wedge d\bar{z}^j \wedge d\bar{z}^\ell$$

$$+ \sum_{i>k} (-1)^{k+i-1} g_{1\ldots\hat{i}\ldots n\bar{j}} h_{1\ldots\hat{k}\ldots n\bar{\ell}}\, dz^\ell \wedge \ldots \wedge d\hat{z}^k \wedge \ldots \wedge d\hat{z}^i \wedge \ldots \wedge dz^n \wedge d\bar{z}^j \wedge d\bar{z}^\ell$$

$$= \sum_{i<k} (-1)^{k+1} (g_{1\ldots\hat{i}\ldots n\bar{j}} h_{1\ldots\hat{k}\ldots n\bar{\ell}} - h_{1\ldots\hat{i}\ldots n\bar{\ell}}\, g_{1\ldots\hat{k}\ldots n\bar{j}}$$

$$\times dz^1 \wedge \ldots \wedge d\hat{z}^i \wedge \ldots \wedge d\hat{z}^k \wedge \ldots \wedge dz^n \wedge d\bar{z}^j \wedge d\bar{z}^\ell .$$

At p ,

$$\partial\psi = \sum_{i<k} (-1)^{i+k} \frac{\partial}{\partial z^s} (g_{1\ldots\hat{i}\ldots n\bar{j}} h_{1\ldots\hat{k}\ldots n\bar{\ell}} - h_{1\ldots\hat{i}\ldots n\bar{\ell}} g_{1\ldots\hat{k}\ldots n\bar{j}})$$

$$\times dz^s \wedge dz^1 \wedge \ldots \wedge d\hat{z}^i \wedge \ldots \wedge d\hat{z}^k \wedge \ldots \wedge dz^n \wedge dz^{\bar{j}} \wedge d\bar{z}^\ell$$

$$= \sum_{i<k} \Bigg( (-1)^{k-1} \frac{\partial}{\partial z^i} (g_{1\ldots\hat{i}\ldots n\bar{j}} h_{1\ldots\hat{k}\ldots n\bar{\ell}}$$

$$- h_{1\ldots\hat{i}\ldots n\bar{\ell}} g_{1\ldots\hat{k}\ldots n\bar{j}} \Bigg) dz^1 \wedge \ldots \wedge d\hat{z}^k \wedge \ldots \wedge dz^n \wedge d\bar{z}^j \wedge d\bar{z}^\ell - (-1)^{i-1}$$

$$\times \frac{\partial}{\partial z^k} (g_{1\ldots\hat{i}\ldots n\bar{j}} h_{1\ldots\hat{k}\ldots n\bar{\ell}} - h_{1\ldots\hat{i}\ldots n\bar{\ell}} g_{1\ldots\hat{k}\ldots n\bar{j}})$$

$$\times dz^1 \wedge \ldots \wedge d\hat{z}^i \wedge \ldots \wedge dz^n \wedge d\bar{z}^j \wedge d\bar{z}^\ell$$

$$= \sum_{\substack{i,k \\ i \neq k}} (-1)^{k-1} \frac{\partial}{\partial z^i} (g_{1\ldots\hat{i}\ldots n\bar{j}} h_{1\ldots\hat{k}\ldots n\bar{\ell}} - h_{1\ldots\hat{i}\ldots n\bar{\ell}} g_{1\ldots\hat{k}\ldots n\bar{j}})$$

$$\times dz^1 \wedge \ldots \wedge d\hat{z}^k \wedge \ldots \wedge dz^n \wedge d\bar{z}^j \wedge d\bar{z}^\ell$$

$$= \sum_{i,k} (-1)^{k-1} \left( g_{1\ldots\hat{i}\ldots n\bar{j}} \frac{\partial h_{1\ldots\hat{k}\ldots n\bar{\ell}}}{\partial z^i} - h_{1\ldots\hat{i}\ldots n\bar{\ell}} \frac{\partial g_{1\ldots\hat{k}\ldots n\bar{j}}}{\partial z^i} \right)$$

$$\times dz^1 \wedge \ldots \wedge d\hat{z}^k \wedge \ldots \wedge dz^n \wedge d\bar{z}^j \wedge d\bar{z}^\ell$$

$$+ \sum_k (-1)^{k-1} \left\{ \left( \sum_i \frac{\partial g_{1\ldots\hat{i}\ldots n\bar{j}}}{\partial z^i} \right) h_{1\ldots\hat{k}\ldots n\bar{\ell}} \right.$$

$$\left. - g_{1\ldots\hat{k}\ldots n\bar{j}} \left( \sum_i \frac{\partial h_{1\ldots\hat{i}\ldots n\bar{\ell}}}{\partial z^i} \right) \right\}$$

$$\times dz^1 \wedge \ldots \wedge d\hat{z}^k \wedge \ldots \wedge dz^n \wedge d\bar{z}^j \wedge d\bar{z}^\ell + [\omega_1, \omega_2] + (\# \partial\omega_1) \wedge \omega_2 - \omega_1 \wedge \#(\partial\omega_2) .$$

Hence, the lemma is proved.

## 4. The proof of Theorem 1

Lemma 4.1 ([GH], p. 149). If $\eta$ is any $\bar\partial$- closed (p,q)-form on X, and $\eta$ is d- , $\partial$- , or $\bar\partial$-exact, then

$$\eta = \partial\bar\partial\gamma$$

for some (p-1,p+1)-form $\gamma$.

This is the well-known $\partial\bar\partial$-lemma in Hodge theory and also crucial in the proof of the following theorem.

Theorem 1. If X is a compact Kähler manifold with $C_1 = 0$, then the local universal deformation space of X is isomorphic to an open set in $H^1(X, T_X)$.

Proof. $K_X$ is trivial, so $i_1 : \Gamma(X, \Omega^{0,1}(T_X)) \overset{\cong}{\to} \Gamma(X, \Omega^{n-1,1})$ and coincides with the usual $\partial$ operator induced by the complex structure over X.

$$H^1(X, \Omega^{n-1}) = \{\phi \in \Gamma(X, \Omega^{n-1,1}) \mid \bar\partial\phi = 0, \ \bar\partial^*\phi = 0\}$$

Since X is Kähler, $\forall\ \phi \in H^1(X, \Omega^{n-2})$, $\partial\phi = 0$ (see [GH]).

To prove the theorem, it suffices to show that (**) is solvable for any $\omega_1 \in H^1(X, \Omega^{n-1}) \cong H^1(X, T_X)$. Note that $\bar\partial\,[\omega_1, \omega_1] = 0$ for such a $\omega_1$.

For $\omega_1 \in H^1(X, \Omega^{n-1})$, $\partial\omega_1 = 0$, by Lemma 3.1, $[\omega_1, \omega_1] = \partial(i^{-1}(\omega_1) \lrcorner\, \omega_1)$, then by Lemma 4.1, $\frac{1}{2}[\omega_1, \omega_1] = \partial\bar\partial\psi$ for certain $\psi \in \Gamma(X, \Omega^{n-2,1})$.

Put $\omega_2 = \partial\psi$, then $\partial\omega_2 = 0$ and $\bar\partial\omega_2 + \frac{1}{2}[\omega_1, \omega_1] = 0$. Now suppose that we already have $\omega_1, \ldots, \omega_{N-1}$ such that

$$\partial\omega_i = 0 \quad \text{for} \quad i \leq N-1$$

$$\bar{\partial}\omega_i + \frac{1}{2}\sum_{j=1}^{i}[\omega_j, \omega_{i-j}] = 0$$

By Lemma 3.1,

$$\frac{1}{2}\sum_{j=1}^{N}[\omega_j, \omega_{N-j}] = \frac{1}{2}\sum_{j=1}^{N}\partial(i_1^{-1}(\omega_j) \lrcorner\, \omega_{N-j}) = \partial\left(\frac{1}{2}\sum_{j=1}^{N} i_1^{-1}(\omega_j) \lrcorner\, \omega_{N-j}\right) .$$

It is straightforward to check that $\bar{\partial}\left(\sum_{j=1}^{N}[\omega_j, \omega_{N-j}]\right) = 0.$

As above, Lemma 4.1 tells us that for certain ,

$$\frac{1}{2}\sum_{j=1}^{N}[\omega_j, \omega_{N-j}] = \partial\bar{\partial}\psi .$$

Set $\omega_N = \partial\psi$ , we have $\partial\omega_N = 0$ ,

$$\bar{\partial}\omega_N + \frac{1}{2}\sum_{i=1}^{N}[\omega_i, \omega_{N-1}] = 0 .$$

Hence, (**) is solvable. //

5. Remarks. If $X$ is a compact Hamiltonian manifold, we can define an isomorphism between $\Gamma(X, \Omega^{0,q}(T_X))$ and $\Gamma(X, \Omega^{1,q})$ by the non-degenerate holomorphic (2.0)-form defining the Hamiltonian structure over X. One can prove that the local universal deformation space of $X$ is isomorphic to $H^{1,1}(X, \mathbb{C})$ by repeating the above arguments except that we use the fact that $\partial H^{0,2}(X, C) = 0$ instead of Hodge theory for Kähler manifold. Thus, one can drop the Kähler condition in this case.

6. Period domain. From now on, we always assume that the compact Kahler manifold X has vanishing first Chern class and the associated Kähler class gives an ordinary polarization $\omega \in H^2(X, Z)$. By the solution of Yau [Y] to Calabi's conjecture, we may also assume that the Kähler metric on X has flat Ricci curvatures. On $H^n(X, C)$, there is a natural quadratic form $Q(\cdot, \cdot)$, i.e. the intersection form. This is a topological invariant, preserving the lattice $H^n(X, Z) \subset H^n(X, C)$.

A straightforward definition of the period domain is the following:

$D' = \{$ complex line $\ell$ in $H^n(X, C) \mid \forall \varphi \in \ell \setminus \{0\},\ Q(\varphi, \bar{\varphi}) > 0,\ Q(\varphi, \varphi) = 0 \}$.

Note that $D'$ depends only on the topological type of X and is a complex homogeneous manifold. But a more sophisticated period domain D was introduced by P. Griffith in [G]. Let us recall his definition. Denote by $H^n(X, C)_o$ the primitive cohomology of degree n, i.e. $H^n(X, C)_o = \ker(L_\omega)$, where $L_\omega: H^n(X, C) \to H^{n+2}(X, C)$, $L_\omega(\eta) = \omega \wedge \eta$. Hodge decomposition gives $V = H^n(X, C)_o = \bigoplus_{p+q=n} \left(H^{p,q}_{\bar{\partial}}(X, C) \cap H^n(X, C)_o\right)$, where $H^{p,q}_{\bar{\partial}}(X, C)$ consists of all harmonic (p, q) forms with respect to the Kähler metric on X. Put $h_i = \dim_{\mathbb{C}} \left(H^{i,n-i}_{\bar{\partial}}(X, C) \cap H^n(X, C)_o\right)$, $0 \le i \le n$. In particular, $h_o = 1$, since $C_1(X) = 0$. Denote $\left[\frac{n-1}{2}\right]$ by t. Define

$$D = \{\text{flags}[S_o, \ldots, S_t] \mid S_o \subset S_1 \subset \cdots \subset S_t \subset H^n(X, C)_o = V,$$

$$\dim_{\mathbb{C}} S_i = h_i,\ Q(S_i, S_i) = 0,\ (-1)^i Q(S_i/S_{i-1}, S_i/S_{i-1}) > 0 \text{ for } 0 \le i \le t\}$$

where $S_i/S_{i-1} = \{\varphi \in S_i \mid Q(\varphi, \bar{S}_{i-1}) = 0\}$. In [Gl], P. Griffith proves that $D = G/H$, where G is the real linear transformation group of $H^n(X, R)_o$ preserving the quadratic form Q, H is the stabilizer in G of one point in D.

H is compact. Certainly, D is a complex homogeneous manifold. In case that X is a K3 surface, $G \cong SO(2,19)$, $H \cong U(1) \times O(19)$, so $D \cong SO(2,19)/U(1) \times O(19)$.

There is a natural projection $\pi: D \to D'$, mapping the flag $[S_o, \ldots, S_t]$ to the line $S_o$.

By Theorem 1 in section 4, we may assume that the local universal deformation space $\Delta$ of X be in $H^1(X, T_X) \cong H^1(X, \Omega^{n-1}) = H_{\bar\partial}^{n-1,1}(X, C)$. Denote by $H^1(X, T_X)_\omega$ the subspace $\{\theta \in H^1(X, T_X) \mid \theta \lrcorner \omega = 0$ in $H^2(X, \Omega^o) = H_{\bar\partial}^{0,2}(X, C)\} \cong H^1(X, \Omega^{n-1}) \cap H^n(X, C)_o = H_{\bar\partial}^{n-1,1}(X, C) \cap H^n(X, C)_o$. Note that $H^1(X, T_X)_\omega$, so $H_{\bar\partial}^{n-1,1}(X, C) \cap H^n(X, C)_o$ is just the tangent space at X of the submanifold $\Delta_\omega = \Delta \cap H^1(X, T_X)_\omega$ of $\Delta$, $\Delta_\omega$ consists of those local deformations of X preserving the polarization. Define period mappings as follows: Each point $\theta$ in $\Delta_\omega$ stands for a compact Kähler manifold $X_\theta$, diffeomorphic to X, with $C_1(X_\theta) = 0$ and the Kähler class $\omega_\theta = \omega$ in $H^2(X_\theta, Z) = H^2(X, Z)$, Hodge theorem [GH] gives a decomposition on the primitive cohomology $H^n(X_\theta, C)_o = \bigoplus_{p+q=n} H_{\bar\partial}^{p,q}(X_\theta, C) \cap H^n(X_\theta, C)_o$.

Take a diffeomorphism $\phi_\theta: X \to X_\theta$, then $\phi_\theta^*(H^n(X_\theta, C)_o) = H^n(X, C)_o$.

Define

$$\Omega(\theta) = \left[\phi^*(H_{\bar\partial}^{n,0}(X_\theta, C) \cap H^n(X_\theta, C)_o), \ldots, \phi^*(H_{\bar\partial}^{t,n-t}(X_\theta, C) \cap H^n(X_\theta, C))\right] \in D,$$

$$\Omega'(\theta) = \pi \cdot \Omega(\theta) \in D'.$$

$\Omega, \Omega'$ are called period mapping. In [G], Griffith proves that $\Omega$, so $\Omega'$, is holomorphic.

7. Peterson-Weil metric on the universal deformation space of $X$.

Let $\mathcal{M}$ be the universal polarized deformation space of X. $\mathcal{M}$ is smooth by Theorem 1. Each $X' \in \mathcal{M}$ has the Kähler class $\omega'$ as its polarization $\omega$ in $H^2(X', Z) = H^2(X, Z)$. By [Y], we may assume that $X'$ be Kähler-Einstein, i.e. the associated Kähler metric $(g'_{\alpha\bar{\beta}})$ is Ricci-flat. The tangent space $T_{X'}\mathcal{M}$ of $\mathcal{M}$ at $X'$ can be identified with $H^1(X', T_{X'})_\omega$. The Peterson-Weil metric $G_{PW}$ on $\mathcal{M}$ is defined by

$$(7.1) \qquad G_{PW}(\varphi, \psi) = \int_{X'} g'^{\alpha\bar{\beta}} g'_{\gamma\bar{\delta}} \varphi^{\gamma}_{\bar{\beta}} \bar{\psi}^{\bar{\delta}}_{\bar{\alpha}} \, dV_{g'}$$

where $\varphi = \varphi^{\gamma}_{\bar{\beta}} \frac{\partial}{\partial z^{\gamma}} d\bar{z}^{\beta}$, $\psi = \psi^{\delta}_{\bar{\alpha}} \frac{\partial}{\partial z^{\delta}} d\bar{z}^{\alpha}$ are in $H^1(X', T_{X'})_\omega$, $g' = g'_{\alpha\bar{\beta}} dz^{\alpha} d\bar{z}^{\beta}$ is the Kähler-Einstein metric on $X$ associated with the polarization $\omega$.

Lemma 7.1. $H^1(X', T_{X'})_\omega =$

$$= \left\{ \varphi = \varphi^{\gamma}_{\bar{\beta}} \frac{\partial}{\partial z^{\gamma}} d\bar{z}^{\beta} \in H^1(X', T_{X'}) \,\middle|\, \varphi^{\gamma}_{\bar{\beta}} = g'^{\gamma\bar{\alpha}} g'_{\delta\bar{\beta}} \varphi^{\delta}_{\bar{\alpha}} \right\}.$$

Proof. By the definition of $H^1(X', T_{X'})_\omega$ and the fact that $\omega'$ is parallel with respect to the Kahler-Einstein metric, $\varphi \lrcorner \, \omega' = 0$ on $X'$. Then the lemma follows from the interpretation of $\varphi \lrcorner \, \omega' = 0$ in local coordinates.

By Lemma 7.1, formula (7.1) can be written as

(7.2) $$G_{PW}(\varphi,\psi) = \int_{X'} \varphi^{\alpha}_{\bar\beta}\, \overline{\psi^{\beta}_{\bar\alpha}}\, dV_{g'}, \quad \text{for } \varphi, \in H^1(X', T_{X'})_{\omega}.$$

Pick up a holomorphic n-form $\Psi_{X'}$ on $X'$. By the well-known Bochner technique and Ricci flatness of $X'$, we conclude that $\Psi_{X'}$ is parallel with respect to the Kahler-Einstein metric on $X'$. Given a point $x \in X'$, choose the local coordinates $(z_1,\ldots,z_n)$ such that $g'_{\alpha\bar\beta} = \delta_{\alpha\bar\beta}$, $dg'_{\alpha\bar\beta} = 0$ and $\Psi_{X'} = c' dz_1 \wedge\ldots\wedge dz_n$ at x. Then, at $x \in X'$,

$$\varphi \lrcorner \Psi_{X'} = c' \sum_{\alpha,\beta} (-1)^{\alpha-1} \varphi^{\alpha}_{\bar\beta}\, dz_1 \wedge\ldots\wedge d\hat z_\alpha \wedge\ldots\wedge dz_n \wedge d\bar z_\beta$$

$$\psi \lrcorner \Psi_{X'} = c' \sum_{\alpha,\beta} (-1)^{\alpha-1} \psi^{\alpha}_{\bar\beta}\, dz_1 \wedge\ldots\wedge d\hat z_\alpha \wedge\ldots\wedge dz_n \wedge d\bar z_\beta$$

where $d\hat z_\alpha$ means that the term $dz_\alpha$ is omitted.

(7.3) $$\begin{aligned}(\varphi \lrcorner \Psi_{X'}) \wedge \overline{(\psi \lrcorner \Psi_{X'})} &= -|c'|^2 \varphi^{\alpha}_{\bar\beta}\, \overline{\psi^{\beta}_{\bar\alpha}}\, dz_1 \wedge\ldots\wedge dz_n \wedge d\bar z_1 \wedge\ldots\wedge d\bar z_n \\ &= (-1)^{n+1} (\sqrt{-1})^n (2\pi)^n |c'|^2 \varphi^{\alpha}_{\bar\beta}\, \overline{\psi^{\beta}_{\bar\alpha}}\, dV_{g'},\end{aligned}$$

Since $\Psi_{X'}$ is parallel, $\Psi_{X'} \wedge \overline{\Psi}_{X'} = c\, dV_{g'}$, for certain constant c. Furthermore, one sees easily that $c = (-2\pi)^n (\sqrt{-1})^n |c'|^2$. Therefore, at $X' \in \mathfrak{M}$,

$$G_{PW}(\varphi,\psi) = -\frac{\int_{X'} (\varphi \lrcorner \Psi_{X'}) \wedge \overline{(\psi \lrcorner \Psi_{X'})}}{\int_{X'} \Psi_{X'} \wedge \bar{\Psi}_{X'}}$$

$$= -\frac{\int_{X} (\varphi \lrcorner \Psi_{X'}) \wedge \overline{(\psi \lrcorner \Psi_{X'})}}{\int_{X'} \Psi_{X'} \wedge \bar{\Psi}_{X'}}$$

Note that in the last equality, we have used the fact that X, X′ have the same differential type.

We will go further to express this Peterson-Weil metric in terms of datas of D or D′ and the period mappings. Let U be the tautological line bundle of the projective space $PH^n(X, C)_o$, U restricts to a G-homogeneous line bundle on D′, also denoted by U. The total space of U is $\{\varphi \in H^n(X, C)_o \mid Q(\varphi,\bar{\varphi}) > 0,\ \varphi \neq 0,\ Q(\varphi,\varphi) = 0\}$, so the intersection form Q( , ) induces a hermitian metric $h_D$ of U on D′, which is G-invariant, since G preserves Q( , ) and is real. Such an invariant metric is unique up to a constant factor.

Theorem 2. Let $\omega_{PW}$ be the Kähler class associated with the Peterson-Weil metric, then $\omega_{PW} = \frac{\sqrt{-1}}{2\pi}\Omega'^{*}(Ric(h_D))$.

Proof. We may assume that X′ = X. Then $T_X\mathcal{M} = T_X\Delta_\omega = H^1(X, T_X)_\omega$. It suffices to prove that for any $\varphi \in H^1(X, T_X)_\omega$, $G_{PW}(\varphi,\varphi) = \Omega'^{*}(Ric(h_D))(\varphi,\bar{\varphi})$. Put $\varphi(s) = s\varphi$, $s \in C$, then by Theorem 1, for s small $\varphi(s)$ defines a complex structure on the differential manifold X, called $X_s$. Let $\Psi_s$ be a family of n-forms on X, parametrized by s in a small disk $\{|s| \le \epsilon\}$ of C, such that each $\Psi_s$ is holomorphic on $X_s$. Then $\Omega'(\varphi(s)) = \Psi_s \in H^n(X, C)_o$. By the definition of Ricci curvature,

$$\Omega'^{*}(\mathrm{Ric}(h_D))(\varphi,\bar{\varphi}) = -\partial_s\bar{\partial}_s \log Q(\Psi_s, \bar{\Psi}_s)\Big|_{s=0}$$

(7.5)

$$= -\left.\frac{Q(\Psi_s,\bar{\Psi}_s)\, Q\left(\frac{\partial\Psi_s}{\partial s}, \overline{\frac{\partial\Psi_s}{\partial s}}\right) - Q\left(\Psi_s, \overline{\frac{\partial\Psi_s}{\partial s}}\right) Q\left(\frac{\partial\Psi_s}{\partial s}, \bar{\Psi}_s\right)}{Q(\Psi_s,\bar{\Psi}_s)^2}\right|_{s=0}$$

Lemma 7.2.

$$\left.\frac{\partial\Psi_s}{\partial s}\right|_{s=0} = c'\Psi_o + \varphi \,\lrcorner\, \Psi_o$$

where $c'$ is a constant.

Proof. At a point $x \in X$, there is a family of local coordinates $(z_1(s), \ldots, z_n(s))$, such that $z_i(s)$ varies holomorphically with the parameter $s$ for $0 \le i \le n$.

$$\Psi_s = v^{i_1\ldots i_n}(s)\, dz_{i_1}(s)\wedge\ldots\wedge dz_{i_n}(s)$$

Hence,

$$\left.\frac{\partial\Psi_s}{\partial s}\right|_{s=0} = \left.\frac{\partial v^{i_1\ldots i_n}(s)}{\partial s}\right|_{s=0} dz_{i_1}\wedge\ldots\wedge dz_{i_n}$$

$$+ \sum_{j=1}^{n} v^{i_1\ldots i_j\ldots i_n}\, dz_{i_1}\wedge\ldots\wedge d\left.\left(\frac{\partial z_{i_j}(s)}{\partial s}\right)\right|_{s=0}\wedge\ldots\wedge dz_{i_n}$$

where $z_i = z_i(0)$.

By the definition of Kodaira-Spencer mapping $\rho$ [G] and $\rho(\partial/\partial s) = \varphi$ in $H^1(X, T_X)_\omega$, choosing $(z_1(s), \ldots, z_n(s))$ properly, we have

$$\varphi = \rho\left(\frac{\partial}{\partial s}\right) = \bar{\partial}\left(\frac{\partial z_\ell(s)}{\partial s}\right)\Big|_{s=0} \frac{\partial}{\partial z_\ell}$$

This results in

$$\frac{\partial \Psi_s}{\partial s}\Big|_{s=0} = \Psi' + \varphi \lrcorner \Psi_o \quad ,$$

where $\Psi'$ is an n-form in $\Gamma(X, \Omega^{n,0})$.

We have known that each $\Psi_s$ is parallel with respect to the corresponding Kähler-Einstein metric on $X_s$, $d\Psi_s = 0$ and $\varphi \lrcorner \Psi_o$ is harmonic once $\varphi$ is. This follows that $d(\Psi' + \varphi \lrcorner \Psi_o) = 0$ and $\varphi \lrcorner \Psi_o \in H^{n-1,1}_{\bar{\partial}}(X, C)$. By Hodge theory [GH], $\partial(\varphi \lrcorner \Psi_o) = 0$. Hence, $\partial\Psi' = 0$, $\bar{\partial}\Psi' = -\partial(\varphi \lrcorner \Psi_o) = 0$, i.e. $\Psi'$ is a holomorphic n-form, so $\Psi' = c'\Psi_o$ for certain constant $c'$. //

Returning to the proof of Theorem 2, by Lemma 7.2,

$$Q(\Psi_o, \bar{\Psi}_o)\, Q\left(\frac{\partial \Psi_s}{\partial s}\Big|_{s=0}, \overline{\frac{\partial \Psi_s}{\partial s}}\Big|_{s=0}\right) - Q\left(\Psi_o, \overline{\frac{\partial \Psi_s}{\partial s}}\Big|_{s=0}\right) Q\left(\frac{\partial \Psi_s}{\partial s}\Big|_{s=0}, \bar{\Psi}_o\right)$$

$$(7.6) \quad = |c'|^2 Q(\Psi_o, \bar{\Psi}_o)^2 + Q(\Psi_o, \bar{\Psi}_o)\, Q(\varphi \lrcorner \Psi_o, \overline{\varphi \lrcorner \Psi_o}) - |c'|^2 Q(\Psi_o, \bar{\Psi}_o)^2$$

$$= Q(\Psi_o, \bar{\Psi}_o)\, Q(\varphi \lrcorner \Psi_o, \overline{\varphi \lrcorner \Psi_o})$$

It follows from (7.4), (7.5), (7.6) that

$$G_{PW}(\varphi,\varphi) = \Omega'^{*}(\mathrm{Ric}(h_D))(\varphi,\bar{\varphi}) \,. \qquad //$$

The first corollary is the Local Torelli theorem, the case for D is already proved in [G].

Corollary 1. The differentials of period mappings $\Omega, \Omega'$ are injective at points of $\mathcal{M}$, i.e. $\Omega, \Omega'$ are holomorphic immersions.

Proof. It follows from the above theorem and the observation that $G_{PW}$ is positive definite on $H^1(X', T_{X'})_\omega$ for $X' \in \mathcal{M}$. //

The pull-back $\pi^* U$ is a G-homogeneous line bundle over D.

$$\omega_{PW} = \frac{\sqrt{-1}}{2\pi}\,\Omega'^{*}(\mathrm{Ric}(h_D)) = \frac{\sqrt{-1}}{2\pi}\,\Omega^{*}\pi^{*}(\mathrm{Ric}(h_D)) \,.$$

Obviously, $\frac{\sqrt{-1}}{2\pi}\pi^* \mathrm{Ric}(h_D)$, as the Chern class of $\pi^* U$, is G-invariant, so it follows

Corollary 2. The Peterson-Weil metric on $\mathcal{M}$ is the pull-back of a G-invariant metric on D by the period mapping $\Omega$.

Remark. $\sqrt{-1}\,\pi^* \mathrm{Ric}(h_D)$ is not a metric on D, but by the G-invariance and the compactness of H, one can find a G-invariant metric on D, which coincides with $\sqrt{-1}\,\pi^* \mathrm{Ric}(h_D)$ along directions tangent to $\Omega(\mathcal{M})$.

Recall $D = G/H$, H is a compact subgroup of G. Let K be the maximal compact subgroup of G, containing H, $P = G/K$, then the fibres in the fibering $D \xrightarrow{p} P$ are compact, complex subvarieties of D. In [G], it is proved that for each $\theta \in D$, there is a unique G-invariant splitting:

$$T_\theta(D) = V_\theta \oplus H_\theta$$

where $V_\theta$ is the tangent space to the fiber of p passing through $\theta \in D$. Moreover, for $\theta = \Omega(X')$, $d\Omega(T_{X'}\mathcal{M}) \subset H_\theta$ [G]. Any $v \in H_\theta$ is called "horizontal direction" of D at $\theta$. In [GS], it is proved that the G-invariant metric on D has negative holomorphic sectional curvatures corresponding to horizontal directions and the curvatures are bounded away from zero. Therefore, the above Corollary 2 implies

<u>Theorem 3</u>. The holomorphic sectional curvatures of the Peterson-Weil metric on $\mathcal{M}$ are negative and bounded away from zero.

References


[B] F. A. Bogomolov, Hamiltonian Kähler manifolds. Dolk. Akad. Nauk SSSR 243 (1978), no. 5, 1101-1104.

[KNS] K. Kodaira, L. Nirenberg and D. C. Spencer. On the existence of deformations of complex analytic structures. Ann. of Math. 68 (1958) 450-459.

[G] P. Griffiths, Periods of integrals on algebraic manifolds, I, II. Amer. J. Math. 90 (1968) 568-626, 805-865.

[GH] P. Griffiths and Harris, Principles of Algebraic Geometry. John Wiley & Sons, New York, Toronto, 1978.

[GS] P. Griffiths and Schmid, Locally homogeneous complex manifolds. Acta Mathematica 123 (1969) 253-302.

[Y] S. T. Yau, On the Ricci curvature of a compact Kähler manifold and the complex Monge-Ampère equation I. Comm. Pure and Appl. Math. 31 (1978) 229-411.

# AN ELEMENTARY INTRODUCTION TO CRITICAL PHENOMENA

Stephen H. Shenker
*Enrico Fermi Institute*
*5640 S. Ellis*
*Chicago, IL 60637*

The material presented in this lecture can be found in a number of review articles, e.g.:

K. Wilson and J. Kogut, "The Renormalization Group and the $\epsilon$-expansion", *Physics Reports* **12C** (1974) 75.

*Phase Transitions and Critical Phenomena* Volume 6, C. Domb and M. S. Green (eds.), Academic Press (1976).

S. H. Shenker, "Field Theories and Phase Transitions", in *Recent Advances in Field Theory and Statistical Mechanics – Les Houches 1982,* J. B. Zuber and R. Stora (eds.), North Holland (1984).

# LIST OF PARTICIPANTS

**T Allen**
Department of Physics
California Institute of Technology
Pasadena CA 91125
USA

**R Akhoury**
Randall Lab of Physics
University of Michigan
Ann Arbor MI 48109-1120
USA

**C Alcalde**
Department of Physics
University of California
Los Angeles CA 90024
USA

**O Alvarez**
Department of Physics
University of California
Berkeley CA 94720
USA

**J Barrett**
Institute of Theoretical Physics
University of Zurich
Zurich
SWITZERLAND

**R Brooks**
Department of Mathematics
University of Southern California
Los Angeles CA 90089
USA

**P Candelas**
Department of Physics
University of Texas
Austin TX 78712
USA

**J Cardy**
Department of Physics
University of California
Santa Barbara CA 93106
USA

**M-C Chang**
Department of Mathematics (253-37)
California Institute of Technology
Pasadena CA 91125
USA

**W S Cheung**
Department of Mathematics
University of Hong Kong
HONG KONG

**K Cho**
Department of Mathematics
University of California
Los Angeles CA 90024
USA

**Paul Chow**
1188 W 28th Street
Los Angeles CA 90007
USA

**M Clements**
Department of Physics
University of Texas
Austin TX 78712
USA

**A Connes**
Department of Mathematics
College de Francais
Paris
FRANCE

**S Cordes**
Department of Physics
Princeton University
Princeton NJ 08544
USA

**L Crane**
School of Mathematics
Institute of Advance Study
Princeton NJ 08540
USA

**M Douglas**
Department of Physics
California Institute of Technology
Pasadena CA 91125
USA

**P Dourmashkin**
24-603 Department of Physics
Massachusetts Inst of Technology
Cambridge MA 02139
USA

**B Driver**
Department of Mathematics
Cornell University
Ithaca NY 14853
USA

**M Dugan**
50A-3115
Lawrence Berkeley Laboratory
Berkeley CA 94720
USA

**S Enguehard**
AMES Department (B-010)
University of California, San Diego
La Jolla CA 92093
USA

**G Faltings**
Department of Mathematics
Princeton University
Princeton NJ 08544
USA

**J Fillmore**
Department of Mathematics
University of California, San Diego
La Jolla CA 92093
USA

**R Finkelstein**
Department of Physics
University of California
Los Angeles CA 90024
USA

**C Fitzgerald**
Department of Mathematics
University of California, San Diego
La Jolla CA 92093
USA

**S Frankel**
Department of Mathematics
Columbia University
New York NY 10027
USA

**T Frankel**
Department of Mathematics
University of California, San Diego
La Jolla CA 92093
USA

**M Freedman**
Department of Mathematics
University of California, San Diego
La Jolla CA 92093
USA

**D Friedan**
Department of Physics
University of Chicago
Chicago IL 60637
USA

**R Fulp**
Department of Mathematics
No. Carolina State University
Raleigh NC 72695
USA

**S Giddings**
Department of Physics
Princeton University
Princeton NJ 08544
USA

**D Gieseker**
Department of Mathematics
University of California
Los Angeles CA 90024
USA

**M Goulian**
Department of Physics
Harvard University
Cambridge MA 02138
USA

**G Green**
Department of Mathematics
University of California
Los Angeles CA 90024
USA

**A Greenspoon**
Mathematical Review
416-4th St
P O Box 8604
Ann Arbor MI 48107
USA

**M Halpern**
Theory Group
Lawrence Berkeley Laboratory
Berkeley CA 94720
USA

**S Hamidi**
Department of Physics
California Institute of Technology
Pasadena CA 91125
USA

**B Hatfield**
Department of Mathematics
University of California, San Diego
La Jolla CA 92093
USA

**G Horowitz**
Department of Physics
University of California
Santa Barbara CA 93106
USA

**S M Huang**
Department of Mathematics
Carnegie-Mellon University
Philadelphia PA 15213
USA

**J Jost**
Math Inst Ruhrunivesitat
Bochum
WEST GERMANY

**M Karoubi**
Department of Mathematics
University of Paris
Paris 7
FRANCE

**D Kazhdan**
Department of Mathematics
Harvard University
Cambridge MA 02138
USA

**E Kiritsis**
Lauritsen Laboratory of Physics
452-48
California Institute of Technology
Pasadena CA 91125
USA

**D Kraines**
Department of Mathematics
Duke University
Durham NC 27706
USA

**R Kreminski**
Department of Mathematics
University of Maryland
College Park MD 20742
USA

**N Kroll**
Department of Physics
University of California, San Diego
La Jolla CA 92093
USA

**R Lazarsfeld**
Department of Mathematics
University of California
Los Angeles CA 90024
USA

**J Lepowsky**
Department of Mathematics
Rutgers University
New Brunswick NJ 08903
USA

**E B Lin**
Department of Mathematics
& Computer Science
University of California
Riverside CA 92521-0135
USA

**R Lipton**
Math Sci Inst
Cornell University
Ithaca NY 14850
USA

**A Lutken**
Department of Physics
University of Texas
Austin TX 78712
USA

**K Mallory**
3610 Nagoles (W203)
W Covina CA 91792
USA

**N Marcus**
50A-3115
Lawrence Berkeley Laboratory
Berkeley CA 94720
USA

**A Marden**
MSRI
1000 Centennial Avenue
Berkeley CA 94720
USA

**M Martellini**
Department of Physics
University of Milano
Milan
ITALY

**M Mayer**
Department of Physics
University of California
Irvine CA 92717
USA

**J Mirahan**
Department of Physics
Princeton University
Princeton NJ 08544
USA

**B Moisheson**
Department of Mathematics
Columbia University
New York NY 10027
USA

**B Moonen**
Max-Planck-Inst fur Math
5300 Bonn 3
WEST GERMANY

**J Morava**
Department of Mathematics
Johns Hopkins University
Baltimore MD 20707
USA

**D Morrison**
Department of Mathematics
Princeton University
Princeton NY 08544
USA

**M Mulase**
Department of Mathematics
University of California
Los Angeles CA 90024
USA

**P Nelson**
Department of Physics
Harvard University
Cambridge MA 02138
USA

**A Niemi**
Lawrence Berkeley Laboratory
Berkeley CA 94720
USA

**R Ojakian**
1400 Carpentier St (#221)
San Leandro CA 94577
USA

**R Penner**
Department of Mathematics
University of Southern California
Los Angeles CA 90089
USA

**Pokonen O**
Lenjus
SF-50100 Mikkeli
FINLAND

**Y S Poon**
Department of Mathematics
State University of New York
Stony Brook NY 11794
USA

**J M Rabin**
Fermi Institute
University of Chicago
Chicago IL 60637
USA

**H Riggs**
Department of Physics
University of Chicago
Chicago IL 60637
USA

**Z Ran**
Department of Mathematics
University of California
Riverside CA 92521-0135
USA

**S S Roan**
2821 Alida Street
Oakland CA 94602
USA

**B Rodin**
Department of Mathematics
University of California, San Diego
La Jolla CA 92093
USA

**S Sabharwal**
Department of Physics
University of California
Los Angeles CA 90024
USA

**A Sagnotti**
50A-3115
Lawrence Berkeley Laboratory
Berkeley CA 94720
USA

**S Sannan**
Department of Physics
University of California
Santa Barbara CA 93106
USA

**L Saper**
Department of Mathematics
Massachusetts Inst of Technology
Cambridge MA 02139
USA

**R Schimmrigk**
Department of Physics
University of Texas
Austin TX 78712
USA

**C Schoen**
MSRI
1000 Centennial Avenue
Berkeley CA 94720
USA

**R Schoen**
Department of Mathematics
University of California, San Diego
La Jolla CA 92093
USA

**J Segert**
Department of Physics
Princeton University
Princeton NJ 08544
USA

**T Shiota**
Department of Mathematics
Harvard University
Cambridge MA 02139
USA

**H Sieburg**
Salk Institute
10010 No. Torrey Pines Road
La Jolla CA 92037
USA

**S-J Sin**
Lawrence Berkeley Laboratory
Building 50A-3115
Berkeley CA 94720
USA

**S Shenker**
Department of Physics
University of Chicago
Chicago IL 60637
USA

**I Singer**
Department of Mathematics
Massachusetts Inst of Technology
Cambridge MA 02139
USA

**R Skora**
Department of Mathematics
University of California, San Diego
La Jolla CA 92093
USA

**H Sonoda**
Lawrence Berkeley Laboratory
1 Cyclotron Road
Berkeley CA 94720
USA

**M Stern**
Department of Mathematics
Duke University
Durham NC 27706
USA

**J Tarski**
Department of Mathematics
University of California
Davis CA 95616
USA

**F Thiess**
Department of Mathematics
University of California, San Diego
La Jolla CA 92093
USA

**Tonoi**
Department of Physics
Johns Hopkins University
Baltimore MD 20707
USA

**R Treger**
Department of Mathematics
University of Toronto
Toronto
Ontario M5S 1A1
CANADA

**K Uhlenbeck**
Department of Mathematics
University of Chicago
Chicago IL 60637
USA

**M Villasande**
Department of Physics
University of California
Los Angeles CA 90024
USA

**B Warr**
Department of Physics
California Institute of Technology
Pasadena CA 91125
USA

**S Webster**
Department of Mathematics
University of Minnesota
Minneapolis MN 55455
USA

**R Williams**
Department of Mathematics
Northwestern University
Evanston IL 60201
USA

**P Windey**
50A-3115
Lawrence Berkeley Laboratory
Berkeley CA 94720
USA

**M Wolf**
Department of Mathematics
Stanford University
Stanford CA 94305
USA

**S Wolpert**
Department of Mathematics
University of Maryland
College Park MD 20742
USA

**B Wong**
Department of Mathematics
University of California
Riverside CA 92521
USA

**C-Y Xu**
Department of Mathematics 2-230
Massachusetts Inst of Technology
Cambridge MA 02139
USA

**S T Yau**
Department of Mathematics
University of California, San Diego
La Jolla CA 92093
USA

**G Zuckerman**
Department of Mathematics
Yale University
New Haven CT 06520
USA